E. P. Zege · A. P. Ivanov · I. L. Katsev

# Image Transfer Through a Scattering Medium

With 88 Figures

Springer-Verlag
Berlin Heidelberg NewYork
London Paris Tokyo
Hong Kong Barcelona

Dr. Eleonora P. Zege
Dr. Arkadii P. Ivanov
Dr. Iosif L. Katsev

Institute of Physics
BSSR Academy of Sciences
70 Leninsky Prospekt
220602 Minsk, USSR

ISBN-13:978-3-642-75288-9 e-ISBN-13:978-3-642-75286-5
DOI: 10.1007/978-3-642-75286-5

Library of Congress Cataloging in Publication Data
Zege, E. P. (Eleonora Petrovna)
[Perenos izobrazheniia v rasseivaiushcheĭ srede. English]
Image transfer through a scattering medium / Eleonora P. Zege,
Arkadii P. Ivanov, Iosif L. Katsev.
Translation of: Perenos izobraheniia v rasseivaiushcheĭ srede.
Includes bibliographical references and index.
ISBN-13:978-3-642-75288-9
1. Image transmission. I. Ivanov, A. P. (Arkadii Petrovich)
II. Katsev, I. L. (Iosif Leibovich) III. Title.
TK5105.2.Z4213 1991


Softcover reprint of the hardcover 1st edition 1991



Typesetting: Asco Trade Typesetting Ltd., Hong Kong;

57/3020-543210 – Printed on acid-free paper.

# Contents

# 1 Introduction

The world around us appears as diverse and beautiful as it does owing to the fact that light is scattered. Scattering plays two roles in the intricate process of image formation. First, it is the means by which we perceive an object. An object which does not scatter light cannot be seen. However, scattering also distorts the image observed. Even in relatively pure air, visibility is limited to a range of a few tens of kilometers owing to aerosol and molecular light scattering. It is reduced to tens of meters in conditions of mist or fog.

What occurs when we look at an object through a scattering medium? For example, when a point diffuse source is being observed, the radiation undergoes multiple scattering along the path to the photodetector. Therefore, the image produced by a source of this kind (i.e., the irradiance distribution in the image plane) appears as a more or less blurred speck. What we have said so far serves as an introduction to the concept of image transfer theory, or the point spread function.

It is clear that the more complex an object is, the greater will be the distortion due to the "overlapping" of the blurred images of the details of that object. The smaller the detail, the more blurred the image will appear. Thus, the scattering medium acts as a low-pass spatial frequency filter. Anyone who has mistaken a post for a person in conditions of mist or fog will be able to vouch for the validity of this scientific statement.

In the present book we shall look at the laws governing image transfer through scattering media by synthesizing the linear system and radiative transfer theories. The image produced by the naked eye as well as the image produced by modern electro-optical vision systems, including laser ones, are analyzed from a unified point of view.

Modern image transfer theory developed at the interface of a variety of disciplines which include: the theories of light scattering and statistical decisions, laser engineering, coherent optics and holography, the optics of the sea and atmosphere, scientific photography etc.

## 1.1 A Brief History

If we were setting out to write a history of, for example, quantum electronics, we should almost certainly begin by stating that the first quantum generators were developed independently under the guidance of C.H. Townes and

A.M. Prokhorov as far back as 1954. The field which we have chosen to discuss does not offer such a clear-cut starting point.

Since time immemorial, man has been blessed with as unique and perfect an optical tool as the eye, and a means of processing visual information. Therefore, some experience of specific features of image transfer in scattering media is involved in forming our everyday picture of the world. The sensitive eye of the artist has often caught the distinctions due to the scattering of light on canvas. Leonardo da Vinci's notebooks show quite clearly that his analytical mind was aware of the concept of a half-plane edge function, albeit formulated in terms consummate with the century in which he lived. Many research workers of the present time can only wonder at and admire many of Leonardo da Vinci's observations on contrast and color transfer. The same can be said of many of the great masters, and especially of the great landscape painters of the impressionist era. One fine example is the unique and stirring effect created by the brilliant depiction of image transfer in haze and drizzle in Pissaro's "Boulevard Montmartre".

However, research into the science of vision in scattering media has been fashioned by rather more prosaic considerations. As is usual, practical requirements were the determining factor. Geophysicists have for some time been attracted by the problem of the range of visibility of remote objects in various scattering media. Until the mid-sixties, the principal advances in this field were made in the theory of horizontal visibility in the atmosphere (Koshmider), the modification of this theory as applied to underwater vision (Dantley, Preisendorfer), and the development of the semi-empirical Sekky disc visibility theory [1.1–4]. It should be recalled that the pioneering experiments, using a disc in the Black sea, were made by Sekky and Chialdi as far back as 1865. In all these cases, the aim was only to study the unaided visibility of small objects in sunlight.

The problems related to image transfer in scattering media accumulated with improved technology. Poor visibility in fog had always been a matter for concern and, with the advent of the automobile and airplane, became a problem which had to be tackled. The problems of all-weather aviation and navigation presented the next challenge.

The 1960s brought about an essential change in the approach to the problem. It became necessary as a matter of urgency to develop underwater laser vision as well as location and navigation systems, and to elaborate remote techniques for exploring the earth and its oceans using modern observation facilities. All this called for a new vision theory capable of providing a unified approach to the analysis of the potential of various observation systems such as: photographic and television systems operated both in sun and artificial light, laser scanning and pulsed systems, etc. In addition, the theory had to allow for various medium effects such as: light scattering and absorption in water and in the atmosphere, the effects of medium stratification and fluctuations in the optical parameters, radiation transmission through the surface of a rough sea, etc.

It would seem that the first attempts to formulate such a theory were made by *Bravo-Zhivotovsky, Dolin, Levin, Luchinin* and *Saveliev* [1.5–9]. This period marked a new stage in the development of vision theory. Since that time, the modern theory of image transfer in scattering media has been applied to the field of instrumentation and, above all, to laser systems.

The theory of image formation in recording media evolved along the same lines. The results, as well as a history of the field, are given in the well-known monograph by *Friezer* [1.10].

The fact that it is the scattering of light in a photographic layer which is the major factor in determining the quality of the image became clear early this century. However, it is the last two decades which have witnessed a wider use of light scattering theory as far as problems connected with scientific photography and high-resolution luminescent screens are concerned [1.12, 13]. New and attractive technological prospects were opened up and the more constructive approaches to describing the sensitometric and information properties of photographic materials were developed.

In this short historical review, we have mentioned only a few of the turning points in the development of image transfer theory. Various specific aspects of the theory have been discussed at length in a number of monographs [1.10–16].

## 1.2 Structure of the Book

The book divides broadly into three parts. The first covers Chaps. 2–4 and, basically, presents the applied scattering theory.

A number of fundamental monographs [1.17–22] have been devoted to the classical radiative transfer theory which gives rigorous solutions to non-trivial model problems. However, experience in solving design problems over many years has pointed to the necessity of using approximate methods to solve the transfer equation and to calculate fields light. It is these methods that constitute the main tool in all applied studies, including image transfer theory. They aim to describe the fields both of heat and laser sources, i.e. sources of natural light and coherent pulsed radiation, while at the same time taking into account such typical properties of real media as strong scattering anisotropy, fluctuations of optical parameters, etc.

An important factor is that in calculating the characteristics of light, the majority of applied problems do not demand a high degree of accuracy. Furthermore, the use of exact but intricate methods cannot be justified because of the initial uncertainties in the parameters of the medium (sea water, clouds, aerosols, and so on). The first part has been written with this in view. To date, no consistent discussion of these methods exists, with the exception of [1.22, 23]. We hope that the first part of our book, which might be called *Applied Radiation Transfer Theory*, makes some contribution to this task.

In Chap. 2, the main optical characteristics of a medium and the photometric characteristics of a radiation field are inroduced, and the radiative transfer equation is formulated. In our discussion of the optical characteristics of a medium, we have not attempted to give a complete representation of the theory of scattering by individual particles. We have only cited simple approximate and asymptotic solutions that are useful in studying the transfer characteristics of a scattering medium as related to the parameters of its microstructure. The similarity and experimental modelling principles are given, as are the optical reciprocity theorem and its corollaries reflecting symmetry properties. They will be used extensively in subsequent chapters.

Chapters 3 and 4 offer useful approaches, approximations and general procedures for calculating light fields due to multiple scattering. We do not cite many of the results of the classical transfer theory. In Chap. 3, which deals with light fields induced by infinite sources, sunlight in sea and in atmosphere, for example, we have drawn only on those classical results which give simple formulas for media having strong scattering anisotropy. They consist mainly of asymptotic solutions and some well-known approximate methods. Two groups of methods are distinguishable. One is based on approximate scattering indicatrix models, while the other relies on the averaging the radiance over scattering angle.

Procedures for calculating light fields induced by narrow laser beams form the background to the applied transfer theory. All the known constructive approaches to the solution of this problem, which cover such complementary methods as the diffusion (DA), small-angle (SAA) and small-angle diffusion (SADA) approximations as used for homogeneous and inhomogeneous, including stochastic media, are found in Chap. 4. Of the recent results presented, the modified small-angle diffusion approximation, which is almost as simple as the ordinary SADA, provides the most exact description of light fields. We also show here the interrelation of various methods, and suggest a common approach to estimating the applicability of different approximate solutions.

A great deal of consideration has been given to nonstationary scattering, i.e. to light fields induced in a scattering medium by pulsed sources. We do not repeat the classical results of the nonstationary theory which refer, for the most part, to the illumination of a medium by infinite sources. This information will be used later in the constructive approach to solving the problem of laser pulse propagation.

The results given in Chaps. 2–4 are, on the whole, a combination of methods and formulas used to calculate light fields due to natural and artificial sources, including laser sources, in real scattering media.

The second part of the book, Chaps. 5–7, is the *Theory of Vision in Scattering Media* proper. It relates the "true" (not distorted by the medium) and "visible" image patterns. Chap. 5 introduces and discussed the main characteristics which determine image quality, such as contrast and signal-to-noise ratio, as well as the optical transfer functions of systems. These quantities are shown to be related to the parameters of a vision system and to the optical properties of a medium. Various passive and active vision systems that have different characteristics in the emitter and photodetector, spatial scanning and image transfer as well as techniques for suppressing backscattering noise are discussed from a single viewpoint. The classification cited embraces all possible types of vision systems, from unaided vision to the latest laser and television systems which use pulsed BSN gating.

Visual perception plays a key role in the vision problem. The fact that human eyesight is one of the components of a complex viewing system makes it all the more difficult to provide a theoretical description. In such a case, one is forced to fall back upon empirical criteria which relate to the visual perception of real objects and of some simple test objects. We have chosen to use Johnson's criteria.

We hope that specialists in the field will be interested in the concept of optimum detection suggested here, as well as in the comparison of different television and laser location systems.

By and large, the theory given in Chap. 5 proposes methods for estimating the potentials of, and optimizing the different systems of viewing and location in scattering media.

Chapter 6 deals with the optical transfer functions (OTF) of scattering media. It presents the general properties of OTF and their characteristics when viewing along oblique observation paths, as well as through inhomogeneous and stochastic media. The set of formulas listed make it possible to calculate OTF's in the media with all possible optical parameters.

In this chapter, special consideration is given to observation through inhomogeneous media. We have endeavoured to present a simple and instructive interpretation of the nonmonotonic image contrast dependence on the scattering layer position along the observation path, for example, on the position of a turbid aerosol layer in space observations of Earth. This problem has recently been the subject of lively debate. However the Carribbean pirates of the 16th century seem to have been able to judge it accurately, producing smoke-screens not in front of their own ship, but close to the enemy during sea battles.

Another attractive aspect, albeit one which has received scant attention in the literature, which is also important for interpreting space observations of Earth is the problem of nonlinear image distortions that may be encountered when viewing through a scattering media which shows appreciable reflection. It appears that in a large number of real situations, it is possible to linearize and use the traditional apparatus of transfer functions.

Early achievements in holography have given rise to the question of its potential for improving image quality and increasing the range of visibility in scattering media.

Chapter 7, which completes the section dealing with general problems in the theory of vision, offers a new approach to estimating the quality of the image formed by a holographic system of observation through scattering media. It relies on the interrelation between photometry and coherence theory. As will become apparent, the methods developed earlier for incoherent systems can be applied to a quantitative representation of holographic vision systems. Profound physical analogies between coherent and incoherent observation methods have been demonstrated. These analogies are important both in practice, and in proposing new systems.

Therefore, there exists a general theory of vision in scattering media (Chaps. 4–7), and we know how to calculate the quantities and functions involved (Chaps. 2–4). This is a tool which can be applied to various problems. An illustration of such an application is given in the third part of the book (Chaps. 8–10) which discusses concrete problems of image transfer in specific media. These are problems relating to *Vision and Location in the Atmosphere and Ocean, Image Formation in Photographic Layers and Luminescent Screens.* The results given in the first two sections serve as the theoretical and methodological background for investigation. For example, if the problem of light propagation in the ocean or atmosphere in which you are interested differs from that discussed in the book, you may attempt to solve it by using the methods developed and the models of the optical properties of media suggested, following the procedure in the given solutions. We believe that this aspect of the book is very important and can be successful in practice, as the Russian edition has already demonstrated.

A solution to specific problems of the theory of vision requires a knowledge of the optical parameters of a medium. Each of Chaps. 8–10, or the appropriate section, begins with an optical model of a medium. All the models rely on the experimental data banks and scattering theory.

Section 8.1 presents concise optical models of aerosol atmospheres and clouds, while simple models of the optical properties of sea water and rough sea surfaces are given in Chap. 9. We draw the reader's attention to some specific features of this set of models, and the potential they offer. Together, they provide a convenient model of the "ocean-atmosphere" system.

We hope that the simple model of the optical properties of sea water will prove convenient in forecasting investigations. Two types of water, such as sea and ocean water, are considered. In both cases, all the characteristics of absorption and scattering necessary for optical studies to be carried out can be calculated, given that the transparency, i.e. the simplest parameter to be measured, is known. The model allows for the interrelation between the different optical properties of sea water well known in optical oceanology.

In Chapters 8 and 9, classical problems (the theory of horizontal visibility in atmosphere, Sekky's disc visibility, etc.) are considered from a single viewpoint, together with questions associated with estimating the ultimate potential of present-day laser systems of vision, location and bathymetry. We have relied on optical atmosphere and ocean models as well as on procedures developed for determining light fields, and on the concept of object detection and identification set out in Chap. 5. Characteristically, even in the case of well-studied classical problems, the general approach used does not only give their modern scientific basis, but also provides additional information which allows one to determine the accuracy and applicability of the known solutions on common principles. We have endeavoured to illustrate these applications. For example, when considering the theory of horizontal visibility range (Koshmider's theory) that assumes the object observed to be of negligible size, we have nevertheless studied the effect of the object size on the visibility range. Rigorous theoretical results are presented alongside semi-empirical data on Sekky's disc visibility in order to prove and specify this well-tried method.

The particular problems associated with observing in conditions of cloud and fog are considered, taking into account the microstructure and stochastic nature of these media. In Chap. 8, we deal with the microstructure effect on the transfer properties of clouds. The theory of signals and noises in viewing through stochastic clouds is presented for the first time. The competition between the improvement of the transfer characteristics on the one hand and, on the other, processes which generate new medium fluctuation noises offers an attractive physical area of investigation. Its analysis both an improvement and a degeneration of the image quality due to the medium's stochasticity has been exposed.

Chapters 8 and 9 discuss the visibility of radiation sources. As early as the 1960s, pioneering experiments on laser radiation propagating in scattering media showed that lasers preserve high brightness contrast in the source direction up to very large optical depths. This fact has been used to design laser navigation systems and laser systems for aircraft landing under conditions of poor visibility. This problem is discussed in Chap. 8 and we suggest convenient procedures for design calculations of a limiting range for light source detection in different media.

In Chapter 9, our attention is focused on less traditional problems, such as those concerned with the scope of television and location systems for viewing underwater objects, and laser bathymetry. Formulas required to calculate background and noise energy in underwater viewing systems are supplied in a convenient form. Taken in conjunction with the detection concept elaborated in Chap. 5, they allow the signal-to-noise ratio and object detection range to be determined for each particular situation.

Among new problems still under development, the theory of image transfer through a rough sea surface, considered mainly in Sect. 9.4, is worth noting. This situation may involve such significant and specific effects as enhanced backscattering, the spatial correlation of fluctuation noises and the nonlocal nature of transfer functions.

We move on to quite a different field in Chap. 10. Here, we deal with photographic layers and luminescent screens, i.e. the media designed for recording and reproducing an image. These problems are also considered from the general viewpoint adopted in this book, i.e. from the standpoint of the theories of scattering and vison. Such an approach proves to be rather beneficial in that. It offers a scientific basis for predicting the sensitometric characteristics of photographic materials and the quality of luminescent screens. It also allows one to outline processes for manufacturing photographic materials and screens with specific properties, and may serve as a scientific background for developing new technologies.

## 1.3 Notes on Terminology

As the reader will already have gathered, the subject of this book embraces several different scientific fields. In addition to the fundamental problems of image and radiation transfer in scattering media, we touch upon a number of aspects of laser engineering, coherent optics and holography, the optics of the sea and atmosphere, scientific photography and luminescent screen technology. Attempts to introduce a common terminology for such diverse fields were not successful and, in fact, the question of terminology has been far from easy to solve.

As far as possible, we have used the terms, acronyms and units adopted by the Commissions of the International Associations on pure and applied physics, radiation and physical oceanography, relying on the usual handbooks [1.24–26]. Information on photometric and radiometric terminology is taken from [1.27], and that on transfer functions from [1.28].

Whenever feasible, we have retained the traditional terminology of the particular fields. Transfer theory proved especially problematic and we reluctantly decided to opt for the terms "radiance" and "scattering indicatrix" in place of the usual "intensity" and "phase function". In our discussion, we have explained the reasons for this change. It is prompted not only by the desire to use the adopted terms, but also the desire to avoid ambiguity due to their connotation in related fields.

# 2 Radiation Field and Scattering Medium Characteristics

In this chapter, we deal with the basic characteristics of a light field as well as with the scattering characteristics of an elementary volume and individual scatterers which affect the mechanism of radiation propagation. We also formulate an equation of transfer whose solutions will be used subsequently, and study some of its general properties. A special section is devoted to the optical reciprocity theorem which represents the most general properties of light fields and is of importance in the image transfer and multiple scattering theories.

## 2.1 Radiation Field

We now consider a radiant flux at the point characterized by a radius-vector $\vec{R}$. For the direction indicated by the solid angle vector $\vec{n}$, the radiant flux $dP$ per solid angle $d\Omega$ per elementary area $d\Sigma$ with the $\vec{n}_0$ normal is

$$dP = I(\vec{R}, \vec{n}) \cos\theta \, d\Sigma \, d\Omega \,, \tag{2.1.1}$$

where

$$\cos\theta = (\vec{n}\,\vec{n}_0) \tag{2.1.2}$$

is the cosine of the angle between the solid angle vectors $\vec{n}$ and $\vec{n}_0$. The quantity $I(\vec{R}, \vec{n})$, a radiant flux per unit solid angle per unit area normal to $\vec{n}$, is referred to as the radiance, $\mathrm{Wm^{-2}sr^{-1}}$.

In radiation transfer theory, this quantity is termed the intensity (luminosity or brightness) [2.1, 2]. We deviate from this tradition and try to keep to the conventional international terminology. In this connection, we use the term brightness for a qualitative description of a radiant pattern, whereas luminance applies to a luminous system [2.3]. In the analysis which follows, only the terms and MKS units for a radiant system are used. The conversion into the luminous system is well known ([2.3] and Sect. 5.6.1).

If the radiance distribution in the $\vec{n}$ directions for each point $\vec{R}$ is given, we say that a light or radiation field is specified. A radiation field is isotropic at this point if the radiance is independent of the direction.

Radiance is a differential characteristic of a light field. It unambiguously determines all integral quantities.

Radiation *density* at a given point of a medium, or the amount of radiation per unit volume, is expressed by the formula

$$w^*(\vec{R}) = \frac{1}{v}\int_{4\pi} I(\vec{R},\vec{n})\,d\Omega\,, \tag{2.1.3}$$

where integration is over the entire solid angle, and $v$ is the light velocity in the medium. Different studies on transfer theory, scattering medium optics and photometry use two more quantities, which differ from density only by constants. These are *scalar irradiance*

$$w(\vec{R}) = \int_{4\pi} I(\vec{R},\vec{n})\,d\Omega = vw^*\,, \tag{2.1.4}$$

and *spherical (average) radiance*

$$\langle I(\vec{R})\rangle = \frac{1}{4\pi}\int_{4\pi} I(\vec{R},\vec{n})\,d\Omega = \frac{v}{4\pi}w^*\,. \tag{2.1.5}$$

Radiation density is measured in $Jm^{-3}$, scalar irradiance in $Wm^{-2}$.

In transfer theory, the quantity $w$ as well as $w^*$ is frequently called radiation density. In the interests of brevity, we shall also use this term below.

An important integral light field characteristic is the *flux density vector* $\vec{F}$ which describes the magnitude and direction of the radiation flux per unit area:

$$\vec{F}(\vec{R}) = \int_{4\pi} I(\vec{R},\vec{n})\vec{n}\,d\Omega\,. \tag{2.1.6}$$

*Flux density in some $\vec{n}_0$ direction* is the vector $\vec{F}(\vec{R})$ projection onto this direction:

$$F(\vec{R},\vec{n}_0) = (\vec{n}_0\vec{F}(\vec{R})) = \int_{4\pi} I(\vec{R},\vec{n})\cos\theta\,d\Omega\,. \tag{2.1.7}$$

The word "density" is often omitted in radiation transfer theory [2.4, 5], and the quantity $\vec{F}$ is simply called a radiation flux.

Consider an area $d\Sigma$ with the $\vec{n}_0$ normal. The flux incident per unit area in the $(-\vec{n}_0)$ direction per solid angle $2\pi$ equal to

$$E(\vec{R},\vec{n}_0) = \int_{2\pi} I(\vec{R},\vec{n})\cos\theta\,d\Omega\,, \quad \cos\theta = (\vec{n}(-\vec{n}_0))\,, \quad 0 \leqslant \theta \leqslant \pi/2 \tag{2.1.8}$$

is called area *irradiance*. Similarly, the *radiant emittance* of a luminous surface is defined as flux density in the $\vec{n}_0$ direction over the solid angle $2\pi$:

$$E(\vec{R},\vec{n}_0) = \int_{2\pi} I(\vec{R},\vec{n})\cos\theta\,d\Omega\,, \quad \cos\theta = (\vec{n}\vec{n}_0)\,, \quad 0 \leqslant \theta \leqslant \pi/2\,. \tag{2.1.9}$$

The quantities $E$ and $\vec{F}$, as well as the scalar irradiance, are measured in $Wm^{-2}$.

The radiation flux per unit angle which is emitted by an area $\Sigma$ in the $\vec{n}$ direction is called the *radiant intensity* (light strength) $I_{em}$, i.e.,

$$I_{em}(\vec{n}) = \int_{\Sigma} I(\vec{R}, \vec{n}) \cos\theta \, d\Sigma \,. \tag{2.1.10}$$

It is measured in $\mathrm{Wsr}^{-1}$. If the radiance emitted by the surface $\Sigma$ is isotropic, the radiant intensity depends on the angle $\theta$ between the $\vec{n}_0$ normal and the $\vec{n}$ direction of observation, in accordance with the cosine law

$$I_{em}(\vec{n}) = \cos\theta \int_{\Sigma} I(\vec{R}) \, d\Sigma \,. \tag{2.1.11}$$

In such cases, we are dealing with an emitting Lambert surface. Such a surface is sometimes referred to as orthotropic or diffuse.

## 2.2 Optical Parameters of a Volume Element

Consider some volume element $dV$ in a scattering and absorbing medium, which is illuminated over a solid angle $d\Omega$ by an incident light beam with radiance $I$. This volume will partly absorb and partly scatter the incident flux. The absorbed flux turns out to be

$$dP_a = \kappa I \, d\Omega \, dV \,, \tag{2.2.1}$$

where $\kappa$ is the proportionality constant called the *absorption coefficient.*

The intensity $dI_{em}$ of the light scattered by a volume element in the direction making an angle $\beta$ to the incident beam is assumed to be

$$dI_{em} = \sigma(\beta) I \, d\Omega \, dV \,. \tag{2.2.2}$$

Here, $\sigma(\beta)$ is the differential scattering coefficient in the $\beta$ direction and is usually termed the *volume scattering function.*

Naturally, the radiant flux scattered in all directions is

$$dP_s = \int_{4\pi} dI_{em} \, d\Omega' = I \, d\Omega \, dV \int_{4\pi} \sigma(\beta) \, d\Omega' = \sigma I \, d\Omega \, dV \,, \tag{2.2.3}$$

where $\sigma = \int_{4\pi} \sigma(\beta) \, d\Omega'$ is the *scattering coefficient of a volume element.*

It is convenient to describe the relative angular distribution of the radiant intensity scattered by a volume element by the function

$$x(\beta) = 4\pi\sigma(\beta)/\sigma \,, \tag{2.2.4}$$

normalized by the relation

$$\frac{1}{4\pi} \int_{4\pi} x(\beta) \, d\Omega = 1 \,. \tag{2.2.5}$$

This characteristic is called the *scattering indicatrix,* or *scattering diagram.* In transfer theory [2.2, 6], the term *phase function* is frequently used for $x(\beta)$. Like the authors of [2.4], we do not believe this term to be appropriate for describing radiance distribution since it suggests the idea of a phase.

If the radiant intensity is equal in all directions (spherical indicatrix), it is said that *isotropic scattering* occurs, and

$$x(\beta) = 1 . \tag{2.2.6}$$

The scattering indicatrices of real scattering media are generally far from isotropic. They may be rather complex functions of the scattering angle.

In multiple scattering theory, the indicatrix is frequently expanded into a series of Legendre polynomials $P_i(\cos\beta)$

$$x(\beta) = \sum_{i=0}^{N} x_i P_i(\cos\beta) . \tag{2.2.7}$$

The expansion coefficients are determined as

$$x_i = \frac{2i+1}{2} \int_{-1}^{1} x(\beta) P_i(\cos\beta)\, d(\cos\beta) , \qquad x_0 = 1 . \tag{2.2.8}$$

The more forward-extended the indicatrix, the greater the number of expansion terms required in (2.2.8) to represent it adequately.

When studying anisotropic scattering in radiation transfer theory, $x(\beta)$ is often described by a simple one–parameter model

$$x(\beta) = \frac{1-g^2}{(1+g^2-2g\cos\beta)^{3/2}} , \tag{2.2.9}$$

where $g$ is some parameter ($-1 \leqslant g \leqslant 1$). The function (2.2.9) is called the Henyey-Greenstein (H-G) indicatrix. It can also conveniently expanded into a series of Legendre polynomials:

$$x_i = (2i+1) g^i . \tag{2.2.10}$$

The function (2.2.9) is a fair approximation of some real scattering indicatrices.

As far as the calculation of integral light field characteristics is concerned, a single number is frequently sufficient to characterize the scattering indicatrix. This may be the first expansion term $x_1$ or the average cosine of the scattering angle,

$$\langle\cos\beta\rangle = g = \frac{x_1}{3} = \frac{1}{2} \int_{-1}^{1} x(\beta) \cos\beta\, d(\cos\beta) , \tag{2.2.11}$$

defined in scattering theory as the asymmetry parameter. In all cases $-1 \leqslant g \leqslant 1$, and for forward-extended scattering indicatrices, $0 \leqslant g \leqslant 1$. The quantity $g = 0$ corresponds to isotropic scattering ($x(\beta) = 1$), $g = 1$ complies with strictly forward scattering after $x(\beta) = \delta(\beta)$, where $\delta(\beta)$ is the Dirac delta function. The asymmetry parameter plays a specific role in multiple scattering theory (see Ref. 2.2 as well as Chaps. 3 and 4). The parameter $F = \int_0^1 x(\beta)\, d(\cos\beta) / \int_{-1}^{1} x(\beta)\, d(\cos\beta)$, that is, the portion scattered into the forward hemisphere, or the so-called forward/back ratio equal to $F/(1-F) = \int_0^1 x(\beta)\, d(\cos\beta) / \int_{-1}^{0} x(\beta)\, d(\cos\beta)$, is frequently used as one more integral characteristic of the scattering indicatrix. The parameters $g$ and $F$ governing the degree of forward extension of the indicatrix increase with increasing particle size (at least, if the refractive index of the particle matter is greater than that of the medium) and as the difference between the refractive indices of the scattering particle and the medium decreases. For a wide class of indicatrices of natural media, $g$ and $F$ are strongly correlated (Sect. 4.7).

If polarization effects are disregarded, it is sufficient to know the scattering and absorption coefficients as well as the scattering indicatrix in order to describe the optical properties of the scattering medium. Sometimes, instead of $\kappa$ and $\sigma$, one considers such quantities as: $\varepsilon = \kappa + \sigma$, the extinction (attenuation) coefficient; $\Lambda = \sigma/\varepsilon$, the single scattering albedo, or $\varepsilon$, $\kappa/\sigma$, the specific absorption. Consider now the dimensionless parameter $\Lambda$. If there is no absorption in the medium, $\Lambda = 1$. This is a case of conservative (pure) scattering. If the medium does not scatter, $\Lambda = 0$. In all cases, $0 \leqslant \Lambda \leqslant 1$. We omit here the enhancing media for which $\Lambda < 0$. It should be noted that, strictly speaking, all the parameters mentioned above apply to monochromatic light. If the spectral behavior of all three independent characteristics are known, we can assess a wide spectral range.

The relations (2.2.1–3) imply that a volume element is one in which the amount of absorbed and scattered light is proportional to its size. Its geometric sizes appear different for different scattering media. For example, in pure atmosphere the linear sizes of such a volume constitute hundreds of meters, while they are micrometer fractions in photographic materials.

The relationship between the scattering medium structure and the characteristics of the volume element depends on the properties of the scatterers which compose the volume element, the distances between them and the incident light wavelength. At low concentrations, when the medium can be regarded as a set of independent scatterers, $\varepsilon$, $\sigma$ and $\kappa$ as well as the indicatrix $x(\beta)$ can be determined by addition of the scattered radiances of different scatterers. Light scattering (diffraction of electromagnetic radiation) on a particle is, in turn, dependent on its shape, sizes (to be specific, size to diffracted radiation wavelength ratio) and relative complex refractive index $m = n - in'$ of the particle material. We refer here to the simplest case of an isotropic and homogeneous particle.

A number of excellent monographs [2.7–11] deal with light scattering by particles. We shall restrict our consideration to a few brief relevant remarks, and refer the reader to these works for comprehensive information.

As far back as the beginning of this century, *Mie* [2.12] proposed a rigorous solution to the simplest, but very important, problem of the diffraction of electromagnetic radiation by a sphere of radius $a$. Mie's theory allows one to calculate the extinction cross-sections $C_{\mathrm{ext}}$, absorption $C_{\mathrm{a}}$ and scattering $C_{\mathrm{s}}$ ($C_{\mathrm{ext}} = C_{\mathrm{a}} + C_{\mathrm{s}}$), having the dimensions of area, as well as the scattering indicatrix $x_{\mathrm{part}}(\beta)$ of a spherical particle. This is discussed comprehensively in all monographs dealing with scattering particles [2.7–11].

If a unit volume contains $N$ equal particles, then the light extinction coefficient is

$$\varepsilon = NC_{\text{ext}}, \tag{2.2.12}$$

for a polydispersed system of spherical particles with the size distribution function $f(a)$, where $a$ is the particle radius,

$$\varepsilon = N \int_{a_1}^{a_2} C_{\text{ext}} f(a)\, da . \tag{2.2.13}$$

Here $a_1$ and $a_2$ are the minimum and maximum particle radii, $\int_{a_1}^{a_2} f(a)\, da = 1$. We can also write the relations similar to (2.2.12, 13) for $\sigma$, $\kappa$ and the scattering indicatrix.

The optical characteristics of various polydispersions calculated by the Mie formulas are cited in [2.13–15]. Figure 2.1 shows the scattering indicatrix of a polydispersion of large spherical water particles, which is in agreement with the Cloud C.1 model [2.15]. We define it, for brevity, as the indicatrix C.1. In Fig. 2.1a, the forward scattering extension typical of large particles is easily seen: $g = 0.853$. At $\beta > \pi/2$, rainbow and glory scattering are observed (Fig. 2.1b). The function $x(\beta)$ is seen in each case to be complex. The Mie theory solutions have been well studied but, unfortunately, are rather clumsy. They do not yield simple analytical relations for the scattering characteristics of spherical particles, their radii and the optical constants of the material.

There exist a number of physical assumptions and approaches aimed at simplifying the problem in particular situations and obtaining convenient analytical solutions. An instructive classification of these approaches using physical

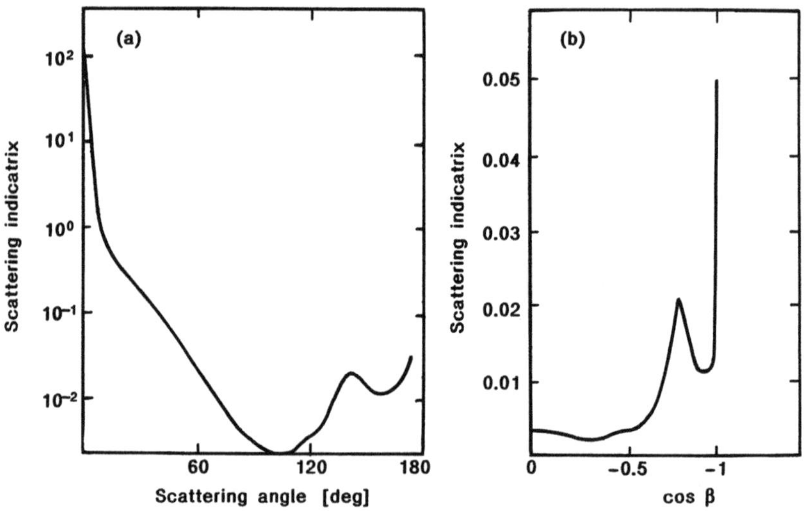


**Fig. 2.1.** Scattering indicatrix of the model water Cloud C.1 with $\lambda = 0.7\ \mu m$. From [2.15]

arguments is given in [2.9]. A more or less comprehensive description can be found in [2.7–11]. Table 2.1 summarize the basic analytical results for the cross-sections of extinction, scattering, absorption ($C_{ext} = C_a + C_s$) and the asymmetry parameter $g$ of a spherical particle. We recall that the optical properties of a scattering spherical particle depend neither on the sphere radius nor on the radiation wavelength $\lambda$, independently, but are functions of their ratio. Therefore, the characteristic parameter of the theory is the diffraction parameter $\rho = 2\pi a/\lambda$.

We now comment on the essential features of the physical approaches whose results are presented in Table 2.1.

1. Light scattering *by a small particle* ($\rho \ll 1$) is of a dipole nature. Formulas for the absorption and scattering cross-sections of small scatterers are given in Table 2.1. At $\rho \leqslant 0.3$, an error in the cross-section is no more than 4% [2.6]. Characteristically, the scattering cross-section is in direct proportion to the squared scatterer volume and in inverse proportion to $\lambda^4$. The scattering indicatrix is of the form

$$x(\beta) = \tfrac{3}{4}(1 + \cos^2 \beta)\,. \qquad (2.2.14)$$

**Table 2.1.** Approximate solutions for scattering characteristics of spherical particles.

| Approximation | $C_{ext}, C_a, C_s, g$ | Notes |
|---|---|---|
| Small particles (Rayleigh scattering) $\rho \ll 1, \lvert m\rvert\rho \ll 1$ | $C_a = \frac{4\pi nn'}{\lambda}\left\lvert\frac{3}{m^2+2}\right\rvert^2 V$ <br> $C_s = \frac{24\pi^3}{\lambda^4}\left\lvert\frac{m^2-1}{m^2+2}\right\rvert^2 V^2$ <br> $g = 0$ | Particle volume $V = \frac{4}{3}\pi a^3$ |
| Large spheres (Geometrical optics) $\rho \gg 1, (n-1)\rho \gg 1$ | $C_{ext} = 2\pi a^2$ <br> $C_a = \left\{1 - \frac{2n^2}{A^2}\left[e^{-A\xi}(1 + A\xi) - e^{-A}(1 + A)\right] + S(n)(1 - e^{-A\xi})^2\right\}\pi a^2$ <br> $g = g_1(n) - \left[g_1(n) - g_0(n) + \frac{\gamma}{\rho}\right]e^{-A\delta(n)}$ | $\xi = \sqrt{1 - n^{-2}}$, $A = 4n'\rho$, $\rho = 2\pi a/\lambda$ See Table 2.2 for the values of $g_1, g_0, \delta$ and $S$ at different $n$ <br> $\gamma = 1.8$ at $n > 1.1$ |
| Anomalous diffraction (Van de Hulst) $n - 1 \ll 1, \rho \gg 1$ $\Delta = 2(n-1)\rho$ | $C_{ext} = \left[2 - \frac{4}{\Delta}\sin\Delta + \frac{4(1-\cos\Delta)}{\Delta^2}\right]\pi a^2$ <br> $C_a = \left\{1 - \frac{2}{A^2}[1 - e^{-A}(1 + A)]\right\}\pi a^2$ | |
| Rayleigh-Gans (Born) approximation; $\rho$ is arbitrary; $\Delta = 2\lvert m-1\rvert\rho \ll 1$, $\lvert m-1\rvert \ll 1$ | $C_s = 2\lvert m-1\rvert^2\rho^2, \rho \gg 1$ <br> $= \frac{32}{27}\lvert m-1\rvert^2\rho^4, \rho \ll 1$ <br> $C_a = \frac{4\pi n'}{\lambda}V$ | For $g$ formulas see [2.16] Values of $g$ are cited in Table 2.3 |

The relation $C_s \sim \lambda^{-4}$ and indicatrix (2.2.14) also characterize molecular scattering. Not wishing to split hairs over details of terminology, we shall use the Rayleigh scattering term, as is often done, both for molecular scattering and for scattering by small particles, and call (2.2.14) the Rayleigh indicatrix. Rayleigh scattering is responsible, in many respects, for atmospheric optical phenomena and must be taken into account when describing the optical properties of sea water and so on.

2. Light scattering *by large particles* ($\rho \gg 1$) is described in terms of geometrical optics, including refraction and reflection at the medium/particle interface and the Fraunhofer diffraction [2.7–11]. As the Mie calculations increase rapidly with the parameter $\rho$, approximate solutions for large particles are especially important in practice.

At $\rho \to \infty$, the extinction efficiency factor $Q_{ext} = C_{ext}/\pi a^2$ is known to tend to its limit of 2, irrespective of the particle shape and internal structure. Formulas for $C_a$ and the indicatrix asymmetry parameter $g$ which are simple, and somewhat more general than those of [2.16], were obtained in [2.17] and cited in Table 2.1. In the formula for $g$, the quantities $g_0(n)$ and $g_1(n)$ are asymptotic at $\rho \to \infty$ values of the asymmetry parameter for nonabsorbing and absorbing spheres. Expressions and detailed tables for the functions $g_0(n)$, $g_1(n)$, $S(n)$ and $\delta(n)$ in the formulas for $g$ and $C_a$ can be found in [2.16, 17]. Relevant data are given in Table 2.2.

Since edge effects have, in part, been taken into account, the formula for $g$ may be applied starting from not very large $\rho$. The formula for $C_a$ goes over into the van de Hulst formulas at $|n-1| \ll 1$, into the Rayleigh-Gans expressions at $|n-1|\rho \ll 1$, $n'\rho \ll 1$ (also given in Table 2.1 and explained below) and into the well-known Bohren formula at $n'\rho \ll 1$ [2.7]. In calculating the absorption cross-sections for polydispersions or for nonmonochromatic medium illumination, the formula for $C_a$ cited in Table 2.1 can be used for either $\rho$. Some quantitative estimates for the accuracy of these formulas as applied to polydispersions are presented in Sect. 8.1 (details are given in [2.17]). The geometrical optics approximation is frequently and successfully used to describe scattering through clouds, fog and large-dispersed aerosols.

3. *The anomalous diffraction* (*van de Hulst*) *approximation* was developed for scattering by soft ($|m-1| \ll 1$), large ($\rho \gg 1$) scatterers [2.9]. The beam is not assumed to deviate at the medium/particle interface, i.e., refraction and reflection

**Table 2.2.** Parameters for $C_a$ and $g$ [2.17].

| $n$ | 1.05 | 1.2 | 1.333 | 1.5 |
|---|---|---|---|---|
| $S$ | 0.0139 | 0.0443 | 0.0644 | 0.0918 |
| $g_0$ | 0.9903 | 0.9341 | 0.8843 | 0.8299 |
| $g_1$ | 0.9981 | 0.9856 | 0.9714 | 0.9520 |
| $\delta$ | 0.388 | 0.653 | 0.747 | 0.816 |

are both neglected. The presence of a nonabsorbing particle changes only the phase of an incident electomagnetic field. Therefore, the field can be found directly from the Huygens principle. The phase shift $\Delta = 2(n-1)\rho$ may be as large as desired. Appropriate formulas for $C_a$ and $C_{ext}$ are given in Table 2.1. It is clear that this method can be applied in full for structural and arbitrarily shaped particles. We take as an example one simple, but practically important, result. The absorption cross-section of soft ($|m-1| \ll 1$, $(n-1)\bar{\rho}$ is arbitrary, $\bar{\rho}$ is the characteristic parameter of a scatterer), weakly-absorbing ($\kappa\bar{\rho} \ll 1$), arbitrarily shaped structural particles is determined by the volume averaged absorption in a particle $\langle n' \rangle = (1/V_{part})\int_{V\,part} n'(\vec{R})\,dV$ ($V_{part}$ is the particle volume, the function $n'(\vec{R})$ defines the absorption distribution inside the particle)

$$C_a = 4\pi\langle n' \rangle V_{part}/\lambda .$$

4. Scattering by soft particles ($|m-1| \ll 1$) with small phase shifts ($(n-1)\rho \ll 1$) and absorption ($n'\rho \ll 1$) can be described by the Rayleigh-Gans approximation (the Born approximation of quantum mechanics). We may now assume that the field inside the scatterer coincides with the incident wave field. Formulas for the scattering and absorption cross-sections are listed in Table 2.1, and the scattering indicatrices are given in [2.6–10]. Since the asymmetry parameter expression is clumsy [2.16], we shall only cite $g(\rho)$ values in Table 2.3. A more detailed tabulation is found in [2.16].

Soft particles of different sizes are generally responsible for scattering through sea water. These also include biological objects (erythrocytes, chloroplasts, biological cells, etc.).

Even a brief description of physical approaches to the above approximations demonstrates their suitability for solving more complex problems such as scattering on arbitrarily shaped, structural and anisotropic particles. A number of such results are presented in [2.7–10, 18].

Rigorous mathematical approaches are also being developed to calculate scattering on radially inhomogeneous and anisotropic spherical particles [2.19], as well as on individual arbitrarily shaped scattered and their ensembles [2.20, 21]. However, these solutions are usually intricate and can rarely be applied directly to design calculations.

So far, we have dealt with a set of independent scatterers. Up to volumetric particle concentrations of less than 0.1–0.2, additive approximation can be applied and the formulas (2.2.12, 13) are valid [2.22]. At high concentrations, account must be taken of the close packing effects, and the relationship between

**Table 2.3.** Asymmetry parameter $g$ for Rayleigh-Gans scattering [2.16].

| $\rho$ | 0.2 | 1.0 | 1.4 | 2 | 4 | 6 | 8 | 10 |
|---|---|---|---|---|---|---|---|---|
| $g$ | 0.0063 | 0.1669 | 0.3298 | 0.6075 | 0.8742 | 0.9345 | 0.9584 | 0.9715 |

**Table 2.4.** Optical characteristics of different scattering media.

| Optical parameter | Atmosphere | | | Sea | Photographic layer | | Luminescent screen | | Model medium | |
|---|---|---|---|---|---|---|---|---|---|---|
| | Cloudless | Cloudy | | | | | | | | |
| | Visible | Visible | IR | | Undeveloped | Developed | Exciting radiation | Visible luminescence | Milk and nigrosine | Latex and nigrosine |
| Single scattering albedo | 0.6–1 | 0.99–1 | 0–1 | 0.5–0.9 | 0–1 | 0–0.9 | 0.1–0.9 | 0.95–1 | 0–1 | 0–1 |
| Asymmetry parameter | 0.1–0.75 | 0.8–0.9 | 0.8–0.98 | 0.7–0.98 | 0–0.8 | 0–0.9 | 0.3–0.97 | 0.7–0.85 | 0.33–0.9 | 0.1–0.9 |

the characteristics of the elementary volume and those of the individual scatterers it includes is more involved [2.22–24].

In the chapters dealing with image transfer in specific scattering media (atmosphere, sea, photographic material, etc.), we shall give the appropriate volume element characteristics for each case. Here we present only the concise Table 2.4 in order to show the range of optical characteristics of sea water, atmospheric aerosols and clouds, as well as of the materials used for image recording. The appropriate data for the model scattering media widely used in laboratory experiments, such as milk or latex emulsions mixed with an aqueous nigrosine dye solution are also tabulated here.

In practice, we often face problems concerned with image formation using luminescent screens, observing certain objects in biological media, through paper, varnish and paint coatings, milky glass, etc. It is difficult to cover all possible situations because of the diversity of the scattering media which play a role in image transfer. It can be assumed, nevertheless, that for these media the scattering, absorption and extinction coefficients lie between $10^{-1}$ and $10^4$ mm$^{-1}$, whereas the geometric thicknesses range from $10^{-1}$ to $10^2$ mm. The scattering indicatrices are generally highly anisotropic. Information on the optical properties of these objects can be gained from [2.25–29].

Having introduced the optical parameters of a volume element for light propagation in scattering media, we shall now formulate the radiation transfer equations.

## 2.3 Radiation Transfer Equation

In free space, radiance is constant along a beam. If absorption, scattering and emission of radiation occur in the medium, then the radiance change along the beam $l$ is described by the integro-differential transfer equation [2.4, 30]

$$\frac{dI}{dl} = -\varepsilon I(\vec{R},\vec{n}) + B_{\text{full}}(\vec{R},\vec{n})$$

$$B_{\text{full}}(\vec{R},\vec{n}) = \frac{\sigma}{4\pi}\iint_{4\pi}(\vec{R},\vec{n}')x(\vec{n},\vec{n}')\,d\vec{n}' + B(\vec{R},\vec{n})\,. \qquad (2.3.1)$$

This is the law of radiant energy conservation for scattering media. The term $(-\varepsilon I(\vec{R},\vec{n}))$ specifies the intensity attenuation per unit beam length. The term $B_{\text{full}}(\vec{R},\vec{n})$ is the total emission coefficient of the scattering medium which defines the radiation power per unit volume at a point $\vec{R}$ per unit solid angle in the $\vec{n}$ direction. The total emission coefficient incorporates the radiation due to medium scattering, which is described by the integral term, and the coefficient $B(\vec{R},\vec{n})$ of proper radiaton from other sources. It is frequently more convenient to use the gradient or divergence operators in the l.h.s. of (2.3.1), rather than the direction derivative

$$\frac{dI(\vec{R},\vec{n})}{dl} = \vec{n}\cdot\nabla_{\vec{R}} I(\vec{R},\vec{n}) = \mathrm{div}[I(\vec{R},\vec{n})\vec{n}]\,. \tag{2.3.2}$$

The derivation of equation (2.3.1) is presented in the majority of monographs dealing with radiation transfer theory [2.4, 30]. In neutron transfer theory, equation (2.3.1) is usually known as the Boltzmann equation.

Equation (2.3.1) must be supplemented by the boundary conditions formulated according to the scattering volume geometry and the properties of the external light sources. Various boundary conditions are described in Chaps. 3 and 4.

We integrate equation (2.3.1) over the total solid angle $4\pi$, using the relation (2.3.2). Then, the flux vector $\vec{F}(\vec{R})$ and the radiation density $w(\vec{R})$ are related as

$$\mathrm{div}\,\vec{F}(\vec{R}) = -\kappa w(\vec{R}) + B_0(\vec{R})\,, \tag{2.3.3}$$

where $B_0(\vec{R}) = \iint_{4\pi} B(\vec{R},\vec{n})\,d\vec{n}$ is the power emitted per unit volume. Formula (2.3.3) expresses the energy conservation law for a volume element. If there are no sources, then we have

$$\mathrm{div}\,\vec{F}(\vec{R}) = -\kappa w(\vec{R})\,. \tag{2.3.4}$$

The relation (2.3.4) is widely used in scattering medium optics. It is used, for example, to determine the medium absorption coefficient [2.31]. For a non-absorbing medium ($\kappa = 0$, $\Lambda = 1$),

$$\mathrm{div}\,\vec{F}(\vec{R}) = 0\,. \tag{2.3.5}$$

The relation (2.3.5) is the conservation law for a radiant flux in the medium with conservative (pure) scattering.

## 2.4 Similarity Principle and Modeling

We first consider the homogeneous medium whose optical properties are independent of the coordinates. By introducing a dimensionless optical radius-vector of the point $\vec{\tau} = \varepsilon\vec{R}$, we pass to the dimensionless optical coordinates. Equation (2.3.1) will then be rewritten as

$$\begin{gathered}\vec{n}\cdot\nabla_{\vec{\tau}} I(\vec{\tau},\vec{n}) = -I(\vec{\tau},\vec{n}) + B'_{\mathrm{full}}(\vec{\tau},\vec{n}) \\ B'_{\mathrm{full}}(\vec{\tau},\vec{n}) = \frac{\Lambda}{4\pi}\iint I(\vec{\tau},\vec{n}')x(\vec{n},\vec{n}')\,d\vec{n}' + B'(\vec{\tau},\vec{n})\,.\end{gathered} \tag{2.4.1}$$

Here,

$$B'(\vec{\tau},\vec{n}) = \frac{B(\vec{R},\vec{n})}{\varepsilon} \tag{2.4.2}$$

is the source function.

Equation (2.4.1) shows that the light field characteristics, as functions of the dimensionless $\vec{\tau}$ coordinates, are independent of $\varepsilon$. This is the reason why the dimensionless coordinates are generally used in theoretical studies.

At the risk of appearing tautological we now express the above statement in another way. Equation (2.4.1) implies that if the scattering indicatrices and the parameter $\Lambda$ are equal in different media, then the radiances are also equal at the points with similar $\vec{\tau}$ coordinates. Naturally, we consider that the boundary conditions in the dimensionless coordinates are the same. We have formulated this rather obvious point in order to explain the concept behind experimental modelling. The free photon path $1/\varepsilon$ in the medium determines only the similarity scale.

Thus, the similarity principle can be used to study general radiation transfer laws for media with a characteristic scale ranging from hundreds of meters to tens of kilometers (atmosphere, sea, etc.) on laboratory models where the appropriate scale is tens of centimeters. On the other hand, in the case of free path lengths below $10^{-2}$ m (photographic materials, luminescent screens, cloth, paper), laboratory models allow the scale to be extended for more convenient study of, for example, light fields inside an object. At such $\varepsilon$, we must, of course, exercise caution and allow for the fact that various close packing effects may occur in an object to be modelled [2.24, 32].

It should be noted that in modelling, all the optical coordinates which characterize the system must be kept equal, both for the object and its model. This also applies to radiation sources and receiver sizes.

Situations are often encountered where the optical parameters of the medium depend on the spatial coordinates. We shall refer to such scattering media as inhomogeneous, as opposed to homogeneous where the optical characteristics at all points of the medium are equal.

For an inhomogeneous scattering medium, the transfer equation can be made dimensionless by introducing, generally speaking, any characteristic scale $1/\varepsilon_0$,

$$\vec{n}\cdot\nabla_{\vec{\tau}} I(\vec{\tau},\vec{n}) = -\frac{\varepsilon(\vec{\tau})}{\varepsilon_0} I(\vec{\tau},\vec{n}) + \frac{\Lambda(\tau)}{4\pi}\iint I(\vec{\tau},\vec{n}')x(\vec{n},\vec{n}')\,d\vec{n}' + B'(\vec{\tau},\vec{n}), \tag{2.4.3}$$

$$\vec{\tau} = \varepsilon_0\vec{R}, \qquad \Lambda(\vec{\tau}) = \frac{\sigma(\vec{\tau})}{\varepsilon_0}. \tag{2.4.4}$$

When realizing the model, it is necessary, in this case, to simulate the relative spatial distribution of the optical characteristics.

We can readily formulate the main modelling principle for a complex medium and experimental geometries: all angular characteristics of the medium and sources must be equal, both in the model and prototype. Such a formulation reveals, for example, the physics of the optical area of sources. This means a solid angle over which a source is seen from a distance equal to the mean free path length of a photon.

The similarity and modelling concepts have proved to be very fruitful [2.31]. Information gained on this basis regarding the laws of narrow beam propagation, as well as the results of model experiments as applied to the problems of hydrooptics and photography, are generalized in monographs [2.32, 33].

## 2.5 Corollaries of the Optical Reciprocity Theorem

Symmetry properties are very important in the study of radiation and image transfer problems and can be determined prior to the mathematical solution. Moreover, this information turns out to be useful when seeking the solution. In many cases, it essentially simplifies the mathematics involved and reduces the number of experimental studies required.

The optical reciprocity theorem and its corollaries make a particular contribution to the study of multiple light scattering and vision theory. In this section, we shall look at some corollaries of the reciprocity theorem which are useful in solving the problems stated in this monograph.

We first formulate the optical reciprocity theorem as applied to a radiation field.

Let a source emitting a flux $P_{\uparrow}(\vec{R},\vec{n})$ in the $\vec{n}$ direction be at the space point $\vec{R}$. Owing to medium scattering, the radiance at some point $\vec{R}'$ in the $\vec{n}'$ direction will be

$$I_{\downarrow}(\vec{R}',\vec{n}') = P_{\uparrow}(\vec{R},\vec{n})I_{\mathrm{n}}(\vec{R},\vec{n};\vec{R}',\vec{n}')\,. \qquad (2.5.1)$$

Here, $I_{\mathrm{n}}(\vec{R},\vec{n};\vec{R}',\vec{n}')$ is the normalized radiance at the point $\vec{R}'$ in the $\vec{n}'$ direction due to the light flux emitted from the point $\vec{R}$ in the $\vec{n}$ direction (normalized radiance definitions are given in Sects. 3.1 and 4.1). Here and below, the sign $\uparrow$ stands for the light field characteristics at the source point; $\downarrow$ denotes the light field characteristics at the observation point. Now, let the source radiating in the $-\vec{n}'$ direction, i.e. a direction opposite to the above, be at the point $\vec{R}'$. The emerging flux is $P_{\uparrow}(\vec{R}',-\vec{n}')$. The radiance at $\vec{R}$ in the $\vec{n}$ direction is then

$$I_{\downarrow}(\vec{R},-\vec{n}) = P_{\uparrow}(\vec{R}',-\vec{n}')I_{\mathrm{n}}(\vec{R}',-\vec{n}';\vec{R},-\vec{n})\,. \qquad (2.5.2)$$

The optical reciprocity theorem states that

$$I_{\mathrm{n}}(\vec{R},\vec{n};\vec{R}',\vec{n}') = I_{\mathrm{n}}(\vec{R}',-\vec{n}';\vec{R},-\vec{n})\,. \qquad (2.5.3)$$

Thus, in particular, if the emerging fluxes are equal, i.e.

$$P_{\uparrow}(\vec{R},\vec{n}) = P_{\uparrow}(\vec{R}',-\vec{n}'), \quad \text{then} \quad I_{\downarrow}(\vec{R}',\vec{n}') = I_{\downarrow}(\vec{R},-\vec{n})\,.$$

The optical reciprocity theorem is, in fact, valid for any media (finite and infinite, homogeneous and inhomogeneous, etc.). It implies that only the condition of beam reciprocity is satisfied. As far as the scattering medium is concerned, this means, in particular, that the scattering indicatrix possesses mirror symmetry.

Consider some corollaries of the reciprocity theorem.

*Corollary 1.* As before, the radiation flux $P_\uparrow(\vec{R},\vec{n})$ emerges from the point $\vec{R}$ in the scattering medium in the $\vec{n}$ direction (Fig. 2.2a). We find the flux $P_\downarrow(\vec{R}',\vec{n}_0')$ over a solid angle $\Omega'$ through an area $d\Sigma'$ with the external normal at $n_0'$. Evidently, this radiant flux will be

$$P_\downarrow(\vec{R}',\vec{n}_0') = d\Sigma'(\vec{n}_0')\iint_{\Omega'} I_\downarrow(\vec{R}',\vec{n}')\cos(\vec{n}',\widehat{\ }-\vec{n}_0')\,d\vec{n}'$$

$$= P_\uparrow(\vec{R},\vec{n})\,d\Sigma'(\vec{n}_0')\iint_{\Omega'} I_\mathrm{n}(\vec{R},\vec{n};\vec{R}',\vec{n}')\cos(\vec{n}',\widehat{\ }-\vec{n}_0')\,d\vec{n}'\,. \quad (2.5.4)$$

We now place at the point $\vec{R}$ a radiation source of area $d\Sigma'(\vec{n}_0')$ which emits the radiance $I_\uparrow(\vec{R}',-\vec{n}')$, over the solid angle $\Omega'$, in the $-\vec{n}'$ direction. Then, at the point $\vec{R}$, the radiance in the $-\vec{n}$ direction will be

$$I_\downarrow(\vec{R},-\vec{n}) = d\Sigma'(\vec{n}_0')\iint_{\Omega'} I_\mathrm{n}(\vec{R}',-\vec{n}';\vec{R},-\vec{n})I_\uparrow(\vec{R}',-\vec{n}')$$

$$\times\cos(-\vec{n}',\widehat{\ }\vec{n}_0')\,d\vec{n}'\,. \quad (2.5.5)$$

If, in the second experimental scheme, the area $d\Sigma'$ emits according to the Lambert law, i.e., $I_\uparrow(\vec{R}',-\vec{n}') = \text{const} = I_\uparrow(\vec{R}')$, from the reciprocity theorem and a comparison of the relations (2.5.4) and (2.5.5), we arrive at

$$\frac{P_\downarrow(\vec{R}',\vec{n}_0')/d\Sigma'(\vec{n}_0')}{P_\uparrow(\vec{R},\vec{n})} = \frac{I_\downarrow(\vec{R},-\vec{n})}{I_\uparrow(\vec{R}')\,d\Sigma'(\vec{n}_0')}\,. \quad (2.5.6)$$

This relation holds for any solid angle $\Omega'$. If $\Omega' = 2\pi$, then $P_\downarrow(\vec{R}',\vec{n}_0')/d\Sigma'(\vec{n}_0') = E_\downarrow(\vec{R}',\vec{n}_0')$ signifies irradiance and, instead of (2.5.6), we have

$$\frac{E_\downarrow(\vec{R}',\vec{n}_0')}{P_\uparrow(\vec{R},\vec{n})} = \frac{I_\downarrow(\vec{R},-\vec{n})}{I_\uparrow(\vec{R}')\,d\Sigma'(\vec{n}_0')}\,. \quad (2.5.7)$$

The ratio $E_\downarrow(\vec{R}',\vec{n}_0')/P_\uparrow(\vec{R},\vec{n}) = S(\vec{R}')$ is the irradiance coefficient [cf (4.1.5, 6)].

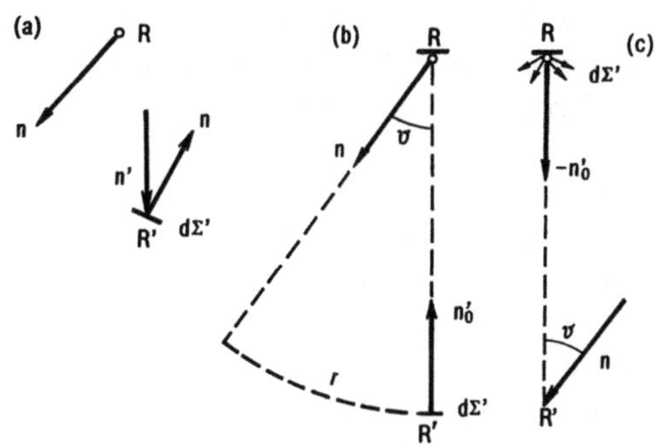


**Fig. 2.2.** Proof of the corollaries of the reciprocity theorem

Expression (2.5.7) is one of the most frequently used corollaries of the reciprocity theorem. It is most widely applied in the theory of vision (for example, Sect. 5.3).

It should be emphasized that in deriving the relations (2.5.6) and (2.5.7), we do not make any additional assumptions, i.e. they follow directly from the reciprocity theorem and can be applied for any media.

*Corollary 2.* We now introduce some restrictions.

Let the scattering medium be infinite and homogeneous. In this instance, in the second experimental scheme the points $\vec{R}$ and $\vec{R}'$ may be interchanged. The $-\vec{n}$ and $\vec{n}_0$ directions must be reversed as $\vec{n}$ and $-\vec{n}_0$. Then,

$$\frac{I_\downarrow(\vec{R}, -\vec{n})}{I_\uparrow(\vec{R}')\, d\Sigma'(\vec{n}_0')} = \frac{I_\downarrow(\vec{R}', \vec{n})}{I_\uparrow(\vec{R})\, d\Sigma'(-\vec{n}_0')},$$

and from the relation (2.5.7), we have

$$\frac{E_\downarrow(\vec{R}', \vec{n}_0')}{P_\uparrow(\vec{R}, \vec{n})} = \frac{I_\downarrow(\vec{R}', \vec{n})}{I_\uparrow(\vec{R})\, d\Sigma'(-\vec{n}_0')}. \tag{2.5.8}$$

Expression (2.5.8) implies that the irradiance distribution $E_\downarrow(\vec{R}', \vec{n}_0')$ (dependent on the coordinate $\vec{R}'$ and the area $d\Sigma'$ orientation determined by $\vec{n}_0'$ due to the source emitting in the $\vec{n}$ direction) can be replaced by the radiance distribution $I_\downarrow(\vec{R}', \vec{n})$ (originated in the scattering medium by the emitting Lambert area $d\Sigma'$ with the external normal vector $-\vec{n}_0'$) and the radiance $I_\uparrow(\vec{R}')$.

This conclusion is also approximately valid for finite (but, it should be emphasized, homogeneous) media if the energy sink through the boundaries is small or, at least, affects only slightly the radiation field at $\vec{R}'$. For example, in the case of the semi-infinite medium, the accuracy of this approximation is therefore greater, the smaller the single scattering albedo and the more forward-extended the scattering indicatrix of the elementary volume. This is because, under such conditions, the intensity of the backscattering, as well as the contribution of multiple scattering to the light field made by the semispace behind the emitter decreases.

Given that $\vec{n}_0' = -\vec{n}$, it is seen from formula (2.5.8) that the axial irradiance from the directional source, normalized to the radiant flux $P_\uparrow(\vec{R}, \vec{n})$, is equal to the radiance in the direction to the diffuse emitter normalized to its intensity $I_\uparrow(\vec{R})\, d\Sigma'$.

Formula (2.5.8) as applied to infinitely extended sources (by integrating with respect to $\vec{R}$) yields more or less apparent results. At any spacial point, the irradiance distribution dependent on the incident angle of external radiation is equivalent to the angular radiance distribution in the scattering medium illuminated by a diffuse source.

*Corollary 3.* We fix the distance $l = |\vec{R}' - \vec{R}|$ between the source and the observation point. Moreover, we consider that the $\vec{n}_0'$ normal of the area $d\Sigma'$ in the first experimental scheme is directed all the time to the source (Fig. 2.2b).

Then, the relation (2.5.8) can be understood as follows. In an infinite homogeneous medium, the irradiance distribution $E_{\downarrow}$ at the points on the circular arc of radius $l$ normalized to the power $P_{\uparrow}$ due to the directional source (Fig. 2.2b) coincides with the angulár radiance distribution $I_{\downarrow}$ normalized to the intensity $I_{\uparrow}(\vec{R})\,d\Sigma'$ at the $-\vec{n}_0'$ axis of the diffuse emitter (Fig. 2.2c).

In addition to imposing the condition of an infinite and homogeneous medium, we now assume that the angle $v$ between the $\vec{n}$ and $-\vec{n}_0'$ directions is small. Therefore, we may approximately consider that all the points $\vec{R}'$ of the radius $l$ arc lie in the same plane. Then, their location may be specified by the distance $r = lv$ in the plane. Hence, the above statement means that the irradiance distribution in the plane normal to the light beam is equivalent to the angular radiance distribution in the scattering media illuminated by a diffuse souce, i.e.

$$\frac{E_{\downarrow}(r = lv)}{P_{\uparrow}(\vec{R}, \vec{n})} = \frac{I_{\downarrow}(v = r/l)}{I_{\uparrow}(\vec{R})\,d\Sigma'}\,. \tag{2.5.9}$$

In particular, for normalized distributions, we have

$$\frac{E_{\downarrow}(r = lv)}{E_{\downarrow}(r = 0)} = \frac{I_{\downarrow}(v = r/l)}{I_{\downarrow}(v = 0)}\,. \tag{2.5.10}$$

The reciprocity theorem, as applied to a radiation field and solutions of the transfer equation, is discussed in monographs [2.1, 9, 30, 34]. Some of its corollaries given in this section and used in vision theory, are formulated in [2.31, 35].

To conclude, we note that the optical reciprocity theorem and its corollaries reflect the most general properties of radiation fields in scattering media. Therefore, their validity must serve as a criterion of the correctness of approximate solutions of the transfer equation, as well as of the experimental data.

# 3 Light Scattering in Semi-Infinite Media and Plane Layers Illuminated by Infinitely Extended Plane Sources

Radiation transfer theory is a well-developed area of mathematical physics. It was devised with the problems of astrophysics, neutron transfer theory, atmospheric optics, hydrooptics and so on in mind. Its basic concepts are put forth in Refs. 3.1–6.

Transfer theory provides rigorous analytical solutions only for simple model problems and, in particular, for a plane homogeneous layer illuminated by an infinitely extended plane source. Relatively simple solutions are only obtained for simple scattering indicatrices: spherical, Rayleigh, $x(\beta) = 1 + x_1 \cos \beta$ [3.1–3, 6]. The analytical theory of anisotropic scattering has been developed over the last twenty years. Its principal results are presented by *Sobolev* and van de *Hulst* [3.4, 5]. However, only some asymptotic solutions of this theory can be used directly to solve applied problems in design calculations.

Thus, problems arising from the study of image transfer, the design of communication and location systems and their optimization, etc. have no rigorous analytic solutions. It has now become clear that such solutions cannot be obtained in a simple form.

At present, numerical methods of calculating light fields are being rapidly developed. A variety of fundamental programs implementing different algorithms for the solution of transfer problems such as finite-difference and iterative procedures [3.7–9], Monte-Carlo simulation [3.10, 11] have been proposed. However, these intricate methods, which require much skill and extensive machine time, can hardly serve as a basis for global design calculations, and especially not in those problems where radiation transfer is only one aspect of a more complex program.

Approximate methods play a key role in providing simple solutions. Preference must, of course, be given to approaches which are genetically bound up with rigorous transfer equations and ensure correct asymptotic solutions. Approximate methods should rely upon numerical procedures to expose the errors they involve and to specify their applications, and be complemented by them when necessary.

Approximate methods and solutions in radiation transfer theory are dealt with in Chap. 3 and 4 of the book. In this chapter, we consider solutions for light fields when a semi-infinite medium with a plane boundary or a plane layer is illuminated by an infinitely extended plane source. These solutions are generally required in order to calculate sun radiation background, understand vision specificities during daytime, etc. In deriving simplified formulas, we shall rely, wherever possible, upon rigorous asymptotic solutions of the radiation transfer equation (Sects. 3.2–5).

Some a priori information on light field properties is required in order to formulate approximate equations. From this point of view, the approaches used can be classified into three major groups. One (Sect. 3.6) is based on the assumption that the properties of multiply scattered light are mainly determined by the integral parameters of the scattering indicatrix (asymmetry parameter, back/forward ratio), and are only weakly dependent on the details of the angular structure of the function $x(\beta)$. Rather than a real indicatrix, we apply a simplified model with the same integral parameter. The transfer equation is then written for this model indicatrix.

For another group, the radiance is averaged over the directions, and differential equations for the angular radiance moments are formulated. A closed system of equations which assumes some relationship between these moments is presented. In Sect. 3.7, a well-known two-stream approximation is described. The diffusion equation and its solutions are not introduced until Chap. 4, since they are generally used to study light fields due to local sources.

The following chapter deals with one more group of approximations which form the basis of the theory of light beam scattering. They include small-angle approaches which assume rapid intensity decrease as a function of the scattering angle. Naturally, in all methods used to describe narrow beam propagation, the appropriate solutions for infinitely extended sources are obtained as a particular case. Chaps. 3 and 4 complement each other and, in general, present the machinery for the applied transfer theory.

## 3.1 Basic Equations

Consider a plane layer of infinite width, scattering and absorbing, and of thickness $z_0$, whose upper boundary is uniformly illuminated. We measure the depth $z$ in the layer from the upper layer boundary. A spherical coordinate system where the polar axis coincides with the $z$ axis is used to define the directions. The polar angle is $\theta$ ($\mu = \cos\theta$), and $\psi$ is the azimuthal angle, i.e., $\vec{n}(\mu, \psi)$ (Fig. 3.1). If we introduce a dimensionless optical depth, then,

$$\tau = \int_0^z \varepsilon(z')\,dz' . \tag{3.1.1}$$

Correspondingly, the optical thickness of a scattering layer is

$$\tau_0 = \int_0^{z_0} \varepsilon(z')\,dz' . \tag{3.1.2}$$

Therefore, the transfer equation in dimensionless coordinates is of the form

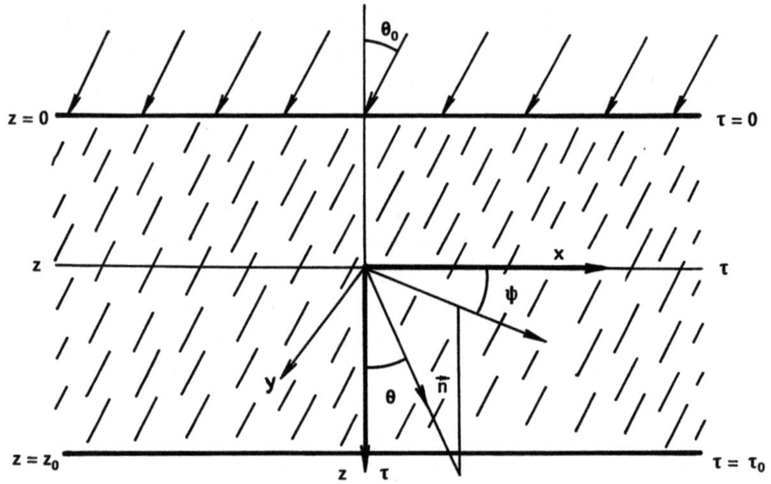


Fig. 3.1. Coordinate system of the layered medium

$$\mu\frac{dI(\tau,\mu,\psi)}{d\tau} = -I(\tau,\mu,\psi) + \frac{\Lambda}{4\pi}\int_0^{2\pi} d\psi' \int_{-1}^{1} I(\tau,\mu',\psi')x(\beta)\,d\mu' + B'(\tau,\mu,\psi)\,, \tag{3.1.3}$$

where

$$\cos\beta = \mu\mu' + \sqrt{1-\mu^2}\sqrt{1-\mu'^2}\cos(\psi-\psi')\,, \tag{3.1.4}$$

and the quantity $B'(\tau,\mu,\psi)$ allows for external sources. If the single scattering albedo and scattering indicatrix are independent of the layer depth, inhomogeneity being determined only by the function $\varepsilon(z)$, then, equation (3.1.3) leads to the obvious but significant conclusion that all light field characteristics in such media depend only on the coordinate $\tau$ and, in this sense, are equal for both homogeneous and inhomogeneous layers.

Now, let a parallel beam be incident on the upper medium boundary, making an angle $\theta_0 = \arccos\mu_0$ with the normal to the boundary, and generate an irradiance $E_0$ on the area perpendicular to the incident beam. Then,

$$B'(\tau,\mu,\psi) = \frac{\Lambda E_0}{4\pi}x(\beta_0)\mathrm{e}^{-\tau/\mu_0}\,, \tag{3.1.5}$$

$$\cos\beta_0 = \mu\mu_0 + \sqrt{1-\mu^2}\sqrt{1-\mu_0^2}\cos\psi\,. \tag{3.1.6}$$

Given such a source function, $I$ represents the scattered light radiance. If we include unscattered light, we obtain total radiance. The zeroth azimuthal plane coincides with the incidence plane, i.e., $\psi_0 = 0$. The boundary conditions are

$$I_0(0,\mu,\psi) = 0 \qquad \text{at } 0 < \mu \leqslant 1$$

$$I(\tau_0,\mu,\psi) = 0 \qquad \text{at } -1 \leqslant \mu < 0\,, \tag{3.1.7}$$

indicating that no diffuse radiation is incident on the layer boundaries from the outside. Equation (3.1.3), (taking into account (3.1.5) and the boundary conditions (3.1.7), provides a complete mathematical statement of the problem.

In the analysis which follows, we frequently use the normalized radiance $I_n(\tau, \mu, \psi)$ per unit incident flux

$$I_n(\tau, \mu, \psi) = I(\tau, \mu, \psi)/E_0\mu_0 . \tag{3.1.8}$$

The quantity $I_n(\tau, \mu, \psi)$ has the dimension $sr^{-1}$. The dimensionless quantity $\pi I_n(\tau, \mu, \psi)$ is the radiance coefficient. We shall explain the physical meaning of this quantity. It is clear that the radiance scattered by an orthotropic absolutely white area at the upper boundary of the scattering layer is equal to $E_0\mu_0/\pi$. Hence, the radiance coefficient is the ratio of the radiance at a depth $\tau$ in the $\vec{n}(\mu, \psi)$ direction to the scattering radiance of the orthotropic white area at the upper boundary of the scattering medium.

This question of determining the radiance in a plane layer illuminated by parallel beams is of particular interest in radiation transfer theory. It is defined as the major problem in [3.4]. Its solution provides the radiance in a layer for any angular incident light distribution for the case of internal radiation sources [3.4], etc. Monographs [3.1, 4, 5, 12] give an analysis and solution of this problem for particular situations, using different methods. In the following four sections, we list and apply asymptotic solutions of this problem to semi-infinite media and layers of finite optical thickness.

## 3.2 Asymptotic Regime in Deep Layers

The problem of a radiation field in a semi-infinite medium is among the most fundamental and well-investigated in radiation transfer theory ([3.4, 6, 13] and references cited therein). Consider now the properties of light fields for deep layers.

At large optical depth ($\tau \gg 1$), an asymptotic regime is developed in the medium to be characterized by separation of the angular and spatial variables, azimuthal independence of the field and exponential attenuation of the radiance with the depth, viz.

$$I_n(\tau, \mu, \mu_0) = \frac{1}{\pi} u(\mu_0)\varphi(\mu)e^{-\gamma\tau} . \tag{3.2.1}$$

Here, $\varphi(\mu)$ is the relative angular radiance distribution for the asymptotic regime and $\gamma$ is the asymptotic attenuation coefficient. The function $u(\mu_0)$ allows for the radiance dependence on the beam incidence angle.

### 3.2.1 Asymptotic Attenuation Coefficient and Angular Radiance Distribution

Substituting (3.2.1) into (3.1.3) shows that the function $\varphi(\mu)$ is determined as a positive eigenfunction in the characteristic transfer equation

$$(1-\gamma\mu)\varphi(\mu)=\frac{\Lambda}{2}\int_{-1}^{1}x(\mu,\mu')\varphi(\mu')\,d\mu'\,, \tag{3.2.2}$$

and conforms to the smallest characteristic number $\gamma$. The kernel of equation (3.2.2)

$$x(\mu,\mu')=\frac{1}{2\pi}\int_{0}^{2\pi}x(\cos\beta)\,d\psi \tag{3.2.3}$$

is the azimuth-averaged scattering indicatrix. The function $\varphi(\mu)$ is frequently normalized by the relation

$$\frac{\Lambda}{2}\int_{-1}^{1}\varphi(\mu)\,d\mu=1\,. \tag{3.2.4}$$

*Ambartsumyan* [3.14] pioneered the study of an asymptotic regime. *Maslennikov* [3.13] made a fundamental study of the characteristic transfer equation for anisotropic scattering.

Nowadays, a number of algorithms are being developed to solve the characteristic equation numerically [3.15–18], and extensive data are tabulated for $\gamma$ and $\varphi(\mu)$ at different $x(\beta)$ and $\Lambda$. The algorithms presented in [3.16, 17] seem to be the most convenient for anisotropic scattering. Although the calculation of $\gamma$ and $\varphi(\mu)$ is now simple, analytical solutions are of special importance in the development of multiple scattering theory and its applications.

In the simplest case of spherical scattering, i.e., at $x(\beta)=1$,

$$\varphi(\mu)=\frac{1}{1-\gamma\mu}\,, \tag{3.2.5}$$

and the asymptotic attenuation coefficient $\gamma$ is determined from the equation

$$\frac{\Lambda}{2\gamma}\ln\frac{1+\gamma}{1-\gamma}=1\,. \tag{3.2.6}$$

Similar simple, although less elegant, formulas for $\gamma$ and $\varphi(\mu)$ using the non-spherical indicatrix $x(\beta)=1+x_1\cos\beta$ and Rayleigh scattering can be found in [3.1, 3].

In the more general case where the indicatrix is given as the Legendre polynomial expansion (2.2.7), the function $\varphi(\mu)$ is sought as

$$\varphi(\mu)=\frac{1}{\Lambda}\sum_{n=0}^{N}(2n+1)b_nP_n(\mu)\,, \tag{3.2.7}$$

where the coefficients $b_n$ are related as in [3.19]

$$nb_n=(2n-1-x_{n-1}\Lambda)b_{n-1}/\gamma-(n-1)b_{n-2}\,;\qquad b_0=1\,. \tag{3.2.8}$$

As seen from (3.2.7, 8), the average cosine of the asymptotic distribution $\langle\mu_{\rm as}\rangle=\int_{-1}^{1}\varphi(\mu)\mu\,d\mu/\int_{-1}^{1}\varphi(\mu)\,d\mu$ can be expressed simply in terms of $\Lambda$ and $\gamma$:

$$\langle \mu_{as} \rangle = \frac{1 - \Lambda}{\gamma} . \tag{3.2.9}$$

This important relation can easily be obtained from equation (3.2.2) by integrating with respect to $\mu$ between $-1$ and 1.

For weak absorption, the quantities $\gamma$ and $\varphi(\mu)$ are described by the known asymptotic relations

$$\varphi(\mu) = 1 + 3q\gamma\mu , \tag{3.2.10}$$

where

$$\gamma = \sqrt{(1 - \Lambda)(3 - x_1)} = \sqrt{(1 - \Lambda)/q} , \tag{3.2.11}$$

$$q = \frac{1}{3 - x_1} = \frac{1}{3(1 - g)} \tag{3.2.12}$$

is the scattering indicatrix parameter. If $\Lambda = 1$, then $\gamma = 0$, $\varphi(\mu) = 1$. However, as the indicatrix extends ($g \to 1$), the applicability range of formulas (3.2.10, 11) is confined to a small vicinity around the point $\Lambda = 1$. In (3.2.10), we retained only the linear term in the $\gamma$ series expansion of $\varphi(\mu)$. The $\gamma$ power series expansion of $\varphi(\mu)$ where allowance is made for the higher-order terms is presented in [3.16, 20], but it is hardly pragmatic. Small absorption applications can be somewhat extended by the exponential approximation

$$\varphi(\mu) = \frac{3q\gamma}{\Lambda \sinh 3q\gamma} e^{3q\gamma\mu} , \tag{3.2.13}$$

which makes exact allowance for the linear term, and only approximately allows for nonlinear ones in the $\gamma$ power series expansion of $\varphi(\mu)$. This approximation, which is accurate to $\leqslant 15\%$, can be used at $q\gamma \leqslant 0.6$ or, which amounts to the same, at $(1 - \Lambda)/(1 - g) \lesssim 1$ since, from (3.2.11), $(1 - \Lambda)/(1 - g) = 3q^2\gamma^2$.

For strongly extended indicatrices and appreciable absorption, the asymptotic angular radiance distribution is very forward-extended. For $\mu > 0$, it can be approximated roughly by [3.21]

$$\varphi(\mu) = \frac{2\gamma}{\Lambda s} \exp\left[ -\frac{\gamma}{s}(1 - \mu) \right] , \tag{3.2.14}$$

where

$$\gamma = 1 - \Lambda + s , \qquad s = \sqrt{\Lambda(1 - \Lambda)(1 - g)} . \tag{3.2.15}$$

As seen from Fig. 3.2, approximation (3.2.14) is sufficient for those problems where it is necessary to know the radiance distribution over the angles where the greater light energy is concentrated. At $\mu \lesssim 0$, the function $\varphi(\mu)$ now closely correlates with $x(\beta)$ at $\beta > \pi/2$. The corresponding estimates, graphs and approximate formulas are given in [3.21]. Expressions (3.2.14, 15) can be applied at

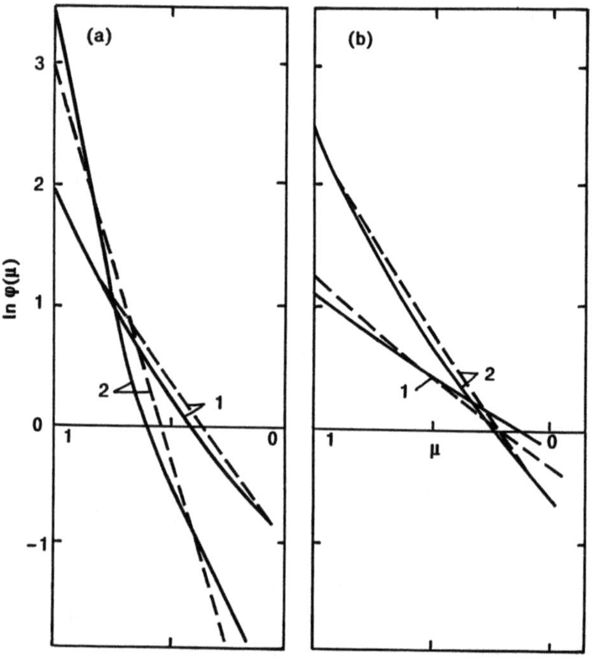


**Fig. 3.2.** Function $\varphi(\mu)$ for sea water scattering indicatrices at (a) $g = 0.953$ and (b) 0.753; (curve 1) $\Lambda = 0.9$ and (curve 2) $\Lambda = 0.6$. *Solid lines*, exact values; *dashed lines*, approximate values from (3.2.14)

$$\alpha = \frac{1 - \Lambda}{\Lambda(1 - g)} \gg 1 . \tag{3.2.16}$$

We have used the parameter $\alpha$ to estimate the applications of the formulas. For weak absorptions, it coincides with the quantity $(1 - \Lambda)/(1 - g)$. Given the parameter $\alpha$, which allows for medium absorption along with scattering anisotropy, we may determine the domains of "weak" and "strong" absorptions. The physics of this quantity as a characteristic parameter in multiple scattering theory will be revealed in Sect. 4.7.

It should be noted that in (3.2.11, 15), which are applicable in two contrasting situations, the impact of the scattering indicatrix on the asymptotic attenuation coefficient $\gamma$ is only exhibited through the asymmetry parameter $g$. This conclusion can be approximately extended to the entire range [3.21]. As noted in [3.5, 22], it is a peculiarity of those light fields with essential multiple scattering discussed in Chaps. 3 and 4. This point, as applied to the asymptotic regime in deep layers, is presented by the data in Table 3.1. A more detailed illustration is given in [3.21]. For anisotropic scattering, [3.21, 22] give the approximation

$$\gamma = 1 - \Lambda + [\Lambda(1 - \Lambda)]^l [3(1 - g)]^m , \tag{3.2.17}$$

where

**Table 3.1.** Values of $\gamma$ for geophysical scattering indicatrices (upper figure in each pair of rows) and Henyey-Greenstein indicatrix (lower figures) for equal $g$ values.

| Media | $g$ | $\Lambda$ | | | | |
|---|---|---|---|---|---|---|
| | | 0.9 | 0.8 | 0.7 | 0.6 | 0.5 |
| Cloud C.1 | 0.853 | 0.231 | 0.351 | 0.457 | 0.554 | 0.645 |
| | | 0.235 | 0.360 | 0.468 | 0.567 | 0.659 |
| Sea water | 0.75 | 0.287 | 0.422 | 0.533 | 0.631 | 0.718 |
| | | 0.287 | 0.422 | 0.533 | 0.631 | 0.718 |
| | 0.83 | 0.221 | 0.341 | 0.447 | 0.544 | 0.636 |
| | | 0.226 | 0.349 | 0.457 | 0.555 | 0.648 |

**Table 3.2.** Values of $\gamma$ for the Henyey-Greenstein indicatrix.

| $\Lambda$ | $g$ | | | | | | | | |
|---|---|---|---|---|---|---|---|---|---|
| | 0.5 | 0.6 | 0.7 | 0.8 | 0.85 | 0.9 | 0.95 | 0.97 | 0.98 |
| 0.2 | 0.9776 | 0.9626 | 0.9424 | 0.9155 | 0.8987 | 0.8783 | 0.8518 | 0.8379 | 0.8294 |
| 0.4 | 0.8911 | 0.8585 | 0.8201 | 0.774 | 0.7467 | 0.7149 | 0.675 | 0.6546 | 0.6422 |
| 0.5 | 0.8279 | 0.7891 | 0.7442 | 0.6919 | 0.6611 | 0.6257 | 0.5818 | 0.5594 | 0.5459 |
| 0.6 | 0.7511 | 0.7079 | 0.6591 | 0.6023 | 0.5695 | 0.5318 | 0.4855 | 0.462 | 0.4479 |
| 0.7 | 0.6578 | 0.6128 | 0.5624 | 0.5042 | 0.4704 | 0.4325 | 0.3857 | 0.362 | 0.3479 |
| 0.8 | 0.5418 | 0.4985 | 0.4502 | 0.3944 | 0.3624 | 0.3258 | 0.2812 | 0.2587 | 0.2453 |
| 0.9 | 0.3856 | 0.3501 | 0.3101 | 0.2639 | 0.2371 | 0.2064 | 0.1686 | 0.1496 | 0.1383 |
| 0.94 | 0.2993 | 0.2702 | 0.2373 | 0.199 | 0.1766 | 0.1508 | 0.1187 | 0.1025 | 0.0928 |
| 0.98 | 0.1731 | 0.1553 | 0.1352 | 0.1114 | 0.0974 | 0.0809 | 0.0601 | 0.0492 | 0.0427 |
| 0.99 | 0.1224 | 0.1097 | 0.0952 | 0.0781 | 0.068 | 0.056 | 0.0407 | 0.0326 | 0.0276 |

$$l = (1 + \Lambda^5)^{-1} ,$$

$$m = 0.6 \qquad \text{at } 1/x(1) \leqslant \Lambda \leqslant 0.82 + 0.18g ,$$

$$= 0.5 \qquad \text{at } \Lambda > 0.82 + 0.18g . \tag{3.2.18}$$

This formula provides $\gamma$ estimates with an error $\lesssim 2\%$ at $g > 0.5$. The restriction $\Lambda > 1/x(1)$ in (3.2.18) excludes the vicinity around the point $\Lambda = 0$, where the function $\varphi(\mu)$ is closely correlated to $x(\mu)$. At $\Lambda \to 1$, expressions (3.2.17, 18) become the asymptotic relation (3.2.11). At $(1 - \Lambda)/(1 - g) \gg 1$, they improve the approximation (3.2.15). Table 3.2 presents the asymptotic attenuation coefficient $\gamma$ for the model Henyey–Greenstein indicatrix (2.2.9). These data are obtained from the numerical solution of (3.2.2), using the methods described in [3.16]. The double line depicts the domain where the error in (3.2.17) is smaller or equal to 1%. In view of this, we can also use the data of Table 3.2 for other scattering indicatrices with the same parameter $g$.

### 3.2.2 The Light Field in Absolute Units

Although light field radiance in deep layers can be calculated in absolute units using (3.2.1), we still have to define the function $u(\mu)$. This function describes the angular radiance distribution at the semi-infinite medium boundary induced by an infinitely remote source deep in the medium [the Milne problem Refs. 3.4, 6, 13]. In radiation transfer theory, it is referred to as the escape function. It depends on the scattering indicatrix and single scattering albedo, and is normalized by the relation

$$2\int_0^1 u(\mu)\varphi(\mu)\mu\,d\mu = 1\,. \tag{3.2.19}$$

Integral equations for the function $u(\mu)$ can be found in [3.4]. We shall cite here only the simple asymptotic and approximate solutions for $u(\mu)$ used in applied problems. Let $u_0(\mu)$ be the function $u(\mu)$ at $\Lambda = 1$. In the region of weak absorptions, $u(\mu)$ is defined by the asymptotic formula

$$u(\mu) = u_0(\mu)\left[1 - 6q\gamma\int_0^1 u_0(\mu)\mu^2\,d\mu\right]. \tag{3.2.20}$$

Analysis of numerical calculations shows [3.18, 23, 24] that, given either of the scattering indicatrices, the linear approximation of the function $u_0(\mu)$

$$u_0(\mu) = \tfrac{3}{7}(1 + 2\mu)\,, \tag{3.2.21}$$

first obtained for isotropic scattering and obeying the normalization condition

$$2\int_0^1 u_0(\mu)\mu\,d\mu = 1\,, \tag{3.2.22}$$

can be used to a sufficient accuracy. In this case, the zeroth ($\int_0^1 u_0(\mu)\,d\mu$) and second ($\int_0^1 u_0(\mu)\mu^2\,d\mu$) angular moments of the approximate function (3.2.21) differ by less than 3% from the real values for different model and geophysical indicatrices. The deviation of more than 5% in the data obtained by (3.2.21) and the exact results is only observed at $\mu < 0.25$ [3.24]. (See also Table 3.6).

For strongly anisotropic scattering and appreciable real absorption $(1 - \Lambda)/\Lambda(1 - g) \geqslant 2$ with insignificant light backscattering by the medium, the angular distribution deep in the medium, as well as that emerging from it, in fact, coincide. Then, we have [3.21, 22]

$$u(\mu) = \frac{\varphi(\mu)}{M}\,, \tag{3.2.23}$$

where

$$M = 2\int_{-1}^1 \varphi^2(\mu)\mu\,d\mu\,. \tag{3.2.24}$$

The relation (3.2.24) results from the normalization condition (3.2.19).

We shall estimate the error in $u(\mu)$ determined by (3.2.23) [3.21]:

$$\frac{\Delta u(\mu)}{u(\mu)} \leqslant \frac{\Lambda}{3q(1-\gamma\mu)} e^{-4\gamma\mu\sqrt{q}} . \qquad (3.2.25)$$

Expression (3.2.25) implies that, at $(1-\Lambda)/\Lambda(1-g) \geqslant 2$ and $\mu > 0.2$, the error in (3.2.23) is no more than 5%. For example, at $\Lambda = 0.7, g = 0.983$ and $\mu = 1$, the error in $\Delta u(1)/u(1)$ is less than 0.4%. The data in Table 3.3 illustrate the accuracy of the approximation (3.2.23) for representative geophysical indicatrices. Substituting (3.2.14) into (3.2.23, 24) gives approximate explicit formulas for $u(\mu)$ and $M$

$$u(\mu) = \frac{\Lambda\gamma}{2\gamma - s} e^{-\gamma(1-\mu)/s} , \qquad (3.2.26)$$

$$M = \frac{2}{\Lambda^2} \frac{2\gamma - s}{s} , \qquad (3.2.27)$$

where $s$ is determined by (3.2.15).

Thus, for absorbing and anisotropically scattering media, it is often superfluous to determine the function $u(\mu)$. Therefore, in the asymptotic regime, the light field is found by solving the characteristic equation (3.2.2):

$$I_n(\tau, \mu, \mu_0) = \frac{\varphi(\mu)\varphi(\mu_0)}{\pi M} e^{-\gamma\tau} . \qquad (3.2.28)$$

Formula (3.2.28) also specifies the radiance for an infinite medium. The physical implication of this statement is clear: if the condition $\alpha = (1-\Lambda)/\Lambda(1-g) \gg 1$ is satisfied, then the reflection is small, and the intensity at a distance $\tau$ from a plane source is equal both for semi-infinite and infinite media. An explicit expression for $I_n(\tau, \mu, \mu_0)$ can be obtained by substituting the approximation (3.2.14, 24) into (3.2.28).

**Table 3.3.** Functions $u(\mu)$ and $\varphi(\mu)/M$ for sea water and cloud C.1 scattering indicatrices at $\Lambda = 0.9$ and 0.7.

| Cloud C.I, $g = 0.853$ | | | | | | Sea water, $g = 0.87$ | | | | | |
|---|---|---|---|---|---|---|---|---|---|---|---|
| $\Lambda$ | | | | | | | | | | | |
| 0.9 | | | 0.7 | | | 0.9 | | | 0.7 | | |
| $\mu$ | $u(\mu)$ | $\varphi(\mu)/M$ | $\mu$ | $u(\mu)$ | $\varphi(\mu)/M$ | $\mu$ | $u(\mu)$ | $\varphi(\mu)/M$ | $\mu$ | $u(\mu)$ | $\varphi(\mu)/M$ |
| 0.107 | 0.12 | 0.15 | 0.1 | 0.021 | 0.023 | 0.1 | 0.010 | 0.120 | 0.1 | 0.015 | 0.016 |
| 0.275 | 0.17 | 0.19 | 0.3 | 0.040 | 0.040 | 0.299 | 0.160 | 0.163 | 0.299 | 0.028 | 0.027 |
| 0.577 | 0.30 | 0.31 | 0.5 | 0.073 | 0.073 | 0.5 | 0.230 | 0.233 | 0.5 | 0.052 | 0.051 |
| 0.667 | 0.35 | 0.37 | 0.676 | 0.130 | 0.130 | 0.676 | 0.330 | 0.330 | 0.676 | 0.100 | 0.099 |
| 0.85 | 0.50 | 0.51 | 0.83 | 0.226 | 0.226 | 0.83 | 0.450 | 0.455 | 0.83 | 0.190 | 0.190 |
| 0.91 | 0.56 | 0.57 | 0.93 | 0.334 | 0.334 | 0.93 | 0.560 | 0.570 | 0.93 | 0.312 | 0.313 |
| 1.0 | 0.67 | 0.68 | 0.9997 | 0.448 | 0.453 | 0.999 | 0.650 | 0.670 | 0.999 | 0.447 | 0.455 |

## 3.3 Reflection from a Semi-Infinite Medium

We now consider diffuse light reflection from a semi-infinite medium illuminated by parallel beams. Let $r(\mu, \mu_0, \psi)$ be the radiance coefficient of the light reflected from a semi-infinite medium, i.e.,

$$r(\mu, \mu_0, \psi) = \pi I_n(\tau = 0, -\mu, \mu_0, \psi) \qquad \text{at } \mu > 0 .$$

As in [3.5], $r(\mu, \mu_0, \psi)$ is called the reflection function. The quantity $\mu$ in the reflected light radiance is set *positive*.

The optical reciprocity theorem implies that the reflection function is a symmetric function of the incidence and reflection angles, i.e., $r(\mu, \mu_0, \psi) = r(\mu_0, \mu, -\psi)$. The integral equations for the function $r(\mu, \mu_0, \psi)$ are given in [3.3–5]. Information on their solution algorithms and programs to realize these algorithms can be found in [3.5, 18].

In classical radiation transfer theory, radiance is generally expressed in terms of auxiliary (Ambartsumyan) functions, for which a system of nonlinear integral equations is formulated in [3.3, 4, 25]. If $N$ terms are sufficient for the series expansion of the indicatrix $x(\beta)$ into Legendre polynomials, then we must find $(N + 1)(N + 2)/2$ such functions. This approach may, to a certain degree, form the basis of a computational procedure. However, in deriving analytical formulas, it is efficient only for simple scattering indicatrices. As for applied problems with strongly anisotropic scattering and $N \gtrsim 100$, an alternative approach is required. Simple approximate formulas for the function $r(\mu, \mu_0, \psi)$ are derived in [3.26–29].

### 3.3.1 Nonabsorbing Media

Let the reflection function of the light reflected by a semi-infinite nonabsorbing medium be $r_0(\mu, \mu_0, \psi)$. The first two moments in the function $r_0(\mu, \mu_0, \psi)$ obey *the relations [3.4]*

$$2\int_0^1 r_0(\mu, \mu_0)\mu\, d\mu = 1 , \tag{3.3.1}$$

$$2\int_0^1 r_0(\mu, \mu_0)\mu^2\, d\mu = \frac{4}{3}u_0(\mu_0) - 1 . \tag{3.3.2}$$

Here, $r_0(\mu, \mu_0) = (1/2\pi)\int_0^{2\pi} r_0(\mu, \mu_0, \psi)\, d\psi$ is the azimuth-averaged reflection function. The relation (3.3.1) is the normalization condition. It expresses a physically evident fact that a semi-infinite nonabsorbing medium reflects all the radiation incident on it.

For not very extended scattering indicatrices ($g < 0.5$), an approximate formula was obtained in [3.26]:

$$r_0(\mu, \mu_0, \psi) = r_0^{(1)}(\mu, \mu_0, \psi) + \Phi(\mu, \mu_0) + \frac{\mu\mu_0\bar{x}}{\mu + \mu_0} . \tag{3.3.3}$$

Here,

$$r_0^{(1)}(\mu,\mu_0,\psi) = \frac{x(\cos\beta_0)}{4(\mu+\mu_0)} \tag{3.3.4}$$

is the reflection function for single scattering, $\cos\beta_0 = \mu\mu_0 + \sqrt{1-\mu^2}\sqrt{1-\mu_0^2}\cos\psi$. The function $\Phi(\mu,\mu_0)$ and the parameter $\bar{x}$ are given in terms of the scattering indicatrix integrals

$$\Phi(\mu,\mu_0) = \frac{\mu_0 F(\mu) + \mu F(\mu_0)}{\mu+\mu_0}, \tag{3.3.5}$$

$$F(\mu) = \frac{1}{4\pi}\int_0^1 d\mu_0 \int_0^{2\pi} x(\mu,\mu_0,\psi)\,d\psi, \tag{3.3.6}$$

$$\bar{x} = 2\int_0^1 [1-F(\mu)]\,d\mu. \tag{3.3.7}$$

In particular, for isotropic and Rayleigh scattering $F(\mu) = 0.5$; $\bar{x} = 1$; for the linear indicatrix $x(\cos\beta) = 1 + x_1\cos\beta$, $F(\mu) = 0.5 + x_1\mu/4$ and $\bar{x} = 1 - x_1/4$. Also, the quantities $r_0(\mu,\mu_0,\psi)$ calculated by (3.3.3) agree to within 5% with the classical solutions [3.1, 3]. In the general case, given the indicatrix $x(\beta)$ expansion into the series of Legendre polynomials (2.2.7), we can find $F(\mu)$ and $\bar{x}$ by

$$F(\mu) = 0.5\left[1 + \sum_n \alpha_n x_n P_n(\mu)\right],$$

$$\bar{x} = 1 - \sum_n \alpha_n^2 x_n,$$

$$\alpha_n = \frac{(-1)^{(n-1)/2}(n-1)!}{(n+1)2^{n-1}([(n-1)/2]!)^2}, \qquad n = 1, 3, 5 \ldots$$

At even $n$, $\alpha_n = 0$. The values of $F(\mu)$ for some indicatrices are listed in Table 3.4. The accuracy of (3.3.3) is illustrated in Fig. 3.3a.

**Table 3.4.** Values of $F(\mu)$ for some model and geophysical scattering indicatrices.

| | Indicatrix | | |
|---|---|---|---|
| $\mu$ | Henyey-Greenstein $g = 0.9$ | Sea water $g = 0.87$ [3.63, 21] | Cloud C.1 $g = 0.853$ [3.64] |
| 1.0 | 0.977 | 0.950 | 0.958 |
| 0.9 | 0.974 | 0.948 | 0.954 |
| 0.8 | 0.970 | 0.945 | 0.950 |
| 0.7 | 0.964 | 0.942 | 0.942 |
| 0.5 | 0.947 | 0.930 | 0.916 |
| 0.3 | 0.906 | 0.9 | 0.866 |

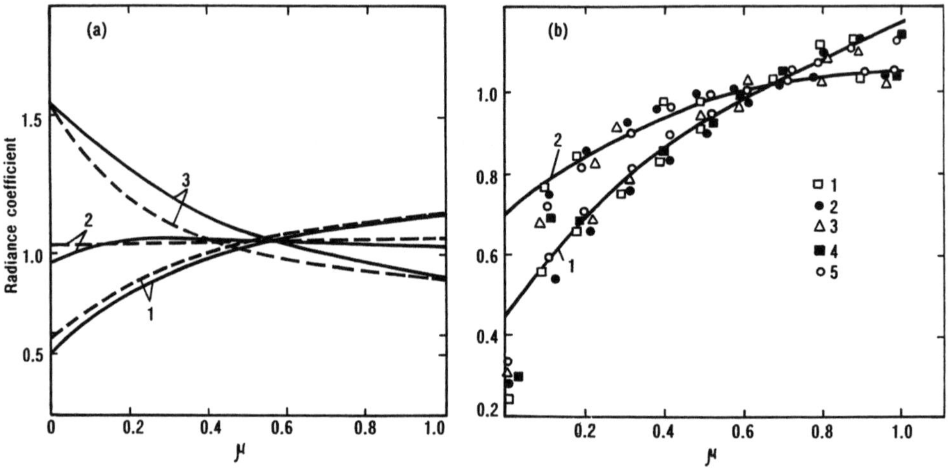


**Fig. 3.3.** Azimuth averaged reflection function $r_0(\mu, \mu_0)$ (radiance coefficient) at (curve 1) $\mu_0 = 1$, (curve 2) 0.7 and (curve 3) 0.54. (a) Exact solutions (*solid lines*) and calculated from (3.3.3) (*dashed lines*) for the Henyey-Greenstein indicatrix at $g = 0.5$. (b) Calculation (*symbols*) for the cloud indicatrix with (curve 1) $g = 0.853$, (curve 2) 0.87, (curve 3) 0.923, (curve 4) 0.958 and Henyey-Greenstein indicatrix at (curve 5) $g = 0.966$ and calculated from (3.3.8) (*lines*)

For strongly anisotropic scattering ($g > 0.8$), and not very oblique incidence and observation angles ($\mu, \mu_0 > 0.2$), the azimuth-averaged reflection function of a semi-infinite medium is described by the simple approximation [3.22, 27]

$$r_0(\mu, \mu_0) = 0.49 \frac{1 + 4\mu\mu_0}{\mu + \mu_0} . \tag{3.3.8}$$

Figure 3.3b compares the radiance coefficient estimated by (3.3.8) with the results of exact calculations for different model and geophysical indicatrices. As mentioned in Sect. 2.2, some of them, the cloud indicatrix C.1, for instance, possess a distinct back-scattering structure (at $|\cos\beta| > 0.6$, glory and rainbow are observed). Nevertheless, we can also use (3.3.8) in such a situation for problems where an error of up to about 10% is possible in determining $r_0(\mu, \mu_0)$.

As seen from (3.3.8) and Fig. 3.3b, for strongly anisotropic scattering, the effect of the indicatrix on the function $r_0(\mu, \mu_0)$ is negligible, except at small $\mu$ and $\mu_0$. In this sense, the function $r_0(\mu, \mu_0)$ becomes universal.

### 3.3.2 Weakly Absorbing Media

For a weakly absorbing medium, radiance is expressed in terms of the known functions $r_0(\mu, \mu_0, \psi)$ and $u_0(\mu)$ [3.24, 30]

$$r(\mu, \mu_0, \psi) = r_0(\mu, \mu_0, \psi) \exp\left[ -4q\gamma \frac{u_0(\mu)u_0(\mu_0)}{r_0(\mu, \mu_0, \psi)} \right] . \tag{3.3.9}$$

At $4q\gamma \ll 1$, relation (3.3.9) becomes the known rigorous asymptotic formula [3.4]:

$$r(\mu, \mu_0, \psi) = r_0(\mu, \mu_0, \psi) - 4q\gamma u_0(\mu)u_0(\mu_0)\,. \tag{3.3.10}$$

We introduce the *reflection coefficient for a directional incident beam* (layer albedo)

$$r(\mu_0) = \frac{1}{\pi}\int_0^{2\pi} d\psi \int_0^1 r(\mu, \mu_0, \psi)\mu\, d\mu\,, \tag{3.3.11}$$

and the *reflection coefficient for diffuse illumination* (spherical albedo)

$$r = 2\int_0^1 r(\mu_0)\mu_0\, d\mu_0\,. \tag{3.3.12}$$

The reflection coefficients of a weakly absorbing semi-infinite medium can be determined by the simple formulas [3.24]

$$r(\mu_0) = \mathrm{e}^{-4q\gamma u_0(\mu_0)}\,, \tag{3.3.13}$$

$$r = \mathrm{e}^{-4q\gamma}\,. \tag{3.3.14}$$

Equations (3.3.13, 14) provide the reflection coefficients with an error of less than 10% at $4q\gamma < 1.2$, and less than 30% at $4q\gamma < 2$. It should be borne in mind that the well-known asymptotic formula $r(\mu_0) = 1 - 4q\gamma u_0(\mu_0)$ [3.4] can, to within 10%, be applied at $4q\gamma \lesssim 0.3$.

### 3.3.3 Absorbing Media with Strongly Anisotropic Scattering

We now consider absorbing media with strongly anisotropic scattering $(1 - \Lambda)/\Lambda(1 - g) \gtrsim 2$. If the scattering indicatrix is a relatively slowly changing function at angles $\beta > \pi/2$, then the reflection function for a semi-infinite medium is given by a simple formula [3.27–29]

$$r(\mu, \mu_0, \psi) = \frac{\Lambda x(\cos\beta_0)}{4(\mu + \mu_0)[1 - \Lambda\Phi(\mu, \mu_0)]}\,. \tag{3.3.15}$$

We shall rewrite (3.3.15), using (3.3.4) for the reflection function of single-scattered light and the relation $r^{(1)}(\mu, \mu_0, \psi) = \Lambda r_0^{(1)}(\mu, \mu_0, \psi)$:

$$r(\mu, \mu_0, \psi) = \frac{r^{(1)}(\mu, \mu_0, \psi)}{1 - \Lambda\Phi(\mu, \mu_0)}\,. \tag{3.3.16}$$

Here, the function $\Phi(\mu, \mu_0)$ is again determined by (3.3.5). For very extended indicatrices, the functions $F(\mu)$ and $\Phi(\mu, \mu_0)$ defined by (3.3.6, 5) have a slight dependence on $\mu$ (Table 3.4). Therefore (3.3.16) implies that the reflection function $r(\mu, \mu_0, \psi)$ is similar to that for single scattering, whereas the radiance is $[1 - \Lambda\Phi(\mu, \mu_0)]^{-1}$ times as large.

The condition $(1-\Lambda)/\Lambda(1-g) \gtrsim 2$ is necessary, but insufficient for the simple and convenient formula (3.3.15) to be applied. It should be supplemented by the "slow change criterion" in the indicatrix at $\beta_0 > \pi/2$ [3.21, 27]:

$$\Lambda\left(1-\frac{x_1}{3}\right)\left|\frac{d\ln x(\cos\beta_0)}{d(\cos\beta_0)}\right| \ll 1\,,$$

$$\frac{\Lambda}{2}\left[\frac{5-x_2}{15}+(\cos\beta_0)^2\left(1-\frac{2x_1}{3}+\frac{x_2}{5}\right)\right]\frac{1}{x(\cos\beta_0)}\left|\frac{d^2x(\cos\beta_0)}{d(\cos\beta_0)^2}\right| \ll 1\,. \tag{3.3.17}$$

If essential angular gradients (for instance, rainbows and glories for large nonabsorbing particles), characterize the backscattering indicatrix, then condition (3.3.17) certainly does not hold. Taking the angular structure of backscattering into account yields the following expression [3.27]:

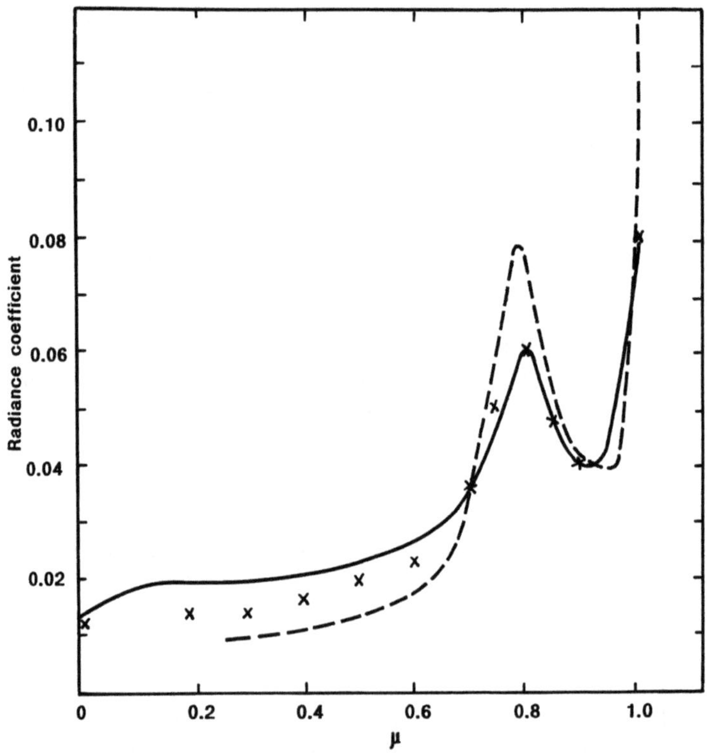


**Fig. 3.4.** Reflection function $r(\mu, 1)$ for the Cloud C.1 indicatrix at $\Lambda = 0.7$. *Solid curve*, exact calculation; *dashed curve*, calculated from (3.3.15); *crosses*, calculated from (3.3.18)

$$r(\mu,\mu_0,\psi) = r^{(1)}(\mu,\mu_0,\psi) + \frac{\Lambda^2}{4(\mu+\mu_0)} \frac{\mu_0\xi(\mu,\mu_0,\psi) + \mu\xi(\mu_0,\mu,-\psi)}{1-\Lambda\Phi(\mu,\mu_0)}, \quad (3.3.18)$$

and the quantity

$$\xi(\mu,\mu_0,\psi) = \frac{1}{4\pi}\int_0^{2\pi} d\psi' \int_0^1 x(\mu,\mu',\psi-\psi')x(-\mu',\mu_0,\psi')\,d\mu'$$

is completely determined by the scattering indicatrix. As seen from Fig. 3.4, the error in (3.3.18) is within 5% at $\mu > 0.7$, and increases to 25% only at $\mu \leqslant 0.2$. Naturally, the greatest error in (3.3.15) is related to the maximum rainbow and glory of single scattering (about 35 and 100%, respectively). We refer the reader to [3.21, 22] for more detailed information about approximate formulas for the function $r(\mu,\mu_0,\psi)$.

Finally, we note that nowadays there are a number of simple asymptotic and approximate solutions to describe light scattering in a semi-infinite medium. They are listed in Table 3.5. The applicability domain of the formulas is such that the error in a given quantity will be no more than 15%. As already noted, the algorithms for computer calculation of the functions $\varphi(\mu)$, $u(\mu)$, $r(\mu,\mu_0,\psi)$ and related quantities written [3.18]. They may be used when the approximate formulas are not sufficiently accurate. We applied these calculations to specify the applicability domain of the approximate formulas. The functions listed are very important in scattering theory and its applications, the theory of light fields in thick layers being a case in point.

## 3.4 Reflection and Transmission by Layers of Finite Thickness

Consider now light fields in layers of finite optical thickness. In applied problems, the solution to describe the radiation diffusely reflected and diffusely transmitted by a layer are of special significance. We introduce the quantities

$$r(\tau_0,\mu,\mu_0,\psi) = \pi I_n(\tau=0,\mu,\psi;\tau_0,\mu_0) \quad \text{at } \pi/2 \leqslant \theta < \pi$$

$$t(\tau_0,\mu,\mu_0,\psi) = \pi I_n(\tau=\tau_0,\mu,\psi;\tau_0,\mu_0) \quad \text{at } 0 \leqslant \theta < \pi/2$$

for the radiance coefficients of the light reflected and transmitted by the layer of optical thickness $\tau_0$, and call them the reflection and transmission functions. Here, $I_n(\tau,\mu,\psi;\tau_0,\mu_0)$ is the normalized radiance at the point $\tau$ in the $(\mu,\psi)$ direction generated in the optical $\tau_0$-thick layer by the incident radiation at an angle cosine $\mu_0$ to the normal, $r(\mu,\mu_0,\psi) = r(\tau_0 \to \infty,\mu,\mu_0,\psi)$. Naturally, the reflection and transmission functions are symmetric relative to the directions of incidence and observation.

It is very often necessary to know the reflection and transmission coefficients of the layer illuminated by directional

**Table 3.5.** Approximate formulas for determining light fields in a semi-infinite medium and in thick layers.

| Function | Conservative scattering | Weak absorption | | Strong absorption |
|---|---|---|---|---|
| | $\Lambda = 1$ | $\frac{1-\Lambda}{1-g} \lesssim 0.02$ | $\frac{1-\Lambda}{1-g} \lesssim 0.2$ | $\frac{1-\Lambda}{1-g} > 2, \mu > 0.3$ |
| $\varphi(\mu)$ | 1 | $1 - 3q\gamma\mu$ | $\frac{3q\gamma}{\Lambda \sinh(3q\gamma)} e^{3q\gamma\mu}$ | $\frac{2\gamma}{\Lambda s} e^{-\gamma(1-\mu)/s}$<br>$s = \sqrt{(1-\Lambda)\Lambda(1-g)}$ |
| $u(\mu)$ | $u_0(\mu) \approx \frac{3}{7}(1 + 2\mu)$ | $u_0(\mu)(1 - 2q\gamma)$ | | $\frac{\varphi(\mu)}{M} \approx \frac{\Lambda\gamma}{2\gamma - s} e^{-\gamma(1-\mu)/s}$ |
| $r(\mu, \mu_0, \psi - \psi_0)$ | $r_0(\mu, \mu_0, \psi - \psi_0)$<br>$= r_1(\mu, \mu_0, \psi - \psi_0) + \Phi(\mu, \mu_0) + \frac{\mu\mu_0\bar{x}}{\mu + \mu_0}, \quad x_1 < 2$<br>$= 0.49\frac{1 + 4\mu\mu_0}{\mu + \mu_0}, \quad x_1 > 2.5$<br>$\mu, \mu_0 > 0.3$ | $r_0(\mu, \mu_0, \psi - \psi_0) - 4q\gamma u_0(\mu)u_0(\mu_0)$ | $r_0(\mu, \mu_0, \psi - \psi_0) \times \exp\left[4q\gamma\frac{u_0(\mu)u_0(\mu_0)}{r_0(\mu, \mu_0, \psi - \psi_0)}\right]$ | $r_1(\mu, \mu_0, \psi) + \frac{\Lambda^2[\mu_0\xi(\mu, \mu_0, \psi) + \mu\xi(\mu_0, \mu, -\psi)}{4(\mu + \mu_0)[1 - \Lambda\Phi(\mu, \mu_0)]}$<br>$\frac{r_1(\mu, \mu_0, \psi)}{1 - \Lambda\Phi(\mu, \mu_0)}$<br>subject to (3.3.17) |
| $r(\mu_0)$ | 1 | $1 - 4q\gamma u_0(\mu_0)$ | $e^{-4q\gamma u_0(\mu_0)}$ | |
| $r$ | 1 | $1 - 4q\gamma$ | $e^{-4q\gamma}$ | |

$$r(\tau_0,\mu_0)=\frac{1}{\pi}\int_0^{2\pi}d\psi\int_0^1 r(\tau_0,\mu,\mu_0,\psi)\mu\,d\mu\,,$$

$$t(\tau_0,\mu_0)=\frac{1}{\pi}\int_0^{2\pi}d\psi\int_0^1 t(\tau_0,\mu,\mu_0,\psi)\mu\,d\mu\,, \tag{3.4.1}$$

and diffuse

$$r(\tau_0)=2\int_0^1 r(\tau_0,\mu_0)\mu_0\,d\mu_0\,,$$

$$t(\tau_0)=2\int_0^1 t(\tau_0,\mu_0)\mu_0\,d\mu_0 \tag{3.4.2}$$

light. A comparison with our earlier notation shows that $r(\mu_0)=r(\tau_0\to\infty,\mu_0)$ and $r=r(\tau_0\to\infty)$. The quantities $r(\tau_0,\mu_0)$ and $r(\tau_0)$ are frequently termed the plane and spherical albedo of the layer, respectively. The reader will recall that if the radiance $I$ is the solution to problem (3.1.3, 5), then the transmission function $t(\tau_0,\mu,\mu_0,\psi)$ and the transmission coefficient $t(\tau_0,\mu_0)$ for directional illumination describe only diffusely scattered light. The total transmission coefficient can be obtained by adding an unscattered component:

$$t_{\text{full}}(\tau_0,\mu_0)=t(\tau_0,\mu_0)+e^{-\tau_0/\mu_0}\,. \tag{3.4.3}$$

Situations are frequently encountered where the scattering layer is adjacent to the reflecting surface. Thus, the Earth's albedo is, on average about 20%, in summer and reaches 80% in winter. In the simplest case of the orthotropic (Lambert) surface with albedo $A$, the reflection $r_A(\tau_0,\mu,\mu_0,\psi)$ and transmission $t_A(\tau_0,\mu,\mu_0,\psi)$ functions of the layer-reflecting surface system are expressed in terms of the relevant functions at $A=0$ [3.4]

$$r_A(\tau_0,\mu,\mu_0,\psi)=r(\tau_0,\mu,\mu_0,\psi)+\frac{At_{\text{full}}(\tau_0,\mu_0)t_{\text{full}}(\tau_0,\mu)}{1-Ar(\tau_0)}\,,$$

$$t_A(\tau_0,\mu,\mu_0,\psi)=t(\tau_0,\mu,\mu_0,\psi)+\frac{At_{\text{full}}(\tau_0,\mu_0)r(\tau_0,\mu)}{1-Ar(\tau_0)}\,. \tag{3.4.4}$$

Equations (3.4.4) follow readily from physical considerations by summing up multiple-scattered radiations through the medium boundary and the underlying Lambert surface.

Simple analytical solutions for the reflection and transmission functions are available for two asymptotic cases, namely, single scattering and high-order scattering in thick layers.

For the optical layers of small thickness and/or those with appreciable absorption (small $\Lambda$), the intensity of the scattered radiation is mainly determined by the first-order scattering. We can easily derive the reflection and transmission functions in a single-scattering approximation by neglecting the integral term in (3.1.3) and solving the differential equation with a source function (3.1.5) [3.3]

$$r^{(1)}(\tau_0, \mu, \mu_0, \psi) = \frac{\Lambda x(\cos\beta_0)}{4(\mu+\mu_0)}[1 - e^{-\tau_0(1/\mu_0+1/\mu)}],$$

$$t^{(1)}(\tau_0, \mu, \mu_0, \psi) = \frac{\Lambda x(\cos\beta_0)}{4(\mu-\mu_0)}[e^{-\tau_0/\mu} - e^{-\tau_0/\mu_0}]. \tag{3.4.5}$$

Equations (3.4.5) are frequently used, for example, to describe scattering in cloudless atmosphere. However, in many problems it is necessary to allow for multiple scattering even for the optical thickness 0.2 to 0.4 realized in the Earth's atmosphere. Some estimates of the accuracy of (3.4.5) for $x(\beta) = 1$ and $\Lambda = 1$ are quoted in [3.3]. For very extended indicatrices, we can transform these relations, by introducing effective parameters, into quasi-single scattering approximation formulas that have a wider applicability domain (Sect. 3.6.2).

Solutions for optically thick layers will be discussed below.

## 3.5 Reflection and Transmission by Optically Thick Scattering Layers

For very thick layers, asymptotic formulas are known to describe the reflection and transmission functions [3.4, 5, 30–33]

$$r(\tau_0, \mu, \mu_0, \psi) = r(\mu, \mu_0, \psi) - \frac{MNe^{-2\gamma\tau_0}}{1 - N^2e^{-2\gamma\tau_0}}u(\mu)u(\mu_0),$$

$$t(\tau_0, \mu, \mu_0, \psi) = \frac{Me^{-\gamma\tau_0}}{1 - N^2e^{-2\gamma\tau_0}}u(\mu)u(\mu_0). \tag{3.5.1}$$

Definitions of the functions $r(\mu, \mu_0, \psi)$, $u(\mu)$ as well as of the quantities $M$ and $\gamma$ were suggested in Sect. 3.2.3, $N = 2\int_0^1 u(\mu)\varphi(-\mu)\mu\, d\mu$. Equations (3.5.1) show that the solution of the problem of the radiation emerging from a thick layer reduces to the theory of scattering in semi-infinite media. All the quantities and functions required are listed in Table 3.5. Integrating (3.5.1) for $\mu$ and $\mu_0$ gives expressions for the reflection and transmission coefficients.

The profound physical significance of (3.5.1) should be emphasized. They stem from the most general symmetry properties of the problem, such as the invariance to displacement [3.6] and the reciprocity theorem [3.2, 6]. We can obtain these formulas irrespective of the radiation transfer equation, basing our considerations only on the symmetry properties. The latter demonstrate their universal nature. Formulas (3.5.1) preserve their structure in regard to reflection at the layer boundaries [3.4, 34, 35], the vector transfer theory (including polarization) [3.36] and even, to a certain degree, in the theory of inhomogeneous media [3.4, 37, 38].

Here, we stress some specific properties of reflection and transmission by thick layers. The azimuthal dependence of reflected light radiance is completely determined by the function $r(\mu, \mu_0, \psi)$, i.e., the specific features of reflection from a semi-infinite medium. The transmitted light radiance is independent of the azimuth, its angular dependence being governed by the function $u(\mu)$.

### 3.5.1 Nonabsorbing Media

For nonabsorbing media, $\Lambda = 1$, $\gamma = 0$. Substituting the asymptotic expansions of the type (3.2.10, 20) into (3.5.1) and setting $\gamma = 0$, we have

$$r(\tau_0, \mu, \mu_0, \psi) = r_0(\mu, \mu_0, \psi) - \frac{4q}{\tau_0 + 4q} u_0(\mu) u_0(\mu_0),$$

$$t(\tau_0, \mu, \mu_0, \psi) = \frac{4q}{\tau_0 + 4q} u_0(\mu) u_0(\mu_0). \tag{3.5.2}$$

The approximate formulas for the functions $r_0(\mu, \mu_0, \psi)$ and $u_0(\mu)$ can be found in Table 3.5. Formulas (3.5.2), together with (3.2.22) and (3.3.1) and the definitions (3.4.1, 2), offer the following relations for the reflection and transmission coefficients

$$r(\tau_0, \mu_0) = 1 - \frac{4q}{\tau_0 + 4q} u_0(\mu_0), \qquad t(\tau_0, \mu_0) = \frac{4q}{\tau_0 + 4q} u_0(\mu_0), \tag{3.5.3}$$

$$r(\tau_0) = \frac{\tau_0}{\tau_0 + 4q}, \qquad t(\tau_0) = \frac{4q}{\tau_0 + 4q}. \tag{3.5.4}$$

In this case, the quantity $t^{-1}$ is the linear function of the layer thickness. A rigorous representation of the asymptotic formulas (3.5.2–4) must include $\tau_0 + 6q\delta$, where $\delta = 2\int_0^1 u_0(\mu)\mu^2\, d\mu$, rather than $\tau_0 + 4q$. As in the majority of formulas used below, in this case also we use the approximation $\delta = 2/3$. The dispersion of the exact $\delta$ values for different scattering indicatrices is seen from Table 3.6.

For an inhomogeneous conservative-scattering layer where the scattering indicatrix $x(\beta, \tau)$ varies along the $\tau$ axis, the authors of [3.37, 38] give solutions which are very similar to (3.5.2–4). They imply that if the weak dependence of the function $u_0(\mu)$ on the scattering indicatrix is neglected (Sect. 3.2.2), (3.5.2–4) can also be used for an inhomogeneous layer by introducing $q = 1/3(1 - \langle g \rangle)$

**Table 3.6.** Values of $\delta = 2\int_0^1 u_0(\mu)\mu^2\, d\mu$ for different scattering indicatrices.

| Indicatrix | $g$ | $\delta$ | Indicatrix | $g$ | $\delta$ |
|---|---|---|---|---|---|
| $1 + 3g\mu$ | $0 \leqslant g \leqslant 1$ | 0.71 | Cloud [3.64] | 0.853 | 0.71 |
| $H - G$ | 0.5 | 0.7 | | 0.87 | 0.71 |
| | 0.8 | 0.73 | | 0.923 | 0.71 |
| | 0.9 | 0.71 | | 0.96 | 0.71 |

where $\langle g \rangle = (1/\tau_0) \int_0^{\tau_0} g(\tau)\, d\tau$ is the average of the asymmetry parameter. Formulas (3.5.2–4), taking into account (3.4.4), can easily be extended to the case of a layer with a reflecting boundary. Equations (3.5.1) and their corollaries (3.5.2–4) were obtained on the assumption that $\tau_0 \gg 1$. We shall now specify their applicability with respect to $\tau_0$. Two approaches, empirical and theoretical, are possible here. Theoretical estimates can be obtained by studying the spectrum of the eigenvalues of the characteristic equation (3.2.2) [3.39, 40]. Following [3.39, 40], at $\tau > 4q$ the relative error in $r$ and $t$ is determined by (3.5.2), i.e. $\Delta r/r \approx \Delta t/t \approx (1/4q)\exp(-\gamma_1 \tau)$, where $\gamma_1$ is the eigenvalue of the characteristic equation closest to $\gamma = 0$. For the model C.1, this results in $\Delta r/r \approx \Delta t/t \leqslant 0.02$ at $\tau \geqslant 4q$. Empirical estimates derived from a comparison of the asymptotic $r$ and $t$ with the data of the corresponding computer solutions (3.1.3–5) yield a similar result.

### 3.5.2 Weakly Absorbing Media

If we wish to obtain the reflection and transmission functions for a weakly absorbing thick layer, it is not sufficient to take into account the linear terms in the series expansion of the asymptotic quantities and functions of (3.5.1). As shown in [3.41], it is convenient to use the representations

$$N = \exp(-6q\gamma\delta)\,,$$

$$Mu(\mu)u(\mu_0) = (4/3\delta)u_0(\mu)u_0(\mu_0)[1 - \exp(-6q\gamma\delta)]\,,$$

which hold, at least, to $\gamma^2$. Then, for the transmitted radiation, we have

$$t(\tau_0, \mu, \mu_0) = \frac{2}{3\delta} \frac{\sinh(6q\delta\gamma)}{\sinh[\gamma(\tau_0 + 6q\delta)]} u_0(\mu)u_0(\mu_0)\,.$$

A similar formula can be written for the reflection. In [3.24], the following simple and convenient formulas, which conform to rigorous asymptotic solutions of the terms about $\gamma^2$ in the approximation $\delta = 2/3$, were obtained for the reflection and transmission coefficients of thick layers:

$$r(\tau_0, \mu_0) = \frac{\sinh X}{\sinh(X + Y)}\,, \qquad t(\tau_0, \mu_0) = \frac{\sinh Y}{\sinh(X + Y)}\,, \tag{3.5.5}$$

where

$$X = \gamma[\tau_0 + 4q(1 - u_0(\mu_0))]\,, \qquad Y = 4q\gamma u_0(\mu_0)\,. \tag{3.5.6}$$

The quantity $\gamma\tau_0$ characterizes the layer absorption. At $\gamma\tau_0 \ll 1$, such absorption can be neglected, and the formulas (3.5.5, 6) become the expressions (3.5.3). Given $\gamma\tau_0 \gg 1$, $t(\tau_0, \mu_0) = 2\exp[-\gamma(\tau_0 + 4q)]\sinh[4q\gamma u_0(\mu_0)]$, i.e. in this case, the layer transmission is an exponential optical thickness function with a decrement $\gamma$. At $X \to \infty$, the first relation in (3.5.5) gives the known (3.3.13) for the reflection coefficient of a semi-infinite medium. In the case of diffuse illumination,

(3.5.5) hold for the reflection and transmission coefficients, but here we must set $X = \gamma\tau_0$ and $Y = 4q\gamma$. Then, (3.5.5) comply with the known formula of Rozenberg [3.30]. At any $\gamma\tau_0$ and $(1 - \Lambda)/(1 - g) \leqslant 0.15$, (3.5.5) provide, to within less than 5%, the same results as the rigorous asymptotic solutions gained from (3.5.1). In the case where $(1 - \Lambda)/(1 - g) \lesssim 0.3$, the error is less than 20%.

Expressions (3.5.5) can also be used to estimate the azimuth-averaged radiance at $\mu$, $\mu_0 > 0.5$ by setting

$$X = \gamma[\tau_0 + 4q(1 - u_0(\mu)u_0(\mu_0))]\,, \qquad Y = 4q\gamma u_0(\mu)u_0(\mu_0)\,. \tag{3.5.7}$$

They can also be used to calculate the reflection and transmission coefficients in the case of a Lambert reflection with albedo $A_1$ at the upper and albedo $A$ at the lower boundary, the arguments $X$ and $Y$ being of the form [3.42]

$$X = \gamma\tau_0 + \frac{4q\gamma A}{1 - A} + 4q\gamma[1 - u_0(\mu_0)]\,,$$

$$Y = 4q\gamma\left[u_0(\mu_0) + \frac{A_1}{1 - A_1}\right]. \tag{3.5.8}$$

A comparison of (3.5.6) and (3.5.8) shows that the lower reflecting boundary (substrate) influences the reflection and transmission coefficients of the layer by an amount $4qA/(1 - A)$ as the optical layer thickness increased. The change in the parameter $Y$ by $4q\gamma A_1/(1 - A_1)$ allows for the impact of the upper illuminated boundary. Now, the reflection and transmission coefficients are calculated with respect to the flux immediately incident on the scattering layer underlying the upper boundary; $\mu_0$ is the incident cosine angle, with allowance for the boundary refraction. The applicability domain of the above formulas at $A_1 \neq 0$ is narrower than that of the relevant expressions for the upper nonreflecting boundary and is estimated as $4q\gamma/(1 - A_1) \lesssim 0.5$. When it is desirable to extend somewhat the applications of the formulas, we may use less compact algebra

$$r(\tau_0, \mu_0, A_1, A) = \frac{r(\tau_0, \mu_0, A)(1 - A_1)}{1 - A_1 r(\tau_0, A)}\,,$$

$$t(\tau_0, \mu_0, A_1, A) = t(\tau_0, \mu_0, A) + \frac{r(\tau_0, \mu_0, A)A_1 t(\tau_0, A)}{1 - A_1 r(\tau_0, A)}\,, \tag{3.5.9}$$

where $r(\tau_0, \mu_0, A)$ and $t(\tau_0, \mu_0, A)$ are again governed by (3.5.5) at $X$ from (3.5.8) and $Y = 4q\gamma$. These relations are valid for any $A_1$ and $4q\gamma \lesssim 1$. In a similar fashion, at $A \approx 1$, formulas (3.4.4), rather than (3.5.5, 8), must be applied.

Finally, it should be noted that in thick weakly absorbing layers under different illumination conditions and different albedos of the reflecting boundaries, the reflection and transmission coefficients depend only on the dimensionless parameters $X$ and $Y$. They can be calculated (3.5.5), using the arguments of Table 3.7. Their applicability domain is determined by the condition $\tau_0 \gtrsim 4q$, $Y \lesssim 1$ [3.24].

**Table 3.7.** Values of $X$ and $Y$ for different boundary conditions.

| Parameter | For reflection and transmission coefficients | | For radiance coefficient at $A_1 = A = 0$ |
|---|---|---|---|
| | Diffuse incidence | Parallel beam incidence at an angle arccos $\mu_0$ to the normal | |
| $X$ | $\gamma\tau_0 + \frac{4q\gamma A}{1-A}$ | $\gamma\tau_0 + 4q\gamma\left[\frac{1}{1-A} - u_0(\mu_0)\right]$ | $\gamma\tau_0 + 4q\gamma[1 - u_0(\mu)u_0(\mu_0)]$ |
| $Y$ | $\frac{4q\gamma}{1-A_1}$ | $4q\gamma\left[u_0(\mu_0) + \frac{A_1}{1-A_1}\right]$ | $4q\gamma u_0(\mu)u_0(\mu_0)$ |

### 3.5.3 Absorbing Media with Strongly Anisotropic Scattering

In this case, at $(1-\Lambda)/\Lambda(1-g) \gtrsim 2$, $N \ll 1$ and the function $u(\mu)$ can be approximately determined by (3.2.23). Then, the transmission function

$$t(\tau_0, \mu, \mu_0) = \frac{\varphi(\mu)\varphi(\mu_0)}{M} e^{-\gamma\tau_0} . \tag{3.5.10}$$

Taking into account the approximate (3.2.14, 27) gives

$$t(\tau_0, \mu, \mu_0) = \frac{2\gamma^2}{s(2\gamma - s)} \exp\left[-\gamma\tau_0 - \frac{\gamma}{s}(1-\mu) - \frac{\gamma}{s}(1-\mu_0)\right] . \tag{3.5.11}$$

It should be borne in mind that the quantity $s$ is found from (3.2.15).

## 3.6 Approximate Indicatrix Model Solutions

Earlier in this chapter, we dealt with asymptotic solutions for semi-infinite media and optically thick layers. If a layer is thin, then the light fields are generated mainly by unscattered and single-scattered light and can easily be calculated. In the general case, for arbitrary optical thickness, computer-aided and different approximate techniques should be used. Some of these approximate procedures are described in this and the following sections. We now consider a group of methods based on approximate indicatrix models. It should be recalled that the scattering indicatrices of real media (sea water, aerosol atmosphere) are very extended in the forward direction. Their description requires dozens of terms in the series expansion in the Legendre polynomials. It is a thankless task to study light fields in such media within the classical transfer theory based on spherical functions. At the same time, we should remember that the light field "forgets" the details of the scattering indicatrix in the course of multiple scattering. Therefore, we can easily derive approximate solutions using different simplified models of the scattering indicatrices.

### 3.6.1 Forward Peak Truncation. Similar Media

The indicatrices of many geophysical objects (sea water, clouds) are distinguished by a distinct "forward peak". The properties of forward scattered light do not differ from those of unscattered radiation. Therefore, it is quite natural to try to find an approximate solution in which some portion of the indicatrix peak applies to the unscattered component. Let us separate a portion of light $\eta$, scattered at very small angles, and represent the scattering indicatrix by the approximate formula

$$x(\beta) = (1-\eta)x_{eq}(\beta) + \eta\delta(\beta)\,, \tag{3.6.1}$$

where $\delta(\beta)$ is the Dirac delta-function; the indicatrix $x_{eq}(\beta)$ approximates $x(\beta)$ when $\beta$ is not small. It is normalized in the usual way,

$$\frac{1}{4\pi}\iint_{4\pi} x_{eq}(\beta)\,d\vec{n} = 1\,.$$

Substituting (3.6.1) into the transfer equation (2.3.1), we obtain

$$\vec{n}\cdot\nabla_{\vec{R}}I(\vec{R},\vec{n}) = -(\varepsilon-\eta\sigma)I(\vec{R},\vec{n}) + \frac{(1-\eta)\sigma}{4\pi}\iint_{4\pi} I(\vec{R},\vec{n}')x_{eq}(\vec{n},\vec{n}')\,d\vec{n}' + B(\vec{R},\vec{n})$$

We now rewrite this equation as

$$\vec{n}\cdot\nabla_{\vec{R}}I(\vec{R},\vec{n}) = -\varepsilon_{eq}I(\vec{R},\vec{n}) + \frac{\sigma_{eq}}{4\pi}\iint_{4\pi} I(\vec{R},\vec{n}')x_{eq}(\vec{n},\vec{n}')\,d\vec{n}' + B(\vec{R},\vec{n})\,, \tag{3.6.2}$$

where

$$\varepsilon_{eq} = \varepsilon(1-\eta\Lambda)\,, \qquad \sigma_{eq} = \sigma(1-\eta)\,. \tag{3.6.3}$$

In this way, we have introduced the characteristics $\sigma_{eq}$, $\varepsilon_{eq}$ and $x_{eq}(\beta)$ of a similar optical medium and reduced the problem to one of determining the light fields in this medium. It is important that the scattering indicatrix of such a medium is less extended. The accuracy of the substitution depends, in many respects, on the choice of the truncation parameter $\eta$. Such an approach, where the diffraction peak is truncated and allowance is made for few terms in the $x_{eq}(\beta)$ expansion in Legendre polynomials frequently forms the basis of a computational procedure [3.43–45] which ensures sufficiently high accuracy of the results. At this point, single scattering may be taken into account in strict conformity with the function $x(\beta)$. We have intentionally used the transfer equation (2.3.1) in order to underline the applicability of such an approach to a wide range of problems, including those where the radiance is a function of the three spatial coordinates. If the simplest functions $x_{eq}(\beta) = 1$, $x_{eq}(\beta) = 1 + x_1\cos\beta$ are used as $x_{eq}(\beta)$, we may apply the wealth of analytical and approximate solutions developed in classical transfer theory [3.1–3, 46, 47]. Recently, the indicatrix approximation (3.6.1) at $x_{eq}(\beta) = 1 + x_1 \cos\beta$ has been successfully applied to calculate cloud radiation ($\delta$-Eddington approximation [3.5, 47]). Nevertheless, it is most convenient, and often expedient, to represent the indicatrix as

$$x(\beta) = 1 - \eta + \eta\delta(\beta) . \tag{3.6.4}$$

At $\eta = g$, it is exactly the solution of the problem in a transport approximation pioneered in neutron transfer theory [3.2, 48].

### 3.6.2 Transport Approximation

Following the representation (3.6.4) at $\eta = g$, the characteristics of the similar medium in a transport approximation are

$$\begin{aligned} &\varepsilon_{eq} = \varepsilon(1 - \Lambda g) , \qquad \tau_{eq} = (1 - \Lambda g)\tau , \\ &\sigma_{eq} = \sigma(1 - g) , \qquad \Lambda_{eq} = \frac{\Lambda(1 - g)}{1 - \Lambda g} , \\ &x_{eq}(\beta) = 1 . \end{aligned} \tag{3.6.5}$$

In neutron transfer theory, the parameters $\varepsilon_{eq} = \varepsilon(1 - \Lambda g)$ and $\sigma_{eq} = \sigma(1 - g)$ are called the transport extinction and scattering coefficients.

Thus, the transport approximation involves passing from the quantities $x(\beta)$, $\Lambda$ and $\tau$ to $\Lambda_{eq}$ and $\tau_{eq}$ for an isotropically scattering medium. In the following, the unknown light field characteristics are found from the isotropic scattering formula [3.1–6, 12, 49].

Consider the simple example of finding the reflection coefficient $r(\mu_0)$ of a semi-infinite anisotropically scattering medium illuminated by an infinitely wide source. Substituting the quantity $\Lambda_{eq}$ for a similar medium into the known solution [3.3] for isotropic scattering gives

$$r(\mu_0) = 1 - \varphi(\mu_0, \Lambda_{eq})\sqrt{1 - \Lambda_{eq}} ,$$

where $\varphi(\mu_0, \Lambda_{eq})$ is the known Ambartsumyan function. Using the approximation [3.50]

$$\varphi(\mu_0, \Lambda) = \frac{1 + 2\mu_0}{1 + 2\mu_0\sqrt{1 - \Lambda}} ,$$

we obtain a simple analytical formula for the reflection coefficient

$$r(\mu_0) = \frac{1 - \sqrt{1 - \Lambda_{eq}}}{1 + 2\mu_0\sqrt{1 - \Lambda_{eq}}} . \tag{3.6.6}$$

The transport approximation provides the correct asymptotic formulas for radiation fluxes in weakly absorbing thick layers illuminated by an infinitely wide source. In this case, the fluxes depend only on the parameters $\gamma\tau$ and $4q\gamma$ (Sect.

3.5). Thus, the exact and approximate formulas will agree, provided the equalities

$$\gamma_{eq}\tau_{eq} = \gamma\tau\,, \qquad 4q_{eq}\gamma_{eq} = 4q\gamma \tag{3.6.7}$$

hold. For isotropic scattering, $q_{eq} = 1/3$. If (3.2.11) is used to determine $\gamma$, then,

$$\begin{aligned}\gamma_{eq}\tau_{eq} &= \tau_{eq}\sqrt{3(1-\Lambda_{eq})} = \tau\sqrt{3(1-\Lambda g)(1-\Lambda)}\\ &\approx \tau\sqrt{3(1-g)(1-\Lambda)} = \gamma\tau\,,\\ 4q_{eq}\gamma_{eq} &= \tfrac{4}{3}\sqrt{3(1-\Lambda_{eq})} = 4\sqrt{(1-\Lambda)/3(1-\Lambda g)}\\ &\approx 4\sqrt{(1-\Lambda)/3(1-g)} = 4q\gamma\,.\end{aligned} \tag{3.6.8}$$

It is seen that at $1 - \Lambda g \approx 1 - g$ (3.6.7) follow from (3.6.5). It is clear that under this condition (3.6.6), which is obtained in a transport approximation, becomes the asymptotic formula

$$r(\mu_0) \approx 1 - (1 + 2\mu_0)\sqrt{1 - \Lambda_{eq}} = 1 - 4q\gamma\frac{\sqrt{3}}{4}(1 + 2\mu_0) \approx 1 - 4q\gamma u_0(\mu_0)\,.$$

As in [3.4, 5], we have called (3.6.5, 7) the similarity relations. It should be remembered that these conditions, as well as the term "similar medium" used above, imply approximate similarity (equivalence).

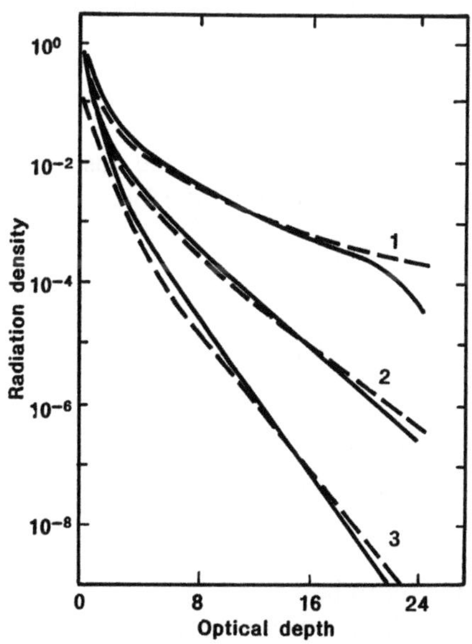


Fig. 3.5. Distribution of radiation density due to a diffuse point source in an infinite medium at $x(\beta) = 1.25/(1.5 - \cos\beta)^2$ and (curve 1) $\Lambda = 0.99$, (curve 2) 0.9 and (curve 3) 0.7. *Solid curves*, numerical finite difference calculation; *dashed curves*, transport approximation

In order to avoid confusion over the terminology, we wish to remind the reader that we have already applied the term "similarity principle" (Sect. 2.4) in a different sense when speaking about the complete identity of light field characteristics in media with equal $x(\beta)$, $\Lambda$ and optical coordinates, but different geometric coordinates and extinction coefficients.

The accuracy of the transport approximation as applied to the calculation of light fields from diffuse point sources is shown in Fig. 3.5, based on the data of [3.51].

Finally, the transport approximation can be recommended for calculating integral angular characteristics. At the same time, this procedure is less advisable when it comes to computing angular functions.

### 3.6.3 Quasi-Single Scattering Approximation

In media with a very extended scattering indicatrix and appreciable absorption $\Lambda_{eq} = \Lambda(1-g)/(1-\Lambda g) \ll 1$, a large body of energy is transferred by the radiation propagating over a small solid angle in the forward direction. The light fields beyond this small angle are generally determined by the single-scattered radiation over large angles. The allowance for the next-order scattering over large angles yields a small correction $\sim \Lambda_{eq}^2$. Such an instructive light field model makes different approximation procedures possible. The quasi-single approximation developed in [3.28, 52, 53] seems to be very simple and convenient. It approximates the scattering indicatrix using (3.6.1). It is assumed that the parameter

$$\eta = F = \frac{1}{2}\int_0^1 x(\cos\beta)\,d(\cos\beta)\,,$$

i.e., the forward-scattered light, is related to the unscattered component. Thus, the parameters of a similar medium are of the form

$$\begin{aligned} x_{eq}(\beta) &= x(\beta) \qquad \text{at } \beta > \pi/2\,, \\ &= 0 \qquad \text{at } \beta < \pi/2\,, \end{aligned}$$

$$\varepsilon_{eq} = \varepsilon(1-\Lambda F)\,, \qquad \Lambda_{eq} = \frac{\Lambda(1-F)}{1-\Lambda F}\,. \tag{3.6.9}$$

Then, the downward flux $S(\tau)$ is determined by the unscattered flux in the equivalent medium and decreases exponentially:

$$S(\tau) = e^{-(1-\Lambda F)\tau}\,. \tag{3.6.10}$$

Here, the incident flux is set to be 1.

Thus, this procedure allows approximately for multiple forward and single backscattering ($\beta > \pi/2$). The light field characteristics beyond a small-angle region are determined by the single scattering formulas, including the effective medium parameters (3.6.9). For example, in such an approximation, the expression for the reflection coefficient in a semi-infinite medium is

$$r(\mu, \mu_0, \psi) = \frac{\Lambda x(\cos \beta_0)}{4(1 - \Lambda F)(\mu + \mu_0)},$$

which complies with (3.3.15), assuming that $\Phi(\mu, \mu_0) = F$. As seen from (3.3.5, 6) and Table 3.4, for the sea and cloud scattering indicatrices, this condition holds at $\mu$, $\mu_0 > 0.5$ with an error of less than 4%.

Any other light field characteristics at large angles can be found in a similar way. The method here is simple: we substitute the similar medium characteristics into the single–scattering formulas.

The accuracy of the quasi-single scattering approximation for calculating the light fields in sea water was found in [3.28, 52, 53] by comparing with experiment and with the Monte-Carlo simulations. In the calculations, $x(\beta)$ was set to a constant for $\beta > \pi/2$. The accuracy is shown to be satisfactory for many practical problems which require that the upward and downward fluxes or angular distribution of the reflected radiation be known. For instance, the error in the reflection function $r(\mu, \mu_0)$ is less than 5%. In [3.21], the domain of applicability of the quasi-single approximation, as compared to rigorous analytical solutions, was estimated. The restrictions are given in Sect. 3.3.3.

In the end, we say that it is the smoothness of the scattering indicatrix at $\beta > \pi/2$ [conditions (3.3.17)] and the validity of the inequality $(1-\Lambda)/\Lambda(1-g) \geqslant 2$ which account for the applicability of the quasi-single approximation. When calculating the angular distributions of reflected radiation, we must impose additional restrictions on the incident and observation angles: $\mu$, $\mu_0 > 0.5$ at $F \geqslant 0.95$.

### 3.6.4 Small-Angle Solution Iteration

The radiance scattered over large angles in strongly anisotropically scattering media can be calculated more consistently and accurately using the approach suggested in [3.54, 55]. The scattering indicatrix is represented as

$$x(\beta) = (1 - a)x'(\beta) + a\,, \tag{3.6.11}$$

where $x'(\beta)$ approximates $x(\beta)$ at small angles. The scattering over large angles is taken to be isotropic and is characterized by $a \ll 1$. The indicatrix $x'(\beta)$ is normalized as usual, i.e.,

$$\frac{1}{4\pi} \iint_{4\pi} x'(\beta)\, d\vec{n} = 1\,.$$

The solution of the transfer equation (2.3.1) is now sought as a minor parameter $a$ power series

$$I(\vec{R}, \vec{n}) = \sum_{k=0}^{\infty} a^k I^{(k)}(\vec{R}, \vec{n})\,. \tag{3.6.12}$$

Substituting (3.6.11, 12) into (2.3.1), we obtain

$$\vec{n}\cdot\nabla_{\vec{R}}I^{(k)}(\vec{R},\vec{n}) = -\varepsilon I^{(k)}(\vec{R},\vec{n}) + \frac{\sigma}{4\pi}\iint_{4\pi} I^{(k)}(\vec{R},\vec{n}')x'(\vec{n},\vec{n}')\,d\vec{n}' + B^{(k)}(\vec{R},\vec{n})\,, \tag{3.6.13}$$

where

$$B^{(k)}(\vec{R},\vec{n}) = B(\vec{R},\vec{n}) \qquad \text{at } k=0\,,$$

$$= \frac{\sigma}{4\pi}\iint_{4\pi} I^{(k-1)}(\vec{R},\vec{n}')[1 - x'(\vec{n},\vec{n}')]\,d\vec{n}' \qquad \text{at } k \geqslant 1\,. \tag{3.6.14}$$

As the function $x'(\vec{n},\vec{n}')$ is very forward-extended, (3.6.13) can be solved in a small-angle or diffuse small-angle approximation. In contrast to other approximations (Sects. 3.6.1–3) which, in fact, disregard the impact of the small-angle peak spread on the light field structure at large angles, this effect is consistently taken into account in this instance. It is essential to take this effect into account in a large number of important problems (laser sounding, backscattering noise in vision and location systems) related mainly to the theory of narrow beam propagation in scattering media. Therefore, we shall again briefly estimate the possibilities which this approximation offers and discuss the results based on the small-angle solutions in Sect. 4.6.

### 3.6.5 Sobolev Approximation. Conservative Scattering

Given an approximate indicatrix of one kind or another, many more types of solution can be provided. We shall complete our list of the most well-known procedures by describing on modification of such an approach developed by Sobolev. Since it makes use of simple and sufficiently exact closed formulas, this approach is frequently used in calculating atmospheric light fields.

As we have argued more than once, the characteristics of multiscattered light are mainly determined by a single indicatrix parameter, i.e., the assymetry parameter $g$, or the first coefficient $x_1 = 3g$ in the $x(\beta)$ series expansion in the Legendre polynomials. In the course of our discussion, we verified the validity of this argument, especially for thick layer scattering. It is exactly this physical assumption which is used to formulate the Sobolev approximation. In the integral term of the transfer equation which allows for multiple scattering, the indicatrix is approximated as $x(\beta) = 1 + x_1\cos\beta$, whereas the integral term of the source function is found in an Eddington approximation. Then, a solution is derived which takes exact account of single scattering and approximately allows for multiple scattering. We have only briefly formulated the concept behind the method; detailed calculations are given in [3.4]. In this approximation, the reflection and transmission functions of a conservative-scattering layer averaged over the azimuth are of the form

$$r_0(\tau_0,\mu,\mu_0,\psi) = 1 - \frac{4q}{\tau_0+4q}u_0(\mu_0,\tau_0)u_0(\mu,\tau_0)$$
$$+ [3(1+g)\mu\mu_0 - 2(\mu+\mu_0) + x(-\mu,\mu_0)]r_0^{(1)}(\tau_0,\mu,\mu_0)\,,$$
$$t_0(\tau_0,\mu,\mu_0,\psi) = \frac{4q}{\tau_0+4q}u_0(\mu_0,\tau_0)u_0(\mu,\tau_0) - \frac{1}{2}(e^{-\tau_0/\mu} + e^{-\tau_0/\mu_0})$$
$$+ [x(\mu,\mu_0) - 3(1+g)\mu\mu_0]t_0^{(1)}(\tau_0,\mu,\mu_0)\,. \tag{3.6.15}$$

Here,

$$x(\mu,\mu_0) = \frac{1}{2\pi}\int_0^{2\pi} x(\beta_0)\,d\psi\,,$$
$$u_0(\mu,\tau_0) = 0.5[1 + 1.5\mu + (1 - 1.5\mu)e^{-\tau_0/\mu}]\,. \tag{3.6.16}$$

The quantities $r_0^{(1)}$ and $t_0^{(1)}$ are the reflection and transmission functions for single scattering at $\Lambda = 1$ and $x(\beta) = 1$ determined by (3.4.5). The solutions for the reflection and transmission functions are, as expected, symmetric with respect to the $\mu$ and $\mu_0$ interchange. It is easily seen that at $\tau_0 \ll 1$, they become rigorous formulas for single scattering. For $\tau_0 \gg 1$, (3.6.15) conform to the asymptotic solutions (3.5.1) entering the following approximate functions

$$u_0(\mu) = 0.5(1 + 1.5\mu)\,, \qquad r_0(\mu,\mu_0) = 0.5 + \frac{3(1+g)\mu\mu_0 + x(-\mu,\mu_0)}{4(\mu+\mu_0)}\,. \tag{3.6.17}$$

The accuracy of (3.6.15) was estimated in [3.56, 57]. The reflection on the underlying surface can readily be taken into account using (3.4.4).

## 3.7 Two-Stream Approximation

In radiation transfer theory, as in the theory of neutron transfer, a number of approximate methods are used which take a direction average of the radiance. They include a group of very similar approximations [Eddington-Milne approximation, $P_1$-approximation of spherical harmonics, diffusion approximation Refs. 3.2, 3, 6] to state the differential equation for the radiation density $w(\tau) = \int_0^{2\pi} d\psi \int_{-1}^{1} I(\mu,\psi)\,d\mu$, assuming the relation $\int_0^{2\pi} d\psi \int_{-1}^{1} I(\mu,\psi)\mu^2\,d\mu = (1/3)\int_0^{2\pi} d\psi \int_{-1}^{1} I(\mu,\psi)\,d\mu$ for the zeroth and second radiance moments. The diffusion approximation as applied to wide and narrow beam propagation problems will be considered in the next chapter.

Another group covers numerous versions of the two-stream approximation. The astrophysicists Schwarzschild and Schuster were the first to apply its simplest variant in 1905 [3.5]. The main idea behind the two-stream approximation is that one may formulate differential equations for the upward and downward fluxes immediately. The equations include the parameters estimated from a priori assumptions about the function $I_n(\tau, \mu)$. Various aspects of the two-stream approximation are detailed in [3.59].

Following [3.59], we shall discuss in this section the key data on this simple and frequently used method.

### 3.7.1 Irradiance Coefficients. Boundary Conditions

In order to characterize the light fields inside a layer, we shall introduce the downward and upward irradiance coefficients of a horizontal area at some depth $\tau$:

$$S_1(\tau, \tau_0, \mu_0) = \int_0^{2\pi} d\psi \int_0^1 I_n(\tau, \mu, \psi; \tau_0, \mu_0)\mu \, d\mu ,$$

$$S_2(\tau, \tau_0, \mu_0) = -\int_0^{2\pi} d\psi \int_{-1}^0 I_n(\tau, \mu, \psi; \tau_0, \mu_0)\mu \, d\mu . \tag{3.7.1}$$

Evidently, $S_1(\tau = \tau_0, \tau_0, \mu_0) = t(\tau_0, \mu_0)$; $S_2(\tau = 0, \tau_0, \mu_0) = r(\tau_0, \mu_0)$. The quantities $S_i(\tau, \tau_0, \mu_0)$ $(i = 1, 2)$ are frequently used in the following analysis. Therefore, for brevity, we generally term the quantity $S_i$ the irradiance. In addition, we omit the arguments if no difficulties arise.

We proceed from the integrodifferential transfer equation (3.1.3) to the system of differential equations for the irradiance coefficients [3.60].

Integrating (3.1.3) with respect to $\mu$ over the upper $(0 \leqslant \mu \leqslant 1)$ and lower $(-1 \leqslant \mu \leqslant 0)$ semispheres, we arrive at the following system of equations for $S_1(\tau)$ and $S_2(\tau)$

$$\frac{dS_1}{d\tau} = -\left[\frac{1 - \Lambda}{\mu_1} + \frac{\Lambda\varphi_1}{\mu_1}\right] S_1(\tau) + \frac{\Lambda\varphi_2}{\mu_2} S_2(\tau) + B_1'$$

$$\frac{dS_2}{d\tau} = \left[\frac{1 - \Lambda}{\mu_2} + \frac{\Lambda\varphi_2}{\mu_2}\right] S_2(\tau) - \frac{\Lambda\varphi_1}{\mu_1} S_1(\tau) + B_2' . \tag{3.7.2}$$

Here,

$$B_1' = B_1'(\tau) = \int_0^{2\pi} d\psi \int_0^1 B'(\tau, \mu, \psi) \, d\mu ,$$

$$B_2' = B_2'(\tau) = \int_0^{2\pi} d\psi \int_{-1}^0 B'(\tau, \mu, \psi) \, d\mu . \tag{3.7.3}$$

The quantities

$$\mu_1 = \mu_1(\tau) = \frac{\int_0^1 I_n(\tau, \mu)\mu \, d\mu}{\int_0^1 I_n(\tau, \mu) \, d\mu} , \qquad \mu_2 = \mu_2(\tau) = \frac{-\int_{-1}^0 I_n(\tau, \mu)\mu \, d\mu}{\int_{-1}^0 I_n(\tau, \mu) \, d\mu} \tag{3.7.4}$$

are the average cosines for the downward and upward light fluxes, respectively; $I_n(\tau, \mu) = (1/2\pi)\int_0^{2\pi} I_n(\tau, \mu, \psi)\, d\psi$ is the azimuth-averaged normalized intensity. It is clear that $0 < \mu_i \leqslant 1$. The quantities

$$\varphi_1 = \varphi_1(\tau) = \frac{1}{2}\frac{\int_0^1 I_n(\tau, \mu)\, d\mu \int_{-1}^0 x(\mu, \mu')\, d\mu'}{\int_0^1 I_n(\tau, \mu)\, d\mu},$$

$$\varphi_2 = \varphi_2(\tau) = \frac{1}{2}\frac{\int_{-1}^0 I_n(\tau, \mu)\, d\mu \int_0^1 x(\mu, \mu')\, d\mu'}{\int_{-1}^0 I_n(\tau, \mu)\, d\mu} \tag{3.7.5}$$

define, correspondingly, the portion of light backscattered by an elementary layer downward ($0 \leqslant \mu \leqslant 1$) and upward ($-1 \leqslant \mu < 0$) irradiated at an angular distribution $I_n(\tau, \mu)$. It is seen from formulas (3.7.5) that in the two limiting cases of isotropic scattering and a very extended indicatrix, the quantity $\varphi_i$ is, in effect, independent of the function $I_n(\tau, \mu)$. In fact, for the inversion center indicatrices, such as the spherical and Rayleigh ones ($x_1 = 0$), formula (3.7.5) yields $\varphi_1 = \varphi_2 = 0.5$. Now, if we assume that the indicatrix has a very forward-extended peak and can be approximately represented as (3.6.4), then $\varphi_1 = \varphi_2 = (1 - \eta)/2$.

The first approximation used to derive the irradiance equations consists in substituting some constants $\mu_i$ and $\varphi_i$ for the unknown functions $\mu_i(\tau)$ and $\varphi_i(\tau)$ at $i = 1, 2$ (the four-parameter variant). Then, (3.7.2) becomes a system of ordinary linear differential equations. The functions $\mu_i(\tau)$ and $\varphi_i(\tau)$ are analysed in [3.59] to show that such an approximate transition is possible.

We must supplement the system (3.7.2) by boundary conditions. Let us consider some of them.

(i) **The boundary $\tau = 0$ of a plane layer is diffusely illuminated.** There are no internal sources in the medium. If the layer boundaries are not reflecting, then we have

$$B_1'(\tau) = B_2'(\tau) = 0\,, \qquad S_1(0) = 1\,, \qquad S_2(\tau_0) = 0\,. \tag{3.7.6}$$

For reflecting boundaries, instead of (3.7.6), we obtain

$$S_1(0) = (1 - A_1') + A_1 S_2(0)\,, \qquad S_2(\tau_0) = A S_1(\tau_0)\,. \tag{3.7.7}$$

Here, $A_1'$ is the reflection coefficient of the downward flux incident on the upper boundary; $A_1$ and $A$ are again the albedo of the surfaces $\tau = 0$ and $\tau = \tau_0$ illuminated by internal scattered radiation. For a semi-infinite medium, we have $S_i(\tau) \to 0$ at $\tau \to \infty$, rather than the second condition in (3.7.7).

(ii) **The layer is illuminated by a parallel beam incident at an angle $\theta_0$ to the normal.** We write down an equation for the scattered component only. Then, for the source function, we obtain

$$B_1'(\tau) = \Lambda F(\mu_0) e^{-\tau/\mu_0}, \qquad B_2'(\tau) = \Lambda[1 - F(\mu_0)] e^{-\tau/\mu_0}, \tag{3.7.8}$$

where the quantity $F(\mu_0)$ can be found from formula (3.3.6). The boundary conditions

$$S_1(0) = 0, \qquad S_2(\tau_0) = 0 \tag{3.7.9}$$

provide for no incident diffuse radiation. More intricate situations involving boundary illumination can be considered in a similar fashion. These are, for example, direct and scattered irradiation of the sea during day-time, etc. The system (3.7.2), together with the boundary conditions, is a complete mathematical statement of the problem. Its solution is very simple. The main difficulty of the two-stream approximation is the a priori definition of the coefficients $\mu_i$ and $\varphi_i$. It is the appropriate choice of these coefficients which determines the application and accuracy of one or other variant of the two-stream approximation.

### 3.7.2 Four-Parameter Variant of the Two-Stream Approximation

By solving the system (3.7.2), subject to the condition (3.7.6), we obtain the following expression for the reflection and transmission coefficients of a diffuse illuminated layer

$$r(\tau_0) = \frac{r[1 - e^{-2(\gamma+b_0)\tau_0}]}{1 - (1+\xi) r^2 e^{-2(\gamma+b_0)\tau_0}},$$

$$t(\tau_0) = \frac{[1 - (1+\xi) r^2] e^{-\gamma\tau_0}}{1 - (1+\xi) r^2 e^{-2(\gamma+b_0)\tau_0}}. \tag{3.7.10}$$

Here,

$$r = \frac{2+\xi}{2(1+\xi)}\left[1 + \frac{\kappa}{\sigma\varphi^*} - \sqrt{\left(\frac{\kappa}{\sigma\varphi^*}\right)^2 + \frac{2\kappa}{\sigma\varphi^*} + \left(\frac{\xi}{2+\xi}\right)^2}\,\right] \tag{3.7.11}$$

is the reflection coefficient of a semi-infinite layer and

$$\gamma = \frac{\Lambda\varphi^*}{\mu^*}\sqrt{\left(\frac{\kappa}{\sigma\varphi^*}\right)^2 + \frac{2\kappa}{\sigma\varphi^*} + \left(\frac{\xi}{2+\xi}\right)^2} - b_0 \tag{3.7.12}$$

is the asymptotic attenuation coefficient in a two-stream approximation;

$$b_0 = \frac{\Lambda\varphi_1\xi}{2\mu_1}\left(1 + \frac{\kappa}{\sigma}\frac{\mu_1 - \mu_2}{\varphi_2\mu_1 - \varphi_1\mu_2}\right), \tag{3.7.13}$$

$$\xi = \frac{\varphi_2\mu_1 - \varphi_1\mu_2}{\varphi_1\mu_2}, \qquad \varphi^* = \frac{\varphi_1\mu_2 + \varphi_2\mu_1}{\mu_1 + \mu_2},$$

$$\mu^* = \frac{2\mu_1\mu_2}{\mu_1 + \mu_2}. \tag{3.7.14}$$

If a layer is illuminated by a parallel beam, then, taking (3.7.8, 9) into account, we have the following solutions:

$$r(\tau_0, \mu_0) = a(\mu_0)r(\tau_0) + b(\mu_0)t(\tau_0)e^{-(1/\mu_0 + 2b_0)\tau_0} - b(\mu_0),$$
$$t(\tau_0, \mu_0) = a(\mu_0)t(\tau_0) + b(\mu_0)r(\tau_0)(1 + \xi)e^{-\tau_0/\mu_0} - a(\mu_0)e^{-\tau_0/\mu_0}. \qquad (3.7.15)$$

The coefficients $a(\mu_0)$ and $b(\mu_0)$ are

$$a(\mu_0) = \frac{\Lambda[F(\mu_0)[(1 - \Lambda)/\mu_2 + 1/\mu_0] + \varphi_2\Lambda/\mu_2]}{\mu_0[(b_0 + 1/\mu_0)^2 - \gamma^2]},$$
$$b(\mu_0) = \frac{\Lambda[(1 - F(\mu_0))[(1 - \Lambda)/\mu_1 - 1/\mu_0] + \varphi_1\Lambda/\mu_1]}{\mu_0[(b_0 + 1/\mu_0)^2 - \gamma^2]}. \qquad (3.7.16)$$

In order that these formulas may be used in practice, we must suggest ways of choosing $\mu_i$ and $\varphi_i$. For very thick optical layers, it is advised in [3.59] to use the asymptotic values of $\mu_i$ and $\varphi_i$. Such an approach is simple, though somewhat tedious. A more convenient variant was implemented in [3.61] for strongly anisotropic scattering when the scattering indicatrix can be approximately represented by (3.6.4). Then, at $\eta = g = x_1/3$, $\varphi_1 = \varphi_2 = (3 - x_1)/6$. For asymptotic, at $\tau \gg 1$, values of $\mu_1$ and $\mu_2$, and making allowance for the numerical and experimental data for the angular distribution $\varphi(\mu)$, [3.61] suggests the semi-empirical formulas

$$\mu_1 = \left(2 - \frac{1 - \Lambda}{\gamma}\right)^{-1}, \qquad \mu_2 = \left(2 + \frac{1 - \Lambda}{\gamma}\right)^{-1}, \qquad (3.7.17)$$

which hold for a wide range of $\Lambda$ and $g$. Then, (3.7.12), on substituting (3.7.17) and $\varphi_1 = \varphi_2 = \varphi = (3 - x_1/6)$, should be considered as an equation for $\gamma$ having the solution

$$\gamma = \sqrt{(1 - \Lambda)[1 - \Lambda(1 - 3\varphi) + 2\sqrt{\varphi\Lambda[1 - \Lambda(1 - 9\varphi/4)]}]}. \qquad (3.7.18)$$

The remaining quantities are expressed in terms of $\gamma$. To be specific, for the reflection coefficient of a diffusely illuminated semi-infinite medium,

$$r = \left(\frac{\gamma + \Lambda - 1}{\gamma + 1 - \Lambda}\right)^2. \qquad (3.7.19)$$

It is clear that the above formulas have correct asymptotics for low absorptions. For $g > 0.96$ and any $\Lambda$, as well as for $\Lambda > 0.7$ and any $g$, the error in the approximate (3.7.18) and (3.7.19) is no more than 3% [3.61].

As an example, we consider the solution of the system (3.72) for a medium with strongly anisotropic scattering and appreciable absorption. Then, for directional medium illumination, we may, in (3.72), approximately set

$$\mu_1 = \mu_0, \qquad \varphi_1 = 1 - F(\mu_0). \qquad (3.7.20)$$

Considering $S_2 \ll S_1$, from (3.7.2) and taking (3.7.20) into account, we have

$$S_1(\tau) = \mathrm{e}^{-[1-\Lambda F(\mu_0)]\tau/\mu_0},$$

$$S_2(\tau) = \frac{\Lambda[1 - F(\mu_0)]}{1 - \Lambda F(\mu_0)} \mathrm{e}^{-[1-\Lambda F(\mu)]\tau/\mu_0}. \tag{3.7.21}$$

Such an approach essentially conforms to the quasi-single approximation (Sect. 3.6.3).

### 3.7.3 Two-Parameter Variant of the Two-Stream Approximation

Light scattering theory and applied problems commonly use the two-parameter variant of the two-stream approximation, assuming

$$\mu_1 = \mu_2 = \tilde{\mu}, \qquad \varphi_1 = \varphi_2 = \tilde{\varphi}. \tag{3.7.22}$$

This can be illustrated by considering the well-known approximations suggested by *Schwarzschild, Schuster* [3.58], *Gurevich, Gershun and Kubelka-Munk* [3.62]. Assuming (3.7.22), formulas (3.7.10–16) are much reduced. Then, for diffuse illumination, we have

$$r(\tau_0) = \frac{r(1 - \mathrm{e}^{-2\gamma\tau_0})}{1 - r^2\mathrm{e}^{-2\gamma\tau_0}}, \qquad t(\tau_0) = \frac{(1 - r^2)\mathrm{e}^{-\gamma\tau_0}}{1 - r^2\mathrm{e}^{-2\gamma\tau_0}}. \tag{3.7.23}$$

The reflection coefficient of a semi-infinite medium $r$ and the asymptotic attenuation coefficient $\gamma$ are determined from the formulas

$$r = 1 + \frac{1 - \Lambda}{\Lambda\tilde{\varphi}} - \sqrt{\left(\frac{1 - \Lambda}{\Lambda\tilde{\varphi}}\right)^2 + 2\frac{1 - \Lambda}{\Lambda\tilde{\varphi}}},$$

$$\gamma = \frac{\Lambda\tilde{\varphi}}{\tilde{\mu}} \sqrt{\left(\frac{1 - \Lambda}{\Lambda\tilde{\varphi}}\right)^2 + 2\frac{1 - \Lambda}{\Lambda\tilde{\varphi}}}. \tag{3.7.24}$$

Given

$$Y = \ln r, \qquad X = \gamma\tau_0, \tag{3.7.25}$$

(3.7.23) coincide with (3.5.5). Such a structural coincidence of approximate, two-stream and rigorous asymptotic solutions is by no means accidental. It displays the symmetry and invariance properties of a plane layer.

The irradiance coefficients for diffuse illumination are found from the relations

$$S_1(\tau, \tau_0) = \frac{\sinh(X - \gamma\tau + Y)}{\sinh(X + Y)}, \qquad S_2(\tau, \tau_0) = \frac{\sinh(X - \gamma\tau)}{\sinh(X + Y)}, \tag{3.7.26}$$

where $X$ and $Y$ are of the form (3.7.25). Equations (3.7.26) imply that, in diffusely illuminated semi-infinite media, the flux attenuates exponentially

$$S_1 = \exp(-\gamma\tau), \qquad S_2 = r\exp(-\gamma\tau).$$

For the layer illuminated by a parallel beam,

$$r(\tau_0, \mu_0) = a(\mu_0)r(\tau_0) + b(\mu_0)t(\tau_0)e^{-\tau_0/\mu_0} - b(\mu_0)\,,$$
$$t(\tau_0, \mu_0) = a(\mu_0)t(\tau_0) + b(\mu_0)r(\tau_0)e^{-\tau_0/\mu_0} - a(\mu_0)e^{-\tau_0/\mu_0}\,, \quad (3.7.27)$$

where the parameters $a(\mu_0)$ and $b(\mu_0)$ are determined by the relations

$$a(\mu_0) = \frac{\Lambda[F(\mu_0)[(1-\Lambda)/\tilde{\mu} + 1/\mu_0] + \tilde{\varphi}\Lambda/\tilde{\mu}]}{\mu_0(1/\mu_0^2 - \gamma^2)}\,,$$
$$b(\mu_0) = \frac{\Lambda[(1-F(\mu_0))[(1-\Lambda)/\tilde{\mu} - 1/\mu_0] + \tilde{\varphi}\Lambda/\tilde{\mu}]}{\mu_0(1/\mu_0^2 - \gamma^2)}\,. \quad (3.7.28)$$

Now, for the irradiance coefficients, we have

$$S_1(\tau, \tau_0, \mu_0) = a(\mu_0)S_1(\tau, \tau_0) + b(\mu_0)S_2'(\tau, \tau_0)e^{-\tau_0/\mu_0} - a(\mu_0)e^{-\tau/\mu_0}\,,$$
$$S_2(\tau, \tau_0, \mu_0) = a(\mu_0)S_2(\tau, \tau_0) + b(\mu_0)S_1'(\tau, \tau_0)e^{-\tau_0/\mu_0} - b(\mu_0)e^{-\tau/\mu_0}\,. \quad (3.7.29)$$

The quantities $S_1(\tau, \tau_0)$ and $S_2(\tau, \tau_0)$ are described by (3.7.26),

$$S_1'(\tau, \tau_0) = \frac{\sinh(\gamma\tau + Y)}{\sinh(X + Y)}\,, \qquad S_2'(\tau, \tau_0) = \frac{\sinh \gamma\tau}{\sinh(X + Y)}\,.$$

The reader is reminded that scattered radiation is understood here. The total irradiance can be obtained by supplementing $S_1$ with the term which includes the unscattered component $\exp(-\tau/\mu_0)$.

Hence, the solutions to a two-parameter two-stream approximation are sufficiently simple and physically instructive. However, the assumptions (3.7.22) used in their derivation have no physical basis. As noted above, the condition $\varphi_1 = \varphi_2$ holds only in some particular cases. In all real situations, $\mu_1 \neq \mu_2$. On account of this, the two-parameter formulas have given rise to reasonable criticism and distrust. Nevertheless, they were re-considered in [3.59]. The assumptions (3.7.22) were shown to be unnecessary in proceeding from the four-parameter (3.7.11, 12) to the two-parameter (3.7.24). Such a procedure is also possible if one takes into account the anisotropy of the angular radiance distribution, i.e., at $\mu_1 \neq \mu_2$. In this case, we should introduce some effective approximation parameters

$$\tilde{\varphi} = \frac{3 - x_1}{8}\,, \qquad \tilde{\mu} = 0.5\,. \quad (3.7.30)$$

We wish to emphasize once again that this is far from being equivalent to the assumption (3.7.22), and the choice of the parameters (3.7.30) allows for $\mu_1 \neq \mu_2$. Naturally, on no account can a physical interpretation of $\mu_i$ and $\varphi_i$ (recalling, for instance, that $\mu_i$ implies the average cosine for an appropriate beam) be extended to the parameters (3.7.30). The introduction of effective parameters alone can probably account for the well-known and significant fact that (3.7.23, 24) have for years been successful both in flux calculations, and as a methodological basis for scattering media spectroscopy.

We now analyse the asymptotic behavior of the two-parameter solutions at $\Lambda \to 1$, choosing the constants from (3.7.30). At $\Lambda \to 1$, (3.7.24), taking (3.7.30) into account, give

$$r = 1 - 4\sqrt{(1-\Lambda)/(3-x_1)} + 8(1-\Lambda)/(3-x_1)\,,$$
$$\gamma = \sqrt{(1-\Lambda)(3-x_1)}\,, \tag{3.7.31}$$

i.e., the asymptotic attenuation coefficient and the reflection coefficient of semi-infinite media conform, at least as accurately as the $(1-\Lambda)^{3/2}$-order terms, to the asymptotic (3.2.11) and (3.3.14). As is revealed by the analysis [3.59], (3.7.24) have wider domains of applicability than expansions (3.7.31).

Consider the dependence of the reflected and transmitted fluxes in a thick layer on the incident angle for the simple case where $\Lambda = 1$. For a thick layer, disregarding the terms $\sim\exp(-\tau_0/\mu_0)$ yields, from (3.7.27, 28),

$$r(\tau_0, \mu_0) = a(\mu_0) r(\tau_0) - b(\mu_0)\,, \qquad t(\tau_0, \mu_0) = a(\mu_0) t(\tau_0)\,, \tag{3.7.32}$$

where

$$a(\mu_0, \Lambda = 1) = F(\mu_0) + \mu_0/4q\,,$$
$$b(\mu_0, \Lambda = 1) = a(\mu_0, \Lambda = 1) - 1\,. \tag{3.7.33}$$

Following (3.7.23, 24), we have

$$r(\tau_0) = \frac{\tau_0}{\tau_0 + 4q}\,, \qquad t(\tau_0) = \frac{4q}{\tau_0 + 4q}\,. \tag{3.7.34}$$

The relations (3.7.34), as expected, comply with the exact asymptotic (3.5.4). At $\Lambda = 1$, we must regard the function $a(\mu_0)$ as an approximate solution for the Milne function $u_0(\mu_0)$. For the simplest scattering indicatrices (isotropic, Rayleigh and $x(\beta) = 1 + x_1 \cos\beta$), we easily obtain the function $a(\mu_0, \Lambda = 1)$ in the form

$$u_0(\mu_0) \approx a(\mu_0, \Lambda = 1) = 0.5(1 + 1.5\mu_0)\,. \tag{3.7.35}$$

A relation similar to Sobolev's approximation (Sect. 3.6) has been derived for $u_0(\mu_0)$. It should be noted that neither the function $a(\mu_0, \Lambda = 1)$ nor $u_0(\mu_0)$ in the rigorous asymptotic theory depends on the parameter $x_1$. It is clear from a comparison of (3.7.35) and (3.2.21) that the difference between them is less than 7%, i.e., the two-stream approximation describes the incident angle dependence to sufficient accuracy.

The curves in Fig. 3.6 demonstrate the accuracy of (3.7.29, 30), and the behavior of the irradiance in a layer.

A thorough analysis of the two-parameter two-stream approximation with parameters chosen from (3.7.30) [3.59] has shown an absolute error of no more

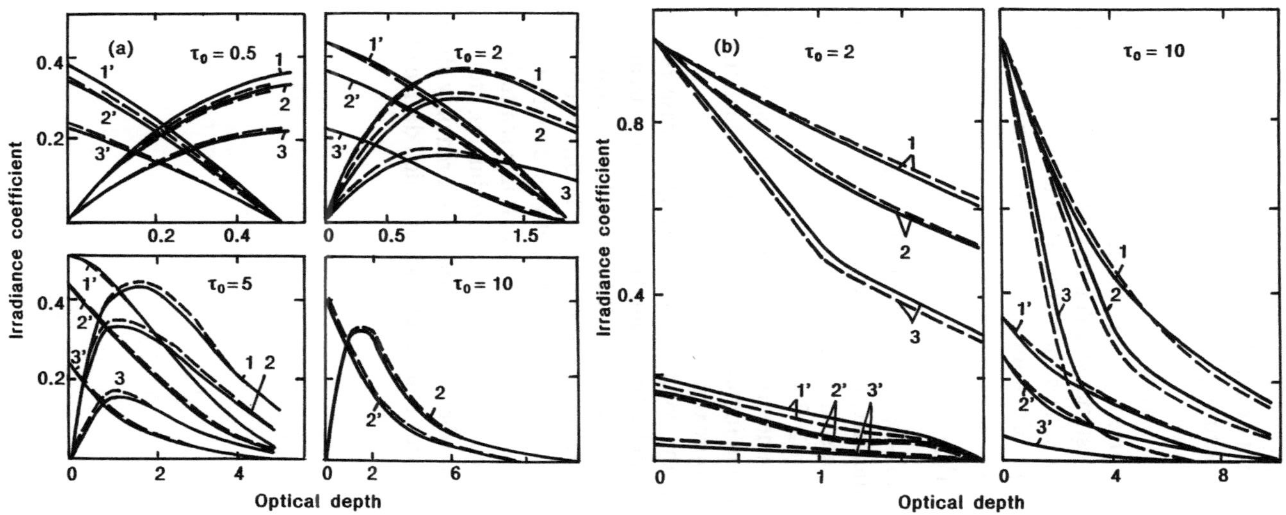


**Fig. 3.6.** $S_1$ and $S_2$ versus $\tau$. (a) $\mu_0 = 1$, $x(\beta) = 1$; (b) diffuse incident light, $x(\beta) = 0.21/(1.1 - \cos\beta)^2$. *Solid curves* are numerical finite-difference calculations and *dashed curves* are calculated from (3.7.26, 29, 30) for $S_1$ (curves 1–3) and $S_2$ (curves $1' - 3'$) at $\Lambda = 0.95$ (curves 1, 1'); 0.9 (curves 2, 2'); 0.7 (curves 3, 3')

than 0.05 in $r(\tau_0)$ and $t(\tau_0)$ calculations. A relative error increases with decreasing $r(\tau_0)$ and $t(\tau_0)$, but does not exceed 10% at $r(\tau_0)$, $t(\tau_0) > 0.1$. This approximation gives a somewhat lower accuracy for directional illumination. In particular, for normal incidence, the relative error in $r(\tau_0, 1)$ and $t(\tau_0, 1)$ is less than 20% at $r$, $t > 0.1$. For small $r(\tau_0, \mu_0)$ and $t(\tau_0, \mu_0)$ ($10^{-3}$ to $10^{-4}$), only the order of magnitude can be estimated correctly. The applicability domain of the formulas with respect to (3.7.30) can be estimated differently. We assume the admissible error in the $r$ determination to be less than 10%. It follows from the data cited that this is realized at $r > 0.1$. Equation (3.7.24) yields constraints for the optical parameters of a medium:

$$(1 - \Lambda)\Lambda\tilde{\varphi} = (1 - r)^2/2r \leqslant 4\,.$$

Otherwise, taking $\tilde{\varphi} = (3 - x_1)/8$, we obtain $(1 - \Lambda)/\Lambda(1 - g) \leqslant 1.5$. Thus, for a spherical scattering indicatrix, the applicability domain of the two-stream approximation with constants chosen from (3.7.30) conforms to $\Lambda \gtrsim 0.4$; for cloud indicatrices at $g \approx 0.82$, $\Lambda \gtrsim 0.8$; for very extended ones at $g \geqslant 0.97$, $\Lambda > 0.95$. The unique choice of parameters (3.7.30) as well as the reasonable accuracy over a wide $\Lambda$ range for different indicatrices of the simple calculation (3.7.23–29) are essential merits of this approximate approach.

We have presented the solution for the simplest case of a homogeneous scattering layer with nonreflecting boundaries. The two-stream approximation offers a simple account of boundary reflection and its extension to the case of a multi-layer medium. The range of problems which can be solved by this approximation is extremely wide. It has been successfully used to study the luminescence of scattering media, specific features of light fields in a scattering medium whose parameters depend of the radiation density, coordinates, etc. It appears that it can reasonably be applied when information is required only about the radiation fluxes from an infinitely wide source, and when a high degree of accuracy is not demanded.

# 4 Radiation Transfer in Scattering Media Illuminated by Localized Sources

The theory of radiative transfer, as was stated in the previous chapter, has accumulated rich stores of rigorous analytical, especially asymptotic, and approximate solutions for calculating light field characteristics in plane layers illuminated by infinitely extended sources. Some of these results were presented in Chap. 3. Appropriate computing algorithms are well-developed. An optimum procedure for obtaining a solution can be chosen according to the problem to be solved, the information available and the accuracy required.

Unfortunately, for more modest progress has been made in the analytical radiative transfer theory as used for localized sources. The numerical methods of solving the relevant boundary-value problems are also far from satisfactory. The only exception is the Monte-Carlo simulations [4.1, 2]. It is no mere chance, therefore, that the general picture of light fields from localized sources was, in many respects, clarified using experimental modelling [4.3–5]. These results were generalized in the monographs [4.6, 7]. For small $\tau$ and $\Lambda$, certain results may be obtained by a series expansion restricted to scattering up to second order [4.8, 9]. However, different approaches are required for strongly scattering media.

The integral transforms and equations for the spatial irradiance distribution moments, which reduce the dimensionality of the problem, are the principal mathematical approaches used to describe the light fields of localized sources in the rigorous and approximate radiative transfer theory. They are discussed in Sect. 4.1. Qualitative studies of the solutions are very important for the development of approximate techniques and effective computation algorithms. Some information on the local properties of solutions of localized source problems and their asymptotic behavior at $\tau \gg 1$ is given in Sect. 4.2.

Diffusion (DA), small-angle (SAA) and small-angle diffusion (SADA) approximations which, fortunately, complement one another are the main approaches used to provide particular design solutions to calculate the light fields originated by localized sources. The reader is referred to Sects. 4.3–5 for these methods and their results. Section 4.5.7 suggests the modified SADA as a means of essentially improving the description of spatial radiance distributions. Some necessary data on backscattering in media with a small angle-type light field generation are given in Sect. 4.6. Section 4.7 deals with different approximate approaches from a single viewpoint and presents a convenient estimate of their applicability as well as solutions for illumination by an infinitely extended source.

So far, we have dealt with stationary radiation sources. The extensive use of pulse laser techniques in vision and location systems has prompted us to include a special section (Sect. 4.8), which describes light fields from localized pulsed sources. In this section, we discuss the general methodology of solving nonstationary problems as well as particular methods and results for pulse backscattering and distortion during its propagation in a scattering medium. The results of this section rely completely upon the approximate methods of solving the stationary transfer equation described in Sects. 4.3–6. It should be noted that the term "narrow beams" is frequently used alongside the term "localized sources".

## 4.1 Transfer Equation for a Narrow Beam. Spatial Radiance and Irradiance Distribution Moments

Let us again write down the initial integro-differential equation (2.3.1) for the medium illuminated by narrow beams

$$\vec{n}\cdot\nabla_{\vec{R}} I(\vec{R},\vec{n}) = -\varepsilon I(\vec{R},\vec{n}) + \frac{\sigma}{4\pi}\int\!\!\int I(\vec{R},\vec{n}')x(\vec{n},\vec{n}')\,d\vec{n}' + B(\vec{R},\vec{n})\,. \quad (4.1.1)$$

We now define the coordinate system depicted in Fig. 4.1 and set the $z$ axis in the Cartesian coordinate system to be normal to the layer boundary and directed inside the layer. The layer boundary is then parallel to the $xy$ plane. As before, the radius vector $\vec{R}(x,y,z)$ characterizes an observation point and the vector $\vec{n}(\theta,\psi)$ is given by the polar angle $\theta$ and the azimuthal angle $\psi$ measured from the $x$ direction in the spherical coordinate system (Fig. 4.1). The differential operator on the l.h.s. of (4.1.1) is of the form:

$$\vec{n}\cdot\nabla_{\vec{R}} = \mu\frac{\partial}{\partial z} + \sqrt{1-\mu^2}\cos\psi\frac{\partial}{\partial x} + \sqrt{1-\mu^2}\sin\psi\frac{\partial}{\partial y}\,. \quad (4.1.2)$$

Here, the radiance depends on two coordinates $\theta$, $\psi$ which determine the $\vec{n}$ direction, and on three coordinates $(x,y,z)$ of the observation point $\vec{R}$. Sometimes,

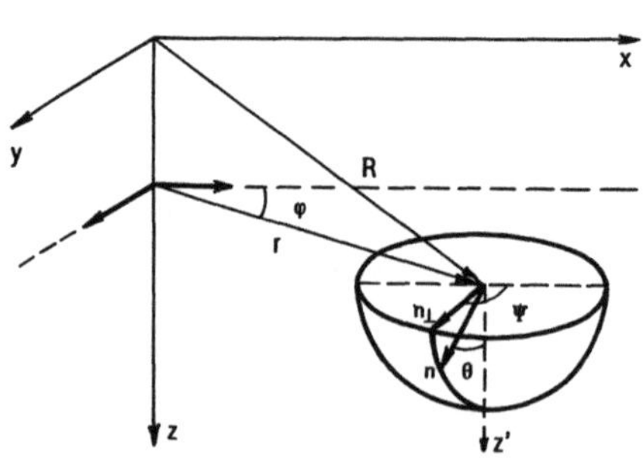


**Fig. 4.1.** Coordinate system for the transfer equation for a narrow beam

problems which involve localized sources are called three-dimensional, in contrast to one-dimensional problems which involve infinitely wide sources and where the radiance depends only on one observation point coordinate. For a homogeneous medium, optical coordinates, for example the optical radius-vector $\vec{\tau} = \varepsilon\vec{R}$ with the components $\tau_x = \varepsilon x$; $\tau_y = \varepsilon y$; $\tau = \varepsilon z$, are often convenient to use.

Let a radiant flux $P_0$ from a localized source be incident on the layer boundary. We include the quantity

$$I_n(x, y, z, \vec{n}) = I(x, y, z, \vec{n})/P_0 \tag{4.1.3}$$

which is the radiance at a point $\vec{R}$ in the $\vec{n}$ direction normalized to unit incident flux. It is understood to be the distribution density, $m^{-2}sr^{-1}$. The integral of this density over the plane $z$ = const

$$I_n(z, \vec{n}) = \iint_{-\infty}^{\infty} I_n(x, y, z, \vec{n})\,dx\,dy \tag{4.1.4}$$

equals the normalized radiance at a depth $z$ in the $\vec{n}$ direction when the medium is illuminated by an infinitely extended source with the same directional diagram. Similarly, we introduce the quantities

$$S_1(x, y, z) = E_1(x, y, z)/P_0\,, \tag{4.1.5}$$

$$S_2(x, y, z) = E_2(x, y, z)/P_0\,, \tag{4.1.6}$$

which are numerically equal to the downward and upward irradiances of the area located at the point $\vec{R}$ and normal to the $z$ axis, provided the incident flux is of unity. They are also measured in $m^{-2}$ and, like the quantity $I_n(x, y, z, \vec{n})$, is understood as the distribution density. The flux transfer coefficient at a depth $z$ is

$$S_1(z) = \iint_{-\infty}^{\infty} S_1(x, y, z)\,dx\,dy = P_1(z)/P_0\,, \tag{4.1.7}$$

$$S_2(z) = \iint_{-\infty}^{\infty} S_2(x, y, z)\,dx\,dy = P_2(z)/P_0\,, \tag{4.1.8}$$

$P_1(z)$ and $P_2(z)$ are the total downward and upward fluxes through the plane $z$ = const. The radiation transfer coefficients are equal to the irradiance coefficients for the layer illuminated by an infinitely extended source. We frequently use the quantities $I_n(x, y, z, \vec{n})$ and $S_i(x, y, z)$. When there is no possibility of confusion, they are called, for brevity, and radiance and irradiance. Moreover, the quantity $S_1$ will be encountered much more often than $S_2$. Therefore, we shall omit the subscript 1, i.e., set $S \equiv S_1$.

The two-dimensional Fourier transformation with respect to the spatial $x$ and $y$ coordinates is a traditional procedure for solving (4.1.1) [4.10, 11]. Let us define the Fourier transform of some function $a(x, y)$:

$$a(\nu_x, \nu_y) = \iint_{-\infty}^{\infty} a(x, y)e^{-i(\nu_x x + \nu_y y)}\,dx\,dy\,. \tag{4.1.9}$$

In order to avoid introducing new notation, here and below the function and its Fourier transform differ only in their arguments.

For the radiance Fourier transform, instead of (4.1.1, 2), we have the following equation:

$$\mu \frac{dI(v_x, v_y, z, \vec{n})}{dz} = -\varepsilon_v I(v_x, v_y, z, \vec{n}) + \frac{\sigma}{4\pi} \iint_{4\pi} I(v_x, v_y, z, \vec{n}')x(\vec{n}, \vec{n}')\, d\vec{n}'$$

$$+ B(v_x, v_y, z, \vec{n}), \tag{4.1.10}$$

where

$$\varepsilon_v = \varepsilon + \mathrm{i}\sqrt{1-\mu^2}(v_x \cos\psi + v_y \sin\psi), \tag{4.1.11}$$

$$B(v_x, v_y, z, \vec{n}) = \iint_{-\infty}^{\infty} B(x, y, z, \vec{n}) \mathrm{e}^{-\mathrm{i}(v_x x + v_y y)}\, dx\, dy\,. \tag{4.1.12}$$

Equation (4.1.10) is formally similar to (3.1.3), i.e., the Fourier transformation reduces a three-dimensional to one-dimensional problem.

Let the angular radiance diagram of the light source be symmetric about the normal to the layer. We introduce the radius-vector $\vec{R}$ and the $\vec{r}$ projection onto the plane $z = \mathrm{const}$, i.e., $\vec{R}(\vec{r}, z)$. In the Cartesian coordinate system, $\vec{r}(x, y)$. As far as problems of axial symmetry are concerned, it is often convenient to use the cylindrical coordinate system. Then, in the polar coordinates, $\vec{r}$ is given by the distance $r = |\vec{r}|$ and the azimuthal angle $\varphi$ taken from the $x$ direction, i.e., $\vec{R}(r, z, \varphi)$ (Fig. 4.1). It should be borne in mind that $\varphi$ is the azimuth of the point $\vec{R}$, in contrast to the $\psi$ azimuth of the $\vec{n}$ direction. Correspondingly, $\tau = \varepsilon z$ and $\vec{\rho} = \varepsilon \vec{r}$ are the optical coordinates. In this case, the spatial-angular radiance distribution has a symmetry axis; the radiance $I(r, z, \theta, \varphi - \psi)$ depends only on 4, rather than on 5 variables; the irradiance distribution is azimuthally symmetric and independent of $\varphi$, and the quantity $S$ depends on two spatial coordinates $S(r, z)$. Here, the Fourier transformation (4.1.9) for the irradiance is equivalent to the one-dimensional Hankel transformation with respect to the $r$ coordinate [4.12]:

$$a(v) = 2\pi \int_0^{\infty} a(r) J_0(vr) r\, dr\,, \tag{4.1.13}$$

where the spatial frequency $v = \sqrt{v_x^2 + v_y^2}$; $J_0(vr)$ is the zero-order Bessel function.

Naturally, the Fourier (or Hankel) transform is also used to solve various approximate transfer equations for a narrow beam (since they include a linear differential operator (Sects. 4.3, 4).

The theory of "narrow beams" often reduces to the solutions in the Fourier region. This is attributed, on the one hand, to the difficulties involved in the inverse Fourier transformation. On the other hand, the information contained in the Fourier transform $a(v_x, v_y)$ is frequently sufficient to solve applied problems. For example, one of the basic characteristics of the image transfer theory (details are found in Chap. 6), is the fact that the optical transfer function (OTF) of the medium is the Fourier transform of the point spread function (PSF). The PSF, in turn, is the spatial distribution of the irradiance due to a monodirectional unit power point source, i.e. the function $S(x, y, z)$ for a monodirectional point source. Thus, the medium OTF is expressed as

$$S(v_x, v_y, z) = \iint_{-\infty}^{\infty} S(x, y, z) \mathrm{e}^{-\mathrm{i}(v_x x + v_y y)} \, dx \, dy \, . \tag{4.1.14}$$

Introducing the dimensionless spatial frequencies

$$\omega_x = v_x/\varepsilon \, , \qquad \omega_y = v_y/\varepsilon \tag{4.1.15}$$

gives, instead of (4.1.14),

$$S(\omega_x, \omega_y, \tau) = \frac{1}{\varepsilon^2} \iint_{-\infty}^{\infty} S(\tau_x, \tau_y, \tau) \mathrm{e}^{-\mathrm{i}(\omega_x \tau_x + \omega_y \tau_y)} \, d\tau_x \, d\tau_y \, . \tag{4.1.16}$$

For the case of normal light incident on the medium, we have an azimuthally symmetric PSF and [4.12]

$$\begin{aligned} S(v, z) &= \int_0^{2\pi} d\psi \int_0^{\infty} S(r, z) \mathrm{e}^{-\mathrm{i} v r \cos \varphi} r \, dr \\ &= 2\pi \int_0^{\infty} S(r, z) J_0(vr) r \, dr \, . \end{aligned} \tag{4.1.17}$$

**Spatial intensity and irradiance distribution moments.** In the theory of narrow beam propagation, the moments of the field characteristic distribution over the plane $z = \text{const}$ are commonly used. We remind the reader that the moments of the two-dimensional distribution of some function $a(x, y)$ are found from

$$M_{n,k} = \iint_{-\infty}^{\infty} x^n y^k a(x, y) \, dx \, dy \, . \tag{4.1.18}$$

In the present case, we often understand by the function $a(x, y)$ the normalized radiance $I_\mathrm{n}(x, y, z, \vec{n})$ and irradiance $S_i(x, y, z)$ coefficients above introduced.

The zeroth moment of the function

$$M_{0,0} = \iint_{-\infty}^{\infty} I_\mathrm{n}(x, y, z, \vec{n}) \, dx \, dy = I_\mathrm{n}(z, \vec{n}) \tag{4.1.19}$$

is the normalized radiance when the medium is illuminated by an infinitely extended source which has the same angular characteristics as a localized source. The zeroth moment of the function $S_i(x, y, z)$ is the flux transfer coefficient through the plane $z = \text{const}$ in the direction of the axis $z$ ($i = 1$), while in the opposite direction, $i = 2$, it is the irradiance due to an infinitely extended source. Equations (4.1.4, 19) relate the fields of infinitely extended and localized sources and give expression to the mathematical formulation of the superposition principle for this case.

If the light field is azimuthally symmetric, then we may consider the radial distribution moments. In particular, for the irradiance moments

$$M^{(i)}_{r,2n} = 2\pi \int_0^{2\pi} S_i(r,z) r^{2n+1}\, dr \,. \tag{4.1.20}$$

The moments $M_{r,2n}$ from (4.1.20) and $M_{n,k}$ from (4.1.18) are related by

$$M_{r,2n} = \frac{2^n n!}{(2n-1)!!} M_{2n,0} \,.$$

It is important that the moments of the spatial irradiance distribution can be expressed in a straightforward way in terms of the OTF [4.13]. In fact, by differentiating (4.1.17) $n$ times with respect to $\nu$ and passing to the limit at $\nu \to 0$, we arrive at

$$M_{r,2n} = (-1)^n \frac{2^n n!}{(2n-1)!!} \lim_{\nu \to 0} \frac{d^{2n} S(\nu)}{d\nu^{2n}} \,. \tag{4.1.21}$$

Here $(2n-1)!! = (2n-1)(2n-3)\dots 1$. The first distribution moments responsible for the variance $V_r$ and kurtosis $\chi_r$ of the spread function are of particular importance:

$$V_r = \frac{M_{2,0}}{M_{0,0}} = \frac{M_{r,2}}{2M_{r,0}} \,, \qquad \chi_r = \frac{M_{4,0}}{M_{2,0}^2} = \frac{3}{2} \frac{M_{r,4}}{M_{r,2}^2} \,. \tag{4.1.22}$$

A number of works [4.14–16] suggest an alternative approach to determine the moments $M_{n,k}$ for the spatial radiance distribution. It provides the equation system for the radiance distribution moments $M_{n,k}$ defined by (4.1.18) by integrating the transfer equation (4.1.1, 2) over the $x$, $y$ coordinates weighted as $x^n y^k$. The equations for $M_{n,k}$ are similar to the transfer equation for the radiance from an infinitely extended source, but have different source functions. Thus, the problem reduces one-dimensional problem which has been thoroughly investigated.

## 4.2 Local and Asymptotic Properties of Transfer Equation Solutions

The qualitative analysis of boundary-value problems is particularly important in radiation transfer theory. Effective computation algorithms and approximate approaches cannot be developed without appropriate information on the local structure of the solutions and their asymptotic behavior.

For the simplest geometries discussed in Chap. 3 (a plane homogeneous layer, infinitely extended source), the solutions have singularities only at the layer boundaries. These singularities are not essential. It is a different matter in problems with localized sources where the singularities in the solutions are essential owing to first order scattering [4.17–19]. There can be found analytically. A consistent discussion of the local properties of the solution to the transfer equation is given in the recently published monograph [4.20].

In the analysis presented below, we only give an outline of the local PSF properties and discuss their significance in the image transfer problem. The local properties of the radiance solution are considered in [4.20].

Let a monodirectional point source of unit power be localized at the coordinate origin at the medium boundary and emit normal to the boundary, i.e.,

$$B(\vec{r}, z, \vec{n}) = \delta(z)\delta(\vec{n})\delta(\vec{r}) . \tag{4.2.1}$$

We represent the PSF as the scattering order series expansion

$$S(r,z) = S^{(0)}(r,z) + S^{(1)}(r,z) + S^{(2)}(r,z) + \sum_{n>2} S^{(n)}(r,z) . \tag{4.2.2}$$

Here, $S^{(0)}(r,z)$ allows for the unscattered light contribution; $S^{(1)}(r,z)$ and $S^{(2)}(r,z)$, respectively, are the contributions made to the PSF by the first- and second-order scattered light terms. The term $\sum_{n>2} S^{(n)}(r,z)$ is a sum of the contributions of the multiple (i.e. higher than second-order) scattering. From (4.1.1, 4.2.1), we may show that at $r \to 0$, the function $S^{(n)}(r,z)$ is characterized by local singularities of the type

$$\begin{aligned} &\delta(r)/r \quad \text{for} \quad S^{(0)}(r,z) \\ &1/r \quad \text{for} \quad S^{(1)}(r,z) \\ &\ln\frac{1}{r} \quad \text{for} \quad S^{(2)}(r,z) . \end{aligned} \tag{4.2.3}$$

The function $\sum_{n>2} S^{(n)}(r,z)$ can be differentiated at all spatial points. It should be noted that the analytical PSF representation is intricate, primarily because of its local properties.

We now consider some asymptotic properties of the transfer equation solutions. It is shown in [4.14] that in deep layers ($\tau \gg 1$) of the semi-infinite medium illuminated by a monodirectional point source, the PSF variance $V_\rho$ linearly increases with the optical depth:

$$V_\rho = d(\gamma)\tau , \tag{4.2.4}$$

the quantity $d(\gamma)$ being determined from the integral equation related to the characteristic radiation transfer equation (3.2.2). We can show that for the spherical scattering indicatrix,

$$d(\gamma) = 1/\gamma . \tag{4.2.5}$$

At $\Lambda \to 1$, the asymptotic coefficient $\gamma \to 0$ and $V_\rho \to \infty$.

The spatial radiance distribution about the beam axis at $\tau \gg 1$ and $\Lambda \neq 1$ for the source, (4.2.1), can be approximated by a normal distribution [4.14]. Then, for the normalized radiance in deep layers, at $\mu_0 = 1$, we obtain

$$I_n(\rho, \tau, \mu, \psi) = \frac{u(1)\varphi(\mu)}{\pi} \frac{e^{-\gamma\tau}}{\tau} \frac{e^{-\rho^2/2d(\gamma)\tau}}{2\pi d(\gamma)} . \tag{4.2.6}$$

As seen from (4.2.6), the radiance about the beam axis ($\rho^2/2d(\gamma)\tau \ll 1$) decreases with the depth as $\exp(-\gamma\tau)/\tau$. It should be noted that such a dependence occurs at large distances from a localized source, irrespective of the directional diagram. It was first obtained for an isotropic point source [4.21]. The angular intensity distribution in the axial region, as seen from (4.2.6), is the same as for an infinitely extended source and obeys the function $\varphi(\mu)$.

Another asymptotic problem, light reflection by a semi-infinite medium, was solved for the particular case of isotropic scattering [4.10]. In this work, an integral equation for the radiance of the light reflected by the semi-infinite medium illuminated by a monodirectional point source was formulated in terms of the invariance principles, and solved using the Fourier transform. The solutions obtained in this way are, however, awkward. They require numerical calculation and may, in general, be used to verify approximate methods.

## 4.3 The Diffusion Equation

The diffusion approximation (DA) was developed and is most frequently used in neutron transfer theory [4.21, 22]. The basic concepts of DA, an analysis of its accuracy and some results have been presented [4.21–24]. These results, however, apply mainly to isotropic sources and media with a spherical or slightly extended scattering indicatrix, whereas optical problems are notable for their concern with monodirectional sources in anisotropically scattering media.

### 4.3.1 Derivation of the Diffusion Equation

We shall now pass from the integro-differential transfer equation (4.1.1) to the differential diffusion equation. Integrating (4.1.1) over all $\vec{n}$ directions gives the known equation (2.3.3):

$$\nabla_{\vec{R}} \cdot \vec{F}(\vec{R}) = -\kappa w(\vec{R}) + B_0(\vec{R}) , \tag{4.3.1}$$

where

$$B_0(\vec{R}) = \iint_{4\pi} B(\vec{R}, \vec{n})\, d\vec{n} \tag{4.3.2}$$

is the zeroth moment of the source function. If we now multiply the l.h.s. and r.h.s. of (4.1.1) by $\vec{n}$ and again integrate the expression so obtained with respect to the solid angle:

$$\iint_{4\pi} \vec{n}(\vec{n} \cdot \nabla_{\vec{R}} I(\vec{R}, \vec{n}))\, d\vec{n} = -\varepsilon(1 - \Lambda g)\vec{F}(\vec{R}) + \vec{B}_1(\vec{R}) . \tag{4.3.3}$$

Here,

$$\vec{B}_1(\vec{R}) = \iint_{4\pi} \vec{n} B(\vec{R}, \vec{n})\, d\vec{n} \tag{4.3.4}$$

is the first moment of the source function. We used the explicit relation

$$\frac{1}{4\pi} \iint_{4\pi} x(\vec{n}, \vec{n}')\vec{n}'\, d\vec{n}' = g\vec{n} \tag{4.3.5}$$

to derive (4.3.3).

The equation system (4.3.1, 3) will be closed provided the integral on the l.h.s. of (4.3.3) is expressed in terms of $w$ and $\vec{F}$. In order for this to be the case, the angular radiance dependence must be set a priori. In the classical diffusion theory, the scattered light radiance is taken to be weakly anisotropic (thick layers, weak absorption) and can be represented as

$$I(\vec{R}, \vec{n}) = \frac{1}{4\pi}[w(\vec{R}) + 3(\vec{F}(\vec{R}) \cdot \vec{n})]\,. \tag{4.3.6}$$

Then,

$$\iint_{4\pi} \vec{n}(\vec{n} \cdot \nabla_{\vec{R}} I(\vec{R}, \vec{n}))\, d\vec{n} = \frac{1}{3} \nabla_{\vec{R}} w(\vec{R})\,. \tag{4.3.7}$$

Equations (4.3.3, 7) imply one more [in addition to (4.3.1)] relation for the radiation density and flux

$$\vec{F}(\vec{R}) = -\frac{1}{3\varepsilon(1 - g\Lambda)}[\nabla_{\vec{R}} w(\vec{R}) - 3\vec{B}_1(\vec{R})]\,, \tag{4.3.8}$$

which is known as the Fick law [4.22, 24]. Substituting (4.3.8) into (4.3.1) and using the optical coordinates $\vec{\tau} = \varepsilon\vec{R}$, we get the known diffusion equation

$$\Delta_\tau w(\vec{\tau}) - \gamma^2 w(\vec{\tau}) = G(\vec{\tau})\,. \tag{4.3.9}$$

Here,

$$\gamma^2 = 3(1 - \Lambda)(1 - \Lambda g)\,, \tag{4.3.10}$$

$$G(\vec{\tau}) = -\frac{3(1 - \Lambda g)}{\varepsilon} B_0(\vec{R}) + \frac{3}{\varepsilon} \nabla_{\vec{\tau}} \vec{B}_1(\vec{R})\,. \tag{4.3.11}$$

This last formula expresses the function $G$ on the r.h.s. of the diffusion equation in terms of the angular moments of the source function. One of the difficulties of the DA is attributable to the indefinite choice of the boundary conditions whose statement also requires a priori informaton. For the nonilluminated boundary of the medium, the DA boundary condition is generally chosen to match the known solution of the Milne problem (Chap. 3) [4.21, 23]. In Fig. 4.2, the behavior of the radiation density about the nonilluminated semi-infinite layer boundary is shown for an infinitely wide radiation source deep in the medium

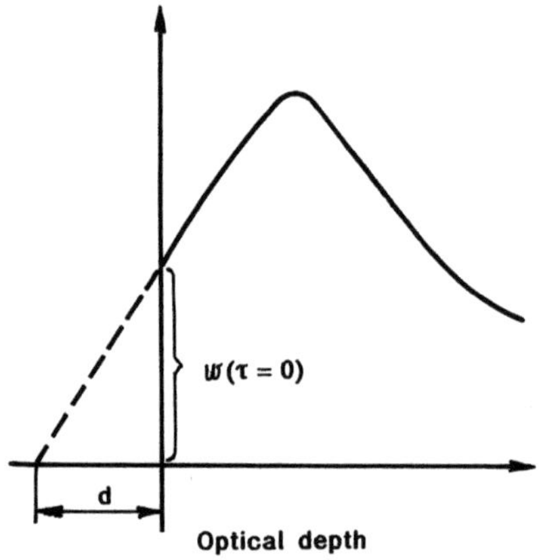


**Fig. 4.2.** Radiation density about the nonilluminate semi-infinite layer boundary (Milne problem)

(solution of the Milne problem for the radiation density [4.22]). We shall linearly extrapolate the function $w(\tau)$ to the domain $\tau < 0$. We may then consider, by convention, that the radiation density becomes zero at $\tau = -d$, and that the optical distance $d$ is the extrapolated length in the Milne problem. Thus, we have the boundary condition

$$w\,(\tau = -d) = 0\,. \tag{4.3.12}$$

The boundary condition might also be set in the form

$$\left.\frac{\partial w(\tau)}{\partial \tau}\right|_{\tau=0} = \frac{w\,(\tau = 0)}{d}\,.$$

We generally use the simpler condition (4.3.12). From the solution of the Milne problem [4.21, 23, 25], $d = 6q\int_0^1 u_0(\mu)\mu^2\,d\mu$ at $\Lambda = 1$ and when there is no reflection at the boundary. With regard to (3.2.21), we find approximately

$$d \approx 2q \qquad \text{at } \Lambda = 1\,. \tag{4.3.13}$$

If $\Lambda \neq 1$, then $d \approx 2q/\Lambda$. The boundary reflection changes the extrapolated length. In particular, as follows from [4.22], the Lambert reflection with albedo $A$ at the boundary can approximately be taken into account by introducing the equivalent extrapolated length

$$d_A = d\frac{1 + A}{1 - A}\,. \tag{4.3.14}$$

The conditions for the function $w(\tau)$ at infinity follow from physical considerations

$$w(\vec{\tau}) \to 0 \qquad \text{at } |\vec{\tau}| \to \infty\,. \tag{4.3.15}$$

Equation (4.3.9), subject to the boundary conditions (4.3.12, 15), is a complete mathematical statement of the problem. The solution for the radiation density $w$ having been found, the total flux follows from Fick's law (4.3.8). The irradiance at $\vec{\tau}$ can also be expressed in terms of the function $w(\vec{\tau})$. Considering (4.3.6, 8), we arrive at

$$S_i(\tau_x, \tau_y, \tau) = \frac{1}{4P_0}\left[w(\tau_x, \tau_y, \tau) + (-1)^i 2q\frac{dw(\tau_x, \tau_y, \tau)}{d\tau} - (-1)^i 6q\frac{B_1(\tau_x, \tau_y, \tau)}{\varepsilon}\right], \quad i = 1, 2\,. \tag{4.3.16}$$

In fact, the derivation of the diffusion equation complies with the $P_1$-approximation of spherical harmonics [4.21, 23] alternatively, the diffusion equation (4.3.9) can be interpreted [without stressing (4.3.10)] as an equation satisfied by the asymptotic solution for the medium points far from the bounaries and sources [4.21, 22, 24]. In this case, the parameter $\gamma$ in (4.3.9) is determined as the smallest eigenvalue in the characteristic equation (3.2.2). Both approaches coincide for weakly absorbing media when the formula $\gamma^2 = (1 - \Lambda)(3 - x_1)$ can be applied.

The problem of the accuracy of the DA has been discussed elsewhere [4.21–23, 26]. In weakly absorbing $[(1 - \Lambda)/(1 - g) \ll 1]$ semi-infinite media and very thick optical layers, the DA provides high enough accuracy in calculating the first two spatial irradiance distribution moments [4.23, 26]. In the case of an infinitely wide beam, DA solutions for the fluxes in a thick layer really coincide with rigorous asymptotic solutions [4.26]. As seen from these studies, it is advisable to use the DA at $\tau_0 > 4q$ and $(1 - \Lambda)/(1 - g) \lesssim 0.3$. (For the applicability of DA, see Sect. 4.7.)

We have used a priori the assumption (4.3.6) for the angular radiance dependence to derive the diffusion equation. Alternative assumptions for the function $I(\vec{\tau}, \vec{n})$ provide further approximate differential equations, which can be regarded as some variants of the diffusion equation, to determine the angular radiance moments. For example, for a strongly extended scattering indicatrix and appreciable absorption, the angular intensity distribution is strongly anisotropic. Therefore, in [4.27, 28], for the case of an infinitely wide beam incident normal to the layer boundary, the relation for the angular moments $w$ and $\vec{F}$ is sought from the a priori intensity distribution:

$$I(\tau, \vec{n}) = \frac{w(\tau)}{4\pi V_\theta(\tau)} e^{-\theta^2/2V_\theta(\tau)}\,.$$

This approximation is essentially similar to the small-angle diffusion approximation that will be considered in Sect. 4.5. Of course, the accuracy of the differential equations obtained in this way and the applicability domain of the relevant approximate solutions are determined by the a priori information used.

### 4.3.2 Solution of the Diffusion Equation for an Isotropic Point Source

Let an isotropic point source of unit power ($P_0 = 1$) lie at some point having the coordinates $\tau = \tau'$ and $\rho = 0$. For the axial symmetry of the light field, we use the cylindrical system of optical coordinates introduced in Sect. 4.1. Then, we have

$$G(\rho,\tau) = -\frac{\varepsilon^2}{q}\delta(\tau-\tau')\frac{\delta(\rho)}{2\pi\rho}, \tag{4.3.17}$$

where $\delta(\tau)$ is the Dirac delta-function. Assuming the possibility of different reflection conditions at the layer boundaries by introducing different extrapolated lengths for the upper ($d_1$) and lower ($d_2$) boundaries of the layers:

$$w(\rho,\tau=-d_1)=0, \qquad w(\rho,\tau=\tau_0+d_2)=0. \tag{4.3.18}$$

Then, for the radiation density, we have [4.29]

$$w(\rho,\tau,\tau_0,\tau') = \frac{\varepsilon^2}{4\pi q}\left[\frac{e^{-\gamma\alpha_{04}}}{\alpha_{04}} - \frac{e^{-\gamma\alpha_{03}}}{\alpha_{03}} - \sum_{k=1}^{\infty}\sum_{i=1}^{4}(-1)^{i+1}\frac{e^{-\gamma\alpha_{ki}}}{\alpha_{ki}}\right], \tag{4.3.19}$$

where

$$\begin{aligned}
\alpha_{ki} &= \sqrt{\beta_{ki}^2+\rho^2}, & \beta_{k1} &= 2k\eta - a_1, \\
a_2 &= \tau-\tau', & \beta_{k2} &= 2k\eta - a_2, \\
a_1 &= \tau+\tau'+2d_1, & \beta_{k3} &= 2k\eta + a_1 \\
\eta &= \tau_0+d_1+d_2, & \beta_{k4} &= 2k\eta + a_2.
\end{aligned} \tag{4.3.20}$$

The structure for the expression $w(\rho,\tau,\tau_0,\tau')$ is physically instructive. The first bracketed term in (4.3.19) determines the radiation density due to an isotropic point source in an infinite medium; the second includes the correction for the radiation sink through the boundary of a semi-infinite medium. The radiation sink through the other boundary in a finite-thickness layer, as well as the combined effect of these two boundaries, are incorporated in the third term of (4.3.19) written as a series. This series converges faster for all $\gamma$, the larger the layer thickness and light absorption in the medium.

As seen from (4.3.19), the solution for a semi-infinite medium is of the form

$$w(\rho,\tau,\tau') = \frac{\varepsilon^2}{4\pi q}\left[\frac{e^{-\gamma\alpha_{04}}}{\alpha_{04}} - \frac{e^{-\gamma\alpha_{03}}}{\alpha_{03}}\right].$$

This formula can be derived simply on the following physical grounds. The condition (4.3.12) for a semi-infinite medium is provided by substituting an infinite medium for a real one, an additional virtual source of negative power being mirror-symmetric to the real source about the extrapolated boundary (Fig. 4.3). Then, the solution will be written for the infinite medium with two sources $\alpha_{04}$ and $\alpha_{03}$ away from the observation point.

We can pursue this idea in order to understand that the solution to the problem for a scattering medium of an arbitrary volume may be expressed through that for an infinite medium having an equivalent combination of hypothetic point sources which will ensure zero radiation density at the extrapo-

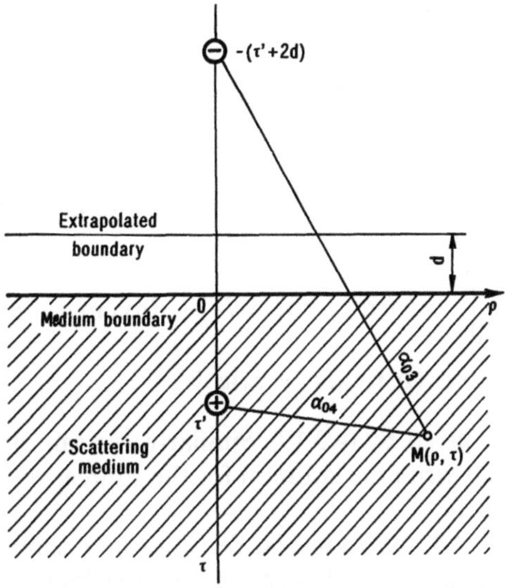


**Fig. 4.3.** Extrapolated boundary and virtual sources in a semi-infinite medium

lated boundary of the volume. In the same way, we can obtain (4.3.19) for the $\tau_0$-thick optical layer, $\alpha_{ki}$ being the distance from the observation point to additional sources. Such a technique was used, in particular, in [4.26, 30].

Expression (4.3.19) is essentially the Green's function of the problem. It suggests a simple solution for any localized source with different directional diagrams.

### 4.3.3 Radial Irradiance Distribution from a Monodirectional Point Source

For the case of a monodirectional source, it is advisable to write the diffusion equation only for the scattered radiation. The unscattered component must be added to the final result to obtain the radiation density. Let a monodirectional point source of unit power ($P_0 = 1$) radiating along the normal to the medium surface be at the point $\rho = 0$, $\tau = 0$ of the $\tau_0$-thick optical layer. Then, in the transfer equation (4.1.1), the source function determined by the single-scattered light is of the form

$$B(\vec{\tau}, \vec{n}) = \frac{x(\mu)}{4\pi} \Lambda \varepsilon^2 e^{-\tau} \frac{\delta(\rho)}{2\pi\rho} [U(\tau) - U(\tau - \tau_0)] , \tag{4.3.21}$$

where $U(\tau)$ is the Heaviside unit function:

$$\begin{aligned} U(\tau) &= 1 , \qquad \tau \geqslant 0 , \\ &= 0 , \qquad \tau < 0 . \end{aligned} \tag{4.3.22}$$

Equations (4.3.21, 22) yield

$$G(\rho,\tau) = -3\Lambda\varepsilon^2 \frac{\delta(\rho)}{2\pi\rho} e^{-\tau}\left[U(\tau) - U(\tau-\tau_0) - \frac{x_1}{3}\delta(\tau) + \frac{x_1}{3}\delta(\tau-\tau_0)\right]. \tag{4.3.23}$$

Now, the problem is solved and reduces to calculating a simple integral since, allowing for (4.3.17, 23),

$$\begin{aligned} w(\rho,\tau,\tau_0) = 3\Lambda q \int_0^{\tau_0} w(\rho,\tau,\tau_0,\tau')e^{-\tau'}d\tau' - \Lambda q x_1 w(\rho,\tau,\tau_0,\tau'=0) \\ + \Lambda q x_1 w(\rho,\tau,\tau_0,\tau'=\tau_0)e^{-\tau_0}, \end{aligned} \tag{4.3.24}$$

where $w(\rho,\tau,\tau_0,\tau')$ is given by (4.3.19). Similarly, we can obtain an expression for the function $G$ and write the solution for the source emitting at some angle to the normal (Sect. 6.3).

As seen from Fig. 4.4, the radiation densities in a semi-infinite weakly absorbing medium calculated by (4.3.24) agree with the data of model experiments at $\tau > 15$. Formula (4.3.24) correctly defines the $w$ dependence on the variables $\rho$ and $\tau$, and the parameter $\Lambda$. The most essential discrepancy between the theory and experiment is observed at small $\tau$. This may be attributed to both the inexact allowance made for the boundary effect on the light field in a diffusion approximation, and the large experimental beam diameter.

We can follow the behavior of the light field originated by a narrow beam in a strongly scattering medium in Fig. 4.4. For example, the growth of the radiation density at small $\rho$ and $\gamma$ can clearly be seen.

Using (4.3.16, 24), we shall write the expressions for the irradiance coefficients in a semi-infinite medium:

$$S_i(\rho,\tau) = 3\int_0^{\infty} S_i(\rho,\tau,\tau')e^{-\tau'}d\tau' - x_1 S_i(\rho,\tau,\tau'=0), \quad i = 1, 2, \tag{4.3.25}$$

$$\begin{aligned} S_i(\rho,\tau,\tau') = \frac{\Lambda\varepsilon^2}{16\pi}\{\rho_1^{-3}e^{-\gamma\rho_1}[\rho_1^2 + (-1)^i 2q(\tau-\tau')(1+\gamma\rho_1)] \\ - \rho_2^{-3}e^{-\gamma\rho_2}[\rho_2^2 + (-1)^i 2q(\tau+\tau'+4q)(1+\gamma\rho_2)]\}, \end{aligned} \tag{4.3.26}$$

$$\rho_1^2 = (\tau-\tau')^2 + \rho^2, \quad \rho_2^2 = (\tau+\tau'+4q)^2 + \rho^2. \tag{4.3.27}$$

Formulas (4.3.25–27) are somewhat awkward, but simple for calculation purposes.

In the axial region ($\rho/\tau \ll 1$) at a large distance from the source and the boundary, the system (4.3.25–27) gives the following asymptotic relationship of the downflux to the spatial $\rho$ and $\tau$ coordinates:

$$S_1(\rho,\tau) = A\frac{e^{-\gamma\tau}}{\tau}e^{-\rho^2/(2\tau/\gamma)}, \tag{4.3.28}$$

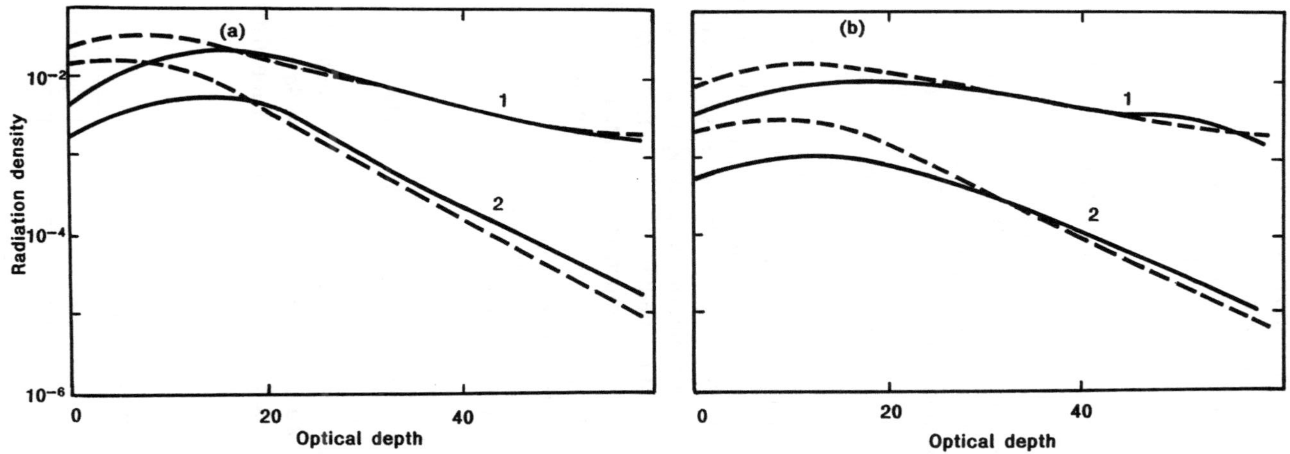


**Fig. 4.4.** Radiation density in a semi-infinite medium. *Solid lines*, model experiment [4.3] and *dashed lines*, calculation by (4.3.24) at (a) $\rho = 9.4$ and (b) 18.8, $q = 2.61$. Curve 1, $\Lambda = 0.998$; curve 2, 0.96. Experimental and theoretical curves are matched for $\Lambda = 0.998$ at $\tau = 30$, $\rho = 9.4$

where $A$ is some constant. A comparison of (4.3.28) and (4.2.6) shows that the asymptotic behavior of the diffusion equation (4.3.9) and transfer equation (4.1.1) result in a similar coordinate dependence of the irradiance. Following (4.3.28), the PSF variance for any scattering indicatrix is

$$V_\rho = \tau/\gamma \,. \tag{4.3.29}$$

The same result was obtained as for the asymptotic behavior of the rigorous equation (4.1.1) for particular cases: isotropic scattering (4.2.4, 5) and arbitrary scattering indicatrix for weak absorptions [4.31].

From (4.3.24, 16) and including (4.3.19), we can formulate a similar solution for the irradiances in layers of finite optical thickness. Since the relevant expressions are awkward, we shall omit them here. In Sects. 6.1, 3 which deal with the OTF of strongly scattering layers, formulas are given for the Fourier transform $S_1(\vec{\omega}, \tau, \tau_0)$ under different layer illumination conditions. These formulas, along with (4.1.21, 22) provide the variance of the radial irradiance distribution. The most simple and instructive expressions have been obtained for thick layers [4.29, 32] (Table 4.1). The formulas listed imply that at $\gamma\tau_0 \gg 1$ (strong absorption in the layer), the variance $V_\rho^{\mathrm{ref}}$ of the distribution $S_2(\rho, \tau = 0, \tau_0)$ of the reflected light is independent of the layer thickness. It is the same as for a semi-infinite medium. In this case, the transmitted radiation variance $V_\rho = \tau_0/\gamma$, i.e., it is equal to that for deep layers of a semi-infinite medium at $\tau = \tau_0$. In a semi-infinite medium at $\Lambda \to 1$ ($\gamma \to 0$), the variance of the function $S_1(\rho, \tau)$ tends to infinity (4.3.29). For a finite thickness layer, the quantities $V_\rho$ and $V_\rho^{\mathrm{ref}}$ are always finite.

If the layer absorption is very weak ($\gamma\tau_0 \ll 1$), then the point spread function variance is primarily determined by the optical layer thickness $\tau_0$. At large $\tau_0$, the variance $V_\rho$ of the transmitted light increases as $\tau_0^2$, whereas the quantity $V_\rho^{\mathrm{ref}}$ increases as $\tau_0$.

For isotropic scattering ($q = 1/3$), at $\Lambda \approx 1$ ($\gamma\tau_0 \ll 1$), the expression for the variance $V_\rho$, equivalent to that cited in Table 4.1, has been obtained in [4.16, 31] as an asymptotic of (4.1.1).

We shall now discuss some inferences of the formulas listed in Table 4.1. Consider a very thick absorbing optical layer ($\tau_0 \gg 4q, \gamma\tau_0 \gg 1$). We write an expression for the variance $V_r$ of the transmitted radiation distribution in dimensionless coordinates:

$$V_r = V_\rho/\varepsilon^2 = z_0/\varepsilon\gamma \,. \tag{4.3.30}$$

**Table 4.1.** Variances $V_\rho$ and $V_\rho^{\mathrm{ref}}$ for thick layers at $\mu_0 = 1$.

| Variance | General case | $\gamma\tau_0 \gg 1$ | $\gamma\tau_0 \ll 1$ |
|---|---|---|---|
| $V_\rho$ | $\dfrac{\tau_0 + 4q}{\gamma}\operatorname{cotanh}[\gamma(\tau_0 + 4q)] - \dfrac{5q}{\gamma}\operatorname{cotanh}(5q\gamma)$ | $\dfrac{\tau_0 + 4q}{\gamma}$ | $\dfrac{(\tau_0 + 4q)^2 - 25q^2}{3}$ |
| $V_\rho^{\mathrm{ref}}$ | $\dfrac{\tau_0 + 4q}{\gamma}\operatorname{cotanh}[\gamma(\tau_0 + 4q)] - \dfrac{\tau_0}{\gamma}\operatorname{cotanh}(\gamma\tau_0)$ | $\dfrac{4q}{\gamma}$ | $\dfrac{8q}{3}(\tau_0 + 2q)$ |

It is seen that when the geometric thickness of the layer is constant, the increase of the extinction coefficient $\varepsilon$ (for example, by an increase in the concentration of scatterers) causes a narrowing of the spread function in the geometric coordinates.

Let us estimate the effective radius $\Delta r_{\text{eff}} = \sqrt{V_r^{\text{ref}}} = \sqrt{V_\rho^{\text{ref}}}/\varepsilon$ of the reflected radiation spot. Accounting for (3.2.11) and the $V_\rho^{\text{ref}}$ formulas presented in Table 4.1, we have

$$\Delta r_{\text{eff}} = \frac{2}{\varepsilon} \frac{q^{3/4}}{(1-\Lambda)^{1/4}} \qquad \text{at } \gamma\tau_0 \gg 1\,,$$

$$= \frac{1}{\varepsilon}\sqrt{\frac{8q}{3}(\tau_0 + 2q)} \qquad \text{at } \gamma\tau_0 \ll 1\,.$$

The effective radius of the reflected radiation spot is seen to be greater, the more extended the indicatrix (greater $q$) due to the increasing light penetration depth. It should be noted that the spread functon dispersions specify the characteristic size of the light spot which is an important scale parameter in transfer problems for stochastically inhomogeneous media (broken clouds), and is required to estimate the parameters of coherent observation systems, etc.

So far, we have considered only homogeneous media. In [4.33], using the DA, solutions were obtained for media with optical parameters dependent on the depth $\tau$. The OTFs of such media are studied in Sect. 6.2.

In this section, only some DA results are presented. Different particular problems concerning narrow beam propagation in scattering media have been solved using the DA and are discussed elsewhere (for example, [4.34–37]).

## 4.4 Small-Angle Approximation

### 4.4.1 Various Approaches

In the previous section, when proceeding from the transfer equation (4.1.1) to the approximate differential diffusion equation (4.3.9), we assumed a relatively small angular light field anisotropy expressed by (4.3.6). Such an assumption is natural for large scattering volumes with weak absorption ($(1-\Lambda)/\Lambda(1-g) \ll 1$). We now consider a contrary case, where a scattering strongly absorbing medium with a very extended scattering indicatrix ($(1-\Lambda)/\Lambda(1-g) \gg 1$) is illuminated by a directional source. Then, the radiance is only significant close to the source $\vec{n}_0$ direction, and rapidly decreases with increasing $|\vec{n} - \vec{n}_0|$. Now, we may set $\vec{n} \approx \vec{n}_0 + \vec{n}_\perp$ ($\vec{n}_\perp$ being the projection onto the plane normal to $\vec{n}_0$, Fig. 4.1) in the differential operator of (4.1.1) and arrive at the small-angle transfer equation [4.38].

It should be emphasized that, in this way, we neglect the light reflected by a scattering medium. This means that at some distance $R^*$ from a point source, the light field is independent of the scattering medium properties at $R > R^*$, i.e., the "layer interaction" is considered to be small. The picture is similar when fast charged particles propagate. The small-angle approximation was first developed in connection with these problems [4.38–42].

Several different versions of small-angle radiation transfer equations are known. One of them [4.15, 39, 43, 44] does not contain any additional assumptions, and the integral term of the transfer equation preserves its form. We shall illustrate these methods for the simplest problem [4.39]. Let a plane-parallel layer be illuminated by an infinitely wide source emitting normal to the layer boundary ($\mu_0 = 1$). Then, assuming $\mu \partial I(\tau, \mu, \psi)/\partial \tau \approx \partial I(\tau, \mu, \psi)/\partial \tau$ in (3.1.3) and integrating (3.1.3) over the azimuth with allowance for the azimuthal homogeneity of the radiance, we arrive at the following boundary-value problem

$$\frac{\partial I_{\mathrm{n}}(\tau, \mu)}{\partial \tau} = -I_{\mathrm{n}}(\tau, \mu) + \frac{\Lambda}{2} \int_{-1}^{1} I_{\mathrm{n}}(\tau, \mu')\, d\mu' \sum_{n=0}^{N} x_n P_n(\mu) P_n(\mu'),$$

$$I_{\mathrm{n}}(\tau = 0, \mu) = \delta(\mu - 1).$$

Here, the indicatrix expansion in terms of the Legendre polynomials (2.2.7) is used.

As in the classical transfer theory, we seek the solution in the form of the radiance expansion into a series of the Legendre polynomials

$$I_{\mathrm{n}}(\tau, \mu) = \sum_{n=0}^{N} b_n(\tau) P_n(\mu). \tag{4.4.1}$$

The number of terms in (4.4.1) is determined by $N$ in the indicatrix expansion (2.2.7). Substituting (4.4.1) into the transfer equation and using the orthogonality of the Legendre polynomials

$$\int_{-1}^{1} P_n(\mu) P_m(\mu)\, d\mu = \frac{2}{2n+1} \delta_{nm},$$

we obtain the following equations for the functions:

$$\frac{db_n(\tau)}{d\tau} = -\left(1 - \frac{\Lambda x_n}{2n+1}\right) b_n(\tau),$$

$$b_n(\tau = 0) = \frac{2n+1}{2},$$

i.e.,

$$b_n(\tau) = \frac{2n+1}{2} \exp\left[-\left(1 - \frac{\Lambda x_n}{2n+1}\right)\tau\right].$$

The boundary condition for the function $b_n(\tau)$ at $\tau = 0$ is written using the well-known expansion $\delta(\mu - 1) = 0.5 \sum_{n=0}^{\infty} (2n+1) P_n(\mu)$. Substituting $b_n(\tau)$ into (4.4.1), we obtain the required solution. By applying the $\delta$-function series in terms of the Legendre polynomials, we rewrite this solution as a sum of the unscattered and scattered components. Finally, we have

$$I_n(\tau,\mu) = e^{-\tau}\delta(\mu - 1) + \frac{1}{2}\sum_{n=0}^{N}(2n+1)e^{-(\Lambda x_n/(2n+1))\tau}P_n(\mu)\,.$$

When an anisotropically scattering medium is illuminated by a narrow beam, such an approach is used to obtain the spatial radiance distribution moments $M_{n,k}$. The solutions for the moments can also be derived as a series of terms which are the product of elementary and associated Legendre functions [4.15]. The number of terms in these series is determined by $N$ in the scattering indicatrix expansion (2.2.7). It is clear that this procedure is inconvenient for strong scattering anisotropy (large $N$) peculiar to the majority of real media.

In this case, especially as used in image transfer problems, it is more efficient to apply a modified small-angle transfer equation where the integral term is approximately substituted by the convolution integral [4.40]. We shall derive this equation in Sect. 4.4.2 and give the resulting solution in Sects. 4.4.3–6. Below, this variant developed in [4.45–49] and used in our work will be called the *small-angle approximation* (SAA).

One more small-angle equation resulting from the approximate substitution of the integral term by a differential one will be considered in Sect. 4.5.

### 4.4.2 Transfer Equation in the Small-Angle Approximation

We shall write down the radiation transfer equation for the case of optical characteristics dependent on the depth $z$:

$$\vec{n}\cdot\nabla_{\vec{R}}I(\vec{R},\vec{n}) = -\varepsilon(z)I(\vec{R},\vec{n}) + \frac{\sigma(z)}{4\pi}\iint_{4\pi} I(\vec{R},\vec{n}')x(\vec{n},\vec{n}',z)\,d\vec{n}' + B(\vec{R},\vec{n})\,. \tag{4.4.2}$$

Let a scattering medium be illuminated by a source in the plane $z = 0$, the axis of the angular emittance diagram being directed along $z$. In the differential operator $\vec{n}\cdot\nabla_{\vec{R}}$, we separate the $z$ derivative

$$\vec{n}\cdot\nabla_{\vec{R}} = \mu\frac{d}{dz} + \vec{n}_\perp\cdot\nabla_{\vec{r}}\,, \tag{4.4.3}$$

where $\vec{n}_\perp$ and $\vec{r}$ are the projections of the vectors $\vec{n}$ and $\vec{R}$ onto the plane $xy$ normal to the beam axis $z$ (Fig. 4.1)

$$\vec{n}_\perp = \vec{i}\sin\theta\cos\psi + \vec{j}\sin\theta\sin\psi\,, \qquad \vec{n} = \vec{k}\cos\theta + \vec{n}_\perp\,,$$
$$\vec{r} = \vec{i}x + \vec{j}y\,, \qquad \vec{R} = \vec{k}z + \vec{r}\,. \tag{4.4.4}$$

Here, $\vec{i}$, $\vec{j}$, $\vec{k}$ are the unit vectors in the $x$, $y$, $z$ directions. We note that the differential $d\vec{n}_\perp$ is related to the differential $d\vec{n}$ as

$$d\vec{n} = d\vec{n}_\perp/\cos\theta\,. \tag{4.4.5}$$

In view of the above assumption that at any medium point the radiance $I(\vec{R}, \vec{n})$ is essential only over the angles about the $\vec{k}$ direction and quickly reduces with increasing $|\vec{n} - \vec{k}|$, we may set $\mu = 1$ in (4.4.3). We also note that the scattering indicatrix depends only on the modulus $|\vec{n}_\perp - \vec{n}'_\perp|$, i.e., no direction is separated in the medium. This is always true of media with a random distribution of scatterers.

The condition of strong extension of the scattering indicatrix allows substitution of the real indicatrix $x(\beta)$ by the function

$$
\begin{aligned}
x'(\beta) &= x(\beta)/F_\beta \qquad \text{at } 0 \leqslant \beta \leqslant \beta_0 \ll 1 \\
&= 0 \qquad \text{at } \beta > \beta_0 ,
\end{aligned}
\tag{4.4.6}
$$

where

$$
F_{\beta_0} = \frac{1}{2} \int_0^{\beta_0} x(\beta) \sin\beta \, d\beta \, . \tag{4.4.7}
$$

The fact $1/F_{\beta_0}$ is introduced to preserve the earlier normalization for the indicatrix $x'(\beta)$: $(4\pi)^{-1} \iint_{4\pi} x'(\beta)\, d\vec{n} = 1$. Then,

$$
\begin{aligned}
\sigma' &= \sigma F_{\beta_0} , \qquad \kappa' = \kappa + \sigma(1 - F_{\beta_0}) , \\
\Lambda' &= \Lambda F_{\beta_0} , \qquad \varepsilon' = \kappa' + \sigma' = \kappa + \sigma = \varepsilon .
\end{aligned}
\tag{4.4.8}
$$

We have now replaced the optical characteristics of the medium by "effective" ones whose use offers an approximate account of the light losses incurred in the "escape from the small-angle region". It should be noted that, following (4.4.6), $x'(\beta)$ is different from zero only within $(0, \beta_0)$. Moreover, we take into account the fact that the function $I(\vec{R}, \vec{n}_\perp)$ decreases rapidly with increasing $|\vec{n}_\perp|$. We assume that if the internal sources are available in the medium, then the function $B(\vec{R}, \vec{n}_\perp) \neq 0$ only for $|\vec{n}_\perp| \ll 1$. Thus, we may formally regard the variable $|\vec{n}_\perp|$ as given on the interval $(0, \infty)$ and going over to infinite integration limits for all integrals containing $I(\vec{R}, \vec{n}_\perp)$ and $x(\beta)$.

In view of the above, we now rewrite (4.4.2) as

$$
\begin{aligned}
&\left( \frac{\partial}{\partial z} + \vec{n}_\perp \nabla_{\vec{r}} + \varepsilon'(z) \right) I(\vec{r}, z, \vec{n}_\perp) \\
&= \frac{\sigma'(z)}{4\pi} \iint_{-\infty}^{\infty} I(\vec{r}, z, \vec{n}'_\perp) x'(|\vec{n}_\perp - \vec{n}'_\perp|) \, d\vec{n}'_\perp + B(\vec{r}, z, \vec{n}_\perp) \, .
\end{aligned}
\tag{4.4.9}
$$

This is the SAA transfer equation. In the analysis which follows, we shall omit the prime on the optical characteristics, but bear in mind that the SAA solutions include the parameters determined by (4.4.6–8). The boundary condition for (4.4.9) is set by the spatial-angular characteristics of the source radiance

$$
I(\vec{r}, z = 0, \vec{n}_\perp) = I_0(\vec{r}, \vec{n}_\perp) \, . \tag{4.4.10}
$$

Equation (4.4.9), subject to the boundary condition (4.4.10), is a complete statement of the SAA problem.

### 4.4.3 Solution of the Small-Angle Transfer Equation

Equation (4.4.9) contains a linear differential operator with respect to $\vec{r}$ and an integral convolution-type term with respect to $\vec{n}'_\perp$. It is, therefore, natural to use the two-dimensional Fourier transformation with respect to the $\vec{r}$ and $\vec{n}_\perp$ coordinates. Let us include the Fourier transform of the unknown function

$$I(\vec{v},z,\vec{p}) = \iint_{-\infty}^{\infty} d\vec{n}_\perp \iint_{-\infty}^{\infty} I(\vec{r},z,\vec{n}_\perp)\mathrm{e}^{-\mathrm{i}(\vec{v}\cdot\vec{r}+\vec{p}\cdot\vec{n}_\perp)}\,d\vec{r}\,. \tag{4.4.11}$$

Here, the angular frequency $\vec{p}$ is used along with the spatial frequency $\vec{v}$. Then, from (4.4.9), we obtain

$$\left(\frac{\partial}{\partial z} - \vec{v}\cdot\nabla_{\vec{p}}\right)I(\vec{v},z,\vec{p}) = [-\varepsilon(z) + \sigma(z)x(p,z)]I(\vec{v},z,\vec{p}) + B(\vec{v},z,\vec{p})\,, \tag{4.4.12}$$

where

$$x(p,z) = \frac{1}{2}\int_0^{\infty} x(\beta,z)J_0(p\beta)\beta\,d\beta \tag{4.4.13}$$

is the zero-order Hankel transform of the scattering indicatrix, $J_0(p\beta)$ is the zero-order Bessel function, and

$$B(\vec{v},z,\vec{p}) = \iint_{-\infty}^{\infty} d\vec{n}_\perp \iint_{-\infty}^{\infty} B(\vec{r},z,\vec{n}_\perp)\mathrm{e}^{-\mathrm{i}(\vec{v}\cdot\vec{r}+\vec{p}\cdot\vec{n}_\perp)}\,d\vec{r}$$

is the Fourier transform of the source function.

Equation (4.4.12) is a quasi-linear first-order partial differential equation. It can easily be solved in a standard fashion, using, for example, the method of characteristics [4.12]. At $B(\vec{v},z,\vec{p}) = 0$, and subject to the boundary condition (4.4.10), the solution (4.4.12) is

$$I(\vec{v},z,\vec{p}) = I_0(\vec{v},\vec{p}+\vec{v}z)\exp\left[-\int_0^z \Phi(z-\xi,\vec{p}+\vec{v}\xi)\,d\xi\right], \tag{4.4.14}$$

$$\Phi(z,\vec{p}) = \varepsilon(z) - \sigma(z)x(p,z)\,. \tag{4.4.15}$$

In order to obtain the spatial-angular distribution of the radiance $I(\vec{r},z,\vec{n}_\perp)$, we must take the inverse Fourier transform

$$I(\vec{r},z,\vec{n}_\perp) = \frac{1}{(2\pi)^4}\iint_{-\infty}^{\infty} d\vec{v} \iint_{-\infty}^{\infty} I(\vec{v},z,\vec{p})\mathrm{e}^{\mathrm{i}(\vec{v}\cdot\vec{r}+\vec{p}\cdot\vec{n}_\perp)}\,d\vec{p}\,. \tag{4.4.16}$$

Formulas (4.4.14–16) provide a complete solution to the problem stated. However, simple analytical solutions can be obtained in some cases only for the Fourier transform of the radiance $I(\vec{v},z,\vec{p})$.

Let a scattering medium be illuminated by a monodirectional unit power point source, i.e.,

$$I_0(\vec{r}, \vec{n}_\perp) = \delta(\vec{r})\delta(\vec{n}_\perp) . \tag{4.4.17}$$

So far as in this case $I_0(\vec{v}, \vec{p}) = 1$,

$$I(\vec{v}, z, \vec{p}) = I_n(\vec{v}, z, \vec{p}) = \exp\left[ -\int_0^z \Phi(z - \xi, \vec{p} + \vec{v}\xi)\, d\xi \right] . \tag{4.4.18}$$

The Fourier transform for the normalized radiance $I_n(\vec{v}, z, \vec{p})$ is essentially the Green's function of the problem. For a random initial radiance distribution, we have

$$I(\vec{v}, z, \vec{p}) = I_0(\vec{v}, \vec{p} + \vec{v}z) I_n(\vec{v}, z, \vec{p}) . \tag{4.4.19}$$

The function $I_0(\vec{v}, \vec{p})$ for some of the simplest and most frequently used initial source radiance distributions is given in Table 4.2, where $2r_s$ and $2\gamma_s$ are the initial cross-section and beam divergence, respectively, $J_1(x)$ the first-order Bessel function and $U(x) = 1$ at $x \leqslant 1$, $U(x) = 0$ at $x > 1$.

Consider some particular case. Set $\vec{p} = 0$. Then, from (4.4.11), we obtain

$$I_n(\vec{v}, z, \vec{p} = 0) = \iint_{-\infty}^{\infty} e^{-i\vec{v}\cdot\vec{r}}\, d\vec{r} \iint_{-\infty}^{\infty} I_n(\vec{r}, z, \vec{n}_\perp)\, d\vec{n}_\perp . \tag{4.4.20}$$

Taking into account (4.4.5), we have

$$\iint_{-\infty}^{\infty} I_n(\vec{r}, z, \vec{n}_\perp)\, d\vec{n}_\perp = S(\vec{r}, z) . \tag{4.4.21}$$

As $S(\vec{r}, z)$ is the irradiance distribution due to a monodirectional unit power point source (PSF),

$$I_n(\vec{v}, z, \vec{p} = 0) = S(\vec{v}, z) . \tag{4.4.22}$$

is the medium OTF.

We now set $\vec{v} = 0$. Formula (4.4.11) then yields

$$I_n(\vec{v} = 0, z, \vec{p}) = \iint_{-\infty}^{\infty} I_n(z, \vec{n}_\perp) e^{-i\vec{p}\cdot\vec{n}_\perp}\, d\vec{n}_\perp , \tag{4.4.23}$$

**Table 4.2.** Function $I_0(\vec{v}, \vec{p})/I_0(0,0)$. From [4.49].

| $I_0(\vec{r}, \vec{n}_\perp)$ | $I_0(\vec{v}, \vec{p})/I_0(0,0)$ |
|---|---|
| $P_0\delta(\vec{r})\delta(\vec{n}_\perp)$ | 1 |
| $\dfrac{P_0}{(\pi r_s^2)(\pi\gamma_s^2)} \exp\left(-\dfrac{r^2}{r_s^2} - \dfrac{n_\perp^2}{\gamma_s^2}\right)$ | $\exp\left(-\dfrac{r_s^2 v^2 + \gamma_s^2 p^2}{4}\right)$ |
| $\dfrac{P_0}{(\pi r_s^2)(\pi\gamma_s^2)} U\left(\dfrac{r}{r_s}\right) U\left(\dfrac{n_\perp}{\gamma_s}\right)$ | $4\dfrac{J_1(r_s v)}{r_s v}\dfrac{J_1(\gamma_s p)}{\gamma_s p}$ |

where $I_n(z, \vec{n}_\perp)$ is the normalized radiance due to an infinitely wide monodirectional source (4.1.4). Thus, the quantity $I_n(\vec{v} = 0, z, \vec{p})$ is the angular Fourier transform of the normalized radiance for the medium illuminated by an infinitely wide source.

Finally, if $\vec{v} = 0$ and $\vec{p} = 0$, then the quantity

$$S(z) = I_n(\vec{v} = 0, z, \vec{p} = 0) = \int\int_{-\infty}^{\infty} I_n(z, \vec{n}_\perp)\, d\vec{n}_\perp = \int\int_{-\infty}^{\infty} S(\vec{r}, z)\, d\vec{r} \qquad (4.4.24)$$

is the radiation flux through the plane $z =$ const due to a narrow unit power beam, or the irradiance for an infinitely extended source. In the next three sections, we shall discuss the characteristics of the light fields originated by wide and narrow beams, based on the solution (4.4.14, 15) and (4.4.22–24).

### 4.4.4 Irradiance and Radiance of a Medium Illuminated by an Infinitely Extended Source

At $\vec{v} = 0$, $\vec{p} = 0$, (4.4.14) gives

$$S(z) = \exp(-\tau_\kappa)\,, \qquad (4.4.25)$$

where

$$\tau_\kappa = \int_0^z \kappa(z')\, dz' \qquad (4.4.26)$$

is the "absorbing" optical thickness. Therefore, absorption alone is responsible for the total flux attenuation in the medium with the permissible effective characteristics. Proceeding to the real optical characteristics and taking into account (4.4.8), we arrive at $S(z) = \exp[-\int_0^z [\kappa(z') + \sigma(z')(1 - F_{\beta_0}(z'))]\, dz'$.

Since the medium is homogeneous,

$$S(z) = \exp[-(1 - \Lambda F_{\beta_0})\varepsilon z]\,, \qquad (4.4.27)$$

where $F_{\beta_0}$ was determined in (4.4.7). The Expression (4.4.27) at $\beta_0 = \pi/2$ is similar to the irradiance formula (3.6.10). It was shown in [4.50, 51], based on experiments and the Monte-Carlo simulations, that (4.4.27) could be applied to estimate the radiation fluxes in sea water up to $\tau$ between 12 and 15.

Setting $\vec{v} = 0$ in (4.4.14) and assuming the optical characteristics to be independent of $z$ gives the following expression for the angular Fourier transform of the radiance $I_n(z, \vec{n}_\perp)$ due to an infinitely wide incident beam:

$$I_n(z, \vec{p}) = \exp[-\varepsilon z + \sigma z x(p)]\,. \qquad (4.4.28)$$

Let us allow for the angular radiance distribution moments

$$M_{\theta, 2n} = 2\pi \int_0^\infty I_n(z, \theta)\theta^{2n+1}\, d\theta\,. \qquad (4.4.29)$$

As shown in Sect. 4.1, the spatial moments $M_{r,2n}$ can be expressed in terms of the derivatives $d^{2n}S(\vec{v})/dv^{2n}$ at $\vec{v} \to 0$ (4.1.21). Similarly, the angular moments are related to the derivatives at $\vec{p} \to 0$ [4.52]. In particular, for the variance of the angular radiance distribution, we have

$$V_\theta = \frac{M_{\theta,2}}{2M_{\theta,0}} = -\lim_{\vec{p}\to 0} \frac{\partial^2 \ln I_n(z,\vec{p})}{\partial p^2} . \tag{4.4.30}$$

Next, we include the angular moments of the scattering indicatrix

$$\beta_{2k} = \frac{\int_0^\infty x(\beta)\beta^{2k+1}\, d\beta}{\int_0^\infty x(\beta)\beta\, d\beta} . \tag{4.4.31}$$

Then, for the variance $V_\theta$ and kurtosis $\chi_\theta$ of the angular radiance distribution, we may write

$$V_\theta = \sigma z \beta_2/2 , \qquad \chi_\theta = \frac{3\beta_4}{2\sigma z \beta_2^2} + 3 . \tag{4.4.32}$$

The characteristics of the angular radiance distribution are seen to be dependent only on the scattering layer thickness $\tau_\sigma = \sigma z$ and the scattering indicatrix. This is typical of the SAA solutions. The distribution kurtosis decreases with increasing depth limiting to the value $\chi_\theta = 3$, and the angular radiance distribution tends to the Gaussian function. Given the set of the angular moments $M_{\theta,2n}$, we can reconstruct the angular radiance distribution $I_n(z,\theta)$ [4.52].

### 4.4.5 The Spread Function and its Moments

The medium OTF and PSF can be obtained by setting $\vec{p} = 0$ in the solution (4.4.18). Taking into account (4.4.22) and the azimuthal field symmetry, $S(\vec{r},z) = S(|\vec{r}|,z) = S(r,z)$ and, correspondingly, $S(\vec{v},z) = S(|\vec{v}|,z) = S(v,z)$, we have

$$S(v,z) = \exp\left\{-\int_0^z [\varepsilon(z-\xi) - \sigma(z-\xi)x(v\xi, z-\xi)]\, d\xi\right\} , \tag{4.4.33}$$

$$S(r,z) = 2\pi \int_0^\infty S(v,z) J_0(vr) v\, dv . \tag{4.4.34}$$

If the optical characteristics are independent of the depth, for the medium OTF, we find

$$S(v,z) = \exp\left[-\varepsilon z + \sigma \int_0^z x(v\xi)\, d\xi\right] . \tag{4.4.35}$$

Equations (4.4.33–35) fail to give simple analytical solutions for the PSF. At the same time, formulas for the integral moments of the spread function prove to be sufficiently simple. We determine the spread function moments $M_{r,2n}$ from the expressions (4.4.35; 4.1.21, 22). In particular, for the variance and kurtosis of the spread function in a homogeneous medium, we have [4.13]

**Table 4.3.** Values of coefficients $d_m$ in (4.4.37).

| $m$ | $d_m$ |
|---|---|
| 0 | 1 |
| 1 | $\frac{3}{4\sigma z}\frac{\beta_4}{\beta_2^2}$ |
| 2 | $\frac{3}{28(\sigma z)^2}\frac{\beta_6}{\beta_2^3}$ |
| 3 | $\frac{1}{64}\left[\frac{1}{(\sigma z)^3}\frac{\beta_8}{\beta_2^4}+\frac{4}{\sigma z}\frac{\beta_4}{\beta_2^2}\right]$ |

$$V_r = \frac{\sigma z^3 \beta_2}{6}, \qquad \chi_r = \frac{2.7\beta_4}{\sigma z \beta_2^2} + 3. \tag{4.4.36}$$

We shall now consider the solutions obtained. The variance of the spread function is proportional to the second angular moment of the indicatrix $\beta_2$ and increases as $z^3$ with the depth, while the kurtosis value decreases.

A simple calculation of the moments sometimes makes it advisable to reconstruct appropriately the PSF in terms of its moments $M_{r,2n}$. In [4.13], the PSF is sought as

$$S(r,z) = \frac{M_0(z)}{2\pi V_r(z)} e^{-r^2/2V_r(z)} \sum_{m=0}^{\infty} d_m L_m\left(\frac{r^2}{2V_r(z)}\right), \tag{4.4.37}$$

where $L_m(x)$ are the Laguerre polynomials on the interval $(0, \infty)$.

The orthogonality property of the Laguerre polynomials provides a simple expression for the coefficients $d_m$ in terms of the $M_{r,2n}$ moments at $n \leqslant m$. It is sufficient to know 3–4 terms in (4.4.37) in order to obtain the PSF to reasonable accuracy (Table 4.3). As the scattering thickness $\sigma z$ grows, the coefficients $d_m$ decrease and the PSF approaches the Gaussian distribution.

### 4.4.6 Light Flux

In practice, it is often necessary to calculate the light flux $P_\Sigma$ through a circular area of size $\Sigma$ symmetrical about the beam axis (Fig. 4.5). Taking into account that $r = z\theta$, we obtain

$$P_\Sigma(z,\theta) = \int_\Sigma S(r',z) r' \, dr' \, d\varphi. \tag{4.4.38}$$

Substituting (4.3.34) into (4.4.38) and integrating with respect to $\varphi$ gives [4.49]

$$P_\Sigma(z,\theta) = 2\pi \int_0^\infty S(v,z) J_1(\theta z v) \, dv. \tag{4.4.39}$$

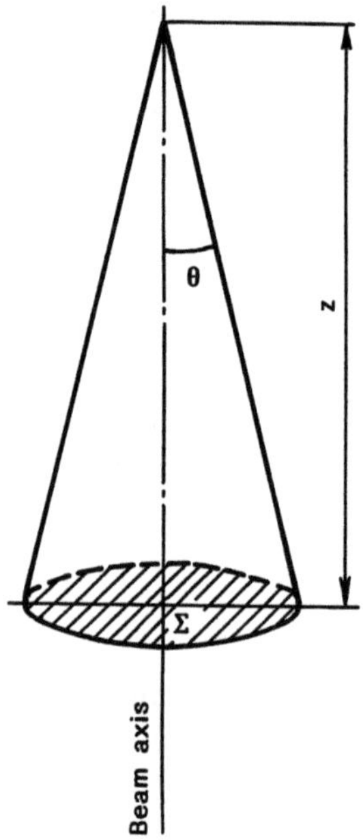


**Fig. 4.5.** Area Σ through which a light flux is determined

We can easily pass from $P_\Sigma(z, \theta)$ to the PSF

$$S(r, z) = \frac{1}{2\pi z \theta^2} \left. \frac{dP_\Sigma(z, \theta)}{d\theta} \right|_{\theta = r/z} .$$

The integral (4.4.39) is more convenient than (4.4.34) for numerical calculations since the subintegrand in (4.4.39) vanishes more rapidly with increasing $v$ than that in (4.4.34). The flux $P_\Sigma(z, \theta)$ was calculated in [4.49] for the scattering indicatrix $x(\beta) = a \exp(-a\beta)/2\pi\beta$. In Fig. 4.6, some theoretical results are compared with experimental data.

The approximation for the indicatrix

$$x(\beta) = \frac{\sqrt{a_1}}{2} \frac{e^{-a_1 \beta}}{(\pi\beta)^{3/2}} \tag{4.4.40}$$

offers a more convenient approximate solution for $P_\Sigma(z, \theta)$ [4.53] i.e.,

$$P_\Sigma(z, \theta) = \frac{[1 + \Phi(\sqrt{2 s_0 \sigma z})]}{2 - (1 - s_0)\sqrt{b/s_0}} e^{-(\varepsilon - \sigma s_0) z} , \tag{4.4.41}$$

where $b = 4 a_1 \sigma z \theta$, and $s_0$ is the root of the equation

$$s - (1 - e^{-\sqrt{bs}}) = 0 , \tag{4.4.42}$$

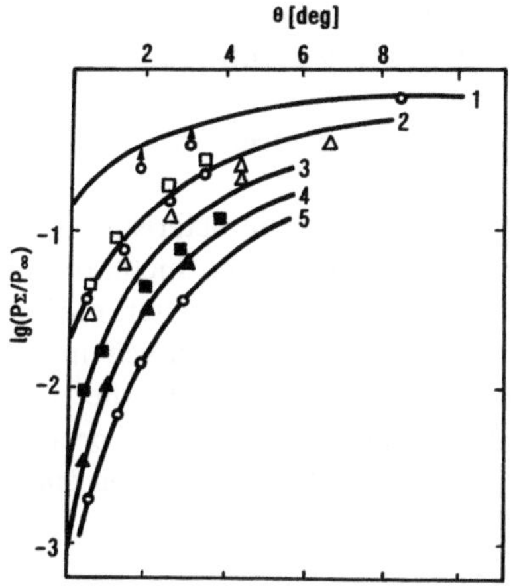


**Fig. 4.6.** Radiant flux through an area Σ as a function of $\theta$ (Fig. 4.5) normalized to the total flux through a plane at $z = \text{const}$. *Solid lines* are calculated by the SAA formula (4.4.39); *symbols* are experimental data for sea water at $\sigma z = 2$ (curve 1), 4 (curve 2), 6 (curve 3), 8 (curve 4) and 10 (curve 5)

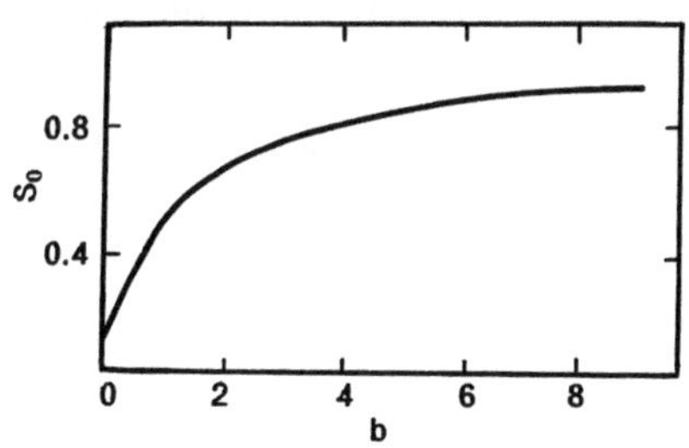


**Fig. 4.7.** Root of (4.4.42) $s_0$ as a function of b

and

$$\Phi(x) = \sqrt{\pi/2} \int_0^x e^{-t^2/2} \, dt \,, \tag{4.4.43}$$

is the probability integral. The root $s_0$ of (4.4.42) vs. the parameter $b$ is plotted in Fig. 4.7. At small $b$ ($b < 0.5$),

$$s_0 = \frac{b}{(1 + b/2)^2} \,. \tag{4.4.44}$$

For $b \to 0$ ($\sigma z \to 0$), $s_0 \to 0$ and, as seen from (4.4.41), the flux attenuates according to the Bourguer law. At $b \to \infty$

$$P_\Sigma(z, \theta) \to P(z) = \exp(-\kappa z) \,.$$

A comparison of $P_\Sigma(z,\theta)$ calculated from (4.4.39, 41) shows that the simple expression (4.4.41) is of satisfactory accuracy for $\theta$ not very large and $\sigma z \lesssim 18$.

### 4.4.7 Radiation Fields in Scattering Media with Fluctuating Optical Parameters

We now consider the radiation field characteristics for an inhomogeneous (stochastic) medium with optical parameters fluctuating in space [4.54, 55]. All natural scattering media (clouds, fog, sea water, etc.) are, to some extent, stochastic. Fluctuations with a spatial scale much in excess of the size of the individual scatterers and commensurate with the photon path length in the medium $\varepsilon^{-1}$ would change the mean radiance and produce radiance fluctuations.

In a small-angle approximation, the radiation transfer equation for a stochastic medium is of the form

$$\left[\frac{\partial}{\partial z} + \vec{n}_\perp \cdot \nabla_{\vec{r}} + \varepsilon(\vec{r}, z)\right] I(\vec{r}, z, \vec{n}_\perp)$$
$$= \frac{\sigma(\vec{r}, z)}{4\pi} \iint_{-\infty}^{\infty} I(\vec{r}, z, \vec{n}'_\perp) x(|\vec{n}_\perp - \vec{n}'_\perp|)\, d\vec{n}'_\perp \,. \tag{4.4.45}$$

Separating $\sigma(\vec{r}, z)$, $\varepsilon(\vec{r}, z)$ and $I(\vec{r}, z, \vec{n}_\perp)$ into the mean and fluctuation components, i.e., assuming

$$\varepsilon(\vec{r}, z) = \langle\varepsilon\rangle + \tilde{\varepsilon}(\vec{r}, z)\,, \qquad \sigma(\vec{r}, z) = \langle\sigma\rangle + \tilde{\sigma}(\vec{r}, z)\,, \tag{4.4.46}$$

$$I(\vec{r}, z, \vec{n}_\perp) = \langle I(\vec{r}, z, \vec{n}_\perp)\rangle + \tilde{I}(\vec{r}, z, \vec{n}_\perp)\,, \tag{4.4.47}$$

we rewrite (4.4.45) as a system of two equations

$$\langle\mathscr{L}\rangle\langle I\rangle + \langle\tilde{\mathscr{L}} I\rangle = 0\,, \tag{4.4.48}$$

$$\langle\mathscr{L}\rangle\tilde{I} + \tilde{\mathscr{L}} I - \langle\tilde{\mathscr{L}} I\rangle = 0\,, \tag{4.4.49}$$

where

$$\langle\mathscr{L}\rangle I(\vec{r}, z, \vec{n}_\perp) = \left(\frac{\partial}{\partial z} + \vec{n}_\perp \cdot \nabla_{\vec{r}} + \langle\varepsilon\rangle\right) I(\vec{r}, z, \vec{n}_\perp)$$
$$- \frac{\langle\sigma\rangle}{4\pi} \iint_{-\infty}^{\infty} I(\vec{r}, z, \vec{n}'_\perp) x(|\vec{n}_\perp - \vec{n}'_\perp|)\, d\vec{n}'_\perp \,, \tag{4.4.50}$$

$$\tilde{\mathscr{L}} I(\vec{r}, z, \vec{n}_\perp) = \tilde{\varepsilon}(\vec{r}, z) I(\vec{r}, z, \vec{n}_\perp)$$
$$- \frac{\tilde{\sigma}(\vec{r}, z)}{4\pi} \iint_{-\infty}^{\infty} I(\vec{r}, z, \vec{n}'_\perp) x(|\vec{n}_\perp - \vec{n}'_\perp|)\, d\vec{n}'_\perp \,. \tag{4.4.51}$$

In (4.4.46–49), the averaging is over the medium realization ensemble. We restrict our consideration to a simpler, but frequently used case of small fluctuations of optical parameters when $\tilde{I} \ll \langle I\rangle$. The possibility of deriving a more general solution from the equations for the random field moments is shown in [4.55].

Below, assuming small fluctuations, we have:

$$\tilde{\mathscr{L}} I - \langle \tilde{\mathscr{L}} I \rangle = \tilde{\varepsilon}(\vec{r}, z)\langle I(\vec{r}, z, \vec{n}_\perp)\rangle - \frac{\tilde{\sigma}(\vec{r}, z)}{4\pi} \iint_{-\infty}^{\infty} \langle I(\vec{r}, z, \vec{n}'_\perp)\rangle x(|\vec{n}_\perp - \vec{n}'_\perp|)\, d\vec{n}'_\perp . \quad (4.4.52)$$

Taking the two-dimensional Fourier transforms of (4.4.49, 52) with respect to the $\vec{r}$ and $\vec{n}_\perp$ coordinates and accounting for the $\vec{r}$ dependence of $\tilde{\varepsilon}$ and $\tilde{\sigma}$, we arrive at the solution to (4.4.49) in the form

$$\tilde{I}(\vec{v}, z, \vec{p}) = \int_0^z [f_2(\vec{v}, z - z', \vec{p} + \vec{v}z') - f_1(\vec{v}, z - z', \vec{p} + \vec{v}z')] I_{\mathrm{n}}^{\mathrm{hom}}(\vec{v}, z', \vec{p})\, dz' , \quad (4.4.53)$$

where

$$I_{\mathrm{n}}^{\mathrm{hom}}(\vec{v}, z, \vec{p}) = \exp\left[ -\langle \varepsilon \rangle z + \langle \sigma \rangle \int_0^z x(\vec{p} + \vec{v}\xi)\, d\xi \right], \quad (4.4.54)$$

$$f_1(\vec{v}, z, \vec{p}) = \frac{1}{4\pi^2} \iint_{-\infty}^{\infty} \tilde{\varepsilon}(\vec{v} - \vec{v}', z)\langle I(\vec{v}', z, \vec{p})\rangle\, d\vec{v}' , \quad (4.4.55)$$

$$f_2(\vec{v}, z, \vec{p}) = \frac{x(p)}{4\pi^2} \iint_{-\infty}^{\infty} \tilde{\sigma}(\vec{v} - \vec{v}', z)\langle I(\vec{v}', z, \vec{p})\rangle\, d\vec{v}' , \quad (4.4.56)$$

$$I(\vec{v}, z, \vec{p}) = \iint_{-\infty}^{\infty} \iint_{-\infty}^{\infty} I(\vec{r}, z, \vec{n}_\perp) \mathrm{e}^{-\mathrm{i}(\vec{v}\cdot\vec{r} + \vec{p}\cdot\vec{n}_\perp)}\, d\vec{r}\, d\vec{n}_\perp . \quad (4.4.57)$$

At $\vec{p} = 0$, the function $I_{\mathrm{n}}^{\mathrm{hom}}(\vec{v}, z, \vec{p} = 0)$ is the Fourier transform of the normalized radiance for a homogeneous medium with average optical characteristics. Formula (4.4.53) defines the fluctuation radiance component in terms of its average value $\langle I(\vec{v}, z, \vec{p})\rangle$ and medium characteristics.

In order to solve (4.4.48) and find $I(\vec{v}, z, \vec{p})$, we must determine $\langle \tilde{\mathscr{L}} I \rangle$. Using (4.4.53–56) and taking simple transforms, (4.4.51) gives

$$\langle \tilde{\mathscr{L}} I(\vec{v}, z, \vec{p})\rangle = \frac{1}{4\pi^2} \int_0^z dz' \iint_{-\infty}^{\infty} \left[ \Gamma_{\varepsilon\varepsilon}(\vec{v} - \vec{v}', z') - 2x(p)\Gamma_{\varepsilon\sigma}(\vec{v} - \vec{v}', z') + x^2(p)\Gamma_{\sigma\sigma}(\vec{v} - \vec{v}', z') \right] \times \langle I(\vec{v}, z - z', \vec{p} + \vec{v}'z')\rangle I_{\mathrm{n}}^{\mathrm{hom}}(\vec{v}', z', \vec{p})\, d\vec{v}' , \quad (4.4.58)$$

$$\Gamma_{\varepsilon\varepsilon}(\vec{v}, z') = \iint_{-\infty}^{\infty} \Gamma_{\varepsilon\varepsilon}(\vec{r}', z')\mathrm{e}^{-\mathrm{i}\vec{v}\cdot\vec{r}'}\, d\vec{r}' , \quad (4.4.59)$$

$$\Gamma_{\varepsilon\varepsilon}(\vec{r}', z) = \langle \tilde{\varepsilon}(\vec{r}, z)\tilde{\varepsilon}(\vec{r} - \vec{r}', z - z')\rangle . \quad (4.4.60)$$

The quantities $\Gamma_{\varepsilon\sigma}(\vec{v}, z')$ and $\Gamma_{\sigma\sigma}(\vec{v}, z')$ are defined in the same way as (4.4.59). Formula (4.4.58) is obtained using

$$\langle \tilde{\varepsilon}(\vec{v}, z)\tilde{\varepsilon}(\vec{v}', z - z')\rangle = 4\pi^2 \Gamma_{\varepsilon\varepsilon}(\vec{v}', z')\delta(\vec{v} + \vec{v}'), \tag{4.4.61}$$

which results from the homogeneity of a random field $\tilde{\varepsilon}(\vec{r}, z)$.

We consider that the correlation radius of the functions $\varepsilon(\vec{r}, z)$ and $\sigma(\vec{r}, z)$ is small in comparison with the scale of the essential variation of the function $\langle I(\vec{r}, z, \vec{n}_\perp)\rangle$ along the $z$ axis. From this assumption and taking into account that $I_n^{\text{hom}}(\vec{v}, 0, \vec{p}) = 1$, from (4.4.58), we obtain

$$\langle \tilde{\mathscr{L}} I(\vec{v}, z, \vec{p})\rangle = \langle I(\vec{v}, z, \vec{p})\rangle [\alpha_{\varepsilon\varepsilon} - 2\alpha_{\varepsilon\sigma} x(p) + \alpha_{\sigma\sigma} x^2(p)], \tag{4.4.62}$$

where

$$\alpha_{\substack{\varepsilon\varepsilon\\ \varepsilon\sigma\\ \sigma\sigma}} = \int_0^\infty \Gamma_{\substack{\varepsilon\varepsilon\\ \varepsilon\sigma\\ \sigma\sigma}}(\vec{r}' = 0, z')\, dz'. \tag{4.4.63}$$

Taking into account (4.4.62), the solution to (4.4.48) is

$$\langle I(\vec{v}, z, \vec{p})\rangle = I_0(\vec{v}, \vec{p} + \vec{v}z)\langle I_n(\vec{v}, z, \vec{p})\rangle, \tag{4.4.64}$$

where

$$\langle I_n(\vec{v}, z, \vec{p})\rangle = \exp\left[-(\langle\varepsilon\rangle - \alpha_{\varepsilon\varepsilon})z + (\langle\sigma\rangle - 2\alpha_{\varepsilon\sigma})\int_0^z x(p + v\xi)\, d\xi \right.$$
$$\left. + \alpha_{\sigma\sigma}\int_0^z x^2(p + v\xi)\, d\xi\right] \tag{4.4.65}$$

is the two-dimensional Fourier spectrum of the average normalized radiance.

We shall analyse some results of (4.4.64). Let a medium be illuminated by a normal infinitely wide monodirectional light beam producing unit irradiance at the boundary, i.e., $I_0(\vec{v}, \vec{p}) = 4\pi^2\delta(\vec{v})$. From (4.4.64), we obtain

$$\langle I(\vec{r}, z, \vec{p})\rangle = \langle I_n(z, \vec{p})\rangle$$
$$= \exp\{[-(\langle\varepsilon\rangle - \alpha_{\varepsilon\varepsilon}) + (\langle\sigma\rangle - 2\alpha_{\varepsilon\sigma})x(p) + \alpha_{\sigma\sigma}x^2(p)]z\}. \tag{4.4.66}$$

Here, $\langle I_n(z, \vec{p})\rangle$ is the Fourier spectrum of the angular distribution of the average normalized intensity for the medium illuminated by an infinitely wide beam.

Setting $\vec{p} = 0$ in (4.4.66) and allowing for the normalization $x(p = 0) = 1$ gives the average irradiance coefficient for the medium illuminated by an infinitely wide beam, that is, the average total flux at a depth $z$ for a monodirectional unit power point source

$$\langle S(z)\rangle \equiv \langle I_n(z, \vec{p} = 0)\rangle = \exp[-(\langle\kappa\rangle - \alpha_{\kappa\kappa})z], \tag{4.4.67}$$

where $\langle\kappa\rangle = \langle\varepsilon\rangle - \langle\sigma\rangle$, $\alpha_{\kappa\kappa} = \alpha_{\varepsilon\varepsilon} - 2\alpha_{\varepsilon\sigma} + \alpha_{\sigma\sigma}$. This means that the irradiance $\langle S(z)\rangle$ in the medium with fluctuating parameters differs from the irradiance $S^{\text{hom}}(z) = \exp(-\kappa z)$ in the homogeneous medium with $\varepsilon = \langle\varepsilon\rangle$, $\sigma = \langle\sigma\rangle$ and $\kappa = \langle\kappa\rangle$ only if the absorption coefficient $\kappa$ fluctuates (a consequence of the small-angle approximation).

Let

$$\Gamma_{\substack{\varepsilon\varepsilon\\ \sigma\sigma\\ \varepsilon\sigma}} = \begin{Bmatrix} V_\varepsilon \\ V_\sigma \\ \langle\tilde{\varepsilon}\tilde{\sigma}\rangle \end{Bmatrix} e^{-R^2/2R_c^2}, \tag{4.4.68}$$

where the correlation radius $R_c$ determines the characteristic scale of the inhomogeneities; $V_\varepsilon = \langle\tilde{\varepsilon}^2\rangle$ and $V_\sigma = \langle\tilde{\sigma}^2\rangle$ are the variances of the fluctuations of the extinction and scattering coefficients. Then,

$$\alpha_{\substack{\varepsilon\varepsilon\\ \sigma\sigma\\ \varepsilon\sigma}} = \begin{Bmatrix} V_\varepsilon \\ V_\sigma \\ \langle\tilde{\varepsilon}\tilde{\sigma}\rangle \end{Bmatrix} R_c\sqrt{\pi/2}\,. \tag{4.4.69}$$

We now find the variance $\langle V_\theta\rangle$ of the angular radiance distribution in a stochastic medium. Equations (4.4.30, 66) give

$$\langle V_\theta\rangle = (\langle\sigma\rangle - 2\alpha_{\sigma\kappa})\beta_2 z/2\,. \tag{4.4.70}$$

Thus, the variance $\langle V_\theta\rangle$ will never exceed the variance $V_\theta^{\text{hom}}$ in a homogeneous medium (4.4.32), and will only differ from it in the case of simultaneous and correlated fluctuations of the absorption and scattering coefficients. It should be noted that, when the scattering coefficient does not fluctuate ($V_\sigma = 0$) even at $V_\kappa \neq 0$, as seen from (4.4.66), the relative angular radiance distributions in homogeneous and inhomogeneous media are equal. This result is typical of the small-angle approximation which disregards the dispersion of the photon paths in a scattering medium. At $V_\kappa = 0$, $V_\sigma \neq 0$, the variances $\langle V_\theta\rangle$ and $V_\theta^{\text{hom}}$ are equal, the angular distributions being, nevertheless, different.

We now consider the case of a medium illuminated by a monodirectional point source, i.e.,

$$I_0(\vec{v}, \vec{p}) = 1\,. \tag{4.4.71}$$

In this case, at $\vec{p} = 0$, we find from (4.4.64, 65) the average OTF for an inhomogeneous medium

$$\begin{aligned}\langle S(\vec{v}, z)\rangle &= \langle I_n(\vec{v}, z, \vec{p} = 0)\rangle \\ &= \exp\left[-(\langle\varepsilon\rangle - \alpha_{\varepsilon\varepsilon})z + (\langle\sigma\rangle - 2\alpha_{\varepsilon\sigma})\int_0^z x(v\xi)\,d\xi\right. \\ &\qquad \left. + \alpha_{\sigma\sigma}\int_0^z x^2(v\xi)\,d\xi\right]. \end{aligned} \tag{4.4.72}$$

### 4.4.8 The Merits and Weaknesses of the Small-Angle Approximation

We have seen that the SAA offers ample scope for calculating and investigating light fields originated in anisotropically scattering media from directional localized sources. It is especially important to the theory of optical image transfer in scattering media. The SAA design formulas are most often used for transfer coefficients and the OTFs of scattering media (Chaps. 5, 6). In the next chapter, we show that the SAA is closely connected to the aspect invariance (Sect. 5.2), which is of significance in the theory of vision systems.

Although the principles of the SAA as used for the solution of optical problems were described as far back as the 60s, it proved to be so convenient and attractive that it still draws the attention of researchers. Recent papers [4.56–59] are devoted to specifying limits of its applicability. The ample scope provided by this approach is implemented in the solution of important and intricate problems. Thus, the SAA was used to develop radiation transfer theory for stochastic media [4.54, 55, 60] (Sect. 4.4.7), multiple scattering theory for moving particles [4.61–63], and coherent light propagation theory for scattering media [4.64–66] (Chap. 7).

However, this approximation has two essential restrictions. First, it disregards the photon path dispersion and, therefore, cannot describe the temporal deformation of a pulse signal propagating in a scattering medium. For the same reason, it cannot be applied at large $\tau$. The modified SAA with $\mu = 1 - n_\perp^2/2$ which, to some extent, allows for the photon path dispersion was developed in [4.48] and suggests appropriate corrections for the integral radiance characteristics. Secondly, the SAA determination of spatial-angular radiance distributions requires computer-aided multidimensional Fourier transformations. In this respect, the small-angle diffusion approximation seems to be more promising since it provides simple formulas for light field characteristics.

## 4.5 Small-Angle Diffusion Approximation

### 4.5.1 Transfer Equation in the Small-Angle Diffusion Approximation

As in the previous section, we shall consider media whose scattering indicatrix is strongly forward-extended and has a sharp maximum at $\beta = 0$. We also assume that the angular radiance distribution is described by smooth functions, so that the variance of the angular radiance distribution is essentially greater than the scattering indicatrix variance, i.e.,

$$V_\theta \gg \beta_2/2 \,. \tag{4.5.1}$$

Then, we can approximately replace the integrand $I(\vec{R}, \vec{n}')$ on the r.h.s. of (4.1.1) by the first terms of its $|\vec{n}_\perp - \vec{n}'_\perp|$ power Taylor series expansion. Restricting ourselves to the quadratic terms of this series expansion at $B(\vec{R}, \vec{n}) = 0$, we obtain [4.67]

$$\left[\mu \frac{\partial}{\partial z} + \vec{n}_\perp \cdot \nabla_{\vec{r}} + \kappa\right] I(\vec{R}, \vec{n}_\perp) = \frac{\sigma\beta_2}{4} \Delta_{\vec{n}_\perp} I(\vec{R}, \vec{n}_\perp) \,. \tag{4.5.2}$$

Equation (4.5.2), without any additional simplifications, was used in [4.67] to solve the classical problem of the asymptotic regime at $\tau \to \infty$ in a medium irradiated by an infinitely wide source. The high accuracy of the solution was demonstrated. We wish to point out that the light field diffusivity, as well as the accuracy of the solution to (4.5.2), increases with decreasing medium absorption (growing $\Lambda$). However, analytical solutions to other problems, and especially to those with a narrow beam, can be obtained only by using supplementary assumptions which provide a small-angle variant of (4.5.2). We shall confine our study to media where the absorption is not small when the angular radiance distribution variance is small at all points in the scattering medium, i.e., $V_\theta \ll 1$. Thus, taking the above assumption (4.5.1) into account, we expect the angular radiance distributions in the medium to satisfy the condition

$$\beta_2/2 \ll V_\theta \ll 1 . \tag{4.5.3}$$

Since this is the case, we shall allow only for the terms $n_\perp^2$ and, as in the SAA, apply the range of the variable $\vec{n}_\perp$ to the whole plane. Setting

$$\mu = 1 - n_\perp^2/2 , \tag{4.5.4}$$

we write the transfer equation in a *small-angle diffusion approximation* (SADA):

$$\left[\left(1 - \frac{n_\perp^2}{2}\right)\frac{\partial}{\partial z} + \vec{n}_\perp \cdot \nabla_{\vec{r}} + \kappa\right] I(\vec{R}, \vec{n}_\perp) = \frac{\sigma\beta_2}{4} \Delta_{\vec{n}_\perp} I(\vec{R}, \vec{n}_\perp) . \tag{4.5.5}$$

The differential equation (4.5.5) is the Fokker-Planck equation. A similar mathematical statement of the problem is used in the theory of propagation of fast charged particles [4.41, 42, 68]. By neglecting the dispersion of the photon path to a given point, i.e., letting $\mu = 1$ rather than (4.5.4), we can easily solve (4.5.5) [4.42]. This solution was then applied to optical problems [4.45, 46, 69]. Their extension to a stratified medium is also easy within the SADA [4.47, 70]. Refined equations with a separated unscattered radiance component were suggested in [4.47, 69]. Asymptotic formulas at $\tau \to \infty$ allowing for the photon path length dispersion were given in [4.69]. Later, more general solutions to equation (4.5.5) were found [4.71–73]. However, the method is rather tedious and requires numerical calculation. An elegant and physically instructive solution to (4.5.5), which also holds for large optical depths, was suggested and analysed in detail in [4.74–77]. We shall discuss it in this section.

Let us supplement (4.5.5) with a boundary condition. We specify that the spatial-angular distribution of the unit-power source radiance is given by

$$I_0(\vec{r}, z = 0, \vec{n}_\perp) = \frac{1}{\Sigma_s \Omega_s} \exp\left[-\frac{\pi r^2}{\Sigma_s} - \frac{\pi n_\perp^2}{\Omega_s}\right] . \tag{4.5.6}$$

Here, $\Sigma_s$ and $\Omega_s$ are the effective cross-section and effective solid angle of the source, respectively. We take it that $\Omega_s \ll 1$. The condition $\Sigma_s \to 0$, $\Omega_s \to 0$ conforms to the case of a monodirectional point source. The situation $\Sigma_s \to \infty$ complies with an infinitely wide source. Equation (4.5.5), subject to the boundary condition (4.5.6), is a complete mathematical statement of the problem. We shall analyse its solutions, starting from the simple case of a medium irradiated by an infinitely wide source.

### 4.5.2 Light Fields Generated by an Infinitely Wide Source

The light fields generated in a scattering medium by an infinitely wide source were studied in [4.27, 28, 75, 76] using the SADA. In this approximation, the solution is sought as a normal angular radiance distribution

$$I_n(\tau, \theta) = \frac{S(\tau)}{2\pi V_\theta(\tau)} e^{-\theta^2/2V_\theta(\tau)} . \tag{4.5.7}$$

The distribution variance $V_\theta$ depends on $\tau$. Substituting (4.5.7) into (4.5.5) yields the differential equations for $V_\theta(\tau)$ and $S(\tau)$, whose solution is of the form

$$S(\tau) = \frac{e^{-(1-\Lambda)\tau}}{\cosh \xi + u_0 \sinh \xi} , \tag{4.5.8}$$

$$V_\theta(\tau) = \frac{s}{1-\Lambda} \cdot \frac{u_0 + \tanh \xi}{1 + u_0 \tanh \xi} . \tag{4.5.9}$$

Here,

$$s = \sqrt{\Lambda(1-\Lambda)\beta_2/2} , \tag{4.5.10}$$

$$\xi = s\tau , \tag{4.5.11}$$

$$u_0 = \Omega_s/2\pi V_\theta^{as} , \tag{4.5.12}$$

$$V_\theta^{as} = s/(1-\Lambda) . \tag{4.5.13}$$

We first consider the laws of irradiance attenuation. Equation (4.5.8) implies that at $\xi \ll 1$, $S(\tau) = \exp[-(1-\Lambda)\tau]$, i.e., the absorption alone is responsible for the total flux attenuation, as is the case with a small-angle approximation.

At large depths, when $\xi \gg 1$, $\sinh \xi = \cosh \xi = \exp(\xi)/2$, the flux is

$$S(\tau) = 2\frac{e^{-\gamma\tau}}{1+u_0} , \tag{4.5.14}$$

i.e., it attenuates exponentially, its asymptotic attenuation coefficient now being

$$\gamma = 1 - \Lambda + s . \tag{4.5.15}$$

The angular radiance distribution variance stabilizes, and its asymptotic quantity $V_\theta^{as}$ at $\xi \to \infty$ is given by (4.5.13). A similar solution for the asymptotic regime was obtained in Sect. 3.2. The expressions for $\gamma$ and $s$ which we consider in the present section coincide with (3.2.15), whereas those for $V_\theta^{as}$ (up to the terms $s^2$) follow from (3.2.14). The quantity $V_\theta^{as} = [\Lambda\beta_2/2(1-\Lambda)]^{1/2}$ is a representative scattering characteristic of the medium. The condition $V_\theta^{as} \ll 1$ is very close to the criterion $(1-\Lambda)/\Lambda(1-g) \gg 1$ used in Chap. 3.

If $\xi \ll 1$, then $V_\theta = s\xi/(1-\Lambda) + \Omega_s/2\pi$ and is equal to the variance obtained in a SAA (4.4.32). The angular variance for small depths increases linearly with $\tau$.

Let us estimate the depth $\tau^{as}$ for the onset of the asymptotic regime for a medium illuminated by a monodirectional source. We take it that the asymptotic regime has developed if the difference between the vertical irradiance attenuation $\gamma_{\text{eff}} = -d[\ln S(\tau)]/d\tau$ or the angular radiance distribution variance $V_\theta$ and their asymptotic values at $\tau \to \infty$ amounts to $\Delta \cdot 100\%$. For $u_0 = 0$ and not small $\xi$ values, (4.5.8, 15) and (4.5.9, 13) give

$$\frac{\gamma_{\text{eff}} - \gamma}{\gamma} = \frac{2s}{1 - \Lambda} e^{-2s\tau}, \tag{4.5.16}$$

$$\frac{V_\theta - V_\theta^{\text{as}}}{V_\theta^{\text{as}}} = 2e^{-2s\tau}. \tag{4.5.17}$$

Given $\Delta$, from (4.5.16, 17), we have

$$\tau_\gamma^{\text{as}} = -\frac{1}{2s} \ln \frac{\Delta(1 - \Lambda)}{2s},$$

$$\tau_\theta^{\text{as}} = -\frac{1}{2s} \ln \frac{\Delta}{2}. \tag{4.5.18}$$

As seen from (4.5.18), $\tau_\gamma^{\text{as}} < \tau_\theta^{\text{as}}$, i.e., the effective attenuation coefficient becomes asymptotic before the angular distribution has stabilized. For the medium with $\beta_2 = 0.05$, which is typical of extended sea indicatrices, and $\Lambda = 0.7$ at $\Delta = 0.05$, from (4.5.18), we have $\tau_\gamma^{\text{as}} \approx 16$ and $\tau_\theta^{\text{as}} \approx 25$. It is clear that at $\Lambda$ between 0.8 and 0.2, $\tau^{\text{as}}$ is a weak function of $\Lambda$ and a stronger function of the indicatrix extension: $\tau^{\text{as}} = \beta_2^{-1/2}$.

It should be noted that the quantities $S(\tau)\exp[(1 - \Lambda)\tau]$ and $V_\theta(\tau)/V_\theta^{\text{as}}$ depend only on the parameter $\xi$. In Fig. 4.8, they are plotted against $\xi$ for different angular source characteristics. At $u_0 = 1$, the asymptotic regime develops at the boundary ($\ln S(\tau) = -\gamma\tau$ for all $\tau$). At small $\tau$, if $\Omega_s/2\pi \leqslant V_\theta^{\text{as}}$ ($u_0 < 1$), the flux attenuates more slowly and at $u_0 > 1$ faster than in the asymptotic regime (4.5.8) (Fig. 4.8a). It is also clear that at $u_0 > 1$, the variance $V_\theta$ decreases with increasing depth and tends to its asymptotic value from above. At $u_0 > 1$, the function $V_\theta(\tau)$ decreases. When $u_0 = 1$, $V_\theta = V_\theta^{\text{as}}$ for all $\tau$ (4.5.9) (Fig. 4.8b).

### 4.5.3 Characteristics of Light Fields Produced by Narrow Beams

Following [4.75, 76], we shall write down the spatial-angular radiance distribution due to a narrow beam as

$$I_{\text{n}}(\vec{\rho}, \tau, \vec{n}_\perp) = \varepsilon^2 S(\tau) S'(\rho, \tau) I'(\vec{\rho}, \tau, \vec{n}_\perp). \tag{4.5.19}$$

Here, the total flux $S(\tau)$ in the plane $\tau = \text{const}$ is found from (4.5.8):

$$S'(\rho, \tau) = \frac{1}{2\pi V_\rho(\tau)} e^{-\rho^2/2V_\rho(\tau)} \tag{4.5.20}$$

is the radial irradiance distribution for which $2\pi \int_0^\infty S'(\rho, \tau)\rho\, d\rho = 1$.

The function

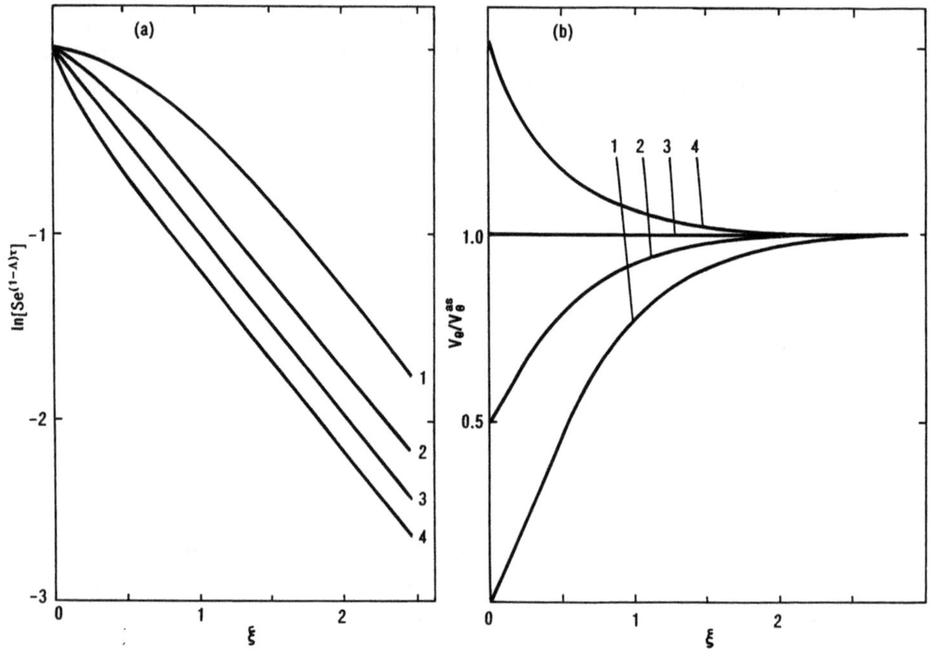


Fig. 4.8. (a) $\ln[S\exp(1-\Lambda)\tau]$ and (b) $V_\theta/V_\theta^{as}$ versus $\xi = s\tau$ at $u_0 = 0$ (curves 1), 0.5 (curves 2), 1 (curves 3) and 1.5 (curves 4)

$$I'(\vec{\rho}, \tau, \vec{n}_\perp) = \frac{1}{2\pi V_{\rho\theta}(\tau)} \exp\left\{-\frac{[\vec{n}_\perp - (\vec{\rho}/\tau)(A(\tau)\tau)/V_\rho(\tau)]^2}{2V_{\rho\theta}(\tau)}\right\} \tag{4.5.21}$$

describes the angular radiance distribution normalized by the condition $\iint_{-\infty}^{\infty} I'(\vec{\rho}, \tau, \vec{n}_\perp)\, d\vec{n}_\perp = 1$. The variance of this distribution is

$$V_{\rho\theta}(\tau) = V_\theta(\tau) - A^2(\tau)/V_\rho(\tau)\,, \tag{4.5.22}$$

where the quantity $V_\theta$ is described by (4.5.9) and

$$V_\rho(\tau) = V_\rho^0 + \frac{\tau}{1-\Lambda}\left[1 - \frac{\tanh\xi}{\xi(1 + u_0\tanh\xi)}\right] \tag{4.5.23}$$

is the variance of the radial irradiance distribution, $V_\rho^0 = \varepsilon^2\Sigma_s/2\pi$ being the effective optical area of the source,

$$A(\tau) = \frac{1}{1-\Lambda}\left[1 - \frac{1}{\cosh\xi + u_0\sinh\xi}\right]. \tag{4.5.24}$$

These formulas for a light field from a Gaussian source (4.5.6) were obtained in [4.76]. At $u_0 = 0$ and $V_\rho^0 = 0$, they provide an explicit solution for a monodirectional point source [4.75].

At small $\xi$ and $u_0 = 0$, $V_\rho^0 = 0$, (4.5.23) yields the known (4.4.36) for the variance of the spread function in a small-angle approximation. The variances of the radial irradiance distribution at $\Omega_s \neq 0$ and $\Sigma_s \neq 0$ obtained for $\xi \ll 1$ both in the SAA and SADA are also equal.

In the asymptotic regime, $\xi \to \infty$, the variance is a linear function of $\tau$

$$V_\rho^{as} = \frac{\tau}{1 - \Lambda} . \tag{4.5.25}$$

As $s \ll 1 - \Lambda$ and making allowance for (4.5.15), this result agrees with the earlier solution (4.2.5) to the rigorous transfer equation, as well as with (4.3.29) which results from the diffusion equation.

We now consider the angular radiance distribution at a point $(\vec{\rho}, \tau)$ described by (4.5.21). At $\vec{\rho} \neq 0$, the radiance distribution is azimuthally inhomogeneous and the radiance maximum is observed in the direction $\vec{n}_\perp^{max} = \vec{\rho}A(\tau)/V_\rho(\tau)$ dependent on $\vec{\rho}$ and $\tau$. The direction of the radiance maximum changes with the depth. We analyse the case $u_0 = 0$, $V_\rho^0 = 0$. As the observation point traverses the plane $\tau =$ const, the radiance diagram axis $n_\perp^{max}$ rotates so as to intersect the axis $\tau$ at some point $\tau^* = \tau(1 - V_\rho/A\tau)$. If $\xi$ is small, then $A\tau/V_\rho = 1.5$ and $\tau^* = \tau/3$. Correspondingly, at $\xi \to \infty$, $A\tau/V_\rho \to 1$, i.e., the direction of the maximum function $I'(\vec{\rho}, \tau, \vec{n}_\perp)$ approaches the direction of the source.

The variance of the angular radiance distribution about the narrow beam axis determined by (4.5.22) is less than the variance $V_\theta$ from a wide beam by $A^2/V_\rho$. In particular, at small $\xi$, (4.5.9, 22–24) imply $V_{\rho\theta} = V_\theta/4$. For a monodirectional point source, the quantities $V_\rho/V_\rho^{as}$, $A\tau/V_\rho$, $V_{\rho\theta}/V_\theta^{as}$, as well as $S(\tau)\exp[(1 - \Lambda)\tau]$ and $V_\theta/V_\theta^{as}$, depend only on the product $\xi = s\tau$. These relationships are depicted in Fig. 4.9. All the formulas required for calculating the light field characteristics, as well as their asymptotics at $\xi \ll 1$ and $\xi \gg 1$, are listed in Table 4.4.

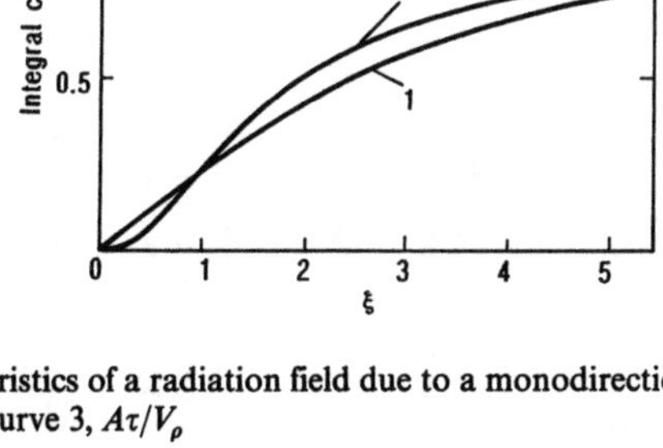


**Fig. 4.9.** Integral characteristics of a radiation field due to a monodirectional point source. Curve 1, $V_{\rho\theta}/V_\theta^{as}$; curve 2, $V_\rho/V_\rho^{as}$; curve 3, $A\tau/V_\rho$

**Table 4.4.** Integral light field characteristics in an absorbing medium with extended scattering indicatrix.

| Quantity | General case | $\xi \ll 1$ | $\xi \gg 1$ | Notes |
|---|---|---|---|---|
| Irradiance coefficient $S(\tau)$ | $\dfrac{e^{-(1-\Lambda)\tau}}{(1+u_0\tanh\xi)\cosh\xi}$ | $e^{-(1-\Lambda)\tau}$ | $\dfrac{2}{1+u_0}e^{-(1-\Lambda+s)\tau}$ | $s=\sqrt{\Lambda(1-\Lambda)\beta_2/2}$ |
| Variance of angular photon deviation $V_\theta$, infinite beam incidence | $\dfrac{s}{1-\Lambda}\dfrac{u_0+\tanh\xi}{1+u_0\tanh\xi}$ | $\dfrac{s}{1-\Lambda}(\xi+u_0)$ | $V_\theta^{as}=\dfrac{s}{1-\Lambda}$ | $\xi=s\tau$ |
| Variance of spread function $V_\rho$ | $V_\rho^0+\dfrac{\tau}{1-\Lambda}\left[1-\dfrac{\tanh\xi}{\xi(1+u_0\tanh\xi)}\right]$ | $V_\rho^0+\dfrac{\tau}{1-\Lambda}\left(\dfrac{\xi^2}{3}+u_0\xi\right)$ | $\dfrac{\tau}{1-\Lambda}$ | $u_0=\dfrac{\Omega_s}{2\pi V_\theta^{as}}$ |
| Variance of angular photon deviation at point $\rho=0$, $V_{\rho\theta}$; narrow beam | $V_\theta-\dfrac{A^2}{V_\rho}$ | $\dfrac{1}{4}\dfrac{s\xi}{1-\Lambda}$ at $\begin{cases}V_\rho^0=0\\u_0=0\end{cases}$ | | |
| $\vec{n}_{max}$ of angular radiance diagram at point $(\vec{\rho},\tau)$ | $\dfrac{\vec{\rho}}{\tau}\dfrac{A\tau}{V_\rho}$ | $1.5\dfrac{\vec{\rho}}{\tau}$ at $\begin{cases}V_\rho^0=0\\u_0=0\end{cases}$ | | |
| Auxiliary quantity $A$ | $\dfrac{1}{1-\Lambda}\left[1-\dfrac{1}{\cosh\xi+u_0\sinh\xi}\right]$ | $\dfrac{\xi^2/2-u_0\xi}{1-\Lambda}$ | $\dfrac{1}{1-\Lambda}$ | |

### 4.5.4 Oblique Medium Illumination

We shall consider the situation where the incident beam axis is in the direction $\vec{n}_0(\theta_0 = b_0, \psi = 0)$, i.e., it makes an angle $b_0$ with the normal and lies in the plane $xz$. Then, $xz$ is the symmetry plane of the problem and the maximum spatial downward irradiation distribution (the beam axis) falls at the $x$ axis, i.e., at the point $\vec{R}(x = x_{\max}, y = 0, z)$. The angular radiance distribution at this point has a maximum in the $\vec{n}_{\max}(\theta = b(z), \psi = 0)$ direction in the symmetry plane $xz$. In contrast to normal incidence which has a symmetry axis, the symmetry here is poorer (only the symmetry plane is available), and the solution is then inevitably more intricate.

Let us include a new local reference system $x', y, z'$ with the $z'(\theta = b(z), \psi = 0)$ axis in the direction of maximum radiance at the beam and $x'$ axes in the $xz$ plane. We also introduce a spherical system with a $z'$ axis, where $\vec{n} = \vec{n}(\theta', \psi')$. The vector $\vec{n}$ projections onto the $x'y$ plane: $n'_x = \sin\theta' \cos\psi' = \sin\theta \cos\psi \cos b - \cos\theta \sin b$, $n_y = \sin\theta' \sin\psi' = \sin\theta \sin\psi$ are small following the assumption of a small angular radiance dispersion. For this case, the small-angle transfer equation with the terms $n_{x'}^2, n_y^2$ was formulated in [4.78] and solved in the SADA. Below, for brevity, we write only the Green function, i.e. the solution for the source $B(\vec{R}, \vec{n}) = \delta(\vec{R})\delta(\vec{n} - \vec{n}_0)$ where $\vec{n}_0(\theta_0 = b_0, \psi = 0)$. The solution for an arbitrary source is given in [4.78]. As in (4.5.19–21), we represent the spatial-angular radiance distribution as

$$I_n(\tau_x, \tau_y, \tau, n_{x'}, n_y) = \varepsilon^2 S(\tau) S'(\tau_x, \tau_y, \tau) I'(\tau_x, \tau_y, \tau, n_{x'}, n_y)\,,$$

$$S'(\tau_x, \tau_y, \tau) = \frac{1}{2\pi\sqrt{V_{\tau_x} V_{\tau_y}}} \exp\left[-\frac{(\tau_x - d)^2}{2V_{\tau_x}} - \frac{\tau_y^2}{2V_{\tau_y}}\right],$$

$$I'(\tau_x, \tau_y, \tau, n_{x'}, n_y) = \frac{1}{2\pi\sqrt{V_{\tau_x, n_{x'}} V_{\tau_y, n_y}}} \times \exp\left\{-\frac{[n_{x'} - A_x(\tau_x - d)/V_{\tau_x}]^2}{2V_{\tau_x, n_{x'}}} - \frac{[n_y - A_y \tau_y / V_{\tau_y}]^2}{2V_{\tau_y, n_y}}\right\}. \quad (4.5.26)$$

Here, $\tau_x = \varepsilon x$, $\tau_y = \varepsilon y$, $\tau = \tau_z = \varepsilon z$ are the optical coordinates of the observation point; $d = \varepsilon x_{\max}$, $o$, $\tau$ are the optical coordinates of the irradiance maximum position, $S(\tau)$ is the total flux through the $z$ plane; $S'(\cdot)$ and $I'(\cdot)$ are the spatial irradiance and angular radiance distributions at the point $\vec{R}$ normalized as in (4.5.19–21); $V_{n_{x'}}$ and $V_{n_y}$ are the variances of the angular radiance distributions in the case of an infinite source; $V_{\tau_x}$ and $V_{\tau_y}$ are the variances of the axial irradiance distribution; $V_{\tau_x, n_{x'}} = V_{n_{x'}} - A_{\tau_x}^2 / V_{\tau_x}$ and $V_{\tau_y, n_y} = V_{n_y} - A_{\tau_y}^2 / V_{\tau_y}$ are the variances of the angular radiance distributions in the case of a point source. Formulas (4.5.26) allow for the light field symmetry $I(\tau_x, \tau_y, \tau, n_{x'}, n_y) = I(\tau_x, -\tau_y, \tau, n_{x'}, -n_y)$. In (4.5.26), distributions are correctly described with regard for the squared spatial-angular photon deviations from the position of a maximum. These solutions, however, disregard the details due to the higher moments, for example, the distribution asymmetry.

The integral characteristics entering into (4.5.26) are determined as

$$V_{n_y} = V_\theta^{\mathrm{as}} \tanh(sl)\,, \qquad V_{\tau_y} = \frac{l(1 - \tanh(sl)/sl)}{1 - \Lambda}\,, \tag{4.5.27}$$

$$A_{\tau_y} = \frac{1 - 1/\cosh(sl)}{1 - \Lambda}\,, \qquad l = \int_0^\tau \frac{d\tau'}{\cos b}\,. \tag{4.5.28}$$

Formulas (4.5.27, 28) differ from the relations for a normal incident beam ($b_0 = 0$) (Table 4.4) only in the effective optical length $l$ substituted for $\tau$ in the case of oblique incidence. Other quantities are determined by the quadratures

$$\sin b = \mathrm{e}^{-p} \sin b_0\,, \qquad V_{n_{x'}} = V_\theta^{\mathrm{as}} V_{n_{x'}}^*\,, \tag{4.5.29}$$

$$V_{n_{x'}}^* = [(1 + \sin^2 b \sin^2 b_0)(1 - \mathrm{e}^{-2p}) - 4p \sin^2 b]^{1/2}/\cos^2 b\,, \tag{4.5.30}$$

$$S(\tau) = \frac{\mathrm{e}^{-(1-\Lambda)l - p/2}}{\sqrt{\cosh(sl)}}\,, \tag{4.5.31}$$

$$d(\tau) = \int_0^p \frac{\sin b - (1 - \Lambda) A_{\tau_x} \tan b}{(1 - \Lambda) V_{n_{x'}}} dp'\,, \tag{4.5.32}$$

$$A_{\tau_x}(\tau) = \frac{1 - \mathrm{e}^{-p} \cos^2 b_0/\cos^2 b + (b - b_0)\tan b}{(1 - \Lambda)\cos b}\,, \tag{4.5.33}$$

$$V_{\tau_x}(\tau) = \int_0^p \frac{2A_{\tau_x}/\cos b - (1 - \Lambda) A_{\tau_x}^2 (1 + 2\tan^2 b)}{(1 - \Lambda) V_{n_{x'}}} dp'\,. \tag{4.5.34}$$

The parameter $p$ entering into these formulas can be found from the equation

$$s\tau = \int_0^p \frac{\cos b}{V_{n_{x'}}^*} dp'\,. \tag{4.5.35}$$

Calculation by (4.5.27–39) is simple. Taking into account (4.5.29, 30) and numerically integrating (4.5.35), we obtain the function $s\tau(p)$. This is a complete solution to the problem since, in (4.5.27–34), all the beam characteristics are expressed in terms of the parameter $p$. The quantities $b$, $p$, $sl$, $sd$, $A_{\tau_x}(1 - \Lambda)$, $A_{\tau_y}(1 - \Lambda)$, $S(\tau)\exp[(1 - \Lambda)\tau]$, as well as the variances derived for their asymptotic magnitudes at $\tau \to \infty$, $V_{n_{x'}(n_y)}^* = V_{n_{x'}(n_y)}/V_\theta^{\mathrm{as}}$ and $V_{\tau_x(\tau_y)} = V_{\tau_x(\tau_y)}/V_\rho^{\mathrm{as}}(V_\theta^{\mathrm{as}} = s/(1 - \Lambda)$, $V_\rho^{\mathrm{as}} = \tau/(1 - \Lambda))$, depend only on $\xi = s\tau$. These relationships are depicted in Figs 4.10–11. They elucidate the main properties of light field transformations in a scattering medium illuminated by oblique light. These are the position of an irradiance maximum (Fig. 4.11a) and the behavior of the spatial variances $V_{\tau_x}^*$ and $V_{\tau_y}^*$ (Fig. 4.11b) in a narrow beam field. The specific features of angular distributions in the case of infinitely wide sources are presented in Fig. 4.10. As seen from Fig. 4.10, for example, if the sun is not at its zenith, then the

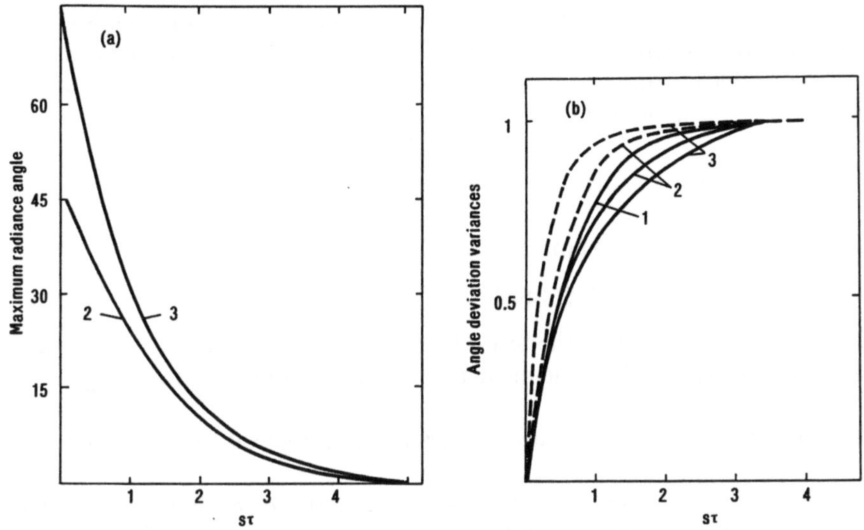


**Fig. 4.10.** (a) Angle $b$ and (b) variances of angular deviation of $n_{x'}$ and $n_y$ axes versus $s\tau$ at $b_0 = 0°$ (curves 1, 1′), 45° (curves 2, 2′), 75° (curves 3, 3′). In (b), *solid curves* show $V^*_{n_{x'}}$ and *dashed curves*, $V^*_{n_y}$

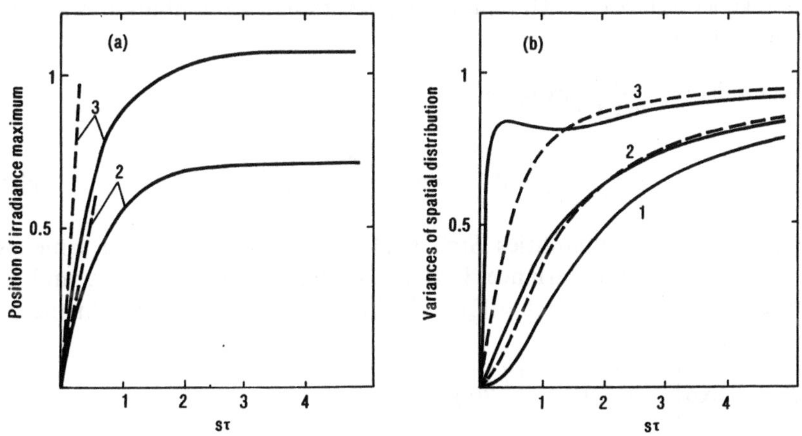


**Fig. 4.11.** Narrow beam incidence at $b_0 = 0°$ (curve 1), 45° (curve 2) and 75° (curve 3). In (a), $sd$ marks the position of the irradiance maximum. *Dashed lines* correspond to $sd = s\tau \tan b_0$. In (b), variances of irradiance distribution along $\tau_x$ and $\tau_y$ axes versus $s\tau$. *Solid lines*, $V^*_{\tau_x}$ and *dashed lines*, $V^*_{\tau_y}$

axis of the angular radiance diagram turns faster in the upper atmospheric layers the greater $b_0$ is, and then slowly tends to the normal at $s\tau \gtrsim 2$. It follows from this figure that the axis of the angular diagram of the scattered sun radiance in the sea is appreciably oblique about the normal ($> 5°$) up to $\tau \approx 20$ if the sun is low above the horizon. Such an estimate was obtained allowing that for sea water

$s \approx 0.1$ and the maximum $b_0 \approx 48°$ due to the boundary refraction. Similarly, other necessary inferences and estimates can be gained from these figures.

The problem simplifies at $b_0 \ll 1$. We shall cite some simple expressions to characterize the light field to $b_0^2$,

$$b = b_0/\cosh\xi\,, \qquad S(\tau) = \frac{e^{-(1-\Lambda)\tau - b_0 \tanh\xi/2V_\theta^{as}}}{\cosh\xi}\,,$$

$$V^*_{n_{x'}} = \tanh\xi + b_0^2(2.5\tanh\xi - 2\xi)/\cosh^2\xi\,,$$

$$V^*_{n_y} = \tanh\xi + b_0^2\tanh\xi/2\cosh^2\xi\,.$$

These solutions, including the terms linear in $b_0$, are given in [4.79]. For additional information on oblique incidence, the reader is referred to [4.80].

### 4.5.5 Light Fields in Media with Depth-Dependent Optical Characteristics

We assume that the optical characteristics of a scattering medium depend on the depth $z$, i.e. are characterized by the functions $\kappa(z)$, $\sigma(z)$ and $\beta_2(z)$. The algorithm for solving the transfer equation for such a medium illuminated by a normal monodirectional point source is developed in [4.77].

It is more convenient to use the dimensional $r$ and $z$ coordinates for an inhomogeneous medium. The spatial-angular radiance distribution is sought, as before, in the form

$$I_n(\vec{r}, z, \vec{n}_\perp) = \frac{S(z)}{2\pi V_r(z)}\exp[-r^2/2V_r(z)] \times \frac{1}{2\pi V_{r\theta}(z)}\exp\left\{-\frac{[\vec{n}_\perp - (\vec{r}/z)(A(z)z/V_r(z))]^2}{2V_{r\theta}(z)}\right\}. \qquad (4.5.36)$$

The physics of the quantities $S(z)$, $V_r(z)$, $V_{r\theta}(z)$ and $A(z)$ is the same as in (4.5.19–21). Here, the variance $V_r(z)$ is the dimensional quantity, and for a homogeneous medium $V_r(z) = V_\rho(\tau)/\varepsilon^2$. We represent a total flux in the plane $z = \text{const}$ as

$$S(z) = \exp\left[-\int_0^z \kappa(z)\,dz\right]\Big/B(z)\,, \qquad (4.5.37)$$

where $B(z)$ is some unknown function. It satisfies the ordinary equation

$$\frac{d}{dz}\left(\frac{1}{\kappa(z)}\frac{dB(z)}{dz}\right) - \frac{\sigma(z)\beta_2(z)}{2}B(z) = 0 \qquad (4.5.38)$$

having the boundary conditions

$$B(z=0) = 1\,, \qquad \left.\frac{dB(z)}{dz}\right|_{z=0} = 0\,. \qquad (4.5.39)$$

The variances $V_\theta$, $V_r$ and parameter $A$ are expressed in terms of the function $B(z)$ as

$$V_\theta(z) = \frac{1}{\kappa(z)} \frac{d}{dz} \ln B(z) , \tag{4.5.40}$$

$$A(z) = \frac{1}{B(z)} \int_0^z \frac{1}{\kappa(z')} \frac{dB(z')}{dz'} dz' , \tag{4.5.41}$$

$$V_r(z) = \int_0^z \left[ A(z') - \frac{\kappa(z')}{4} A^2(z') \right] dz' . \tag{4.5.42}$$

We note that for a homogeneous scattering medium, the solution to (4.5.38) is simple. For a monodirectional point source $B(z) = \cosh(z\sqrt{\sigma\kappa\beta_2/2})$, and from (4.5.40–42), formulas (4.5.9, 23, 24) immediately follow at $u_0 = 0$, $V_\rho^0 = 0$.

An analytical solution to (4.5.38) may be found only in some particular cases. For example, in [4.81] such a solution is obtained for the exponential $\kappa$ and $\sigma\beta_2$ dependence on $z$. A computer solution, however, presents no difficulties. Formulas (4.5.38–42) give a simple algorithm for calculating the integral characteristics for stratified media.

For small depths ($z\sqrt{\sigma(z)\kappa(z)\beta_2(z)/2} \ll 1$), the SADA simply allows for stratification [4.47, 70]. The quantities entering into (4.5.36) are determined by the following expressions:

$$S(z) = \exp\left[ -\int_0^z \kappa(z')\, dz' \right] = \exp(-\tau_\kappa) , \tag{4.5.43}$$

$$V_\theta(z) = \frac{1}{2} \int_0^z \sigma(z')\beta_2(z')\, dz' , \tag{4.5.44}$$

$$A(z) = \frac{1}{2} \int_0^z \sigma(z')\beta_2(z')(z - z')\, dz' , \tag{4.5.45}$$

$$V_r(z) = \frac{1}{2} \int_0^z \sigma(z')\beta_2(z')(z - z')^2\, dz' . \tag{4.5.46}$$

Characteristically, the variances $V_\theta$, $V_r$ and $V_{r\theta}$ depend only on the stratification of the medium scattering properties [the product $\sigma(z)\beta_2(z)$], and are independent of $\kappa(z)$. We emphasize that (4.5.43–46) for the integral characteristics are the same as the appropriate SAA solutions.

### 4.5.6 The Scope of the Small-Angle and Small-Angle Diffusion Approximations

Equation (4.5.3) is the condition for the SADA to be applicable. We shall discuss the restrictions it imposes on the medium parameters. Consider a medium illuminated by a monodirectional point source. Then, the variance of the angular radiance distribution is highest in the asymptotic regime. Therefore, the necessary condition for the SADA to be applied to describe light fields in the medium is the fulfillment of the inequality $\beta_2/2 \ll V_\theta^{as} \ll 1$. Substituting $V_\theta^{as} = s/(1 - \Lambda)$ (Table 4.4) gives the following condition imposed on the optical characteristics of the medium:

$$2/\beta_2 > \Lambda/(1-\Lambda) > \beta_2/2\,. \tag{4.5.47}$$

In the case of sea water, when $\beta_2 \approx 0.1$, the inequality (4.5.47) suggests the following conditions: $\Lambda > 0.05$, $1-\Lambda > 0.05$. The condition (4.5.47) is necessary but insufficient. At small $\tau$, the variance $V_\theta(\tau) < V_\theta^{\mathrm{as}}$ can be compared with the mean-square angle single scattering deviation $\beta_2/2$, i.e., the SADA may not be successful at small $\tau$.

We shall consider independently the problems of the application SADA to calculate integral characteristics (variances), and to describe spatial-angular distributions. Let the condition (4.5.47) be real. Then, the SADA applicability at large $\tau$ is not in doubt. At small $\tau$, the SAA solutions may be the accuracy criteria for the SADA formulas. At small $\xi = s\tau$, the SAA and SADA formulas for the fluxes and spatial-angular distribution variances are similar. Therefore, since the condition (4.5.47) is fulfilled, the SADA formulas suggest a correct description of these integral characteristics for all $\tau$.

The case is more intricate as far as spatial-angular distributions are concerned. At small $\tau$, real light fields are far from Gaussian, the spatial-angular radiance distributions are closely correlated with the scattering indicatrix and are determined by all its moments. The local singularities of the light field characteristics for the $\delta$-source are described in Sect. 4.2. As $\tau$ increases, all distributions approach the normal distribution. The range of optical depths obeying a normal law valid for approximating the angular radiance distribution is estimated from (4.4.32)

$$\Lambda\tau \gg 3\beta_4/2\beta_2^2\,. \tag{4.5.48}$$

Thus, for $\tau$ satisfying (4.5.48), the radiance at a point $(\vec{\rho},\tau)$ for $\rho/\tau \ll 1$ in the $\vec{n}(\vec{n}_\perp \ll 1)$ direction can be calculated by the SADA formulas. For smaller $\tau$, the SAA or modified SADA solutions presented in the next section should be used. Naturally, for the medium irradiated by a source with the Gaussian directional diagram at $1 \gg \Omega_s \gg \pi\beta_2$, the radiance in the scattering medium is described correctly enough by SADA for any $\tau$.

We have applied the SAA solutions as a means of specifying the SADA applicability at low $\tau$. However, the SADA can be used for large $\tau$ to specify the $\tau$ range where the SAA can be applied. In fact, we shall consider that since (4.5.47) is fulfilled, the variances $V_\theta$ and $V_\rho$ described are exactly enough by the SADA for all $\tau$. As is known, the appropriate SAA formulas for $V_\theta$ and $V_\rho$ can be applied in a limited $\tau$ range due to the neglect of the photon path dispersion. If $\Delta$ is the limiting relative error in this case, then the range of optical depths where the error in the SAA determination of $V_\theta$ is less than $\Delta$ can be found from the inequality

$$[V_\theta^{\mathrm{SAA}}(\tau) - V_\theta(\tau)]/V_\theta(\tau) \lesssim \Delta\,. \tag{4.5.49}$$

Here, $V_\theta^{\mathrm{SAA}}(\tau) = s^2\tau/(1-\Lambda)$ is the mean-square deviation in a SAA, for $V_\theta(\tau)$ see Table 4.4. Expanding the function $\tanh\xi$ in a Taylor series, from (4.5.49) we obtain the estimate $\xi^2/3 \leqslant \Delta$. A similar estimate has been obtained for the applicability range of the $V_\rho^{\mathrm{SAA}}(\tau)$ expression $2\xi^2/5 \leqslant \Delta$. Thus, we may consider that the angular and radial distribution variances are defined by the SAA formulas with an error of less than $\Delta$ for the $\tau$ range that satisfies

$$\tau \leqslant \frac{\sqrt{3\Delta}}{s} \approx \sqrt{\frac{3\Delta}{\Lambda(1-\Lambda)(1-g)}}\,. \tag{4.5.50}$$

For example, at $\Lambda = 0.7, g = 0.9$ and $\Delta = 0.3$ (i.e., the error is no more than 30%), from (4.5.50), we have $\tau \leqslant 7$. It is clear that the estimate (4.5.50) holds for $\Lambda$ limited by the inequalities (4.5.47). For other $\Lambda$, and for $\Lambda \to 0$ and $\Lambda \to 1$ in particular, the SAA applicability range must be estimated from other considerations.

Therefore, if the optical parameters of a medium satisfy (4.5.47), then the SAA and SADA conveniently complement each other, thereby providing solutions for light fields in a small-angle region at small and large thickness. For weakly absorbing media ($(1-\Lambda)/\Lambda(1-g) \ll 1$) and large optical thicknesses, the diffusional approximation described in Sect. 4.3 must be used. The conventional applicability diagram in Fig. 4.12 illustrates different approximate solutions to determine downward radiation characteristics. It corresponds to an indicatrix of $g = 0.9$. The vertical dashed line $[\Lambda(1-g) = 1-\Lambda]$ divides the diagram into regions of weak and strong absorptions. The SAA applicability region bounded by the $AB$ curve from below is defined here as the region where the error in the SAA determination of the variances $V_\rho$ and $V_\theta$ is not more than 30%. Similarly, the $EF$ curve is the upper bound of the domain where the diffusion approxima-

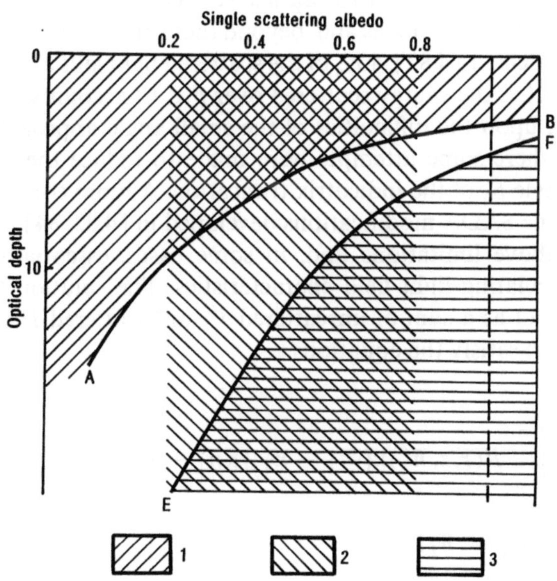


**Fig. 4.12.** Applicability domains of approximation methods: (1) small-angle approximation; (2) small-angle diffusion approximation; (3) diffusion approxmation

tion provides the radiation flux and density to the same maximum error. The applicability of the SADA solutions for integral light field characteristics is limited to the vertical lines (4.5.47). Depending on the angular source diagram, the upper boundary of the SADA applicability to describe spatial angular radiance distribution may shift from $\tau = 0$ to the $EF$ curve. It should be noted that the DA and SADA applicability domains overlap at large $\tau$. It is also illustrated in Table 8.6 and commented upon in detail in Chap. 8 where we quote the limiting depths of the self-luminous source detection calculated in these two approximations. We also wish to point out that the SADA and DA overlap is approximately the interval in which spatial-angular radiance distributions can be described by the normal law, even in the case of monodirectional sources, i.e., here also, the SADA correctly describes spatial-angular distributions for any boundary conditions.

### 4.5.7 Modified Small-Angle Diffusion Approximation

The solutions discussed in Sects. 4.5.2–4 determine the integral characteristics for any $\tau$ to sufficient accuracy. They are convenient for various design calculations and estimations. Nevertheless, real light fields, especially those not far removed from local directional sources, may essentially differ from Gaussian ones (Sect. 4.2). In particular, the condition $V_\theta \gg \beta_2/2$ may not always hold. Moreover, the SADA does not allow for the dependence of the light field structure on the scattering indicatrix properties, the data for which appear in the solution only as a single parameter, the mean square scattering angle $\beta_2$. This situation is extremely critical for the optical scattering indicatrices which have a "sharp peak" at very small angles and, beyond these limits, decrease slowly. Thus, for the cloud model C.1 [4.82], 40% of the scattered light falls at scattering angles $\beta \lesssim 3°$; in sea water, the energy scattered over $\beta \leqslant 6°$ accounts for some 60%.

We touched upon the difficulties of the small-angle approaches in the previous section, where some attempts were made to reduce them somewhat by choosing the angle $\beta_0$ of the indicatrix truncation. A more consistent approach is provided by the SADA modifications developed in [4.83, 84], which partially eliminate the above limitations and, in some cases, essentially improve the accuracy in calculating light field characteristics. The solution to the transfer equation is sought as a sum of unscattered $I^0$ and scattered $I^s$ light radiances, $I = I^0 + I^s$ which satisfy the equations

$$\hat{D}I^0 = 0\,. \tag{4.5.51}$$

$$\hat{D}I^s = \frac{\sigma}{4\pi}\iint_{-\infty}^{\infty} I^s(\vec{r}, z, \vec{n}'_\perp)x(|\vec{n}_\perp - \vec{n}'_\perp|)\,d\vec{n}'_\perp + B(\vec{r}, z, \vec{n}_\perp)\,,$$

$$\hat{D} = \left(1 - \frac{n_\perp^2}{2}\right)\frac{\partial}{\partial z} + \vec{n}_\perp \nabla_{\vec{r}} + \varepsilon\,. \tag{4.5.52}$$

In the equation for $I^s$, the source function $B$ describes single-scattered light

$$B(\vec{r}, z, \vec{n}_\perp) = \frac{\sigma}{4\pi} \iint_{-\infty}^{\infty} I^0(\vec{r}, z, \vec{n}'_\perp) x(|\vec{n}_\perp - \vec{n}'_\perp|) \, d\vec{n}'_\perp \,, \tag{4.5.53}$$

i.e., a single iteration has been performed here, as in the classical transfer theory for wide beams (3.1.3–7).

The transfer equation in the form of (4.5.51, 52) did not advance our analysis. However, the condition (4.5.3) of the SADA applicability, as well as the assumption of the Gaussian spatial-angular radiance distribution, are much better fulfilled for the scattered component $I^s$ than for the total radiance $I$. This specifies higher accuracy of the solutions of (4.5.51, 52) in [4.83].

Let us take the next step. We represent the scattering indicatrix as [4.84]

$$x(\beta) = \eta x_1(\beta) + (1 - \eta) x_2(\beta) \,, \tag{4.5.54}$$

where $x_1(\beta)$ describes the peak for small angles, $x_2(\beta)$ stands for the remaining portion of the indicatrix in the forward semi-sphere, and

$$\frac{1}{2} \int_0^{\infty} x_i(\beta) \beta \, d\beta = 1 \,, \qquad i = 1, 2$$

$$\beta_2^{(i)} = \int_0^{\infty} \beta^2 x_i(\beta) \beta \, d\beta \bigg/ \int_0^{\infty} x_i(\beta) \beta \, d\beta \,, \qquad \beta_2^{(1)} \ll \beta_2^{(2)} \ll 1 \,.$$

In conformity with (4.5.54), we shall seek the scattered light radiance $I^s$ in the form

$$I^s(\vec{r}, z, \vec{n}_\perp) = I_1^s(\vec{r}, z, \vec{n}_\perp) + I_2^s(\vec{r}, z, \vec{n}_\perp) \,, \tag{4.5.55}$$

where $I_1^s$ obeys the equation

$$\hat{D} I_1^s(\vec{r}, z, \vec{n}_\perp) = \frac{\sigma_1}{4\pi} \iint_{-\infty}^{\infty} I_1^s(\vec{r}, z, \vec{n}'_\perp) x(|\vec{n}_\perp - \vec{n}'_\perp|) \, d\vec{n}'_\perp + B_1(\vec{r}, z, \vec{n}_\perp) \tag{4.5.56}$$

at

$$B_1(\vec{r}, z, \vec{n}_\perp) = \frac{\sigma_1}{4\pi} \iint_{-\infty}^{\infty} I^0(\vec{r}, z, \vec{n}'_\perp) x_1(|\vec{n}_\perp - \vec{n}'_\perp|) \, d\vec{n}'_\perp \,. \tag{4.5.57}$$

The formulas (4.5.52–57) imply that $I_2^s$ is the solution of the equation

$$\begin{aligned} \hat{D} I_2^s(\vec{r}, z, \vec{n}_\perp) = {} & \frac{\sigma(1-\eta)}{4\pi} \iint_{-\infty}^{\infty} I_2^s(\vec{r}, z, \vec{n}'_\perp) x_2(|\vec{n}_\perp - \vec{n}'_\perp|) \, d\vec{n}'_\perp \\ & + \frac{\sigma\eta}{4\pi} \iint_{-\infty}^{\infty} I_2^s(\vec{r}, z, \vec{n}'_\perp) x_1(|\vec{n}_\perp - \vec{n}'_\perp|) \, d\vec{n}'_\perp + B_2(\vec{r}, z, \vec{n}_\perp) \,, \end{aligned} \tag{4.5.58}$$

where

$$\begin{aligned} B_2(\vec{r}, z, \vec{n}_\perp) = {} & \frac{\sigma(1-\eta)}{4\pi} \iint_{-\infty}^{\infty} [I^0(\vec{r}, z, \vec{n}'_\perp) + I_1^s(\vec{r}, z, \vec{n}'_\perp)] \\ & \times x_2(|\vec{n}_\perp - \vec{n}'_\perp|) \, d\vec{n}'_\perp \,. \end{aligned} \tag{4.5.59}$$

It should be emphasized that (4.5.56, 58) are exact results of (4.5.52) for arbitrary representation of the indicatrix (4.5.54), and may be used to build different approximate analytic solutions and exact numerical algorithms for solving the transfer equation.

We next use the property of the representation (4.5.54) at $\beta_2^{(1)} \ll \beta_2^{(2)}$. In this case, $\beta_2^{(1)}$ is negligible in comparison with the mean square scattering angle of the component $I_2^s(\vec{r}, z, \vec{n}_\perp)$. Then, the kernel $x_1(|\vec{n}_\perp - \vec{n}_\perp'|)$ of the second integral term on the r.h.s. of (4.5.58) can be regarded as the $\delta$-function, and we may approximately set $(\sigma\eta/4\pi)\iint_{-\infty}^{\infty} I_2^s(\vec{r}, z, \vec{n}_\perp')x_1(|\vec{n} - \vec{n}_\perp'|)\,d\vec{n}_\perp' \approx \sigma\eta I_2^s(\vec{r}, z, \vec{n}_\perp)$. Equation (4.5.58) will be written in a similar way to (4.5.56), i.e., as an ordinary transfer equation including the optical parameters $\sigma_2 = \sigma(1-\eta)$ and $\varepsilon_2 = \varepsilon(1 - \Lambda\eta)$. Now, in conformity with the SADA concept and as in (4.5.19), cast the scattered radiance component in the form

$$I_{n,i}^s(\vec{\rho}, \tau, \vec{n}_\perp) = \varepsilon^2 S_i(\tau) S_i'(\rho, \tau) I_i'(\vec{\rho}, \tau, \vec{n}_\perp)\,, \tag{4.5.60}$$

where

$$S_i'(\rho, \tau) = \frac{1}{2\pi V_{\rho,i}(\tau)} \exp\left[-\frac{\rho^2}{2V_{\rho,i}(\tau)}\right], \tag{4.5.61}$$

$$I_i'(\vec{\rho}, \tau, \vec{n}_\perp) = \frac{1}{2\pi V_{\rho\theta,i}(\tau)} \exp\left\{-\frac{[\vec{n}_\perp - (\vec{\rho}/\tau)(A_i(\tau)\tau/V_{\rho,i}(\tau))]^2}{2V_{\rho\theta,i}(\tau)}\right\}, \tag{4.5.62}$$

$$V_{\rho\theta,i}(\tau) = V_{\theta,i}(\tau) - A_i^2(\tau)/V_{\rho,i}(\tau)\,. \tag{4.5.63}$$

For the case of a monodirectional source, approximate expressions for the flux $S_1(\tau)$, variances $V_{\rho,1}(\tau)$ and $V_{\theta,1}(\tau)$, and the quantity $A_1(\tau)$ are [4.84]

$$S_1(\tau) = \frac{e^{-(1-\Lambda_1)\tau}}{\cosh s_1\tau} - e^{-\tau} = T_1 \frac{e^{-(1-\Lambda_1)\tau}}{\cosh s_1\tau}\,, \tag{4.5.64}$$

$$V_{\rho,1}(\tau) = \frac{\tau}{1-\Lambda_1}\left(1 - \frac{\tanh s_1\tau}{s_1\tau}\right)\frac{1}{T_1}\,, \tag{4.5.65}$$

$$V_{\theta,1}(\tau) = \frac{s_1}{1-\Lambda_1}\,\frac{\tanh s_1\tau}{T_1}\,, \tag{4.5.66}$$

$$A_1(\tau) = \frac{1}{1-\Lambda_1}\left(1 - \frac{1}{\cosh s_1\tau}\right)\frac{1}{T_1}\,, \tag{4.5.67}$$

$$T_1 = 1 - e^{-\Lambda_1\tau}\cosh s_1\tau\,, \tag{4.5.68}$$

$$s_1 = \sqrt{\Lambda_1(1-\Lambda_1)\beta_2^{(1)}/2}\,, \qquad \Lambda_1 = \Lambda\eta\,. \tag{4.5.69}$$

The quantities which determine the component $I_{n,2}^s$ are found from (4.5.64–69) in terms of the substitutions

$$\tau \to (1-\Lambda_1)\tau\,, \qquad \vec{r} \to (1-\Lambda_1)\vec{r}\,, \qquad \Lambda_1 \to \Lambda_2\,, \qquad \beta_2^{(1)} \to \beta_2^{(2)}\,.$$

For example,

$$S_2(\tau) = T_2 \frac{e^{-(1-\Lambda_2)(1-\Lambda_1)\tau}}{\cosh[s_2(1-\Lambda_1)\tau]}, \tag{4.5.70}$$

$$V_{\rho,2}(\tau) = \frac{(1-\Lambda_1)\tau}{1-\Lambda_2}\left[1 - \frac{\tanh[s_2(1-\Lambda_1)\tau]}{s_2(1-\Lambda_1)\tau}\right]\frac{1}{T_2}, \tag{4.5.71}$$

$$V_{\theta,2}(\tau) = \frac{s_2}{1-\Lambda_2}\frac{\tanh[s_2(1-\Lambda_1)\tau]}{T_2}, \tag{4.5.72}$$

$$A_2(\tau) = \frac{1}{1-\Lambda_2}\left[1 - \frac{1}{\cosh[s_2(1-\Lambda_1)\tau]}\right]\frac{1}{T_2}, \tag{4.5.73}$$

$$T_2 = 1 - e^{-\Lambda_2(1-\Lambda_1)\tau}\cosh[s_2(1-\Lambda_1)\tau], \tag{4.5.74}$$

$$s_2 = \sqrt{\Lambda_2(1-\Lambda_2)\beta_2^{(2)}/2}, \qquad \Lambda_2 = \Lambda(1-\eta)/(1-\Lambda\eta). \tag{4.5.75}$$

For a source with a finite angular and spatial aperture, the corresponding expressions are somewhat more tedious [4.84]. It should be noted that the two-term representation (4.5.54) was not chosen by chance. Experience shows that at $\beta < \pi/2$, the scattering indicatrices of sea water, atmospheric aerosols and cloud dispersions allow a fair approximation of the sum of two exponents of the type $\exp(-\alpha\beta)$ or $\exp(-\alpha\beta^2)$.

The efficiency of such a modified approach is illustrated in Fig. 4.13 which gives the scattered light radiance for a cloud as calculated in the SAA (chosen as a standard at small $\tau$), as well as in ordinary and modified SADA approximations. The scattering indicatrix C.1 has been approximated by the function

$$x(\beta) = \frac{4\eta}{\beta_2^{(1)}} e^{-\beta^2/\beta_2^{(1)}} + 32(1-\eta)e^{-4\beta}, \qquad \beta \leqslant \pi/2 \tag{4.5.76}$$

at $\eta = 0.42$, $\beta_2^{(1)} = 0.001$. The losses due to backscattering are included by effective absorption, so that $\Lambda = 0.96$. As seen from the figure, the comprehensive allowance for the scattering indicatrix form and scattering radiance representation as a sum of two components essentially improves the accuracy of the calculation. It is important that such an improvement in the accuracy has been achieved without complicating the SADA procedure, the solutions having a simple analytic form as before. At large $\tau$, where the light fields become, in fact, Gaussian due to multiple scattering, the SADA and its modification yield similar results. The estimates in [4.84] are evidence that in comparison with the ordinary SADA, its modification offers essentially higher accuracy in calculating the irradiance up to $\tau \lesssim 12$ and the radiance up to $\tau \lesssim 20$–25.

The approach used in this section to formulate the SADA modification allows a promising generalization. Depending on the indicatrix properties and problem to be solved, we may choose an indicatrix representation such as $x(\beta) = \sum_{i=1}^{N} \eta_i x_i(\beta)$, which will result in substitution of the equation system by simpler kernels for the initial transfer equation. Each of the equations in the system can be solved using the appropriate numerical algorithm or approximate method [4.85].

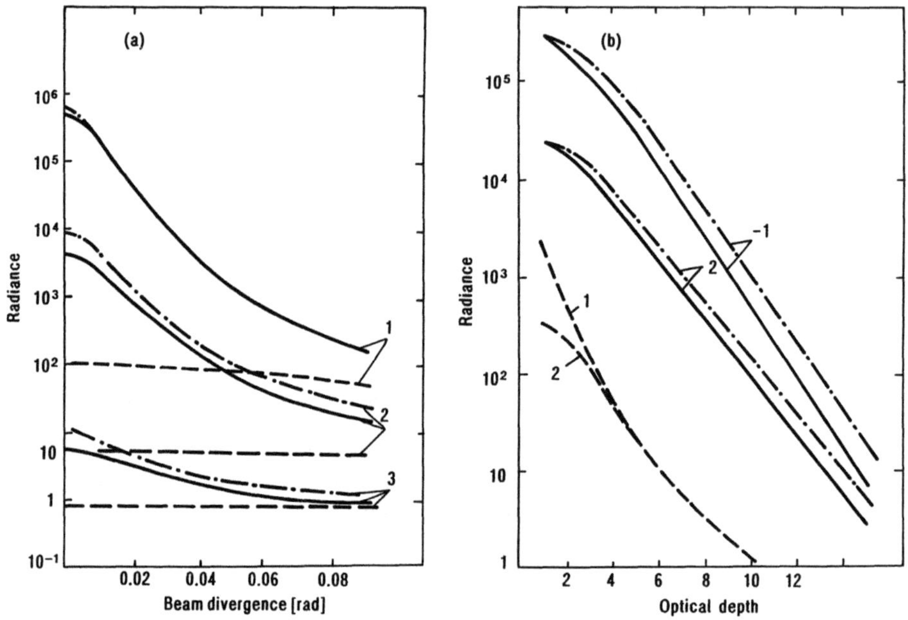


**Fig. 4.13.** Radiance of scattered light in the forward direction. (a) $I^s(\tau, \vec{n}_\perp = 0)$ as a function of the incident beam divergence $\sqrt{V_\theta^0/\pi}$ at $\tau = 1$ (curve 1), 5 (curve 2) 10 (curve 3) due to an infinitely wide source. (b) Radiance $I_n^s(\vec{\rho} = 0, \tau, \vec{n}_\perp = 0)$ at the narrow beam axis as a function of $\tau$ at $V_\theta^0 = 7.8 \times 10^{-5}$, $V_\rho^0 = 0.01$ (curve 1); $V_\theta^0 = 7.8 \times 10^{-3}$, $V_\rho^0 = 0.1$ (curve 2). *Dash-dot lines* correspond to SAA; *dashed lines* to SADA, and *solid lines* to modified SADA

## 4.6 Notes on Multiple Backscattering

This small section should be regarded as a continuation of Sect. 3.6.

It is frequently imperative to know the radiance at large angles. Such a problem arises in calculating the natural background and backscattering noises in active vision or location systems, the echo-signal in remote sensing, etc. Of all the approximate methods presented in this chapter, only the diffusion approximation describes backscattering (neglected both in the SAA and SADA). For strongly anisotropic single- and multiple-scattering media, the DA solution is not correct. Backscattering in such media may be conventionally estimated through approximate approaches using simplified indicatrix models (Sects. 3.6.2–4). We comment on these methods as used to solve the narrow-beam backscattering problem.

One can hardly expect a significant advance due to the application of the transport approximation. The point is that this approach reduces the problem of beam propagation in a strongly anisotropic medium to one of isotropic scattering. The isotropic scattering problem is well developed for wide beams, while for narrow ones progress in the analytic theory has been poor, even in such a simple case. Nevertheless, some possibilities do exist and are recommended in [4.10, 21].

For media with $1-g \ll 1$, provided that $(1-\Lambda)/\Lambda(1-g) \gg 1$ or $\Lambda(1-g)\tau < 1$ is obeyed, backscattering may be estimated to a quasi-single scattering approximation using the known single scattering formulas for a narrow beam [4.8, 9]. Thus, in [4.86], formulas have been suggested for backscattering in the sea at different distances from the axis of a beam emitted by a collimated point source. Unfortunately, one must pay dearly for a simple solution to the quasi-single scattering approximation by neglecting the effect of the narrow-beam spread on the light field beyond the small-angle region.

The consistent account of this effect in the methods described in Sect. 3.6.4 offers more exact results. Mathematically, this approach consists of a SAA solution of the system of equations (3.6.13, 14). We confine ourselves to one iteration, and take $\kappa = 1$ in (3.6.12–13). This means that only the photons single-scattered over large angles are allowed for, as is the case with the quasi-single scattering approximation. In [4.87], expressions are obtained to this approximation for the radiance reflected from a medium illuminated by a narrow-beam collimated source. In the general case, however, the solutions are expressed in terms of multiple integrals and are of little use for design calculations. In some important particular cases, they essentially reduce. Let the receiver and source axes be parallel, the receiver of the area $\Sigma_{\text{rec}}$ having a sufficiently large angular aperture (much in excess of the source aperture). Then, the recorded backscattered flux $P(r)$ is

$$P(r) = \frac{P_0 \sigma x(\pi) \Sigma_{\text{rec}} \Omega_{\text{rec}}}{2\pi} \int_0^\infty S(2r, z, \Omega_{\text{s}} = \Omega_{\text{rec}}, \Sigma_{\text{s}} = 4\Sigma_{\text{rec}})\, dz \,, \tag{4.6.1}$$

where $S(r, z, \Omega_{\text{s}}, \Sigma_{\text{s}})$ is the irradiance originated at a depth $z$ and $r$ units away from the beam axis due to a unit power source with a solid angle $\Omega_{\text{s}}$ and cross section $\Sigma_{\text{s}}$; $x(\pi)$ is the indicatrix in the backward direction. As used in image transfer problems and vision system calculations, the quantity $P(t)$ denotes the backscattering noise (BSN). Using the simple SADA solutions (4.5.8, 20, 21), formula (4.6.1) can be successful in calculating the BSN power dependent on the medium optical parameters and receiver-source system characteristics, as well as on the distance $r$ between the receiver and the source. The simplification is most essential in the case of pulse sources. These solutions and the quasi-single scattering BSN approximation in the case of a pulsed source will be considered in Sect. 4.8.

## 4.7 Generalized Multiple Scattering Theory Parameters and Applicability of Approximate Solutions

In this and previous chapters, we have considered different approximate methods to describe light fields in media with essentially extended scattering indicatrices. Each of them has its own system of optical medium characteristics expressed in terms of the single scattering albedo $\Lambda$, extinction coefficient $\varepsilon$ and some integrals of the scattering indicatrix $x(\beta)$ (asymmetry parameter $g$, mean square scattering angle $\beta_2$, the fraction of backscattered light $1 - F$, etc.). The diversity of these parameters obscures the picture and impedes the comparison and matching of the results in different approximations as well as the evaluation of the applicability of the approximate technique. In this section, we show that all these characteristics are expressed only by two generalized scattering medium parameters, one defining the regions of weak and strong absorption, the other specifying those of small and large optical medium sizes.

We introduce the specific absorption $\alpha = (1 - \Lambda_{eq})/\Lambda_{eq}$ and scattering thickness $\zeta = \Lambda_{eq}\varepsilon_{eq}z$ of an equivalent medium in a transport approximation (Sect. 3.6.2). As follows from (3.6.5),

$$\alpha = \frac{1-\Lambda}{\Lambda(1-g)}, \qquad \zeta = \Lambda(1-g)\tau . \tag{4.7.1}$$

The absorption thickness is the same for the real and equivalent media, i.e., $\kappa z = (1 - \Lambda)\tau = (1 - \Lambda_{eq})\tau_{eq} = \alpha\zeta$.

1. The optical parameters of scattering media appear in the asymptotic solutions (3.5.5, 6) for the reflection and transmission coefficient of a thick weakly absorbing layer only in terms of $4q\gamma$ and $\gamma\tau_0$. Taking into account (3.2.11, 12), we shall write down

$$4q\gamma = \frac{4}{\sqrt{3}}\sqrt{\frac{1-\Lambda}{1-g}}, \qquad \gamma\tau_0 = \sqrt{3}(1-g)\tau_0\sqrt{\frac{1-\Lambda}{1-g}} . \tag{4.7.2}$$

We emphasize that the parameter $4q\gamma$ defines the most important characteristics of scattering media, such as the semi-infinite layer reflection coefficient (3.3.13) and asymptotic angular radiance distribution in a semi-infinite medium at $\tau \gg 1$ (3.2.13). In the asymptotic theory, the parameters $(1 - \Lambda)/(1 - g)$ and $(1 - g)\tau$ are equal to the quantities $\alpha$ and $\zeta$ accurate to terms of order $(1 - \Lambda)$, i.e., we have

$$4q\gamma = \sqrt{\alpha/3}, \qquad \gamma\tau_0 = \zeta_0\sqrt{3\alpha},$$
$$\gamma(\tau_0 + 4q) = (\zeta_0 + 4/3)\sqrt{3\alpha} . \tag{4.7.3}$$

The "weak absorption" concept is very important in transfer theory. As noted in Chaps. 3 and 4, the range of the weak absorption must be specified by some parameter which combines the medium absorption and the extension of the scattering indicatrix. The quantity $\alpha$ is exactly this parameter. It also complies with the physics of $\alpha$ as the specific absorption of the equivalent medium. In Sect. 4.5.6, we described the weak absorption range conventionally by the inequality $\alpha \lesssim 1$, which is roughly $\Lambda \gtrsim g$ for essentially extended indicatrices.

An analysis of numerical calculations demonstrates that at $\Lambda = 1$, the layer with isotropic scattering can be regarded as thick even at $\tau_0 \gtrsim 1$ to 2. This agrees with the estimate $\tau_0 > 2d$ ($d$ being the extrapolated length) which follows from the diffusion theory (for a nonabsorbing isotropically scattering medium, $d \approx 2/3$). Therefore, the quantity $\zeta_0 + 4/3 = \zeta_0 + 2d$ in the relation (4.7.3) is the layer thickness within the extrapolated boundaries, whereas the "thick layer" concept can be evaluated by the condition $\zeta_0 > 4/3$, i.e., $\tau_0 > 4q$.

2. The parameters $4q\gamma$ and $\gamma\tau_0$, and hence, $\alpha$ and $\zeta_0$ are essentially the diffusion approximation parameters. Now, consider, for example, the mean square deviation from the beam axis (PSF variance) at large $\tau_0$ (Table 4.1).

$$\frac{V_\rho}{\tau_0^2} = \frac{(\gamma\tau_0 + 4q\gamma)\operatorname{cotanh}(\gamma\tau_0 + 4q\gamma) - 5q\gamma\operatorname{cotanh}(5q\gamma)}{(\gamma\tau_0)^2} . \tag{4.7.4}$$

It is exactly this quantity, $V_\rho/\tau_0^2$, i.e., the variance $V_\rho$ in the angular measure, which is defined in terms of $\gamma\tau_0$ and $4q\gamma$. This conforms to the similarity theorem which imply that the angular characteristics are invariant against the variations of the extinction coefficient $\varepsilon$ (Sect. 2.4). Similarly, the parameters $4q\gamma$ and $\gamma\tau_0$, and hence, $\alpha$ and $\zeta_0$ may appear in the DA formulas for other quantities which are invariant to $\varepsilon$.

3. In a similar way, the two-stream approximation solutions (Sect. 3.7.3) depend on the parameters $(1 - \Lambda)/\Lambda\tilde{\varphi}$ and $\gamma\tau$, the quantity $\gamma$ being defined by formula (3.7.24). In the version of the two-stream approximation with (3.7.30),

$$(1 - \Lambda)/\Lambda\tilde{\varphi} = 8\alpha/3 , \qquad \gamma\tau = \zeta\sqrt{3\alpha + 4\alpha^2} .$$

4. Within the small-angle and small-angle diffusion approximations (Sects. 4.4, 5), the radiant fluxes, as well as the variances of the angular and spatial (in the angular measure) distributions, are functions only of the quantities $\xi = \tau\sqrt{\Lambda(1 - \Lambda)\beta_2/2}$ and $(1 - \Lambda)\tau$. For essentially extended indicatrices, when $\beta_2/2 = 1 - g$ (the case for which small-angle procedures were originally developed), we have

$$\xi = \tau\sqrt{\Lambda(1 - \Lambda)(1 - g)} = \zeta\sqrt{\alpha} , \qquad (1 - \Lambda)\tau = \alpha\zeta . \tag{4.7.5}$$

Again, the integral light field characteristics are expressed in terms of $\alpha$ and $\zeta$.

It should be noted that the relation $\beta_2/2 = 1 - g$ holds only when scattering at large angles can be neglected. This condition is obeyed, for example, for the scattering of fast charged particles and, though less satisfactorily, when optical radiation scatters off very soft large scatterers. For real geophysical media (sea water, clouds, etc.), in spite of essential indicatrix extension, the portion of the light scattered at large angles is not small (for example, $(1/2)\int_{\pi/4}^{\pi} x(\beta)\sin\beta\, d\beta = 0.1 - 0.25$). In line with the physics of small-angle approximations, we must regard such radiation as absorbed (it does not contribute to the light field in the small-angle region) and, correspondingly, redetermine the elementary volume parameters. In view of this, we introduce (Sect. 4.4.2) an effective medium with elementary volume characteristics (4.4.6–8) dependent on the angle $\beta_0$. Here,

$$\beta_2' = \frac{\displaystyle\int_0^{\beta_0} x(\beta)\beta^3\, d\beta}{\displaystyle\int_0^{\beta_0} x(\beta)\beta\, d\beta} . \tag{4.7.6}$$

There is an arbitrary element in the choice of the angle $\beta_0$. For better agreement between the SADA and exact solutions to be achieved, the quantity $\beta_0$ must depend not only on $\tau$, but also on which of the light characteristics is being calculated (see, for example, [4.88]). It shows different SAA limitations for media with optical indicatrices. Nevertheless, taking into account that the relations $\sin\beta \approx \beta$, $\cos\beta \approx 1 - \beta^2/2$ hold satisfactorily at $\beta \lesssim 40$–$45°$, we may assume that $\beta_0 \approx 40$–$45°$. This recommendation is given in [4.89] to calculate light fields in sea water.

It is still a debatable question whether it is possible to consider the quantities $V_\rho$ and $V_\theta$, calculated for the effective medium, as the true variances of the $\rho$ and $\theta$ distributions in the real medium. The problem is that the tails of the $\rho$ and $\theta$ light field distributions, which are incompletely allowed for by the effective parameters in (4.4.6–8), may essentially contribute to the true variances $V_\rho$ and $V_\theta$. It seems to be more correct to regard the quantities $V_\rho$ and $V_\theta$ as specifying the width of the appropriate distributions at the level $\exp(-1)$.

Matching different approximate solutions (for example, in the DA and SADA) requires relating $F_{\beta_0}$ and $\beta_2'$, calculated by the angle range $0 \leqslant \beta \leqslant \beta_0$, to the indicatrix parameter $g$. An analysis of $F_{\beta_0}$ and $\beta_2'$ computed for a number of scattering indicatrices of sea and ocean water by the data of [4.90] and the cloud models [4.82] has shown that at $\beta_0 = 45°$, the correlation

$$1 - F_{\beta_0} \approx 2(1-g)/3\,, \qquad \beta_2' \approx (1-g)/2 \tag{4.7.7}$$

holds well. More accurate correlations between $1 - F_{\beta_0}$ and $g$ for the scattering indicatrices of sea and ocean water at $\beta_0 = 45°$ are given in [4.89]. At $\beta_0 = 90°$ ($F_{\beta_0} \equiv F$),

$$1 - F \approx (1-g)/3\,, \qquad \beta_2' \approx 1 - g\,. \tag{4.7.8}$$

The accuracy of (4.7.8′) is illustrated by the data in Fig. 4.14. The proportion between $\beta_2'$ and $1 - g$ allows the system of the generalized parameters $\alpha$ and $\xi$ to be valid even upon reevaluation of the optical parameters in the SADA and SAA solutions. It only changes the numerical coefficients in the SADA and SAA formulas for the integral light field characteristics expressed in terms of the generalized parameters.

5. We now consider the quasi-single scattering approximation (Sect. 3.6.3), which is commonly used to define the reflection characteristics of a strongly absorbing medium, or of a comparatively thin scattering layer [4.91, 92]. To this approximation, the radiance coefficient of the light reflected by a $\tau_0$-thick layer is

$$r(\tau_0, \mu, \mu_0, \psi) = \frac{\Lambda x(\beta_0)}{4(\mu + \mu_0)(1 - \Lambda F)}\left[1 - e^{-\tau_0(1-\Lambda F)(1/\mu + 1/\mu_0)}\right]. \tag{4.7.9}$$

Let us introduce the mean single-backscattering probability $\langle x \rangle = \int_{-1}^{0} x(\cos\beta)d(\cos\beta)$ and $\tilde{x}(\beta) = x(\beta)/\langle x \rangle$. Then, taking into account (4.7.8′),

$$\langle x \rangle = 2(1-F) \approx 2(1-g)/3\,, \qquad x(\beta) \approx 2(1-g)\tilde{x}(\beta)/3\,,$$

$$1 - \Lambda F = (1-\Lambda) + \Lambda(1-F) = (\alpha + 1/3)\Lambda(1-g)$$

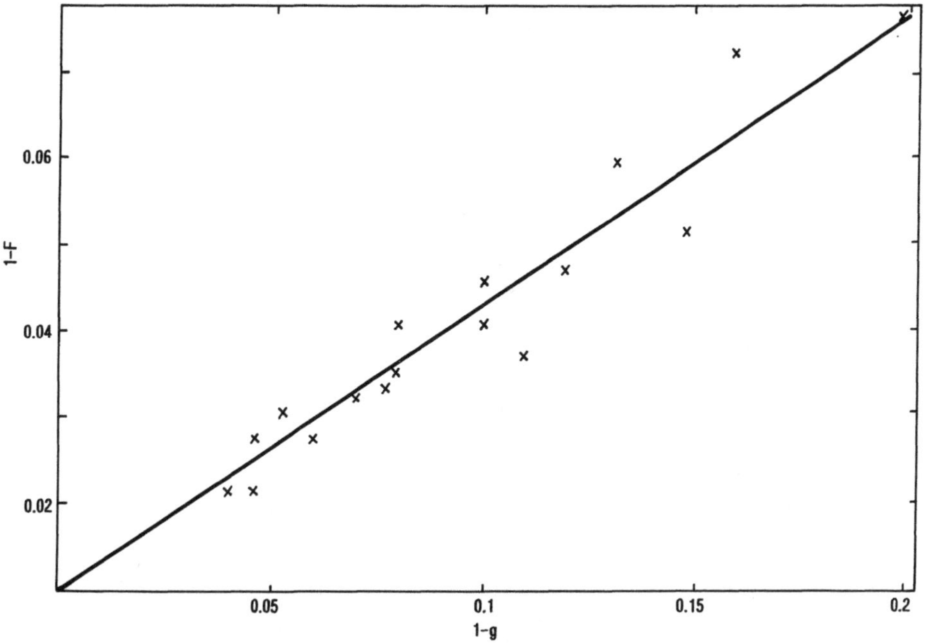


**Fig. 4.14.** Relationship between $1 - F$ and $1 - g$ for real sea water and cloud scattering indicatrices, shown by the *crosses*. The *line* shows $1 - F = (1 - g)/3$

and, instead of (4.7.9), we have

$$r(\tau_0, \mu, \mu_0, \psi) = \frac{\tilde{x}(\beta_0)}{(\mu + \mu_0)(6\alpha + 2)}[1 - e^{-\zeta_0(\alpha+1/3)(1/\mu+1/\mu_0)}] . \tag{4.7.10}$$

In the back semi-sphere, ($\beta > \pi/2$) $\tilde{x}(\beta) \approx 1$. Therefore, (4.7.10) implies that the reflected radiance is also expressed in terms of the generalized parameters $\alpha$ and $\zeta$. For a thin layer, from (4.7.10), we have

$$r(\tau_0, \mu, \mu_0, \psi) \approx \zeta_0/6\mu\mu_0 . \tag{4.7.11}$$

Correspondingly, for the reflection coefficient of diffuse light (3.4.2), we approximately have

$$r \approx 2\zeta_0/3 . \tag{4.7.12}$$

For $\mu$, $\mu_0 \gtrsim 0.5$, the transition from (4.7.10) to (4.7.11, 12) is in error by no more than 10 to 15% at $(\alpha + 1/3)\zeta_0 \lesssim 0.1$. For media with essential absorption as well as for weakly absorbing media, this results in the following thin layer conditions:

$$\begin{aligned} &\alpha\zeta_0 \lesssim 0.1 , \quad \text{i.e., } \tau_0 \lesssim 0.1/(1 - \Lambda) \quad \text{at } \alpha \gtrsim 1 \\ &\zeta_0/3 \lesssim 0.1 , \quad \text{i.e., } \tau_0 \lesssim 0.3/(1 - g) \quad \text{at } \alpha \ll 1 . \end{aligned} \tag{4.7.13}$$

**Table 4.5.** Generalized parameters $\alpha = (1 - \Lambda)/\Lambda(1 - g)$ and $\zeta = \Lambda(1 - g)\tau$ related to transfer theory approximation parameters[a].

| Approximation | Approximation parameters | In terms of $\alpha$ and $\zeta$ |
|---|---|---|
| Transport | $\Lambda_{eq} = \dfrac{\Lambda(1-g)}{1-\Lambda g}$ | $\dfrac{1-\Lambda_{eq}}{\Lambda_{eq}} = \alpha$ |
| | $\tau_{eq} = (1 - \Lambda g)\tau$ | $\Lambda_{eq}\tau_{eq} = \zeta$ |
| Diffusion, asymptotic solutions | $4q\gamma = 4\sqrt{\dfrac{1-\Lambda}{3(1-g)}}$ | $4q\gamma = 4\sqrt{\alpha/3}$ |
| | $\gamma\tau = \tau\sqrt{3(1-\Lambda)(1-g)}$ | $\gamma\tau = \zeta\sqrt{3\alpha}$ |
| Two-stream | $\dfrac{1-\Lambda}{\Lambda\tilde{\varphi}}$ | $\dfrac{1-\Lambda}{\Lambda\tilde{\varphi}} = \dfrac{8}{3}\alpha$ |
| | $\gamma\tau = \dfrac{\Lambda\tilde{\varphi}}{\tilde{\mu}}\sqrt{\left(\dfrac{1-\Lambda}{\Lambda\tilde{\varphi}}\right)^2 + 2\dfrac{1-\Lambda}{\Lambda\tilde{\varphi}}}$ | $\gamma\tau = \zeta\sqrt{3\alpha + 4\alpha^2}$ |
| Small angle (SAA, SADA) | $\xi = \sqrt{\Lambda(1-\Lambda)\beta_2/2}$ | $\xi = \zeta\sqrt{\alpha}$ |
| | $(1-\Lambda)\tau$ | $(1-\Lambda)\tau = \alpha\zeta$ |
| Quasi-single scattering | $\dfrac{\Lambda(1-F)}{1-\Lambda F}$ | $\dfrac{\Lambda(1-F)}{1-\Lambda F} = \dfrac{1}{1+3\alpha}$ |
| | $(1-\Lambda F)\tau$ | $(1-\Lambda F)\tau = \left(\alpha + \dfrac{1}{3}\right)\zeta$ |

[a] We set in SAA and SADA $\beta_2/2 = 1 - g$; in the quasi-single scattering approximation $1 - F = (1 - g)/3$; in the two-stream approximation $\tilde{\varphi} = (3 - x_1)/8$, $\tilde{\mu} = 0.5$.

Compare the last condition with a similar evaluation gained from the asymptotic (3.5.5). For diffuse light, (3.5.5) can be applied even at moderate $\tau_0$. At $\alpha \ll 1$ and $\tau_0/4q \lesssim 0.2$ (thin layer condition), from (3.5.5) we have

$$r \approx 0.75\tau_0(1 - g) = 0.75\zeta_0 \qquad \text{at } \tau_0 \lesssim \frac{0.3}{1 - g} . \tag{4.7.14}$$

The formulas (4.7.12–14), which were derived using contrasting approaches, are similar.

Finally, the integral light field characteristics obtained to different approximations are expressed in terms of the generalized parameters $\alpha$ and $\zeta$. The relations for $\alpha$ and $\zeta$ and the optical medium characteristics which appear in the various approximations are listed in Table 4.5.

## 4.8 Nonstationary Radiation Field from Localized Pulsed Sources

We have so far considered light fields originated by stationary radiation sources. The use of pulsed lasers in vision, location and information transfer systems frequently requires a knowledge of a nonstationary light field in a scattering medium.

The nonstationary radiation field theory was originally developed for problems in astrophysics [4.93]. Here, as well as in the radiative transfer theory as a whole, simpler problems of a medium irradiated by infinitely wide sources have been solved in the first instance. The main results have been generalized in [4.94–96].

Given the nonstationary light field originated by a localized light source, we can solve the applied problems. These are the questions which occupy us in this section.

### 4.8.1 The Nonstationary Transfer Equation

We take it that the photon residence time in a medium is determined by the photon mean free time $t_2 = 1/\varepsilon v$ ($v$ being the light velocity in the medium) between two scattering events. In the theory of nonstationary scattering, the characteristic time $t_1$ of a photon in an absorbed state is also considered [4.95, 97]. However, as far as the problems of interest are concerned, we may assume that $t_1 = 0$.

If the time of essential changes in the light source intensity is much in excess of the mean photon residence time in the medium, then radiative equilibrium develops at each medium point, that is, the field "follows" the source power variations and is quasi-stationary. Otherwise, the light field appears to be essentially nonstationary and is determined by solving the nonstationary transfer equation

$$\left[\frac{\partial}{\partial u} + \vec{n}\cdot\nabla_{\vec{\tau}} + \varepsilon\right] I(\vec{\tau},\vec{n},u) = \frac{\Lambda}{4\pi}\iint_{4\pi} I(\vec{\tau},\vec{n}',u)x(\vec{n},\vec{n}')\,d\vec{n}' + B'(\vec{\tau},\vec{n},u)\,. \tag{4.8.1}$$

Here, we have introduced the dimensionless time $u = t/t_2 = \varepsilon v t$.

Basically, the nonstationary light scattering problem consists of determining the medium response $I_\delta(\vec{\tau},\vec{n},u)$ to an instantaneous excitation $B'(\vec{\tau},\vec{n},u) = B'(\vec{\tau},\vec{n})\delta(u)$, where $\delta(u)$ is the Dirac function. For the other source function $B'(\vec{\tau},\vec{n},u) = B'(\vec{\tau},\vec{n})I_0(u)$, the radiance may be found by integration,

$$I(\vec{\tau},\vec{n},u) = \int_0^u I_\delta(\tau,\vec{n},u')I_0(u-u')\,du'\,. \tag{4.8.2}$$

The mathematician would prefer to term $I_\delta(\vec{\tau},\vec{n},u)$ the Green's function of the nonstationary transfer equation; the researcher dealing with applied problems would call it the propagating pulse spread function, whereas the physicist would refer to it as the distribution of the photon time-of-arrival fluctuations, or the photon path distribution function. The physical terms reflect the mechanism of the phenomenon: fluctuations of the photon time-of-arrival at a point of observation (photon path length), due to multiple scattering.

It is important that in the majority of cases, it is sufficient to determine the Green's function $I_\delta(\vec{\tau},\vec{n},u)$ only for the conservative-scattering medium, since the radiances in the absorbing $I_\delta(\vec{\tau}, \vec{n}, u, \Lambda)$ and nonabsorbing $I_\delta(\vec{\tau}, \vec{n}, u, 1)$ media are simply related [4.96, 97]:

$$I_\delta(\vec{\tau}, \vec{n}, u, \Lambda) = \Lambda e^{-(1-\Lambda)u} I_\delta(\Lambda\vec{\tau}, \vec{n}, \Lambda u, \Lambda = 1) . \quad (4.8.3)$$

**Laplace Transformation.** Take the Laplace transform of (4.8.1) with respect to the time $u$. Then, the Laplace transform of the radiance

$$I_\delta(\vec{\tau}, \vec{n}, p) = \int_0^\infty I_\delta(\vec{\tau}, \vec{n}, u) e^{-pu} du . \quad (4.8.4)$$

obeys the equation which is similar to (2.4.1) upon substitution

$$\tau \to \tau(1 + p) , \qquad \Lambda \to \frac{\Lambda}{1 + p} . \quad (4.8.5)$$

Therefore, without solving (4.8.1) for each particular case once again, we can easily find the Laplace transform $I_\delta(\vec{\tau}, \vec{n}, p)$. For this purpose, in the solution $I(\vec{\tau}, \vec{n})$ of the corresponding stationary problem, it is sufficient to make the substitutions (4.8.5) [4.95, 98] or, which amounts to the same,

$$\varepsilon \to \varepsilon(1 + p) . \quad (4.8.5')$$

In order to obtain the final solution, it only remains to take the inverse Laplace transform. For the substitution (4.8.5′), we must take into account that $\vec{\tau} = \varepsilon\vec{R}$, $\Lambda = \sigma/\varepsilon$.

The Laplace transformation is most often used to solve the problems of pulse propagation.

However, the complete solution for describing a nonstationary radiation field for any time instants $u$ was not successful. This is attributable to the absence of exact analytic solutions to the stationary problem which are valid within the entire range of optical medium parameters and, in particular, for all $\Lambda$. Nevertheless, the known properties of the Laplace transform suggest some important results in the theory of nonstationary light scattering [4.95] and, for example, asymptotic solutions for $u \ll 1$ and $u \gg 1$. It is very important that asymptotic transfer theory formulas can be used in this case. Thus, in order to find a nonstationary light field $u \gg 1$, it suffices to know the function $I_\delta(\vec{\tau}, \vec{n}, p)$ at $p \ll 1$ and, hence, the function $I(\vec{\tau}, \vec{n})$ for $1 - \Lambda \ll 1$ [4.12, 26]. In this way, a large number of results, later generalized in [4.94, 95], have been obtained in the theory of the nonstationary scattering of infinitely wide light beams.

**Method of Moments.** A different approach to studying nonstationary light fields is possible [4.99]. It is frequently sufficient to know only some integral characteristics of a pulse time spread. These may be time moments of the function $I_\delta(\vec{\tau}, \vec{n}, u)$

$$M^\delta_{u,k} = \int_0^\infty u^k I_\delta(u) du , \quad (4.8.6)$$

which may be expressed in a straightforward manner in terms of $I(\vec{\tau},\vec{n})$. Allowing for (4.8.4–6), we find

$$M_{u,k}^{\delta} = (-1)^k \lim_{p\to 0} \frac{d^k I_\delta(\vec{\tau},\vec{n},p)}{dp^k} = (-1)^k \varepsilon^n \frac{d^k I(\vec{\tau},\vec{n})}{d\varepsilon^k}\,. \tag{4.8.7}$$

In differentiating with respect to $\varepsilon$, we should remember that $\vec{\tau} = \varepsilon\vec{R}$, $\Lambda = \sigma/\varepsilon$.

Thus, the time integral characteristics of the time spread function $I_\delta(\vec{\tau},\vec{n},u)$ are expressed using the solution of an appropriate stationary problem, stationary solution differentiation being sufficient to determine $M_{u,k}^{\delta}$. It is worth emphasizing that information regarding the appropriate stationary problem solution for a comparatively narrow range of optical parameters is often quite enough to allow one to find the most important first integral moments determined by low-order derivatives. With this in mind, we can use the approximate and asymptotic stationary solutions discussed, for example, in this chapter, as well as the experimental data for stationary sources.

Knowing that $M_{u,k}^{\delta}$ at $k \leqslant 2$ gives, in particular, such important nonstationary light scattering characteristics as the mean time $\langle u_\delta \rangle$ and variance $V_u^\delta$ of photon paths to the given point $\vec{\tau}$ in the $\vec{n}$ direction

$$\langle u_\delta \rangle = \frac{M_{u,1}^{\delta}}{M_{u,0}^{\delta}} = -\varepsilon \frac{d \ln I(\vec{\tau},\vec{n})}{d\varepsilon}\,, \tag{4.8.8}$$

$$V_u^\delta = \frac{M_{u,2}^{\delta}}{M_{u,0}^{\delta}} - \left(\frac{M_{u,1}^{\delta}}{M_{u,0}^{\delta}}\right)^2 = \varepsilon^2 \frac{d^2 \ln I(\vec{\tau},\vec{n})}{d\varepsilon^2}\,. \tag{4.8.9}$$

If the scattering medium is irradiated by a pulse $I_0(u)$ of arbitrary shape and duration, then the mean time $\langle u \rangle$ and the variance $V_u$ in the medium are

$$\langle u \rangle = \langle u_0 \rangle + \langle u_\delta \rangle\,, \qquad V_u = V_u^0 + V_u^\delta\,. \tag{4.8.10}$$

Here, $\langle u_0 \rangle$ and $V_u^0$ are the mean time and variance of $I_0(u)$. Below, we shall confine our consideration to the Green's function, i.e., medium response to a $\delta$-pulse. Therefore, for brevity, we omit $\delta$.

The entire set of the moments $M_{u,k}$ being known, we can reconstruct the function $I(\vec{\tau},\vec{n},u)$. If there is some a priori information regarding the pulse time spread, then it can be reconstructed approximately only in terms of the first moments [4.100]. It seems that the reconstruction may be made more complete by using, as additional information, the asymptotic solutions at $u \to 0$ and $u \to \infty$ derived, for instance, in terms of the asymptotic Laplace transform.

Thus, the appropriate approximate and asymptotic solutions to the stationary transfer equation, as well as the procedures mentioned above (Laplace transformation or method of moments), provide the required data on the nonstationary light field for the majority of situations.

One further comment on the applicability of the nonstationary light field solutions is important. The approximate nature of the stationary solutions used imposes constraints on the time range where the appropriate nonstationary solutions can be used. We will clarify this statement. As already noted, the nonstationary light field characteristics at $u \ll 1$ are determined by the corresponding stationary solution at $\Lambda \to 0$. On the other hand, the light field in a semi-infinite medium and very thick optical layers at $u \gg 1$ is found from the stationary solution in the vicinity of the point $\Lambda = 1$ [4.26]. Therefore, we can correctly describe the temporal radiance attenuation at large time intervals if the initial stationary solutions are applicable for small absorptions. For this reason, for example, the diffusion approximation can be used here. For the time range at which the major portion of the pulse energy falls, the nonstationary solutions are always correct in the same situations for which the initial stationary solution is valid. When speaking about "the same situation", we assume that in both cases the optical and geometric parameters of the medium and source-receiver system are the same, and that the initial stationary and subsequent nonstationary solutions correspond to stationary and pulsed sources. These remarks provide an a priori judgement regarding the nonstationary field data which can be derived from one or other approximate solution to the stationary problem.

### 4.8.2 Pulse Propagation in Optically Thick Media

Consider a nonstationary light field generated by a pulsed source in a strongly scattering semi-infinite medium or optically thick layer. We shall rely upon the solutions to the stationary diffusion equation presented in Sect. 4.3.2.

For the radiation density generated at the point $(\vec{\rho}, \tau)$ of the semi-infinite medium by the diffuse point source located at the point $(\rho = 0, \tau')$, we have

$$w(\vec{\rho}, \tau, \tau') = \frac{\varepsilon^2}{4\pi q}\left[\frac{e^{-\gamma\sqrt{(\tau-\tau')^2+\rho^2}}}{\sqrt{(\tau-\tau')^2+\rho^2}} - \frac{e^{-\gamma\sqrt{(\tau+\tau'+2d_1)^2+\rho^2}}}{\sqrt{(\tau+\tau'+2d_1)^2+\rho^2}}\right]. \tag{4.8.11}$$

Formula (4.8.11) is (4.3.19) at $\tau_0 \to \infty$. Using the substitutions (4.8.5) in (4.8.11) yields the Laplace transform $w(\vec{\rho}, \tau, \tau', p)$. Inverting the Laplace transform, for the time dependence of the radiation density at $\Lambda = 1$ and $u \gg 1$, we obtain [4.101]

$$w(\vec{\rho}, \tau, \tau', u) = \frac{\varepsilon^2 e^{-\rho^2/4qu}}{8\sqrt{q\pi^3}\,u^{3/2}}\left[e^{-(\tau-\tau')^2/4qu} - e^{-(\tau+\tau'+2d_1)^2/4qu}\right]. \tag{4.8.12}$$

At $u \gg [(\tau + \tau' + 2d_1)^2 + \rho^2]/4q$, from (4.8.12), we have

$$w(\vec{\rho}, \tau, \tau', u) = \frac{\varepsilon^2(\tau + d_1)(\tau' + d_1)}{8q(\pi q)^{3/2} u^{5/2}}. \tag{4.8.13}$$

Formula (4.8.13) describes the asymptotic radiation damping in a semi-infinite nonabsorbing medium illuminated by a diffuse pulsed point source. The radiation density at $\Lambda \neq 1$ may be found from (4.8.3).

The solution, similar to (4.8.12), for the temporal distribution of the radiation density from a point source in a finite-thickness layer is [4.26, 102]

$$w(\vec{\rho}, \tau, \tau', \tau_0, u) = \frac{\varepsilon^2}{2\pi qu(\tau_0 + d_1 + d_2)} e^{-\rho^2/4qu} \times \sum_{k=1}^{\infty} e^{-k^2 vu} \sin\left(\frac{\tau + d_1}{\tau_0 + d_1 + d_2} k\pi\right) \sin\left(\frac{\tau' + d_1}{\tau_0 + d_1 + d_2} k\pi\right), \tag{4.8.14}$$

where

$$v = \frac{\pi^2 q}{(\tau_0 + d_1 + d_2)^2}. \tag{4.8.15}$$

When there is no reflection at the layer boundaries, $d_1 = d_2 = 2q$. In the layer with reflecting boundaries, $d_1$ and $d_2$ are specified by formula (4.3.14). For large $u$, we may restrict ourselves to one term in (4.8.14). Thus, the asymptotic radiation density variation in the layer will be

$$w(\vec{\rho}, \tau, \tau', \tau_0, u) = \frac{\varepsilon^2 e^{-vu-\rho^2/4qu}}{2\pi q(\tau_0 + d_1 + d_2)u} \sin\left(\frac{\tau + d_1}{\tau_0 + d_1 + d_2} \pi\right) \times \sin\left(\frac{\tau' + d_1}{\tau_0 + d_1 + d_2} \pi\right). \tag{4.8.16}$$

For the time instants $u \ll 2\tau_0 - \tau' - \tau$ with no boundary $\tau = \tau_0$ effect, the radiation density for a semi-infinite medium and finite-thickness layer will be the same. It is shown in [4.26, 102] that the entire $w(u)$ curve for a finite-thickness layer can be calculated in terms of the solution (4.8.12) for a semi-infinite medium at comparatively small time instants, and of the asymptotic (4.8.16) at large time instants.

We can analyse the time variation of the radiation density at different medium points using (4.8.14–16). As seen from (4.8.14), the reflection at the boundaries with a small albedo $A$ affects the radiation density as the equivalent increase of the optical layer thickness by the quantity $4qA/(1 - A)$.

The above solutions are the Green's function for the boundary-value problems under consideration. Therefore, for a monodirectional source, the solution can be obtained by integrating the above solutions for an isotropic point source. For a finite-thickness layer illuminated normal to the surface by a monodirectional pulse source of unit power, it is of the form

$$w(\vec{\rho}, \tau, \tau_0, u) = 3q \int_0^{\tau_0} w(\vec{\rho}, \tau, \tau', \tau_0, u - \tau') e^{-\tau'} d\tau' - x_1 q[w(\vec{\rho}, \tau, \tau' = 0, \tau_0, u) - w(\vec{\rho}, \tau, \tau' = \tau_0, u - \tau_0) e^{-\tau_0}]. \tag{4.8.17}$$

Assuming that $\tau_0 \to \infty$ in (4.8.17), we obtain the solution for a semi-infinite medium.

The radiation density expressions are particularly simple in the asymptotic regime at $u \to \infty$. For example, for a semi-infinite medium illuminated by a monodirectional pulsed point source located at the upper medium boundary, from (4.8.13, 17), we have

$$w(\vec{\tau}, u) = \frac{5\varepsilon^2(\tau + 2q)}{8(\pi q)^{3/2} u^{5/2}} e^{-\rho^2/4qu}, \qquad u \to \infty . \tag{4.8.18}$$

Knowing the radiation density, the irradiances $S_1$ and $S_2$ can easily be found using (4.3.16). To be specific, from (4.8.18) and (4.3.16), we find

$$S_i(\vec{\tau}, u) = \frac{5\varepsilon^2 e^{-\rho^2/4qu}}{32(\pi q)^{3/2} u^{5/2}} [\tau + 2q + (-1)^i 2q] . \tag{4.8.19}$$

The time variation of the radial downward flux distribution (the point spread function) is of key interest in vision problems. It follows from (4.8.19) that at $u \to \infty$, the irradiance distribution is normal along the $\rho$ coordinate. The distribution variance $2qu$ increases linearly with time. The same variance for $\tau \gg 1$ has been obtained more exactly in [4.103].

Allowing for the general light field properties in the asymptotic regime at $u \to \infty$ [4.26, 104] provides a comprehensive picture of the nonstationary radiation field, i.e., the time variation of the angular radiance distribution at any medium point with an arbitrary direction for the incident beam. Thus, if a monodirectional pulsed source illuminates a semi-infinite medium at an angle $\arccos \mu_0$ to the layer normal, then the radiance in the medium is described by [4.26]

$$I_n(\vec{\tau}, u) = \frac{5\varepsilon^2(\tau + 2q)e^{-\rho^2/4qu}}{16\pi u(\pi q u)^{3/2}} \frac{\mu_0 \Psi_M(\mu_0, 0)\Psi_M(-\mu, \tau)}{\Psi_M(1,0) \int_{-1}^{1} \Psi_M(\mu, \tau)\mu \, d\mu} . \tag{4.8.20}$$

Here, $\Psi_M(\mu, \tau)$ is the angular radiance distribution in a semi-infinite layer with no sources at $\Lambda = 1$ (solution of the conservative Milne problem [4.25]. At $\tau = 0$,

$$\Psi_M(\mu, \tau = 0) \equiv u_0(\mu) \approx \frac{3}{7}(1 + 2\mu) . \tag{4.8.21}$$

We have only cited here the asymptotic solutions at $u \gg 1$ since these are the most simple and instructive. If we do not restrict ourselves to the asymptotic Laplace transform at $p \ll 1$, we can find the solutions that are applicable in a wider $u$ range [4.101, 102]. Using the stationary solution of the transfer equation, formulas have been derived in [4.101] for the first three time moments to study the light field at the time close to the instant of maximum pulse spread.

More comprehensive studies of light fields from point sources of various angular divergency in semi-infinite media, finite-thickness layers and complex geometry media, as well as results as applied to problems of vision, pulse location and remote laser sensing can be found elsewhere [4.26, 101–109]. Later, these results were supplemented through a thorough examination of light fields at a moderate dimensionless times $u$ [4.110–112]. Moreover, in [4.112], the nonstationary DA solutions were compared with experiments [4.113] on laser pulse propagation in clouds and it was found that there is good agreement between them. Studies of light fields from narrow-beam pulsed sources are summarized briefly in [4.114].

In conclusion, we shall comment on the applicability of nonstationary DA solutions. It is known [4.21, 23] that the diffusion approximation as applied to stationary problems ensures exact solutions for the regions far from the radiation sources and sinks. In conformity with the closing remarks of Sect. 4.8.1, the nonstationary DA solutions for the same cases properly describe the variation of the light field characteristics for the time interval containing the main signal energy. In all cases, the accuracy of the solutions increases with $u$. For very thick optical layers, the DA exactly specifies the laws of asymptotic radiance attenuation with $u \to \infty$ [4.26, 101, 102]. The reasonable accuracy of the DA solutions was verified in a number of situations by comparison with experiment and the Monte-Carlo simulations [4.26, 107, 108, 112, 113].

### 4.8.3 Pulse Reflection from a Semi-Infinite Scattering Medium

Consider the BSN estimation in the case of a pulsed source. At $\tau = 0$, (4.8.20) yields an expression for the BSN power on a photodetector when allowance is made for the particular experimental geometry:

$$P(u) = \frac{5\varepsilon^2 W_0 \Sigma_{\mathrm{rec}} Q}{8\pi^{5/2} q^{1/2} u^{5/2}} \mathrm{e}^{-\rho^2/4qu}, \qquad u \to \infty, \tag{4.8.22}$$

where $W_0$ is the pulse energy and $\Sigma_{\mathrm{rec}}$ the receiving entrance pupil area;

$$Q = \frac{1}{\Omega_s u_0(1)} \iint_{2\pi} \mu u_0(\mu) \varphi_{\mathrm{rec}}(\vec{n})\, d\vec{n} \iint_{2\pi} \mu' u_0(\mu') \varphi_s(\vec{n}')\, d\vec{n}', \tag{4.8.23}$$

$\varphi_s(\vec{n})$ and $\varphi_{\mathrm{rec}}(\vec{n})$ are the angular directional diagrams of the source and receiver. The functions $\varphi_s(\vec{n})$ and $\varphi_{\mathrm{rec}}(\vec{n})$ satisfy the conditions $\varphi_s(\vec{n} = \vec{n}_s) = 1$, $\varphi_{\mathrm{rec}}(\vec{n} = \vec{n}_{\mathrm{rec}}) = 1$,

$$\iint_{2\pi} \varphi_s(\vec{n})\, d\vec{n} = \Omega_s, \qquad \iint_{2\pi} \varphi_{\mathrm{rec}}(\vec{n})\, d\vec{n} = \Omega_{\mathrm{rec}}, \tag{4.8.24}$$

where $\vec{n}_s$ and $\vec{n}_{\mathrm{rec}}$ are the source and receiver axis directions; $\Omega_s$, $\Omega_{\mathrm{rec}}$ the solid angles of the source and the receive. For the system with small angular aperture source and receiver, expression (4.8.23) including (4.8.21, 24), yields

$$Q = \frac{9}{7}\Omega_{\mathrm{rec}} = \frac{9}{7}\pi\gamma_{\mathrm{rec}}^2,$$

where $2\gamma_{\mathrm{rec}}$ is the receiver aperture angle. For the system with a collimated source and an irradiance receiver, $Q = \pi$.

Formula (4.8.22) is written for $\Lambda = 1$. Since, in practice, it is important to know the BSN, we shall give the quantity $P(t)$ for arbitrary medium absorption as a function of the real time $t$ and the geometric coordinates. Allowing for (4.8.3, 22), we arrive at

$$P(t) = \frac{5\sigma^3 v W_0 Q \Sigma_{\text{rec}}}{8\pi^{5/2} q^{1/2} (\sigma v t)^{5/2}} e^{-\kappa v t - \sigma r^2/4qvt}, \qquad t \to \infty , \tag{4.8.25}$$

where $v$ is the light velocity in the medium and $r$ is the geometric distance between the source and the receiver. It should be noted that at large $t$, the detected signal does not depend on the pulse shape and is determined only by the total pulse energy. The accuracy of (4.8.25) is illustrated by the data in Fig. 4.15a. The figure gives the reflected irradiance calculated by (4.8.25) and the Monte Carlo simulations. The vertical lines indicate the dispersion in the Monte Carlo simulations.

Formula (4.8.25) gives a light field at large $t$ for any medium parameters and any experimental geometry. At small $t$, similar characteristics can be computed to a single- or double-order scattering approximation [4.115, 116], and to a quasi-single or small-angle approximations for very extended indicatrices [4.86, 87]. The signal power at intermediate time instants can be found by interpolating between the above two cases [4.26, 107].

We shall illustrate this procedure for a particular situation (Fig. 4.15b). Let us plot the asymptotic curve $P^{\text{as}}(t)$ using (4.8.25) (curve 1) and the single-scattered

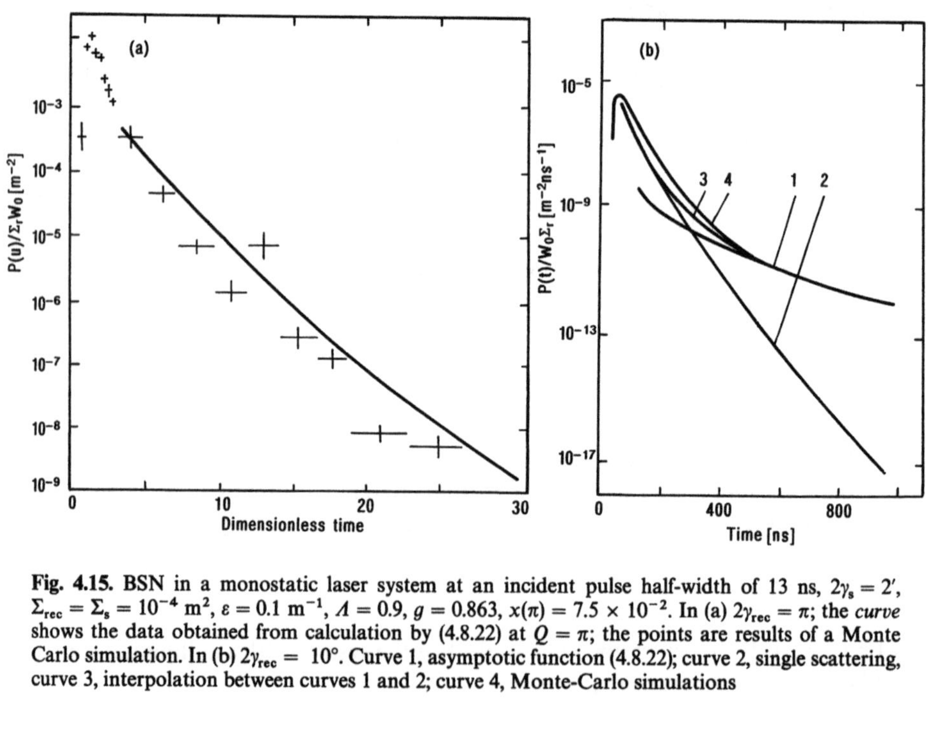


**Fig. 4.15.** BSN in a monostatic laser system at an incident pulse half-width of 13 ns, $2\gamma_s = 2'$, $\Sigma_{\text{rec}} = \Sigma_s = 10^{-4}$ m$^2$, $\varepsilon = 0.1$ m$^{-1}$, $\Lambda = 0.9$, $g = 0.863$, $x(\pi) = 7.5 \times 10^{-2}$. In (a) $2\gamma_{\text{rec}} = \pi$; the *curve* shows the data obtained from calculation by (4.8.22) at $Q = \pi$; the points are results of a Monte Carlo simulation. In (b) $2\gamma_{\text{rec}} = 10°$. Curve 1, asymptotic function (4.8.22); curve 2, single scattering, curve 3, interpolation between curves 1 and 2; curve 4, Monte-Carlo simulations

light power (curve 2). Consider a time instant $t_0$ when the curves 1 and 2 intersect. At this moment, the contributions of the multiple $P^{(m)}(t)$ and first-order $P^{(1)}(t)$ scattering are approximately equal. At $t \gg t_0$, the first-order scattering is small, and we may estimate the signal from the asymptotic curve. For $t < t_0$, when $P^{(1)}(t) \gg P^{(m)}(t)$, $P(t)$ is, on the contrary, a complete function of single scattering. At intermediate time instants, the signal reflected by the medium can be found by interpolating between the above two limiting curves. The function $P(t)$ so obtained is plotted as curve 3 in Fig. 4.15b. For comparison, a Monte Carlo simulation is also presented for the same case (curve 4). Naturally, the interpolation between curves 1 and 2 includes an arbitrary element. For other receiving aperture angles, the accuracy is the same as in the case considered ($2\gamma_{rec} = 10°$).

We point out that the time instant for the equality

$$P^{(1)}(t_0) = P^{(m)}(t_0)\,, \tag{4.8.26}$$

suggests an estimation of the applicability of the single-scattering and asymptotic solutions. The time $t_0$ dependence on the optical medium parameters and experimental geometry can be found analytically. Let, for example, the medium be illuminated by a monodirectional pulsed source normal to the boundary, the receiver and source axes be parallel, and the receiver possess a small angular aperture ($\sin\gamma_{rec} \approx \gamma_{rec}$) and constant sensitivity within the receiving aperture angle. Starting from some time instants [4.26, 109],

$$P^{(1)}(t) = \frac{vx(\pi)\sigma^3 W_0 \Sigma_{rec}}{2\pi(\sigma vt)^2}\,\mathrm{e}^{-\varepsilon vt}\,. \tag{4.8.27}$$

Substituting (4.8.25, 27) at $Q = 9\pi\gamma_{rec}^2/7$ into (4.8.26) yields the following equation for $t_0$:

$$\sigma vt_0 = \frac{1}{2}\ln(\sigma vt_0) - \ln\gamma_{rec}^2 + \frac{\sigma^2 r^2}{4q\sigma vt_0} + \ln\left(\frac{28}{45}x(\pi)\sqrt{\pi q}\right). \tag{4.8.28}$$

It is seen from (4.8.28) that the range of applicability of the single-scattering approximation and asymptotic solution is independent of $\varepsilon$, and is determined rather by the scattering coefficient $\sigma$. The dependence of $\sigma vt_0$ on $\gamma_{rec}$ in an isotropically scattering medium and one of the sea water type with the same optical characteristics as shown in Fig. 4.15 is plotted in Fig. 4.16.

For very extended scattering indicatrices and appreciable absorption in the medium, the BSN for moderate instants can be assumed using the quasi-single scattering approximation or iterations of the small-angle solutions presented in Sects. 3.6.3, 4 and 4.6. If the optical axes of the receiver and source are parallel, the angular aperture of the receiver is large ($\gamma_{rec} \gg \gamma_s$) and the source emits a $\delta$-pulse, then (4.6.1) gives [4.87]

$$P(t) = \frac{W_0\sigma x(\pi)v\Sigma_{rec}\Omega_{rec}}{2\pi}\,S(2r, z = vt, \Omega_s', \Sigma_s')\,, \tag{4.8.29}$$

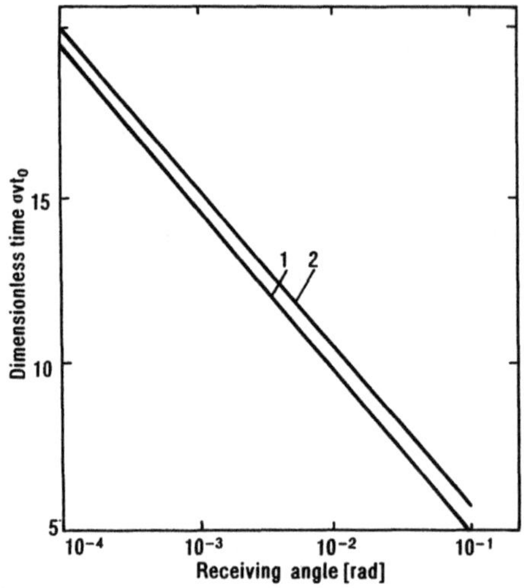


**Fig. 4.16.** $\sigma v t_0$ versus aperture angle $\gamma_{rec}$ at $r = 0$ for the medium with a spherical scattering indicatrix (curve 1) and $g = 0.86$ (curve 2)

where $S(2r, z, \Omega_s', \Sigma_s')$ is the irradiance due to a stationary source with a solid angle $\Omega_s' = \Omega_s + \Omega_{rec}$ and area $\Sigma_s' = 4\Sigma_{rec}$ at depth $z = vt$ and $2r$ away from the beam axis. Substituting the SADA solution (4.5.20) for $S(2r, z, \Omega_s', \Sigma_s')$ at $\xi \ll 1$ (Table 4.4) into (4.8.29) gives for the axial region ($r^2/2V_r \ll 1$)

$$P(t) = \frac{W_0 \varepsilon^3 \Lambda x(\pi) v \Sigma_{rec}}{2\pi} \frac{e^{-(1-\Lambda)\varepsilon v t}}{(\varepsilon v t)^2 + \Lambda \beta_2 (\varepsilon v t)^3 / 3\gamma_{rec}^2 + (2\rho/\gamma_{rec})^2} . \tag{4.8.30}$$

A similar formula for the medium with the indicatrix $x(\beta) = 2a \exp(-a\beta)/\beta(\beta_2 = 2/a^2)$ is cited in [4.87]. The solutions are simpler in the quasi-single scattering approximation. To be specific, for the same case of parallel receiver and source axes, $\gamma_{rec} = \gamma_s$, [4.87] suggests the following expression

$$P(t) = \frac{W_0}{2\pi} \Sigma_{rec} \varepsilon^3 \Lambda v x(\pi) \frac{e^{-(1-\Lambda F)\varepsilon v t}}{(\varepsilon v t)^2 + \rho^2} \qquad \text{at } t \geqslant \frac{r}{v} \operatorname{cotan} \frac{\gamma_{rec}}{2}$$

$$= 0 \qquad \text{at } t < \frac{r}{v} \operatorname{cotan} \frac{\gamma_{rec}}{2} . \tag{4.8.31}$$

The receiver is assumed to have constant sensitivity over the whole aperture angle of $2\gamma_{rec}$. As seen from Fig. 4.17, at $\varepsilon v t \geqslant 1$, the single-scattering approximation underestimates the results. Equations (4.8.25, 30, 31) give similar BSN quantities in a wide time range.

### 4.8.4 Forward Pulse Spread in a Strongly Anisotropic Scattering Medium

If a directional source illuminates an absorbing medium with an extended scattering indicatrix, the scattered radiation field preserves sharp anisotropy up

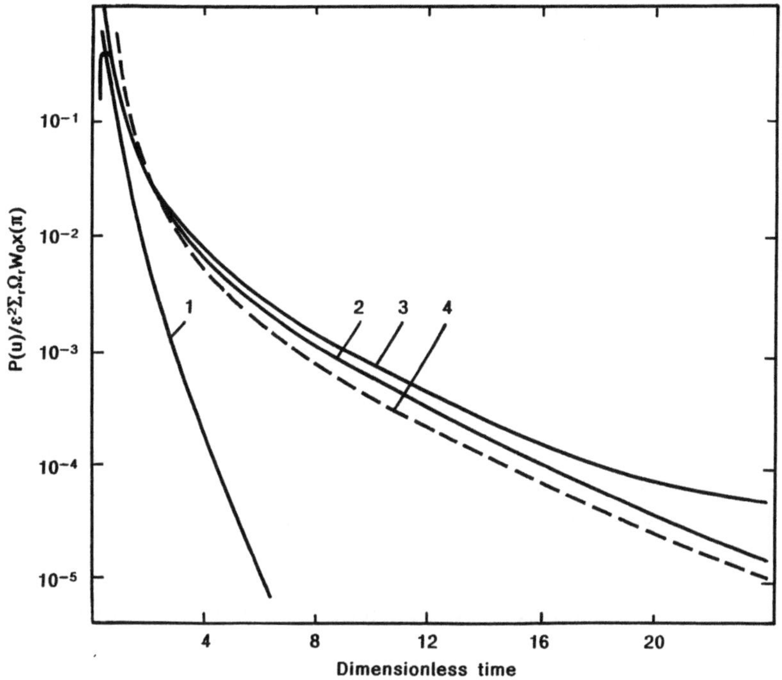


**Fig. 4.17.** BSN at $\Lambda = 0.9$, $\gamma_{rec} = 30°$, $\varepsilon r = 0.1$, $\beta_2 = 0.04$. Curve 1 is single scattering; curve 2, quasi-single scattering at $F = 0.98$ calculated using (4.8.31); curve 3, SADA by (4.8.30); curve 4, asymptotic formula (4.8.22)

to relatively large depths. Then, the diffusion approximation does not offer a satisfactory solution to the stationary problem and, hence, does not allow one to describe the radiance time spread at the time range which includes the instant of a maximum pulse.

The small-angle approximation cannot be used either since it disregards the photon time spread. Therefore, small time instants are described by using the first iteration of the SAA solutions [4.116, 117] or SAA modification [4.48], including the expansion $\mu = \cos\theta \approx 1 - \theta^2/2$ in the transfer equation (4.1.1, 2). The simple SADA solutions to the stationary problem elaborated in Sect. 4.5 are most promising for the study of light fields in such media at arbitrary optical depths. The moments of the time spread function are simple to estimate ([4.27, 83] and Sect. 4.8.5). Now, following [4.77], we shall discuss time-space radiance and irradiance distributions in a scattering medium.

Ordinary substitutions (4.8.5) in the SADA formulas (4.5.19–24) at $V_\rho^0 = 0$ provide the time $u$ Laplace transform of the relevant nonstationary solution. The inverse Laplace transform was taken at $p \gg 1$ and gave the following expression for the time-dependent radiance in the axial region within small forward angles originated by a monodirectional point pulsed source of unit energy

$$I_\delta(\vec{\rho}, \tau, \vec{n}, u) = \frac{\varepsilon^2 (s\tau)^3 f^4(\vec{\rho}, \tau, \vec{n}_\perp)}{32\pi^{5/2}(1-\Lambda)^{3/2}(u')^{9/2}} \times \exp\left[-(1-\Lambda)u - \frac{2(1-\Lambda)}{s^2\tau}\left(\frac{\vec{\rho}}{\tau} - \frac{\vec{n}_\perp}{2}\right)^2 - \frac{(s\tau)^2}{4(1-\Lambda)u'} f^2(\vec{\rho}, \tau, \vec{n}_\perp)\right], \quad (4.8.32)$$

where

$$s = \sqrt{\Lambda(1-\Lambda)\beta_2/2}\,,$$

$$f(\vec{\rho}, \tau, \vec{n}_\perp) = 1 + \frac{1-\Lambda}{s^2\tau}\left[\frac{n_\perp^2}{4} + \left(\frac{\vec{\rho}}{\tau} - \frac{\vec{n}_\perp}{2}\right)^2\right]. \quad (4.8.33)$$

The quantity $u' = u - \sqrt{\tau^2 + \rho^2}$ is the difference between the dimensionless time $u$ and the time of the shortest photon path $\sqrt{\tau^2 + \rho^2}$ from the source to the given medium point $(\vec{\rho}, \tau)$. The solution (4.8.32) can be applied at

$$u' \lesssim 0.1(s\tau)^2/(1-\Lambda) = 0.1\Lambda(1-g)\tau^2\,.$$

The time-space irradiance distribution due to a downward flux is described by the formula

$$S(\vec{\rho}, \tau, u) = -\varepsilon^2 \frac{\partial}{\partial \rho^2}\left\{\frac{s\tau}{\pi^{3/2}(1-\Lambda)^{1/2}(u')^{3/2}} \times \exp\left[-(1-\Lambda)u - \frac{\rho^2}{\sigma V_\rho} - \frac{(s\tau)^2}{4(1-\Lambda)u'}\left(1 - \frac{\rho^2}{\sigma V_\rho}\right)\right]\right\} \quad (4.8.34)$$

at

$$u' \ll 0.2(s\tau)^2/(1-\Lambda) = 0.2\Lambda(1-g)\tau^2\,.$$

Here, $V_\rho = s^2\tau^3/3(1-\Lambda)$ is the PSF variance at $s\tau \ll 1$ in the stationary problem (Table 4.4).

We now consider the applicability of these solutions, bearing in mind the general concluding remark of Sect. 4.8.1. The SADA formulas do not work at $\Lambda \to 0$ and $\Lambda \to 1$. Therefore, the relevant nonstationary solutions cannot be used to describe the pulse shape either at the very first moments of photon arrival at the given point of the scattering medium, or at $u \to \infty$. In the case where the SADA describes the light field from a stationary source (Sect. 4.5.6), (4.8.32, 34) correctly define the pulse propagation for the time interval in which the major portion of the pulse energy falls. Next, we shall discuss some important physical inferences of these solutions. It follows from (4.8.34) that the irradiance distribution along the $\rho$ coordinate at a fixed time instant is not essentially Gaussian. In fact, the shortest optical path to the point $\tau$ is $\sqrt{\tau^2 + \rho^2}$. Therefore, for the time $u$, the photons approach only some limited range $\rho \ll \sqrt{u^2 - \tau^2}$. Figure 4.18

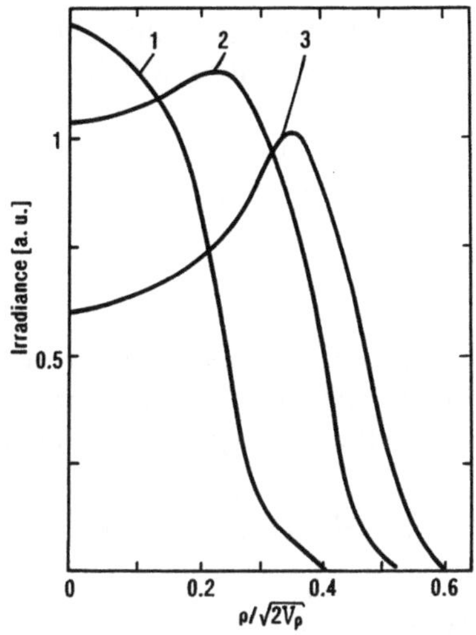


**Fig. 4.18.** Irradiance distribution across the beam at depth $\tau$ and time instants $u = \tau + 2aV_\rho/\tau$ for $a = 0.09$ (curve 1), 0.135 (curve 2) and 0.18 (curve 3). From [4.77]

gives the point spread function $S(\rho, \tau = \text{const}, u = \text{const})$ at fixed instants of time. As seen, at small time instants, the point spread function is bell-shaped with a maximum at $\rho = 0$. Starting from the time $u = 0.14(s\tau)^2/(1 - \Lambda)$, the PSF has a maximum at $\rho_{\max}/\sqrt{2V_\rho} \approx 1.33\sqrt{\tau(u - \tau)/2V_\rho - 0.105}$. Thus, the PSF changes essentially with growing $u$. This fact may be of significance in vision systems using pulsed lasers. We shall now proceed to analyze the integral time characteristics.

### 4.8.5 Mean Time and Variance of Photon Paths

We analyze the integral time characteristics of the pulse spread and find the mean time $\langle u \rangle$ and variance $V_u$ of the photon path to the point $(\vec{\rho}, \tau)$ with an incident $\delta$-pulse.

The first photons arrive at the point $(\vec{\rho}, \tau)$ at the time instant $\sqrt{\rho^2 + \tau^2}$ (the shortest path from the source to the receiver). Therefore, it is more appropriate to characterize the pulse spread by the mean delay time of the photons $\langle u \rangle$-$\sqrt{\rho^2 + \tau^2}$ with respect to their shortest path time $\sqrt{\rho^2 + \tau^2}$. In the following analysis, we shall use the term "mean delay time".

If we want to use the language which is more commonly employed by experimenters and emphasize the meaning of the quantities $\langle u \rangle$ and $V_u$ through their measurements, then we say that $\langle u \rangle$ and $V_u$ are the centre of gravity coordinate and the variance of the pulse spread function which, in the case of a small-angle receiver, is the radiance time distribution, which is the irradiance time distribution for the receiver with a $2\pi$ solid angle.

We shall use (4.8.8, 9) to calculate $\langle u\rangle$ and $V_u$ in terms of the appropriate stationary solution. The SADA solutions will be employed for media with essential absorption and strongly anisotropic scattering. Let a scattering medium be illuminated by a monodirectional point source. The irradiance coefficient at the point $(\vec{\rho}, \tau)$ is found from (4.5.19, 20) as

$$S(\vec{\rho}, \tau) = \frac{\varepsilon^2 S(\tau)}{2\pi V_\rho(\tau)} e^{-\rho^2/2V_\rho(\tau)} . \tag{4.8.35}$$

The quantitites $S(\tau)$ and $V_\rho(\tau)$ are correspondingly determined by (4.5.8, 23). Expressions (4.8.8, 9, 35) suggest that the mean time $\langle u(\vec{\rho}, \tau)\rangle$ and variance $V_u(\vec{\rho}, \tau)$ of the photon path to the point $(\vec{\rho}, \tau)$ at a solid angle $2\pi$ have the form

$$\langle u(\vec{\rho}, \tau)\rangle = \langle u^{\mathrm{w}}(\tau)\rangle - \langle \Delta u(\tau)\rangle + \frac{\rho^2}{2V_\rho(\tau)}\langle \Delta u(\tau)\rangle ,$$

$$V_u(\vec{\rho}, \tau) = V_u^{\mathrm{w}}(\tau) - \Delta V_u(\tau) + \frac{\rho^2}{2V_\rho^{(\tau)}}[\Delta V_u(\tau) - \langle \Delta u(\tau)\rangle^2] , \tag{4.8.36}$$

Here, each term has a clear physical significance. The quantities $\langle u^{\mathrm{w}}(\tau)\rangle$ and $V_u^{\mathrm{w}}(\tau)$ determine the mean time and variance of the time irradiance distribution at a depth $\tau$ when a medium is illuminated by an infinitely wide monodirectional $\delta$-pulse source. The second terms $\langle \Delta u(\tau)\rangle$ and $\Delta V_u(\tau)$ define the decrease in the mean time and variance at the axial point of the beam from the monodirectional $\delta$-pulse point source as compared with $\langle u^{\mathrm{w}}(\tau)\rangle$ and $V_u^{\mathrm{w}}(\tau)(\langle \Delta u(\tau)\rangle > 0$, $\Delta V_u(\tau) > 0)$. The dependence of $\langle u(\vec{\rho}, \tau)\rangle$ and $V_u(\vec{\rho}, \tau)$ on the $\vec{\rho}$ coordinate is described by the third term of the sums in (4.8.36). In the axial region, where $\rho < \sqrt{2V_\rho}$, $\langle u(\vec{\rho}, \tau)\rangle < \langle u^{\mathrm{w}}(\tau)\rangle$, i.e., the pulse from the point source arrives in a shorter time than that from the wide one. At $\rho = \sqrt{2V_\rho}$, we have $\langle u(\vec{\rho}, \tau)\rangle = \langle u^{\mathrm{w}}(\tau)\rangle$. If $\rho > \sqrt{2V_\rho}$, then $\langle u(\vec{\rho}, \tau)\rangle > \langle u^{\mathrm{w}}(\tau)\rangle$.

As in (4.8.36), we shall define the mean time $\langle u(\tau, \vec{n}_\perp)\rangle$ and variance $V_u(\tau, \vec{n}_\perp)$ of the radiance distribution, when the medium is illuminated by an infinitely wide monodirectional $\delta$-pulse source. In fact, having applied (4.5.7) for the angular radiance distribution and using (4.8.8, 9), we obtain

$$\langle u(\tau, \vec{n}_\perp)\rangle = \langle u^{\mathrm{w}}(\tau)\rangle - \langle \Delta u_1(\tau)\rangle + \frac{n_\perp^2}{2V_\theta(\tau)}\langle \Delta u_1(\tau)\rangle ,$$

$$V_u(\tau, \vec{n}_\perp) = V_u^{\mathrm{w}}(\tau) - \Delta V_{u,1}(\tau) + \frac{n_\perp^2}{2V_\theta(\tau)}[\Delta V_{u,1}(\tau) - \langle \Delta u_1(\tau)\rangle^2] , \tag{4.8.37}$$

where the quantity $V_\theta(\tau)$ (mean square angle deviation in the stationary problem) is expressed by (4.5.9).

All the quantities required to calculate the mean time and variance using (4.8.36, 37) are listed in Table 4.6. We note that the quantities $2(1-\Lambda)(\langle u^{\mathrm{w}}\rangle - \tau)$, $2(1-\Lambda)\langle \Delta u\rangle$, $2(1-\Lambda)\langle \Delta u_1\rangle$ as well as $4(1-\Lambda)^2 V_u^{\mathrm{w}}$, $4(1-\Lambda)^2 \Delta V_u$ and $4(1-\Lambda)^2 \Delta V_{u,1}$ depend only on the parameter $\xi = s\tau (s = \sqrt{\Lambda(1-\Lambda)\beta_2/2})$.

**Table 4.6.** Parameters for calculation of the mean time $\langle u\rangle$ and variance $V_u$ of pulse spread.

| Quantity | General case | $\xi \ll 1$ | $\xi > 2$ |
|---|---|---|---|
| $2(1-\Lambda)(\langle u^w\rangle-\tau)$ | $\xi\tanh\xi$ | $\xi^2$ | $\xi$ |
| $2(1-\Lambda)\langle\Delta u\rangle$ | $B=3-\dfrac{\xi\tanh^2\xi}{\xi-\tanh\xi}$ | $\frac{4}{5}\xi^2$ | $2-\dfrac{1}{\xi-1}$ |
| $2(1-\Lambda)\langle\Delta u_1\rangle$ | $B_1=\dfrac{\xi\tanh^2\xi-\xi+\tanh\xi}{\tanh\xi}$ | $\frac{2}{3}\xi^2$ | $1-4\xi e^{-2\xi}$ |
| $4(1-\Lambda)^2V_u^w$ | $B_1\xi\tanh\xi$ | $\frac{2}{3}\xi^4$ | $\xi$ |
| $4(1-\Lambda)^2\Delta V_u$ | $2B-\xi\dfrac{dB}{d\xi}$ | $\frac{344}{525}\xi^4$ | $4-\dfrac{2\xi-1}{(\xi-1)^2}$ |
| $4(1-\Lambda)^2\Delta V_{u,1}$ | $2B_1-\xi\dfrac{dB_1}{d\xi}$ | $\frac{28}{45}\xi^4$ | $2(1-4\xi^2e^{-2\xi})$ |
| $\dfrac{1-\Lambda}{\tau}V_\rho$ | $1-\dfrac{\tanh\xi}{\xi}$ | $\dfrac{\xi^2}{3}$ | $1-\dfrac{1}{\xi}$ |
| $\tau(1-\Lambda)V_\theta$ | $\xi\tanh\xi$ | $\xi^2$ | $\xi$ |

**Table 4.7.** Relative error in the approximation of small $\xi$.

| $\xi$ | $\langle u^w\rangle$ | $\sqrt{V_u^w}$ | $\langle u(\rho=0,\tau)\rangle$ | $\sqrt{V_u(\rho=0,\tau)}$ | $\langle u(\tau,n_\perp=0)\rangle$ | $\sqrt{V_u(\tau,n_\perp=0)}$ |
|---|---|---|---|---|---|---|
| 0.3 | 0.03 | 0.04 | 0.003 | 0.02 | 0.01 | 0.01 |
| 0.5 | 0.08 | 0.10 | 0.006 | 0.02 | 0.02 | 0.02 |
| 1.0 | 0.31 | 0.40 | 0.028 | 0.05 | 0.07 | 0.09 |
| 1.5 | 0.66 | 0.88 | 0.061 | 0.11 | 0.14 | 0.21 |
| 2.0 | 1.07 | — | 0.108 | 0.36 | 0.24 | 0.43 |

The applicability of the expressions in the last column of the table is determined by $\xi$, for which we can assume $\tanh\xi \approx 1$. We note also that at $\xi \geqslant 2$, the quantity $1-\tanh\xi < 0.04$.

In practice, it is useful to estimate the accuracy of the very simple asymptotic formulas ($\xi \ll 1$) in the third column of Table 4.6. The relative errors in these formulas, compared to the general relations in the second column of Table 4.6, are listed in Table 4.7. This approximation can be applied, with an error of less than 10%, to calculate $\langle u^w\rangle$ and $\sqrt{V_u^w}$ at $\xi \approx 0.5$–$0.7$, as well as to find $\langle u(\vec{\rho}=0,\tau)\rangle$ and $\sqrt{V_u(\vec{\rho}=0,\tau)}$ at $\xi \approx 1.5$–$2$.

Figure 4.19 is a comparison of $\langle u^w\rangle$ and $V_u^w$ calculated in the SADA for small $\xi$ and the results of Monte Carlo simulations [4.118] and the experimental data [4.119] for different model media. In the SADA, $4(\langle u^w\rangle-\tau)/\Lambda\beta_2=\tau^2$, $\sqrt{24V_u^w}/\Lambda\beta_2=\tau^2$; the Monte Carlo simulation data for the scattering media at $\Lambda=1$ are approximated by $\langle u^w\rangle-\tau=0.62(1-g)^{0.94}\,\tau^{1.94}$, $\sqrt{V_u^w}=0.64(1-g)^{0.81}\,\tau^{1.81}$, where $g$ is the asymmetry parameter of the scattering indicatrix. This figure

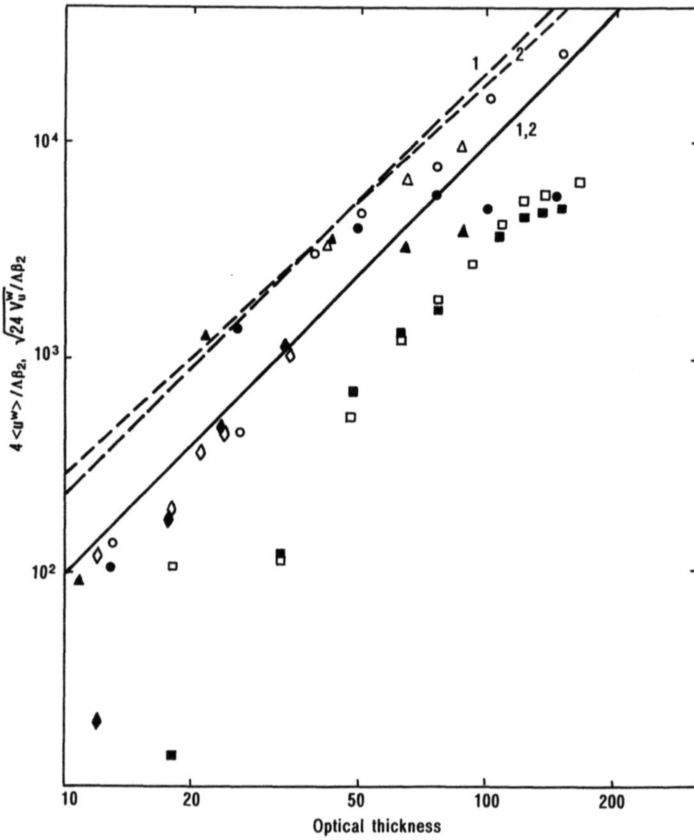


**Fig. 4.19.** Mean time and variance as a function of $\tau$. SADA results (*solid lines*) are compared with Monte Carlo simulations (*dashed lines*) and experimental data (*symbols*). Curves 1 and *light symbols* correspond to mean time, curves 2 and *dark symbols* show the variance

shows that the agreement between the SADA results, the Monte Carlo simulations and experimental data in a wide range of optical thicknesses is satisfactory. It should be noted, however, that for this experiment ($1 - \Lambda \approx 5.10^{-4}$, $1 - g \approx 0.15$), one of the main SADA conditions, $V_\theta \ll 1$, is violated at $\tau > 15$. Therefore, the coincidence of the results at large $\tau$ in Fig. 4.19 seems to be strange.

Let us ascertain in what way the mean time $\langle u^w \rangle$ and variance $V_u^w$ of the photon path to the plane $\tau$ at a solid angle $2\pi$ depend on the thickness $\tau$ of an optically thick weakly absorbing layer. To do this, we use the stationary asymptotic solution for radiant fluxes at $\tau \gg 1$ and $(1 - \Lambda)/\Lambda(1 - g) \ll 1$. In the usual way (4.8.8, 9), using (3.5.5) at $\tau(1 - g) \gg 1$, we obtain

$$\langle u^w \rangle - \tau = 0.5(1 - g)\tau^2 , \qquad \sqrt{V_u^w} = 0.55(1 - g)\tau^2 . \tag{4.8.38}$$

We should not forget that these estimates can be useful in studying light pulse propagation in clouds. The true solution of the problem, correct at all $\tau$, must

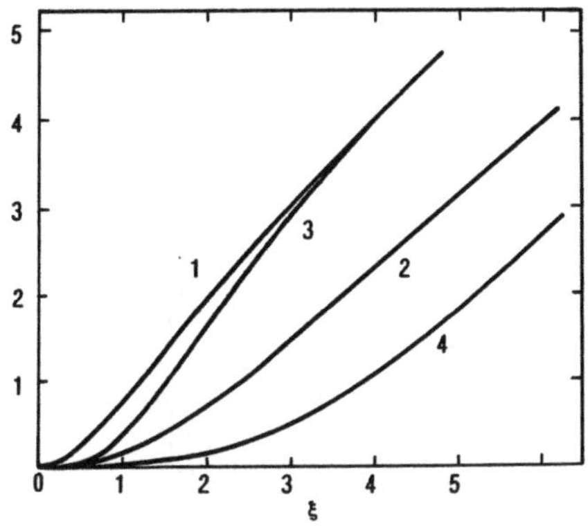


**Fig. 4.20.** Mean time $\langle u^w \rangle$ and variance $V_u^w$ versus $\xi = s\tau$ for the medium illuminated by an infinite or a point source: curve 1, $2(1-\Lambda)\langle u^w \rangle$; curve 2, $2(1-\Lambda)\langle u(\rho=0,\xi)\rangle$; curve 3, $4(1-\Lambda)^2 V_u^w$; curve 4, $4(1-\Lambda)^2 V_u(\rho=0,\xi)$

be close both to the asymptotic estimation (4.8.38) for $\tau(1-g) \gg 1$ and to the SADA estimation at $\tau(1-g) \leqslant 1$. As seen from comparing (4.8.38) and the approximate formulas in Table 4.6, these estimates are close to each other and to the above Monte-Carlo simulation. This is, perhaps, the reason for the surprising coincidence of the curves in Fig. 4.19.

At the narrow beam axis, the mean photon delay time $\langle u \rangle - \tau$ is 5 times as small as that for a wide source. In the case of a point source, the variance of the photon paths is only about $0.01 V_u^w$. As seen from Fig. 4.20, at $\xi = s\tau \leqslant 1.5$ (for example, for sea water with $s \approx 0.1$, which complies with $\tau \approx 15$), the variance of the photon paths $V_u(\vec{\rho} = 0, \tau)$ at the beam axis is negligible, i.e., up to very large optical thicknesses, the pulse at the beam axis is slightly distorted. This accounts, for example, for the wide applicability range of the approximate solutions (4.8.29–31) for the BSN in which the pulse spread in the axial region is disregarded. In all cases, the photon delay time and the variance of the photon paths at the given medium point decrease with increasing $(1-\Lambda)$ and the indicatrix extension. In fact, the increase of the indicatrix extension promotes narrowing of the beams. The absorption results in an effective escape of the photons delayed for large times. Such behavior is evident and was first discussed in [4.108] where it was obtained using Monte Carlo simulations and asymptotic solutions. They are also supported by experimental data [4.120–122]. The SADA formulas have complemented these results by providing an analytical description of the entire pulse propagation process.

Consider now $\langle u(\vec{\rho},\tau)\rangle - \tau$ and $\sqrt{V_u(\vec{\rho},\tau)}$ as functions of the $\vec{\rho}$ coordinate, plotted in Fig. 4.21. The quantity $\sqrt{V_u}$ characterizes the half-width of the pulse time spread. In the figure, typical sea water parameters are given as $\Lambda = 0.8$, $\beta_2 = 0.04$. We note that, at $\varepsilon = 0.3\ \mathrm{m}^{-1}$, for example, the quantity $u = \varepsilon v t = 1$ corresponds to the real time $t \approx 15$ ns. As seen from Fig. 4.21, the pulse half-width is a weaker function of the $\vec{\rho}$ coordinate than the mean delay time $\langle u(\vec{\rho},\tau)\rangle$. We

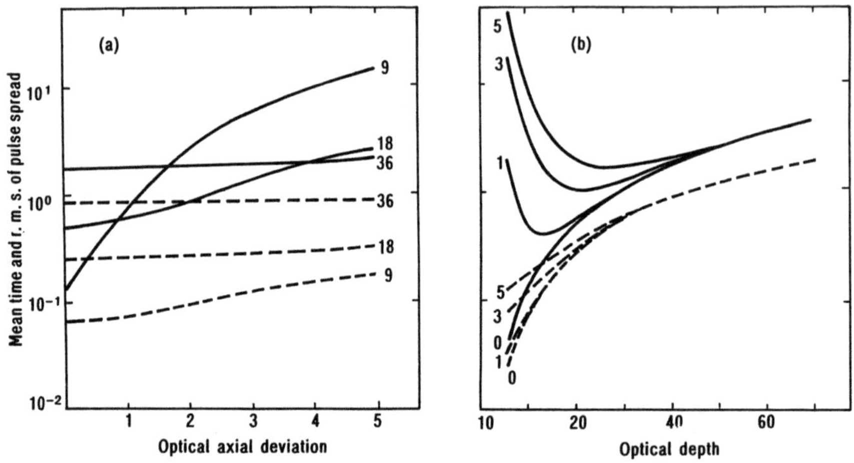


**Fig. 4.21.** Pulse spread mean time $\langle u(\rho, \tau)\rangle - \tau$ (*solid lines*) and half-width $[V_u(\rho, \tau)]^{1/2}$ (*dashed lines*) at $\Lambda = 0.8$, $\beta_2 = 0.04$. The numbers at the curves give the values of $\tau$ and $\rho$

may represent such a time variation of the light field in the plane $\tau =$ const as a pulse traveling along the $\vec{\rho}$ coordinate, its distortion being relatively weak. It agrees with the characteristic behavior of the curve of the radial irradiance distribution at the moment $u$ in the plane $\tau =$ const induced by a pulsed source (Fig. 4.18).

The simple structural formulas (4.8.36, 37) may be derived using only the Gaussian approximations (4.8.35 and 4.5.7) for the functions $S(\vec{\rho}, \tau)$ and $I(\tau, \vec{n}_\perp)$. Therefore, at large $\tau$ when the error in (4.8.35, 4.5.7) is small (see the estimations in Sect. 4.5.6), the above simple formulas can be used for $\langle u(\vec{\rho}, \tau)\rangle$ and $V_u(\vec{\rho}, \tau)$. At small $\tau$, these solutions give only a rough description of the dependence of the integral time characteristics of the optical medium parameters, the accuracy being essentially lower in this case. To obtain convenient solutions for moderate $\tau$ requires using the SADA modifications discussed in Sect. 4.5.7. We now consider, for example, the time characteristics of the irradiance originated by a pulsed narrow-beam source. Take a scattered component of the irradiance in the stationary case and once more give its spatial distribution as

$$S^{s}(\vec{\rho}, \tau) = \frac{\varepsilon^2 S^{s}(\tau)}{2\pi V_\rho^{s}(\tau)} e^{-\rho^2/2V_\rho^s(\tau)}\,, \tag{4.8.39}$$

where $S^s(\tau) = S(\tau) - \exp(-\tau)$. The variance $V_\rho^s(\tau)$ for the scattered component is expressed in terms of the variance $V_\rho(\tau)$ by the relation

$$V_\rho^s = V_\rho/\eta^s\,, \qquad \eta^s = S^s(\tau)/S(\tau)\,, \tag{4.8.40}$$

where $\eta^s$ is the fraction of the scattered radiation in the total flux $S(\tau)$. For the scattered component, equations (4.8.39, 40) yield relations similar to (4.8.36)

$$\langle u^s(\vec{\rho},\tau)\rangle = \langle u^{w,s}\rangle - \langle \Delta u^s\rangle + \frac{\rho^2}{2V_\rho^s}\langle \Delta u^s\rangle\,,$$

$$V_u^s(\vec{\rho},\tau) = V_u^{w,s} - \Delta V_u^s + \frac{\rho^2}{2V_\rho^s}[\Delta V_u^s - \langle \Delta u^s\rangle^2]\,, \tag{4.8.41}$$

where

$$\langle u^{w,s}\rangle - \tau = \frac{\langle u^w\rangle - \tau}{\eta^s}\,,$$

$$V_u^{w,s} = \frac{V_u^w}{\eta^s} - \left(\frac{\langle u^w\rangle - \tau}{\eta^s}\right)^2(1-\eta^s)\,,$$

$$\langle \Delta u^s\rangle = \langle \Delta u\rangle - \frac{\langle u^w\rangle - \tau}{\eta^s}(1-\eta^s)\,,$$

$$\Delta V_u^s = \Delta V_u - (1-\eta^s)\left[\frac{V_u^w}{\eta^s} - \left(\frac{\langle u^w\rangle - \tau}{\eta^s}\right)^2\right]. \tag{4.8.42}$$

Similar formulas can also be obtained for $\langle u^s(\tau,\vec{n}_\perp)\rangle$, $V_u^s(\tau,\vec{n}_\perp)$, $\langle \Delta u_1^s\rangle$ and $\Delta V_{u,1}^s$.

Somewhat more intricate SADA solutions for the scattered radiation component, disregarding the approximation (4.8.39), were derived in [4.83]. A comparison of the results of these two approaches exhibited their agreement [4.123].

As in Sect. 4.5.7, we proceed to the next step and divide the scattered radiance component $I^s(\vec{\rho},\tau,\vec{n}_\perp)$ into two parts, $I_1^s$ and $I_2^s$, which is relevant in dividing the scattering indicatrix (4.5.54). It is not difficult to obtain from (4.5.60) the expressions for the integral temporal characteristics of each component similar in structure to (4.8.41, 42). Such a division of the scattered light into two components first of all essentially improves the estimation of the integral pulse spread characteristics and, in addition, provides an adequate physical description of the real two-model structure of the scattered radiation pulse [4.124] in reconstructing the shape of the temporal pulse spread [4.100].

In the asymptotic regime at very high $\tau$, expressions can be obtained for $\langle u(\vec{\rho},\tau)\rangle$ and $V_u(\vec{\rho},\tau)$ which are applicable for any optical medium characteristics. Formula (4.2.6) yields that the irradiance distribution in the $\tau$ plane at large $\tau$ in the medium illuminated by a narrow beam is described by the expression similar to (4.8.35) in which, however, we must set (4.2.4, 5)

$$S(\tau) = u(\mu=1)e^{-\gamma\tau}\int_0^1 \varphi(\mu)\mu\,d\mu\,, \qquad V_\rho(\tau) = \tau/\gamma\,. \tag{4.8.43}$$

Then, for the mean time $\langle u\rangle$ and variance of the photon paths $V_u$, (4.8.36) also hold, the quantities $(\langle u^w\rangle, V_u^w, \langle \Delta u\rangle, \Delta V_u)$ they include being different. Their values, obtained by (4.8.8, 9) using (4.8.35, 43), are listed in Table 4.8. They incorporate $\Lambda$ derivatives of the asymptotic attenuation coefficient: $\gamma' = d\gamma/d\Lambda$

**Table 4.8.** Parameters for the calculation of the mean time and variance of the pulse spread in deep scattering medium layers.

| Quantity | Notation | General case | Specific case of $(1-\Lambda)/(1-g) \ll 1$ |
|---|---|---|---|
| Mean time | $\langle u^w \rangle$ | $\gamma\tau\left(1-\frac{\Lambda\gamma'}{\gamma}\right)$ | $\tau\sqrt{\frac{3(1-g)}{4(1-\Lambda)}}$ |
| | $\langle \Delta u \rangle$ | $1-\frac{\Lambda\gamma'}{\gamma}$ | $\frac{1}{2(1-\Lambda)}$ |
| Variance | $V_u^w$ | $-\gamma\tau\Lambda^2\frac{\gamma''}{\gamma}$ | $\frac{3(1-g)}{4(1-\Lambda)}\tau$ |
| | $\Delta V_u$ | $-\Lambda^2\frac{\gamma''}{\gamma}+\left(1-\frac{\Lambda\gamma'}{\gamma}\right)^2$ | $\frac{1}{4(1-\Lambda)^2}$ |

**Table 4.9.** Derivation of $-\gamma'$ (above the line) and $-\gamma''$ (below the line) for the Henyey-Greenstein indicatrix.

| | *g* | | | | | | | | |
|---|---|---|---|---|---|---|---|---|---|
| $\Lambda$ | 0.5 | 0.6 | 0.7 | 0.8 | 0.85 | 0.9 | 0.95 | 0.97 | 0.98 |
| 0.2 | 0.288 | 0.39 | 0.5 | 0.619 | 0.685 | 0.758 | 0.845 | 0.889 | 0.914 |
| 0.4 | 0.566 | 0.638 | 0.711 | 0.786 | 0.825 | 0.868 | 0.917 | 0.941 | 0.954 |
| 0.5 | 0.697 | 0.751 | 0.804 | 0.858 | 0.886 | 0.915 | 0.947 | 0.962 | 0.972 |
| 0.6 | 0.844 | 0.876 | 0.906 | 0.935 | 0.95 | 0.964 | 0.979 | 0.986 | 0.988 |
| 0.7 | 1.031 | 1.034 | 1.034 | 1.032 | 1.03 | 1.026 | 1.018 | 1.014 | 1.000 |
| 0.8 | 1.314 | 1.272 | 1.227 | 1.176 | 1.148 | 1.115 | 1.076 | 1.056 | 1.014 |
| 0.9 | 1.908 | 1.779 | 1.642 | 1.486 | 1.4 | 1.305 | 1.195 | 1.14 | 1.109 |
| 0.94 | 2.481 | 2.278 | 2.053 | 1.802 | 1.658 | 1.499 | 1.314 | 1.225 | 1.174 |
| 0.98 | 4.329 | 3.906 | 3.425 | 2.874 | 2.558 | 2.193 | 1.745 | 1.527 | 1.401 |
| 0.99 | 6.098 | 5.464 | 4.808 | 3.968 | 3.484 | 2.924 | 2.227 | 1.866 | 1.658 |
| 0.2 | 1.634 | 1.523 | 1.333 | 1.073 | 0.912 | 0.719 | 0.472 | 0.348 | 0.266 |
| 0.4 | 1.286 | 1.114 | 0.931 | 0.72 | 0.6 | 0.469 | 0.305 | 0.221 | 0.169 |
| 0.5 | 1.361 | 1.163 | 0.956 | 0.734 | 0.611 | 0.473 | 0.305 | 0.222 | 0.172 |
| 0.6 | 1.615 | 1.368 | 1.113 | 0.845 | 0.698 | 0.537 | 0.347 | 0.251 | 0.193 |
| 0.7 | 2.201 | 1.861 | 1.504 | 1.128 | 0.927 | 0.708 | 0.453 | 0.328 | 0.245 |
| 0.8 | 3.712 | 3.145 | 2.547 | 1.9 | 1.55 | 1.169 | 0.742 | 0.533 | 0.379 |
| 0.9 | 9.96 | 8.546 | 7.057 | 5.315 | 4.343 | 3.253 | 2.012 | 1.431 | 1.101 |
| 0.94 | 21.1 | 13.38 | 15.33 | 11.83 | 9.733 | 7.34 | 4.509 | 3.182 | 2.42 |
| 0.98 | 109.2 | 96.62 | 81.9 | 65.04 | 55.19 | 43.25 | 27.46 | 19.3 | 14.51 |
| 0.99 | 303.56 | 273.3 | 236.1 | 188.03 | 160.6 | 128.09 | 85.32 | 60.96 | 46.48 |

and $\gamma'' = d^2\gamma/d\Lambda^2$. Table 4.9 contains $\gamma'$ and $\gamma''$ for different $\Lambda$ and the Henyey-Greenstein indicatrices (2.2.9) of the different extensions [4.125]. The appropriate $\gamma$ values are cited in Table 3.2. The data of Table 4.9 can be used to estimate $\gamma'$ and $\gamma''$ for real media, although their scattering indicatrices may essentially differ from the Henyey-Greenstein function. In this case, the tabulated $\gamma'$ and $\gamma''$ values must be taken at the $g$ valid for this indicatrix. For geophysical scattering

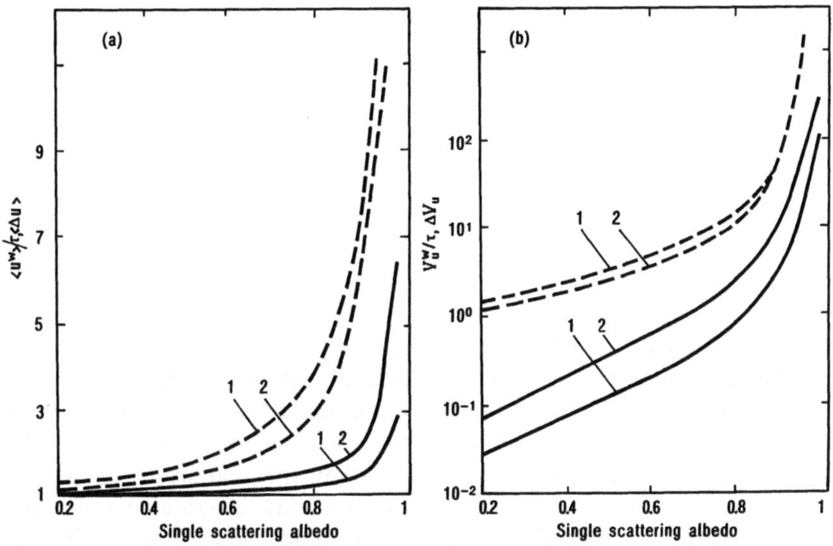


**Fig. 4.22.** Quantities needed for calculation of mean time (a) and variance (b) under asymptotic conditions at $g = 0.9$ (curve 1) and $g = 0.5$ (curve 2). *Solid lines* are $\langle u^w \rangle/\tau$, $V_u^w/\tau$, and *dashed lines* are $\langle \Delta u \rangle$ and $\Delta V_u$

indicatrices (different sea water and cloud models), the error in $\gamma'$ is no more than 3%, that in $\gamma''$ being no more than 20% [4.125].

For weak absorptions, when (3.2.11) gives $\gamma$ as a function of $\Lambda$, the expressions for $\langle u \rangle$ and $V_u$ are significantly simplified (Table 4.8). At $\Lambda \to 1$, the variance $V_u$ goes to $\infty$. The quantities $\langle u^w \rangle/\tau$, $V_u^w/\tau$ and $\langle \Delta u \rangle$, $\Delta V_u$ are plotted vs. $\Lambda$ in Fig. 4.22. The asymptotic values of $\langle u^w \rangle$ and $V_u^w$ have been studied in [4.100].

# 5 Elements of Vision Theory

The theory of image transfer in scattering media is based upon the theories of scattering and linear optical systems. The previous two chapters discussed in detail the potential, methods and results of the applied radiative transfer theory, which allows one to calculate all the required photometric characteristics of the radiation field, taking into account absorption and scattering effects in the medium. This chapter deals with some of the problems encountered in the theory of linear systems as applied to vision theory, and classifies the main vision systems.

Following its success in radio engineering, communication and information theories, the theory of linear systems has been applied in optics since the late 1930s. The fundamental properties of linearity and invariance are characteristic of all these systems. These have also been displayed with the same mathematical apparatus as used for their description, or Fourier analysis. The theory of linear systems for image transfer in a scattering medium represents the medium as an element of the entire system of vision. The influence of other elements, of the optical receiver system, for example, on the image quality may be considered irrespective of the scattering medium.

The principle of Fourier optics in image theory is discussed in several monographs [5.1–4]. This theory has been extended to scattering media and, especially, to the problems of imaging by means of active observation systems with external lighting [5.5–9].

We shall also consider some points which bear no direct relation to the theory of linear systems but which, nevertheless, cannot be ignored as far as the vision problem is concerned. These are image quality characteristics, details of image processing and visual perception, etc. Finally, we shall discuss the concept of object detection and suggest methods of evaluating the limiting range of visibility in a scattering medium, as well as compare related television and location detection systems.

## 5.1 Image Quality Characteristics

We see an image. This may be composed of a TV or motion picture frame and the image formed on the retina of the eye.

Let us ignore for the moment the picture content and consider the image as a combination of a large number of individual cells. In television, the size and number ($\sim 600^2$) of cells is fixed. These quantities are determined by the number of frame lines. It is more difficult to measure exactly the size of the retina cells, and this will be discussed below. However, it is clear that more complex the content of the picture, the greater the amount of information it carries and, hence, the greater the number of cells required to represent the image completely accurately.

### 5.1.1 Contrast and Signal-to-Noise Ratio

We shall consider one of the image cells and term it the test cell. Assume that the difference between the number of photons incident on the test cell and the mean number $\langle N \rangle$ of photons reaching surrounding areas of the same size during the observation time is $\Delta N$. This difference distinguishes the test cell from its neighbours, i.e., $\Delta N$ is a signal carrying valid information regarding the object observed. For this to be the case, the quantity $\Delta N$ must obey certain conditions. First, the valid signal must exceed some minimum value (photodetector perception threshold), otherwise it will not be recorded. This condition is, however, insufficient.

For various reasons, some of which will be discussed below, the real number $N$ of photons arriving at the given resolution cell fluctuates about some mean value. These fluctuations manifest themselves as spatial noise against whose background the valid signal may be lost. The quantitative measure of the fluctuations is their variance $V_N$, or mean-square deviation $\sqrt{V_N}$. Naturally, the test cell can reliably be distinguished from its neighbours provided the regular deviation $\Delta N$ is essentially in excess of the random noise deviations. These quantities determine the most important image quality characteristics, notably the signal-to-noise ratio(SNR)

$$\delta = \frac{\Delta N}{\sqrt{V_N}} . \tag{5.1.1}$$

Various factors are responsible for fluctuation noise. Fluctuations of the optical (for example, atmospheric) characteristics of the image transfer channel are among them. Any recording, expecially photoelectronic and photographic, is accompanied by so-called dark noise which is independent of the number of incident photons and exists even when there is no radiation. In electronic devices, it is due to nonuniform field emission, whereas in photographic systems it is due to nonuniform photomaterial fog. Furthermore, since the incident light power is constant, the photon number fluctuates owing to the quantum nature of the light, i.e. the discrete and random distribution of photons. In photorecorders, fluctuations are displayed as shot noise. This noise is most often encountered in visual range devices and will form the main subject of the following analysis. In contrast to dark noise, its level depends on the number of incident photons. For incoherent radiation, the variance $V_N = \langle N \rangle$. Therefore, the signal-to-noise ratio is

$$\delta = \frac{\Delta N}{\sqrt{\langle N \rangle}}. \tag{5.1.2}$$

We recast the relation (5.1.2) as

$$\delta = \frac{\Delta N}{\langle N \rangle} \frac{\langle N \rangle}{\sqrt{\langle N \rangle}} = k\sqrt{\langle N \rangle}. \tag{5.1.3}$$

The quantity

$$k = \frac{\Delta N}{\langle N \rangle}, \tag{5.1.4}$$

termed the contrast, is one of the basic image quality characteristics.

None of the photorecorders uses all the incident radiation and transforms only some portion of it with efficiency $\eta$. Therefore, in the relation (5.1.3), the quantity $\langle N \rangle$ is understood as the number of photons used. It is worth emphasizing that we can only speak about signal-to-noise ratio when the image cell size and its recording time are known. The larger the cell area and the longer the image recording time, the greater the number of photons used and the higher $\delta$. Therefore, both the space and time integraton improve the signal-to-noise ratio. Failure to take this factor into account may cause misunderstanding. To be specific, since the test-objects being used are long narrow strips, the resolution of the photographic film is essentially overestimated [5.10]. Really, such a test-object with $d$-spaced strips can be much better seen than an individual element, a $d$-side square being an example. When observing a set of long strips, the photon number $\langle N \rangle$ in (5.1.3) is determined by the area of the whole strip, or an appreciable portion of it, and essentially exceeds the number of photons per square. The SNR will, naturally, be higher when viewing the strips.

Space and time integrations are interchangeable since the SNR depends solely on th total number of photons used. Human vision is peculiar for space integration over an angle of 0.5° [5.11, 12] and time integration over the interval $\Delta t = 0.1 - 0.2$ s [5.10, 13]. The number of elements in a TV frame is fixed. In this case, the eye may realize the space integration when perceiving the image received. The time integration must also be considered in this case. As the eye integration time is larger than the frame exposure time, the SNR in the visual image accumulated from several TV frames is higher than for a single frame.

The relation (5.1.3), written on heuristic grounds, is rather general and its applications are diverse, even though it allows only for shot noise. With the exception of some threshold effects associated, for example, with the diffraction resolution limit, threshold vision contrast and so on, the fluctiation theory gives a fairly accurate description of image quality in all observation systems, including television, photography and human vision. Moreover, (5.1.3) can sometimes be applied in the case of an image constructed by the coherence methods, and for a holographic image in particular (Chap. 7).

We shall consider, for example, an expression for the signal-to-noise ratio in photoelectronic recording. In this case [5.9],

$$\delta = k\sqrt{\frac{S_\lambda \langle P \rangle}{2eB\Delta\Omega}} = k\sqrt{\frac{S_\lambda \langle P \rangle t_{\rm el}}{eB}}\,, \tag{5.1.5}$$

where $\langle P \rangle$ is the mean radiation power incident on the image cell, $S_\lambda$ the spectral response of the photocathode, $e$ the electron charge, $B$ the gain in the fluctuation noises due to secondary emission, $\Delta\Omega$ the frequency band transmitted by the detector, and $t_{\rm el} = 1/2\Delta\Omega$ the recording time of the image cell. The shorter $t_{\rm el}$, the wider the frequency band $\Delta\Omega$ required and the higher the shot noise level. The product $\langle P \rangle t_{\rm el}$ is the signal energy arriving at the image cell during the frame recording time. The quantity $\langle P \rangle t_{\rm el} S_\lambda / e$, being the mean number of photoelectrons per image cell, operates as the number $\langle N \rangle$ of photons used, and (5.1.5) is similar to (5.1.3).

### 5.1.2. Threshold Contrast

Formula (5.1.4) introduces the second image quality charactetistic. The higher the contrast $k$, the better the image quality. The test element can be distinguished from its neighbours provided its contrast is higher than some threshold value $k^*$, charctteristic of the given observation system. For example, for human vision, $k^* \approx 0.02$, while for TV and thermal imagers $k^* \approx 0.1$.

Here, we may will question whether both image quality characteristics ($k$ and $\delta$) are necessary in practice, or whether one of them suffices — for example, the signal-to-noise ratio which, as seen from (5.1.3), is a generalized parameter which includes the dependence both on the mean image irradiance and on the contrast.

At sufficiently high irradiance (large $\langle N \rangle$ and $\Delta N$), the object can, nevertheless be indiscernible because of low contrast ($k < k^*$). On the other hand, in viewing with a signal energy gap (low $\langle N \rangle$ and $\Delta N$), even the high-contrast image may "sink in noise". Thus, objects around us are less distinct in twilight than in the daylight, although the contrast is the same. Therefore, it is often said that the resolved contrast depends on the irradiance. Thus, since the SNR is determined by (5.1.2) (i.e., only shot noise is taken into account), both characteristics ($k$ and $\delta$) are required since at $k \gg k^*$ the image quality is determined by the SNR, whereas at $\delta \gg \delta^*$ (discussed below), it is determined by the contrast $k$.

However, an alternative, more general approach is possible which requires only one characteristic. We shall proceed from the basic definition (5.1.1) for the quantity $\delta$. In the general case, the noise variance $V_N$ can be expressed as the series

$$V_N = \alpha_0 + \alpha_1 \langle N \rangle + \alpha_2 \langle N \rangle^2 + \alpha_3 \langle N \rangle^3 + \cdots\,, \tag{5.1.6}$$

where the first terms allow a simple physical interpretation. Thus, the quantities $\alpha_0$ and $\alpha_1 \langle N \rangle$ imply the variance of the dark and shot photodetector noises, respectively. For the Poisson statistics of the distribution of photons used, $\alpha_1 = 1$.

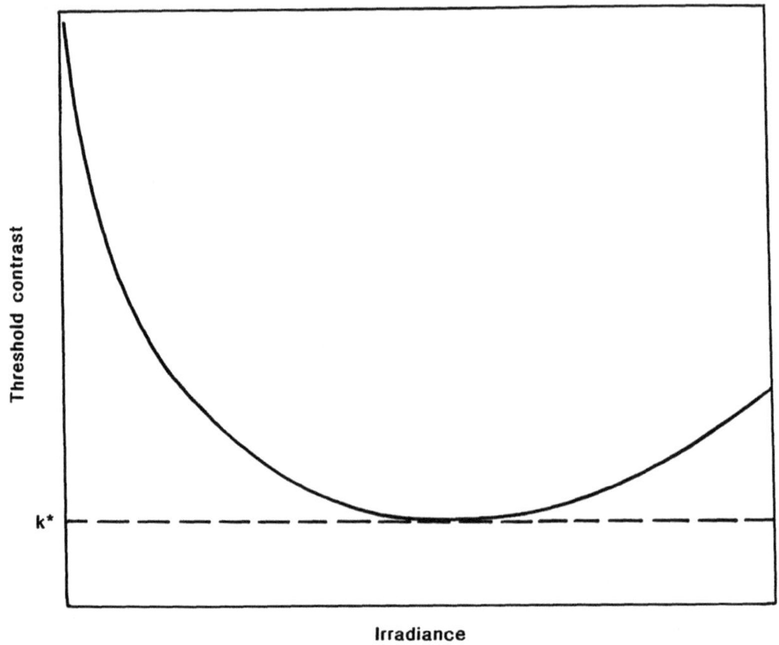


Fig. 5.1. Threshold contrast of human sight $k_E^*$ as a function of object irradiance

The term $\alpha_2\langle N\rangle^2$ may include different noises. They are dependent on the instability of the power of the radiation source and fluctuations of the optical characteristics in the image transfer channel [5.14], and on internal noises in the receiver system (nonuniformity of the detector sensitivity, gain fluctuations, etc). We shall refer to them as the transport noises of the receiving system. It is clear from the physics of the coefficients $\alpha_i$ that at least $\alpha_0$, $\alpha_1$ and $\alpha_2$ are positive.

The relations (5.1.1, 6) allow one to determine the threshold contrast $k^*$ in terms of $\delta^*$, rather than include it as an independent parameter. Moreover, at positive $\alpha_i$ and taking into account four terms in (5.1.6), equations (5.1.1, 6) define both the existence of the threshold contrast in the image forming system and the known dependence of the human vision threshold contrast $k_E^*$ on the object irradiance (Fig. 5.1). In fact, assuming that the threshold contrast restriction of vision ($k \geqslant k_E^*$) physically based on the SNR $\delta$ decreases to its threshold value $\delta^*$, from (5.1.1, 6) and including (5.1.4), we obtain

$$k_E^* = \delta^* \sqrt{\alpha_0/\langle N\rangle^2 + \alpha_1/\langle N\rangle + \alpha_2 + \alpha_3\langle N\rangle}\ . \tag{5.1.7}$$

The relation (5.1.7) correctly describes the function $k_E^*$ over the entire irradiance range (Fig. 5.1). The minimum $(k_E^*)_{min} = k^*$ which is responsilbe for the vision threshold under optimum irradiance conditions is generally a function of $\alpha_2$. It is shown [5.13] that for human vision, the term $\alpha_2\langle N\rangle^2$ in (5.1.6) applies to the transport noises of the neuron system which is a physical basis of the known

Weber-Fechner law. Thus, when the medium characteristics do not fluctuate in the image transfer channel, $\alpha_2$ is the parameter of the video system itself.

We shall consider in more detail the physical significance this quantity. Assuming the number $\langle N\rangle$ of photons used not to be large, we disregard the terms of order $\langle N\rangle^3$ and greater in (5.1.6). In this case, the video system parameter $\alpha_2$ uniquely relates the threshold values $\delta^*$ and $k^*$, and, as seen from (5.1.7), is determined by

$$\alpha_2 = (k^*/\delta^*)^2 \,. \tag{5.1.8}$$

Another physical significance can be assigned to the parameter $\alpha_2$. It defines an ultimate accuracy of signal measurement of the system. In fact, from (5.1.6), the relative error in the photon number measurement is

$$\sqrt{V_N}/\langle N\rangle = \sqrt{\alpha_0/\langle N\rangle^2 + \alpha_1/\langle N\rangle + \alpha_2}$$

and at $\langle N\rangle \to \infty$

$$\sqrt{V_N}/\langle N\rangle = \sqrt{\alpha_2} \,. \tag{5.1.9}$$

Assuming, for example $\delta^* = 5$, at $k^* = 0.1$ (a characteristic of many TV systems) $\sqrt{V_N}/\langle N\rangle = 0.02$. For $k^* = 0.02$ (the threshold contrast of human vision), $\sqrt{V_N}/\langle N\rangle = 0.004$. It should be noted that relation (5.1.8) provides a real estimate of $\alpha_2$ from the known system parameters $k^*$ and $\delta^*$.

### 5.1.3 General Image Quality Criterion

We shall return to the question of vision object criteria. An analysis of (5.1.7, 8) has shown that the threshold contrast $k^*$ can be considered as a manifestation of he photodetector transport noises. Therefore, if we define the SNR by (5.1.1, 6), confining ourselves to the $\langle N\rangle^2$ order terms, then the condition $\delta \geqslant \delta^*$ is the general criterion of object vision. Consider the case where dark noises are insignificant and the SNR is defined by

$$\delta = \frac{\Delta N}{\sqrt{\langle N\rangle + \alpha_2\langle N\rangle^2}} \,. \tag{5.1.10}$$

At low irradiance, shot noises are dominant and (5.1.10) becomes (5.1.2). Otherwise, (5.1.10, 8) yields the relation $\delta = k\delta^*/k^*$ and the vision condition $\delta \geqslant \delta^*$ is equivalent to $k \geqslant k^*$.

It is worth noting that, in principle, both approaches characterize the image quality. The use of the contrast $k$ as an independent characteristic and of the inequality $k > k^*$ as an independent condition of visionis physically instructive. However, the alternative approach based on (5.1.10) seems to be more consistent. Its merits are most evident where the problem of optimizing the vision system parameters and estimating its ultimate potential is concerned (Sect. 5.4.2). In these cases, the use of the general vision criterion offers a more simple and unambiguous solution.

### 5.1.4 Threshold Signal-to-Noise Ratio

What is the minimum $\delta$ at which image cell resolution is possible? The condition $\delta \geqslant 1$ is frequently regarded as the vision criterion. In fact, the condition of equal signal and noise is insufficient to observe an object. Moreover, the threshold quantity $\delta^*$ is by no means some fixed characteristic of the vision system. On the contrary, it depends on the requirements laid down by the system when solving one particular problem or other. It is determined in terms of possible false alarm $P_f$ and detection $P_{det}$ probabilities.

We now consider, in particular, the problem of remote TV or visual detection of an object in a scattering medium. We compare the signals of two image cells, one corresponding to the object center and the other determining the background level. Let $\langle N_1 \rangle$ and $\langle N_2 \rangle$ denote the mean number of photons used per image cell in the background and object image domains respectively. Then, assuming Gaussian statistics, the probability of $N$ photons arriving at the $i$-th element (Fig. 5.2), is

$$P(N/\langle N_i \rangle) = \frac{1}{\sqrt{2\pi V_{N_i}}} \exp\left[-\frac{(N - \langle N_i \rangle)^2}{2V_{N_i}}\right], \quad i = 1, 2, \tag{5.1.11}$$

where $V_{N_i}$ is the variance of the photon number fluctuation in the object and background image.

We assume that the observer decides that an object is present at $(N - \langle N_1 \rangle)/\sqrt{V_{N_1}} \geqslant \theta^*$. The quantity

$$\theta^* = \frac{N^* - \langle N_1 \rangle}{\sqrt{V_{N_1}}} \tag{5.1.12}$$

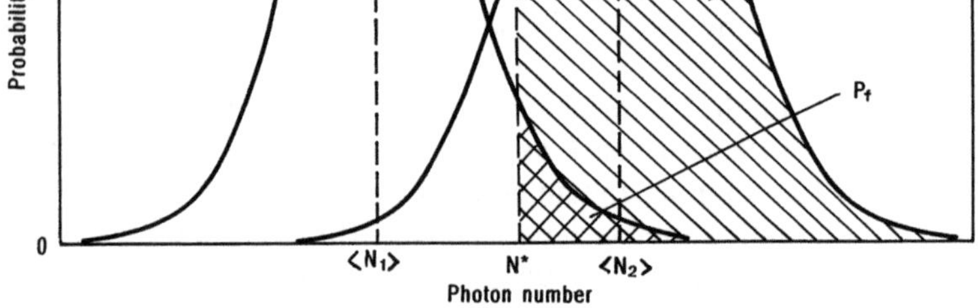


**Fig. 5.2.** Detection $P_{det}$ and "false alarm" $P_f$ probabilities

is the decision-making threshold, and $N^*$ is some discrimination level. The probability of a false alarm $P_f$, i.e., the probability that $N > N^*$ without an object, and the detection probability $P_{det}$, i.e., the probability that in the presence of an object $N > N^*$, are

$$P_f = \int_{N^*}^{\infty} P(N/\langle N_1 \rangle) dN \,, \qquad P_{det} = \int_{N^*}^{\infty} P(N/\langle N_2 \rangle) dN \,. \tag{5.1.13}$$

Then, taking (5.1.11) into account, we have

$$\begin{aligned} P_f &= 0.5 \operatorname{erfc}(\theta^*/\sqrt{2}) \,, \\ P_{det} &= 0.5 \operatorname{erfc}(\eta^*/\sqrt{2}) \,, \end{aligned} \tag{5.1.14}$$

where

$$\eta^* = \frac{N^* - \langle N_2 \rangle}{\sqrt{V_{N_2}}} \,. \tag{5.1.15}$$

As seen from (5.1.14), the decision-making threshold $\theta^*$ is determined by the given $P_f$. As, by definition,

$$\delta = \frac{\langle N_2 \rangle - \langle N_1 \rangle}{\sqrt{V_{N_1}}} \,, \tag{5.1.16}$$

taking into account (5.1.12, 5), we find for the threshold SNR

$$\delta^* = \theta^* - \eta^* \sqrt{V_{N_2}/V_{N_1}} \,. \tag{5.1.17}$$

At $\Delta N << \langle N_1 \rangle$, when $V_{N_2} \approx V_{N_1}$, we have

$$\delta^* = \theta^* - \eta^* \,. \tag{5.1.18}$$

This simple and most frequently encountered case if of common concern [5.13, 15]. For this case, the values of $\delta^*$ at different $P_f$ and $P_{det}$ are presented in Table 5.1. We can easily find $\delta^*$ from Table 5.1 or by solving (5.1.14, 17).

It should be noted that the Gaussian distribution (5.1.11) may be used in place of the Poisson distribution, which really describes the functions $P(N/\langle N_i \rangle)$ only

**Table 5.1.** Signal-to-noise ratio $\delta^*$ as a function of $P_f$ and $P_{det}$.

| | $P_{det}$ | | | | | |
|---|---|---|---|---|---|---|
| $P_f$ | 0.5 | 0.9 | 0.95 | 0.99 | 0.995 | 0.999 |
| $10^{-7}$ | 5.3 | 6.9 | 6.9 | 7.5 | 7.8 | 8.3 |
| $10^{-6}$ | 4.9 | 6.1 | 6.4 | 7.1 | 7.3 | 7.8 |
| $10^{-5}$ | 4.4 | 5.6 | 5.9 | 6.6 | 6.8 | 7.3 |
| $10^{-4}$ | 3.9 | 5.0 | 5.8 | 6.0 | 6.2 | 6.7 |
| $10^{-3}$ | 3.3 | 4.4 | 4.7 | 5.3 | 5.6 | 6.1 |
| $10^{-2}$ | 2.6 | 3.6 | 3.9 | 4.6 | 4.8 | 5.3 |
| $10^{-1}$ | 1.6 | 2.6 | 2.9 | 3.6 | 3.8 | 4.4 |

at $\langle N_i \rangle \gg 1$. This substitution is impossible at very small $P_f$ and $1 - P_{det}$ where these functions are determined by the "tails" of the distributions $P(N/\langle N_i \rangle)$.

Thus, the object vision criterion is the condition $\delta > \delta^*$. In practice, different values are used for this threshold. We shall analyse them in he light of the above results.

The greater the number of cells in an image, the higher the probability that false signals will appear in it and the lower should be the $P_f$ provided for an individual cell. For some high-quality imaging systems (human vision. TV tubes, etc.), the typical number of cells is about $10^5$. To avoid false signals which could easily occur with such a large number of picture elements requires that $P_f$ be less than $10^{-5}$ and, hence, at $P_{det} = 0.5$, $\delta \approx 4 - 5$ (Table 5.1).

However, it is not always necessary to adopt such stringent precautions against false signals. According to modern psychophysiological research, the human vision system (eye + brain) is able to encompass about $10^3$ picture elements at one time. In this case ($P_f \approx 10^{-3}$), as seen from Table 5.1, the SNR, which allows false signals to be avoided at $P_{det} = 0.5$, must be $\delta^* \approx 3$. This value agrees well with the data obtained by Rozell and Wilson [5.16] who measured the probability of detecting some small rectangular test-object against a homogeneous background. They found that the SNR $\delta = 3$ ensures the detection probability $P_{det} = 0.5$.

On the other hand, the detection probability $P_{det} = 0.5$ is insufficient in practice, whereas $P_{det} \approx 0.95$ to $0.99$ is acceptable. In this case, $P_f \approx 10^{-3}$, $P_{det} \approx 0.95$ to $0.99$, $\delta^* \approx 5$. This is presumably why $\delta^* = 5$ is often taken as the threshold signal-to-noise ratio for television and human vision systems [5.10].

### 5.1.5 Signal-to-Noise Ratio in a Medium with Fluctuating Optical Parameters

We have so far supposed that the main noises in a photodetector are due to photon number fluctuations dependent on the mean signal power (shot noise), as well as to transport noises of the system which determine the threshold contrast of the system. Under real conditions, however, a noise component due to random signal modulation caused by fluctuations of the medium parameters may appear in the image. In some cases, this contribution is comparable to, or even greater than the contribution of shot noises. Such a situation occurs, for example, in observation through stochastic clouds, rough sea surfaces, etc. It is characteristic and novel that these noises can be correlated even for the remote image cells under comparison, since the spatial correlation scale of image noises conforms to the correlation scale of macroinhomogeneities in clouds or sea waves, and may even exceed the frame sizes. Such a correlation requires us to generalize some of the results discussed in Sect. 5.1.4.

We assume, as before, that the decision on the presence or absence of an object is made by comparing the number of photons reaching a test image cell with the number of photons on a background cell of the same area $\Sigma_{el}$ averaged over some area $\Sigma$. The correlation radius for the noises due to fluctuations of the optical medium parameters can be comparable to, or greater than, the linear size of the averaging area $\Sigma$. Therefore, the $\Sigma$-mean number of photons on a background image cell

$$N_{1f} = \frac{1}{\Sigma} \iint_{\Sigma} N_1(\vec{r}) d\vec{r} , \tag{5.1.19}$$

to which a test element signal is in fact compared, is different from $\langle N_1 \rangle$, the mean number of photons on the background image cell averaged over all realizations. The subscript "f" in (5.1.19) underlines that averging is taken over a sample of the area $\Sigma$ in the given frame.

Let a signal $N_i$ on a test cell in the absence of an object ($i = 1$), or in the presence of an object ($i = 2$), exceed the sample mean $N_{1f}$ by the value

$$\Delta N_i = N_i - N_{1f} , \qquad i = 1, 2 . \tag{5.1.20}$$

By definition, the false alarm probability $P_f$ and the detection probability $P_{det}$ are, correspondingly, the probabilities of $\Delta N_1 > \Delta N^*$ and $\Delta N_2 > \Delta N^*$, where $\Delta N^*$ is the decision-making threshold in the presence of an object. Assuming the Gaussian statistics in the $\Delta N_i$ distribution, we can write

$$P_f = \int_{\Delta N^*}^{\infty} \frac{1}{\sqrt{2\pi V_{\Delta N_1}}} \exp\left[ -\frac{(\Delta N_1 - \langle \Delta N_1 \rangle)^2}{2V_{\Delta N_1}} \right] d(\Delta N_1) , \tag{5.1.21}$$

$$P_{det} = \int_{\Delta N^*}^{\infty} \frac{1}{\sqrt{2\pi V_{\Delta N_2}}} \exp\left[ -\frac{(\Delta N_2 - \langle \Delta N_2 \rangle)^2}{2V_{\Delta N_2}} \right] d(\Delta N_2) . \tag{5.1.22}$$

Here

$$\langle \Delta N_i \rangle = \langle N_i - N_{1f} \rangle = \langle N_i \rangle - \langle N_1 \rangle , \tag{5.1.23}$$

$$V_{\Delta N_i} = \langle (\Delta N_i - \langle \Delta N_i \rangle)^2 \rangle , \qquad i = 1, 2 . \tag{5.1.24}$$

We now introduce the quantities

$$\theta^* = \frac{\Delta N^*}{\sqrt{V_{\Delta N_1}}} , \tag{5.1.25}$$

$$\eta^* = \frac{\Delta N^* - (\langle N_2 \rangle - \langle N_1 \rangle)}{\sqrt{V_{\Delta N_2}}} , \tag{5.1.26}$$

and the signal-to-noise ratio

$$\delta = \frac{\langle N_2 \rangle - \langle N_1 \rangle}{\sqrt{V_{\Delta N_1}}} . \tag{5.1.27}$$

The threshold signal-to-noise ratio $\delta^*$ can then be written as

$$\delta^* = \theta^* - \eta^* \sqrt{V_{\Delta N_2} / V_{\Delta N_1}} , \tag{5.1.28}$$

where, following (5.1.21, 22) and allowing for (5.1.25, 26), $\theta^*$ and $\eta^*$ are found from the equation system (5.1.14).

Formulas (5.1.25–28) coincide in their structure with the appropriate (5.1.12, 15–17) and are their natural generalization including the correlation between the number of photons in a test cell and the mean number $N_{1f}$ of photons in a reference cell. It is worth noting that instead of $V_{N_1}$ and $V_{N_2}$, (5.1.27, 28) incorporate $V_{\Delta N_i}$ $(i = 1, 2)$ equal to $V_{N_i}$ at $N_{1f} = \langle N_1 \rangle$.

We try to relate $V_{\Delta N_i}$ and $V_{N_i}$ to the image noise correlation function. From (5.1.24) and accounting for (5.1.19, 20), we have

$$V_{\Delta N_i} = \left\langle \left[ N_i(\vec{r}_0) - \frac{1}{\Sigma} \iint_{\Sigma} N_1(\vec{r})\, d\vec{r} \right]^2 \right\rangle - (\langle N_i \rangle - \langle N_1 \rangle)^2 , \qquad (5.1.29)$$

Here, the coordinate $\vec{r}_0$ characterizes the location of the test element, the coordinate origin being taken at some center of the area $\Sigma$. To describe the averaging area $\Sigma$, we introduce the function

$$\begin{aligned} U(\vec{r}, \Sigma) &= 1 , \qquad \vec{r} \in \Sigma \\ &= 0 , \qquad \vec{r} \notin \Sigma , \end{aligned} \qquad (5.1.30)$$

and denote its Fouries-spectrum as $U(\vec{v}, \Sigma)$. Then, from (5.1.29), we have

$$V_{\Delta N_i} = V_{N_i} + \frac{1}{(2\pi)^2} \iint_{-\infty}^{\infty} \Gamma(\vec{v}) [V_{N_1} U_0^2(\vec{v}) - 2 V_{\sqrt{N_1 N_i}} U_0(\vec{v}) e^{-i\vec{v}\cdot\vec{r}_0}]\, d\vec{v} ,$$

$$i = 1, 2 , \qquad (5.1.31)$$

where

$$U_0(\vec{v}) = \frac{U(\vec{v}, \Sigma)}{U(\vec{v} = 0, \Sigma)} , \qquad (5.1.32)$$

$$V_{\sqrt{N_1 N_i}} = \langle N_1(\vec{r}) N_i(\vec{r}) \rangle - \langle N_1 \rangle \langle N_i \rangle , \qquad (5.1.33)$$

and $\Gamma(\vec{v})$ is the Fourier-spectrum of the spatial noise correlation function $\Gamma(\vec{r})$ normalized by

$$\Gamma(\vec{r} = 0) = \frac{1}{(2\pi)^2} \iint_{-\infty}^{\infty} \Gamma(\vec{v})\, d\vec{v} = 1 . \qquad (5.1.34)$$

Expression (5.1.31) is inconvenient for calculations because of the uncertainly inherent in the choice of the averaging area $\Sigma$ and, correspondingly, the spectrum $U_0(\vec{v})$. Therefore, we consider some particular cases.

Let the averaging area be large as compared to the correlation area of the medium noises. Then, $U_0(\vec{v}) \to 0$ and, from (5.1.31), we have

$$V_{\Delta N_i} = V_{N_i} , \qquad (5.1.35)$$

i.e., the variances of the random values of $\Delta N_i$ and the SNR (5.1.27, 28) are independent of the spatial correlation of the medium noises. It should be noted that, as in (5.1.31), we refer here to the entire set of noises, including shot, dark and transport noises of the system as well as those noises due to the medium parameter fluctuations.

If the averaging area $\Sigma$ is small ($U_0(\vec{v}) \approx 1$) and the noise correlation function $\Gamma(\vec{r})$ varies slightly over it, then, from (5.1.31), we get

$$V_{\Delta N_i} = V_{N_i} + V_{N_1} - 2V_{\sqrt{N_1 N_i}}\Gamma(\vec{r}_0). \tag{5.1.36}$$

At a large distance $\vec{r}_0$ between the image cells being compared, $\Gamma(\vec{r}_0) \to 0$ and $V_{\Delta N_i} \approx V_{N_i} + V_{N_1}$.

In practice, this case if of key interest where the averaging area $\Sigma$ is large compared to the range of correlation of shot and transport noises of the system, but small as compared to the region of correlation of the medium fluctuation noises. In this case, from (5.1.31), we obtain

$$V_{\Delta N_i} = V_{N_i} + V_{N_1}^{\text{fl}} - 2V_{\sqrt{N_1 N_i}}^{\text{fl}}\Gamma^{\text{fl}}(\vec{r}_0), \qquad i = 1, 2, \tag{5.1.37}$$

where $V_{N_1}^{\text{fl}}$ and $V_{\sqrt{N_1 N_i}}^{\text{fl}}$ are the corresponding variances of the medium fluctuation noises whose correlation function $\Gamma^{\text{fl}}(\vec{r}_0)$ is normaized by the relation (5.1.34). At $|\vec{r}_0| \to \infty$, (5.1.37) yields

$$V_{\Delta N_i} = V_{N_i} + V_{N_1}^{\text{fl}}. \tag{5.1.38}$$

Now, let the distance $|\vec{r}_0|$ between the cells being compared be much less than the medium noise correlation radius and $\langle N_2 \rangle \approx \langle N_1 \rangle$. Bearing in mind that the total variance of the shot and transport system noises is $\langle N \rangle + \alpha_2 \langle N \rangle^2$, from (5.1.37), we have

$$V_{\Delta N_1} = V_{N_1} - V_{N_1}^{\text{fl}} = \langle N_1 \rangle + \alpha_2 \langle N_1 \rangle^2,$$

$$V_{\Delta N_2} = V_{N_2} + V_{N_1}^{\text{fl}} - 2V_{\sqrt{N_1 N_i}}^{\text{fl}} \approx \langle N_2 \rangle + \alpha_2 \langle N_2 \rangle^2. \tag{5.1.39}$$

As seen from (5.1.39), in this case the variances $V_{\Delta N_i}$ and, consequently, the signal-to-noise ratio are independent of the fluctuation of the optical medium parameters. Such a situation may occur, for example, when observing through rough sea surfaces: long-period swell waves whose scale is comparable to or greater than the frame size do not distort the image of an underwater object, but only cause it to swing slowly.

## 5.2 Image Transfer Characteristics

We have, so far, considered the basic parameters of image quality and now proceed to analyse the characteristics which define the effect of the medium and optical vision system on image transfer, i.e., those characteristics which relate the observed contrast $k$ and irradiance in the image plane to the real object contrast $k_0$ and its radiant emittance.

A diffusely luminous surface having a radiant emittance distribution $E_0(\vec{r})$, where $\vec{r}$ is the radius-vector of a point in the object plane, is in view. Along with various real light sources, a great number of objects observed in the infrared on Earth can be regarded as self-luminous. Also, many other image transfer problems reduce to one of viewing a self-luminous object. Observation of a diffusely reflecting object with the albedo distribution $A(\vec{r})$ in sunlight can be reduced to the problem of viewing a self-luminous object with the radiant emittance distribution $E_0(\vec{r}) = E_s A(\vec{r})$, where $E_s$ is the object surface irradiance due to the sun.

We write $E(\vec{r}\,')$ for the irradiance distribution in the image plane $\vec{r}\,'$ of an optical system. It is clear that the image transfer problem consists of finding agreement between the object radiant emittance $E_0(\vec{r})$ and the image irradiance $E(\vec{r}\,')$, making allowance for the effect of the medium along the object-photodetector path and the parameters of the receiving system.

### 5.2.1 Point Spread Function. Optical Transfer Function

In view of the infinite variety of objects which can be observed, it makes no sense to consider the question of image transfer for each particular case. It seems more reasonable to choose a procedure which is capable of expressing the image quality parameters of any object in terms of the image transfer characteristics of some simple test-objects. Such an approach, based on the theory of linear systems, is commonly used in vision theory [5.1–4, 17]

Any self-luminous diffuse object can be considered as a combination of elementary point light sources. In a linear optical system, its image can be otained by superposing the images of individual sources, i.e.,

$$E(\vec{r}\,') = \iint_{-\infty}^{\infty} E_0(r) a(\vec{r}\,', \vec{r})\, d\vec{r}\,, \tag{5.2.1}$$

where the function $a(\vec{r}\,', \vec{r}_1)$, termed the point spread or point scattering function (PSF), defines the irradiance distribution in the image of a diffuse source having the radiant emittance $E_0(\vec{r}) = \delta(\vec{r} - \vec{r}_1)$.

In image transfer problems, it is usual to distinguish a wide and very important class of invariant (isoplanatic) systems whose point spread function is independent of the source coordinate $\vec{r}_1$ and is determined only by the difference $\vec{r}\,' - \vec{r}_1$, i.e., $a(\vec{r}\,', \vec{r}_1) = a(\vec{r}\,' - \vec{r}_1)$. For such systems, discussed below, (5.2.1) is replaced by the convolution integral

$$E(\vec{r}\,') = \iint_{-\infty}^{\infty} E_0(\vec{r}) a(\vec{r}\,' - \vec{r})\, d\vec{r}\,. \tag{5.2.2}$$

We shall apply the two-dimensional Fourier transform over the coordinate $\vec{r}\,'$ to (5.2.2). Then, for the Fourier domain, we obtain the relation

$$E(\vec{v}) = E_0(\vec{v}) a(\vec{v})\,, \tag{5.2.3}$$

where

$$E(\vec{v}) = \iint_{-\infty}^{\infty} E(\vec{r}\,') e^{-i\vec{v}\cdot\vec{r}\,'}\, d\vec{r}\,'\,, \tag{5.2.4}$$

$$a(\vec{v}) = \iint_{-\infty}^{\infty} a(\vec{r}\,')e^{-i\vec{v}\cdot\vec{r}\,'}\,d\vec{r}\,', \tag{5.2.5}$$

$\vec{v}$ being the spatial frequency. To avoid introducing a large number of new symbols, we shall distinguish between the functions $E(\vec{r}\,')$, $E_0(\vec{r})$, $a(\vec{r}\,')$ and their Fourier transforms, as in Chap. 4, only by the argument.

Expressions (5.2.2, 3) seem to be consistent with the main idea of applying the theory of linear systems and, in particular, the Fourier transform, to the image transfer problem. Image determination for the entire variety of objects is reduced to the question of determining a single function, the PSF $a(\vec{r}\,')$ or its Fourier transform $a(\vec{v})$, independent of the properties of the object. Here, the image $E(\vec{r}\,')$ is determined by (5.2.2), or expressed by the inverse Fourier transform in terms of the spectrum $E(\vec{v})$:

$$E(\vec{r}\,') = \frac{1}{4\pi^2}\iint_{-\infty}^{\infty} E(\vec{v})e^{i\vec{v}\cdot\vec{r}\,'}\,d\vec{v}\,. \tag{5.2.6}$$

The function $a(\vec{v})$ is the optical transfer function (OTF) of a system. It is preferable to the PSF since it involves simple multiplication of the spectra in the Fourier space rather than the intricate integration (5.2.2) required to determine the structure of the image $E(\vec{r}\,')$.

If the spread function is symmetric about an azimuth, $a(\vec{r}\,') = a(|\vec{r}\,'|) = a(r')$, then the OTF also possesses circular symmetry, $a(\vec{v}) = a(|\vec{v}|) = a(v)$, and is related to the spread function by the Hankel transform [5.3, 18]:

$$a(\vec{v}) = \int_0^{2\pi} d\varphi \int_0^{\infty} a(r')e^{-ivr'\cos\varphi}r'\,dr' = 2\pi\int_0^{\infty} a(r')J_0(vr')r'\,dr'\,, \tag{5.2.7}$$

$$a(r') = \frac{1}{2\pi}\int_0^{\infty} a(v)J_0(vr')v\,dv\,, \tag{5.2.8}$$

where $J_0(vr')$ is the Bessel function.

The function

$$a_n(\vec{v}) = \frac{a(\vec{v})}{a(\vec{v}=0)} \tag{5.2.9}$$

is called the normalized OTF. It is correlated (in terms of the direct and inverse Fourier transform) with the frequently used normalized scattering function $a_n(\vec{r}\,') = a(\vec{r}\,')/\iint_{-\infty}^{\infty} a(\vec{r}\,')\,d\vec{r}\,'$ which is seen to obey the relation

$$\iint_{-\infty}^{\infty} a_n(\vec{r}\,')\,d\vec{r}\,' = 1\,. \tag{5.2.10}$$

In the general case, the sytem OTFs $a_n(\vec{v})$ and $a(\vec{v})$ are complex functions, i.e.

$$a_n(\vec{v}) = |a_n(\vec{v})|e^{i\varphi(\vec{v})}\,. \tag{5.2.11}$$

The normalized OTF modulus $|a_n(\vec{v})| \equiv T(\vec{v})$ is termed te modulation transfer function (MTF). The dependent of the phase $\varphi(\vec{v})$ on the frequency $\vec{v}$ is known as the phase transfer function (PTF). That the phase $\varphi(\vec{v})$ is other than zero points to the asymmetric behavior of the point spread function. In optics, the PSF is generally symmetric. In this case, the normalized system OTF coincides with the MTF. Nevertheless, it is sometimes necessary to deal with the asymmetric PSF and take account of the PTF. It occurs, for example, at certain aberrations in optical systems, when observing through clouds and fog along inclined paths [5.19, 20] (Sect. 6.3), in using some scanning techniques in IR imagers [5.11] and on imaging by luminescent screens (Chap. 10), etc.

The relation (5.2.5) formally defines the OTF as the Fourier spectrum of the point spread function. Let us clarify its physical significance. Consider a diffuse luminous test object whose spatial radiant emittance distribution is harmonic[1]

$$E_0(\vec{r}) = \langle E_0 \rangle (1 + k_0 e^{i\vec{v}\cdot\vec{r}}) . \tag{5.2.12}$$

Here, $\langle E_0 \rangle$ is the mean radiant emittance of the harmonic test object, and

$$k_0 = \frac{E_0^{max} - E_0^{min}}{E_0^{max} + E_0^{min}} \tag{5.2.13}$$

defines the true object contrast $E_0^{max} = \langle E_0 \rangle (1 + k_0)$ and $E_0^{min} = \langle E_0 \rangle (1 - k_0)$ are, correspondingly, the maximum and minimum radiant emittances of the test object and $\vec{v}$ stands for the spatial frequency, so that the space between ligth (or dark) strips on the object $\Delta r = 2\pi/v$. In the following analysis, such a test object will be termed the harmonic test object.

Substituating (5.2.12) into (5.2.2) and taking into account (5.2.5), we arrive at

$$E(\vec{r}') = \langle E \rangle [1 + k(\vec{v}) e^{i(\vec{v}\cdot\vec{r}' + \varphi(\vec{v}))}] , \tag{5.2.14}$$

where

$$\langle E \rangle = \langle E_0 \rangle \iint_{-\infty}^{\infty} a(\vec{r}') d\vec{r}' = \langle E_0 \rangle a(\vec{v} = 0) , \tag{5.2.15}$$

$$k(\vec{v}) = k_0 |a_n(\vec{v})| . \tag{5.2.16}$$

Expression (5.2.14) implies that the irradiance distribution $E(\vec{r}')$ in the test object image, as well as the distribution of the radiant emittance $E_0(\vec{r})$, is harmonic and has the same spatial frequency $\vec{v}$[2]. This immediately follows from the linearity of the system. The quantities $\langle E \rangle$ and $k(\vec{v})$ are the mean image irradiance and the contrast observed, respectively.

---

[1] It is obvious that the radiant emittance is a real quantity $\mathrm{Re}\{E_0(\vec{r})\} = \langle E_0 \rangle [1 + k_0 \cos(\vec{v}\vec{r})]$. However, the exponential form is mathematically more convenitent.

[2] We omit here possible linear-scale variations of the spatial frequency with the magnifying power of an optical system. Thus, the harmonic test object period for a photographic image differs, in fact, from the period of an original test object. Their proportionality factor is, however, independent of the frequency $\vec{v}$.

Thus, the system MTF is the ratio of the contrast observed in the image of a diffusely luminous object having harmonically distributed radiant emittance to the initial contrast $k_0$, which depends on the test object frequency $\vec{v}$. The system PTF defines a phase shift in the harmonic test object image. In some cases (see, for example, Chap. 10), a wider significance is added to the OTF and MTF concepts, without restricting to a diffuse nature for the harmonic test object emittance.

It is clear from the condition $|a_n(\vec{v})| < 1$ that, in any linear sytem, the contrast observed is lower than the true one, i.e., $|k(\vec{v})| < k_0$. The contrast of the harmonic test object image, which depends on its spatial frequency $\vec{v}$, is defined by the sytem MTF $T(\vec{v})$. The lower contrast is due to the superposition of light fluxes from different object areas which occurs in imaging due to the imperfect nature of the system, to scattering by the medium, for example. Generally, the greater the harmonic test object frequency $\vec{v}$, the shorter the interval between the function $E_0(\vec{r})$ maxima and the wider the point spread function $a(\vec{r}')$, the more distinct the superposition of fluxes and the more important the contrast degrading. The optical system MTF is usually a monotonic function which decreases faster, the wider the PSF. Here, $T(\vec{v} = 0) = 1$. It is obvious that the system for which the MTF $a(\vec{r}') = \delta(\vec{r}')$, and $T(\vec{v}) = 1$ for all $\vec{v}$, would be ideal for transfer. The narrower the spread function, i.e., the slower the decrease in the MTF with increasing $\vec{v}$, the greater the spatial frequencies (and, correspondingly, the smaller the object elements) which can be resolved by such a system.

The radiant emittance $E_0(\vec{r})$ of any object can be represented in terms of the Fourier series (5.2.6) as a superposition of the harmonics. In a linear system, the transfer of individual components (harmonics) can be considered independently. In this case, the spatial harmonic amplitude and phase transfer coefficients dependent on their frequency are described by the system OTF. Therefore, knowing the OTF as well as the PSF of the system allows one to predict the structure of any image. The OTF, naturally, does not provide any new information, in contrast to the point spread function. The problem of the advantages of one or other characteristic in different situations generally determines how instructive the description will be.

### 5.2.2 Aspect Invariance of a System

We shall now consider one of the requirements which the imges transfer system must obey to be reliably described by the apparatus of transfer functions (OTF, MTF and PTF).

For convenience, an inverted imaging scheme will be introduced in the following analysis in place of the direct technique. Locate a diffuse point source with a radiant emittance $E_0(\vec{r}_1') = \delta(\vec{r}' - \vec{r}_1')$ in the optical system image plane. We use $a(\vec{r}, \vec{r}')$ for the irradiance distribution $E(\vec{r})$ due to this source in the subject plane $\vec{r}$. The optical reciprocity theorem implies that (Sect. 2.5)

$$a(\vec{r}, \vec{r}') = a(\vec{r}', \vec{r}) , \quad (5.2.17)$$

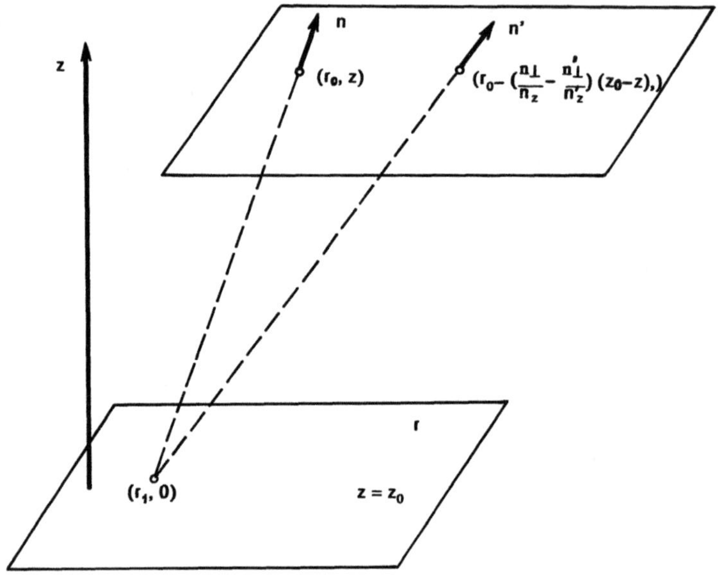


**Fig. 5.3.** Aspect invariance

where $a(\vec{r}', \vec{r})$ is the point spread function in the direct imaging scheme where the diffue source is at the point $\vec{r}$. The relation (5.2.17) means that the contribution of the isotropic source at the point $\vec{r}$ to the irradiance $E(\vec{r}')$ in the direct technique is the same as the contribution of the same source from the point $\vec{r}'$ to the irradiance $E(\vec{r})$ in the inverted scheme. We note that due to the relation (5.2.17), the inverted scheme is often used to estimate the PSF.

Let us divide, by convention, the entire image transfer path into (1) a scattering medium and (2) an optical system (a lens, in the simplest case). By the term scattering medium we understand any medium which distorts the image along the path from an object to the optical receiving system.

First, we shall consider the optical system to be ideal, i.e. in the absence of a scattering medium, the point source image being a point, $a(\vec{r}', \vec{r}_1) = C\delta(\vec{r}' - \vec{r}_1')$. Here, $C$ is the proportionality factor; the points $\vec{r}_1$ and $\vec{r}_1'$ are optically conjugate. Also assume that in a scattering media the field from the diffusely luminous object located in the plane $z = z_0$ (Fig. 5.3) obeys the following condition [5.21], i.e.,

$$I(\vec{r}_0, z, \vec{n}) = I[\vec{r}_0 - (\vec{n}_\perp/n_z - \vec{n}_\perp'/\vec{n}_z')(z_0 - z), z, \vec{n}'] \,. \tag{5.2.18}$$

Here, as in Chap. 4, $\vec{n}$ is the unit directional vector; $\vec{n}_\perp$ and $n_z$ are the vector $\vec{n}$ projection onto the plane $z = 0$ and axis $z$, respectively; $I(\vec{r}_0, z, \vec{n})$ is the radiance at the point $(\vec{r}_0, z)$ in the $\vec{n}$ direction. Here and below we say that the system, for which the relation (5.2.18) holds, possesses the *property of aspect invariance*. Strictly speaking, the radiation field obeys the condition of aspect invariance in a free space alone. The applicability of the assumption (5.2.18) to a scattering medium will be discussed below.

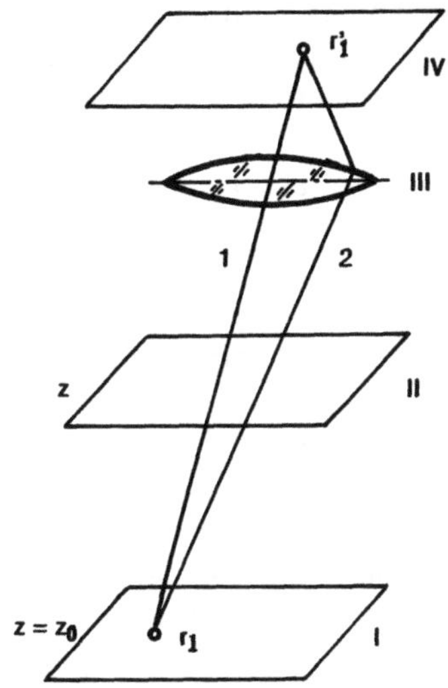


**Fig. 5.4.** Beam path in an ideal optical system: I, object plane; II, plane of a scatterer; III, optical system; IV, image plane

With reference to the optical reciprocity theorem, the aspect invariance property of the medium and the ideality of the optical system have the following consequences.

1. If a scattering medium is illuminated by a monodirectional point source of unit power located at a point $(\vec{r}_0, z)$ and emitting in the $-\vec{n}$ direction (Fig. 5.3), then the transmitted irradiance distribution $E(\vec{r})$ in the plane $z = z_0$ is independent of the source coordinate $\vec{r}_0$ and of its $-\vec{n}$ direction. On the contrary, it is completely determined only by the coordiates $(\vec{r}_1, z = z_0)$ of the point at which the beam is directed. Here, $E(\vec{r}) = S(\vec{r} - \vec{r}_1)$, $S(\vec{r})$ being the medium PSF. This consequence does not apply to the optical system and may be the validity criterion for the aspect invariance condition in the medium.

2. The radiance of all the beams from a diffuse source incident at some fixed point $\vec{r}_1'$ in the image plane (for example, beams 1 and 2 in Fig. 5.4) is equal. Therefore, given the radiance $I(\vec{n})$ at the center of the entrance pupil of the optical system (beam 1 in Fig. 5.4), the irradiance $E(\vec{r}_1')$ and energy flux $P(\vec{r}_1')$ through a picture element of area $\Delta\Sigma$ can be calculated. The flux incident on the picture element is

$$P(\vec{r}_1') = t_{\mathrm{opt}}\Sigma_{\mathrm{rec}} \int_{\Omega_{\mathrm{rec}}} I(\vec{n})\, d\vec{n} \,, \tag{5.2.19}$$

where $t_{\mathrm{opt}}$ is the transmission of the optical system, $\Sigma_{\mathrm{rec}}$ the entrance pupil area, $\Omega_{\mathrm{rec}}$ the solid receiving angle corresponding to the resolution cell $\Delta\Sigma$.

For the simplest optical system, a lens, $\Sigma_{\mathrm{rec}} = \pi D^2/4$, $\Omega_{\mathrm{rec}} = \Delta\Sigma/l^2$, $D$ is the lens diameters, and $l$ refers to the distance between the lens and image planes. We emphasize that (5.2.19) is rigorous and determines the energy flux through the area $\Delta\Sigma$ in the image plane itself rather than in the focal objective plane as is frequently assumed. Correspondingly, the quantity $l$, which determines the solid receiving angle $\Omega_{rec}$, is equal to the focal objective distance when infinitely remote objects are observed.

3. The point spread function $a(\vec{r}, \vec{r}')$ can be determined according to the inverted scheme simply by directing one beam through the entrance pupil center, rather than locating a diffuse source in the image plane. In this case, the PSF for the entire system $a(\vec{r}, \vec{r}_1') = CS(\vec{r} - \vec{r}_1)$ (points $\vec{r}_1$ and $\vec{r}_1'$ are conjugate) i.e., it is equal to the medium PSF, accurate to the coefficient, and the system as a whole possesses isoplanacity.

We now consider a non-ideal optical receiving ststem. The deviation from the ideal can be due, for example, to the fact that the planes of the image and recording are different.

Let a fictitious diffuse source $E_0(\vec{r}') = \delta(\vec{r}')$ located in the photodetector plane (the inverted scheme) generate the irradiance distribution $E(\vec{r}) = a_{opt}(\vec{r})$ in the absence of a medium. Then, according to the first consequence of aspect invariance, the irradiance distribution $E(\vec{r})$ in the presence of a medium is of the form

$$E(\vec{r}) = \int\!\!\int_{-\infty}^{\infty} a_{opt}(\vec{r}'')S(\vec{r} - \vec{r}'')d\vec{r}'' . \tag{5.2.20}$$

Applying the Fourier transform to (5.2.20), we find the OTF of the entire system

$$a(\vec{v}) = a_{opt}(\vec{v})S(\vec{v}) , \tag{5.2.21}$$

i.e., the OTF of the entire system can be represented as a product of the OTFs of its components. This property is very important and accounts for the fact that the linear system approach is so widely applied to the image transfer problem. It enables one to divide the complicated image transfer path which consists of several subsystems with different optical and image transfer characteristics (for example, sea water, atmosphere, media interface, optial receiver system, photo-recording medium, etc.) into a number of simple constituents. Then, one may study the image transfer characteristics independently and find the OTF of the entire system as a product of the individual OTFs. It will be recalled that, in contrast to the OTF, the spread function of the entire system is expresed in a more intricate fashion as the convolution (5.2.20) of the PSFs of its components.

The relations (5.2.20, 21) are widely used in practice. As has been seen, their validity is postulated by the absence of feedback between the system components. The output of the first subsystem is the input of the second, the output of the second system serves as the input for the third, etc. This condition is violated when an image is transferred in strongly scattering media (Sect. 6.4).

Finally, we shall emphasize one important feature which is frequently overlooked. In estimating the spread function $a_i(\vec{r})$ and OTF $a_i(\vec{v})$ of individual subsystems by the inverted scheme, all of them must be examined in one and the same (subject) plane, rather than at the input of the following subsystem.

### 5.2.3 Image Recording Techniques

Various receiving techniques can be used for imaging by recording the radiance $I(\vec{r}_0, z, \vec{n})$ which escapes from the scattering medium. The following three are the most important [5.21].

1. Imaging is achieved by recording the angular radiance distribution at a fixed space point. The eye, camera and TV camera tube can be regarded as obeying this principle of operation.

2. The spatial radiance distribution in one and the same fixed direction over the plane $z$ = const is recorded. This technique can be applied to nearly all the systems, includng those which operate according to the first scheme, when the scattering medium and the observation system are spaced far apart.

3. An image is constructed using a two-dimensional screentype system. An object is scanned by moving the receiver itself along one coordinate (longitudinal scanning) and by oscillating the directional diagram axis of the optical receiver along the other (across the motion). Such a system is used, for example, in one of the meteorological satellites of the USA Department of Defense [5.22], as well as in submarine TV systems (Sect. 9.3.1).

These techniques are seen to be equivalent in their transfer functions when the aspect invariance condition is obeyed. What are the consequences when this condition is violated?

First of all, it is clear that the isoplanarity of the system is violated in the first and third techniques, i.e., (5.2.20,21) are no longer valid. It should be note that in the second case, the isoplanarity is preserved only owing to the homogeneity of the medium.

The violation of isoplanarity is, in fact, responsible for the loss, in a practical sense, of the medium PSF and OTF concepts. Moreover, the applications of the OTF of the entire system are essentially limited [5.3]. In fact, from (5.2.1) and allowing for the Parseval equality, we write

$$E(\vec{r}') = \frac{1}{4\pi^2} \iint_{-\infty}^{\infty} E_0(\vec{v}) a^*(\vec{r}', \vec{v}) d\vec{v}, \qquad (5.2.22)$$

where

$$a^*(\vec{r}', \vec{v}) = \iint_{-\infty}^{\infty} a(\vec{r}', \vec{r}) e^{i\vec{v}\vec{r}} d\vec{r}.$$

A comparison of (5.2.22) and (5.2.3, 6) shows that for a nonisophlanar system, the OTF

$$a(\vec{v}) = \frac{E_0(\vec{v})}{E_0(\vec{v})} = \iint_{-\infty}^{\infty} a(\vec{r}', \vec{r}) e^{-i\vec{v}\cdot(\vec{r}'-\vec{r})} d\vec{r} \qquad (5.2.23)$$

is a function of $\vec{r}'$ in the image plane. At $a(\vec{r}', \vec{r}) = a(\vec{r}' - \vec{r})$, the relation (5.2.23) is the same as (5.2.5).

### 5.2.4 Aspect Invariance Applicability

In view of the extreme importance of the aspect invariance condition (5.2.18) to the problem of image transfer in scattering media, we shall discuss briefly its range of applicability.

As seen from (5.2.18), in order that the aspect invariance condition be obeyed for a scattering medium, the same requirements must be met as for the small-angle approximation (Sect. 4.4). For example, in both cases the radiation diffusely reflected by the medium is disregarded, and the PSF is independent of the incident radiation direction. Therefore, the range of applicability of the aspect invariance condition for scattering media, as well as that of the SAA, is specified by the relation [5.21]

$$|\vec{n}_\perp|, |\vec{n}'_\perp| << \theta_0(z) = \min\left\{\sqrt{\frac{2}{(\langle\sigma\rangle + \langle\kappa\rangle)z}}, \sqrt{\frac{3}{\beta_2\langle\sigma\rangle\langle\kappa\rangle^2 z^3}}\right\}, \tag{5.2.24}$$

where $\langle\sigma\rangle$ and $\langle\kappa\rangle$ are the mean scattering and absorption coefficients for a medium layer, and $\beta_2$ is the second angular moment of the scattering indicatrix. The theoretical estimate (5.2.24) has been verified by the Monte-Carlo simulations [5.23, 24].

We shall illustrate this statement by calculating the OTF of a scattering medium subject only to the aspect invariance condition.

Assume that an elementary $\Delta z$-thick layer at a depth $z$ is illuminated by a monodirectional unit power source. Taking into account that the PSF and OTF of all individual susbsystems must be considered in one and the same object plane $z = z_0$, for the PSF of a particular elementary layer, we obtain

$$S(\vec{r}, \Delta z) = [1 - \varepsilon(z)\Delta z]\delta(\vec{r}) + \frac{\sigma(z)\Delta z x(r/(z_0 - z), z)}{(z_0 - z)^2}, \tag{5.2.25}$$

where $x(\beta, z)$ is the scattering indicatrix at a depth $z$, and $\beta$ is the scattering angle. Correspondingly, the OTF of an elementary layer is

$$S(\vec{v}, \Delta z) = 1 - \Delta z[\varepsilon(z) - \sigma(z)x(v(z_0 - z), z)] \tag{5.2.26}$$

or

$$\ln S(\vec{v}, \Delta z) = -[\varepsilon(z) - \sigma(z)x(v(z_0 - z), z)]\Delta z, \tag{5.2.27}$$

where $x(p, z)$ is the Hankel transform of the function $x(\beta, z)$ with respect to the angle $\beta$.

Representing the entire scattering layer as a sum of elementary layers, from (5.2.21), we obtain

$$\ln S(\vec{v}, z) = -\int_0^{z_0} [\varepsilon(z) - \sigma(z)x(v(z_0 - z), z)]\, dz \tag{5.2.28}$$

or, finally,

$$S(\vec{v}, z_0) = \exp\left\{-\int_0^{z_0} [\varepsilon(z) - \sigma(z)x(v(z_0 - z), z)]\, dz\right\}. \tag{5.2.29}$$

Formula (5.2.29) coincides with the solution of the small-angle equation (4.4.33). The conclusion is physically instructive. Moreover, it demonstrates the generality of the aspect invariance and small-angle approximation condition.

### 5.2.5 PSF and OTF Measurements

We consider briefly the PSF and OTF estimation for the entire image transfer system and its individual elements.

First, we note that direct measurement of the OTF using test objects of the type (5.2.12) is intricate because of the large number of such test objects required [5.25]. In practice, therefore, one technique or other is applied to measure the PSF, and then the OTF is found in terms of the Fourier transform using a computer or, for example, spectrum analyser. In theoretical studies, the situation is usually quite the opposite. The point is that in the majority of cases, the equation of radiative transfer from finite sources is solved in terms of the Fourier transform (Chap. 4). Therefore, an immediate result of the solution is the medium OTF which is then used, along with (5.2.8), to estimate the PSF.

Straightforward measurement of the point spread function, by definition, consists of recording the spatial irradiance distribution generated in the receiver image plane by an isotropic point source at $\vec{r} = 0$. Such a technique, frequently used in practice, has an inherent limitation. Since the field of vision of the receiver system is limited, we fail to measure the spread function $a(\vec{r}')$ for high enough $\vec{r}'$ which, as shown in [5.26], may yield essentially erroneous MTF estimates. It is seen from formula (5.2.5) that the truncation of the spread function results in an underestimation of the OTF and distortion at small frequencies. The evaluation given in [5.26] has shown that the error in the system MTF may be essential.

Another PSF measurement technique is based on the inverted scheme, which is most frequently used in studying the PSF and OTF of scattering systems. According to this technique, the medium is illuminated by a narrow monodirectional radiation beam and the irradiance distribution recorded. For systems obeying the aspect invariance condition, it relies on the third consequence of the aspect invariance. Nevertheless, in some systems that do not obey the aspect invariance, isoplanacity may be satisfied. The second imaging technique whereby the radiance is recorded in one and the same direction at different locations in the plane is a case in point. This procedure relies upon the second corollary of the reciprocity theorem [Sect. 2.5, (2.5.8)].

Basically, the PTF and OTF of the medium can be found using a source with any known distribution $E_0(\vec{r})$. Indeed, from (5.2.3), we have

$$a(\vec{v}) = E(\vec{v})/E_0(\vec{v}), \tag{5.2.30}$$

i.e., the medium OTF is found as the ratio of the Fourier transforms of the spatial distribution of the irradiances arriving at, and escaping from the scattering medium.

Different test objects such as a point source, luminous line, luminous half-plane, variable-thickness light and dark strips, etc are used in different problems, depending on their specific features and measuring traditions. A diffuse source must be used in straightforward PSF measurements, whereas a monodirectional source is preferable in the inverted techniques.

An infinitely long narrow line with radiant emittance $E_0(x, y) = \delta(x - x_1)$, along with a point source is most frequently used as an elementary test object. It will be recalled that $x$ and $y$ are the Cartesian coordinates in the object plane $\vec{r}$, $x'$ and $y'$ the Cartesian coordinates in the conjugate recording plane $\vec{r}'$. Then, in the image plane, the irradiance distribution $a(x', x_1)$, termed the line spread function (LSF), is observed. For the invariant system, $a(x', x_1) = a(x' - x_1)$. The LSF is seen to define the image of the object whose irradiance $E_0(x)$ is a function of a single coordinate:

$$E(x') = \int_{-\infty}^{\infty} E_0(x)a(x' - x)\,dx\,. \tag{5.2.31}$$

The functions $a(\vec{r}')$ and $a(x')$ are related as

$$a(x') = \int_{-\infty}^{\infty} a(\vec{r}')\,dy'\,. \tag{5.2.32}$$

It should be noted that inverse transition from $a(x')$ to $a(\vec{r}')$ is possible only for the PSF with azimuthal symmetry, when $a(\vec{r}') = a(|\vec{r}'|)$. In this case [5.4]

$$a(r') = -\frac{1}{2\pi r'}\frac{da(x')}{dx'}\bigg|_{x'=r'}\,. \tag{5.2.33}$$

We write the system OTF in terms of the line spread function $a(x')$. As $v = \sqrt{v_x^2 + v_y^2}$, at $v_y = 0$ formula (5.2.5) gives

$$\begin{aligned} a(v_x) &= \int_{-\infty}^{\infty} a(x', y')\mathrm{e}^{-\mathrm{i}(v_x x' + v_y y')}\,dx'\,dy' \\ &= \int_{-\infty}^{\infty} a(x')\mathrm{e}^{-\mathrm{i}vx'}\,dx'\,. \end{aligned} \tag{5.2.34}$$

The function $a(x')$ provides a reconstruction of the one-dimentional object image given by (5.2.31). In the case of the PSF having azimuthal symmetry, the OTF (5.2.34) obtained from it for $v_x \to v$ can be used to find the images of any objects.

To do so, (5.2.3, 6), for example, can be employed. It is clear that such a possibility stems from (5.2.33) for the PSF and LSF.

When a luminous half-plane

$$
\begin{aligned}
E_0(x) &= 1 \quad && \text{at } x > 0 \\
&= 0 \quad && \text{at } x < 0 \\
&= 1/2 \quad && \text{at } x = 0 ,
\end{aligned} \tag{5.2.35}
$$

is used as a test object, the irradiance distribution in the light-dark interface image, i.e., the boundary-value function $a_b(x')$ must be measured. Its relationship to the LSF $a(x')$ is given in Table 5.2 which also contains additional information required when using other test objects.

Light or dark strips of variable thickness are often used as test objects [5.27, 28] to estimate the transfer characteristics of photographic layers and luminescent screens. In this case, the strip contrast function, defined as the normalized irradiance in the image center ($x' = 0$) of a strip $b$-wide, is measured, depending on $b$:

$$
c(b) = E(x' = 0, b)/E(x' = 0, b \to \infty) . \tag{5.2.36}
$$

The contrast function $c'(b)$ of a dark strip is used for the diffusometric estimation of the quality of photographic materials [5.29–31]. The measuring methodology is detailed in monographs [5.27, 32].

## 5.3 Active Vision Systems

We have so far confined our consideration to image transfer in passive vision systems with a self-luminous object, or with an object uniformly illuminated by natural sources. We shall now analyse imaging details in vision systems which include an artificial source of illumination. Such active systems are employed to observe nonluminous objects. Lasers have provided a sound technological basis for modern active systems. The theory of image transfer in active systems was first elaborated in detail in [5.5–9, 33, 34], whose data form the basis of this section.

### 5.3.1 Basic Relations

We shall assume an object with a diffusely reflecting surface having a reflection coefficient distribution $A(\vec{r})$ located in the plane $z = z_0$ (Fig. 5.5). The image is formed in the plane $z = 0$ by scanning with the directional diagram of the source $\varphi_s((\vec{r} - \vec{r}_1)/z_0)$, receiver $\varphi_{rec}((\vec{r} - \vec{r}_1)/z_0)$ or of both at any one time. We take it that the source and receiver axes $\vec{n}_s$ and $\vec{n}_{rec}$ intersect at the point $\vec{r}_1$ of the object, and that the directional diagrams obey the conditions $\varphi_s(0) = 1$, $\varphi_{rec}(0) = 1$. Then, following (5.2.19) at $t_{opt} = 1$, the power of the valid signal arriving at an image cell at the point $\vec{r}_1'$ conjugate with $\vec{r}_1$ is

**Table 5.2.** Parameters for the estimation of image transfer characteristics. Symmetric spread function.

| Test object | Test object radiant emittance distribution | Functions | | OTF |
|---|---|---|---|---|
| Point source | $E_0(r) = \frac{\delta(r)}{2\pi r}$ | Point spread $a(r')$ | $a(r) = a(\sqrt{x^2 + y^2})$ | $a(v) = 2\pi \int_0^{\infty} a(r) J_0(vr) r\, dr$ |
| Luminous line | $E_0(x) = \delta(x)$ | Line spread $a(x')$ | $a(x) = \int_{-\infty}^{\infty} a(x, y)\, dy$<br>$a(r) = -\frac{1}{2\pi r} \frac{da(x)}{dx}\Big\vert_{x=r}$ | $a(v) = \int_{-\infty}^{\infty} a(x) \cos vx\, dx$ |
| Luminous half-plane | $E_0(x) = 1, x > 0$<br>$= 0, x < 0$<br>$= \frac{1}{2}, x = 0$ | Half-plane boundary spread $a_b(x')$ | $a_b(x) = \int_{-b}^{x} a(x')\, dx'$<br>$a(x) = \frac{da_b(x)}{dx}$ | $a(v) = a_b(x \to \infty)$<br>$- 2v \int_0^{\infty} [a_b(x \to \infty)$<br>$- a_b(x)] \sin vx\, dx$ |
| Light strip of width $b$ | $E_0(x) = 1, \vert x\vert < b/2$<br>$= 0, \vert x\vert > b/2$<br>$= \frac{1}{2}, \vert x\vert = b/2$ | Irradiance in the light strip center $E(b) \equiv E(x' = 0, b)$ | $E(b) = 2 \int_0^{b/2} a(x)\, dx$<br>$a(x) = \frac{dE(b)}{db}\Big\vert_{b=2x}$ | $a(v) = E(b \to \infty)$<br>$- 2v \int_0^{\infty} [E(b \to \infty)$<br>$- E(2b)] \sin vb\, db$ |
| Dark strip of width $b$ | $E_0(x) = 0, \vert x\vert < b/2$<br>$= 1, \vert x\vert > b/2$<br>$= \frac{1}{2}, \vert x\vert = b/2$ | Irradiance in the dark strip center $E'(b)$ | $E'(b) = E(b \to \infty) - E(b)$<br>$a(x) = -\frac{dE'(b)}{db}\Big\vert_{b=2x}$ | $a(v) = E'(b = 0)$<br>$- 2v \int_0^{\infty} E'(2b) \sin vb\, db$ |
| Harmonic test object | $E_0(x) = 1 + k_0 \cos vx$ | Contrast<br>$k(v) = \frac{E_{max} - E_{min}}{E_{max} + E_{min}}$<br>Average irradiance<br>$\langle E \rangle = \frac{E_{max} + E_{min}}{2}$ | $k(v) = k_0 \frac{\int_{-\infty}^{\infty} a(x) \cos vx\, dx}{\int_{-\infty}^{\infty} a(x)\, dx}$<br>$\langle E \rangle = \int_{-\infty}^{\infty} a(x)\, dx$<br>$a(x) = \frac{\langle E \rangle}{2\pi k_0} \int_{-\infty}^{\infty} k(v) \cos vx\, dv$ | $a(v) = \frac{k(v)}{k_0} \langle E \rangle$ |

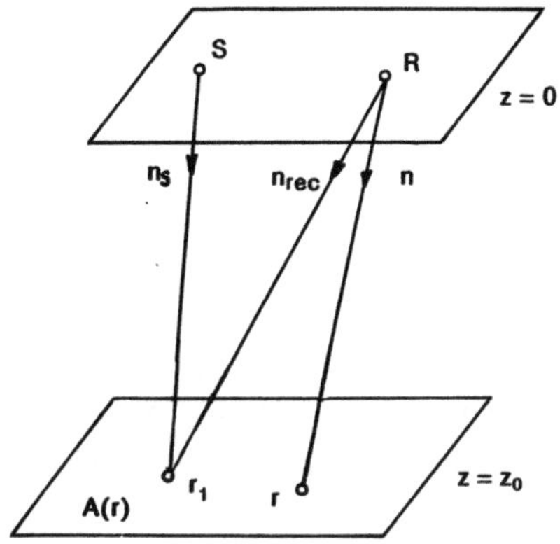


**Fig. 5.5.** Scheme of active observation: *S*, light source; *R*, radiation receiver

$$P_{VS}(\vec{r}_1') = \sum_{rec} \iint_{-\infty}^{\infty} I_{rec}(-\vec{n})\varphi_{rec}(\vec{n} - \vec{n}_{rec})\, d\vec{n}\,. \tag{5.3.1}$$

Here, $I_{rec}(-\vec{n})$ is the angular radiance distribution over the receiver entrance pupil, the vectors $\vec{n}$ and $\vec{n}_{rec}$ being, correspondingly, directed to the points $\vec{r}$ and $\vec{r}_1$.

As before, let the function $S(\vec{r})$ be the irradiance distribution in the object plane illuminated by a monodirectional unit power point source through the scattering medium. The methodologies and particular expressions for estimating the medium PSF $S(\vec{r})$ are discussed in Chap. 4. The irradiance distribution due to the unit power source having the directional diagram $\varphi_s((\vec{r} - \vec{r}_1)/z_0)$ is of the form

$$E_s(\vec{r} - \vec{r}_1) = \frac{1}{z_0^2\Omega_s} \iint_{-\infty}^{\infty} \varphi_s\left(\frac{\vec{r}\,'' - \vec{r}_1}{z_0}\right) S(\vec{r} - \vec{r}\,'')\, d\vec{r}\,''\,, \tag{5.3.2}$$

where the solid source angle $\Omega_s = \iint_{-\infty}^{\infty} \varphi_s(\vec{n} - \vec{n}_s)\, d\vec{n}$.

For the source power $P_0$, the radiance diffusely reflected by the object

$$I_0(\vec{r}) = \frac{P_0}{\pi} A(\vec{r}) E_s(\vec{r} - \vec{r}_1)\,. \tag{5.3.3}$$

In conformity with the optical reciprocity theorem, the angular radiance distribution at the receiver entrance is

$$I_{rec}(-\vec{n}) = \iint_{-\infty}^{\infty} I_0(\vec{r}\,'') S(\vec{r} - \vec{r}\,'')\, d\vec{r}\,''\,. \tag{5.3.4}$$

Taking into account (5.3.1), for the valid signal power on the image cell, we arrive at [5.6]

$$P_{VS}(\vec{r}_1') = \sum_{rec} \Omega_{rec} \iint_{-\infty}^{\infty} I_0(\vec{r}'') E_{rec}(\vec{r}'' - \vec{r}_1)\, d\vec{r}''$$

$$= \frac{P_0 \sum_{rec} \Omega_{rec}}{\pi} \iint_{-\infty}^{\infty} A(\vec{r}) E_s(\vec{r} - \vec{r}_1) E_{rec}(\vec{r} - \vec{r}_1)\, d\vec{r}$$

$$= P_0 \eta_{VS} \iint_{-\infty}^{\infty} A(\vec{r}) a_n(\vec{r} - \vec{r}_1)\, d\vec{r}\,. \qquad (5.3.5)$$

Here, the receiving solid angle $\Omega_{rec} = \iint_{-\infty}^{\infty} \varphi_{rec}(\vec{n} - \vec{n}_{rec})\, d\vec{n}$, while the function $E_{rec}(\vec{r} - \vec{r}_1)$, given as (5.3.2), is the irradiance distribution from some fictitious unit power source which replaces a real receiver and has the directional diagram $\varphi_{rec}((\vec{r} - \vec{r}_1)/z_0)$. The quantity

$$\eta_{VS} = \frac{\sum_{rec} \Omega_{rec}}{\pi} \iint_{-\infty}^{\infty} E_s(\vec{r}) E_{rec}(\vec{r})\, d\vec{r} \qquad (5.3.6)$$

is the power transfer coefficient in observing the uniform surface with $A(\vec{r}) = 1$.

As seen from (5.3.5), the function

$$\eta_{VS} a_n(\vec{r}) = \frac{\eta_{VS} E_s(\vec{r}) E_{rec}(\vec{r})}{\iint_{-\infty}^{\infty} E_s(\vec{r}) E_{rec}(\vec{r})\, d\vec{r}} \qquad (5.3.7)$$

signifies the point spread function for the entire active observation system. It describes the contribution of the emittance of the point $\vec{r}$ of the object with $A(\vec{r}) = 1$ to the signal power on the photodetector, whose axis is directed at the point $\vec{r} = 0$. Such a PSF describes scattering through the medium along the source-object-receiver path

### 5.3.2 Classification of Vision System

The spread function $a_n(\vec{r})$ and, hence, the MTF of the system depend on both the optical medium properties and the source and receiver directional diagrams. As for the directional characteristics, the observation systems can conventionally be classified into three main types (Fig. 5.6) [5.6].

1. The illuminating sourse has a narrow directional diagram while the receiving solid angle is sufficiently large, subtending, in fact, the entire field of vision, i.e., $\Omega_s \ll \Omega_{rec}$, $\varphi_{rec}(\vec{r}/z_0) = 1$ (narrow-wide or system 1). The combination of a narrow-beam laser source with an EM-photodetector is a case in point.

2. The solid angle of the source is sufficiently large, thus providing uniform and simultaneous illumination of the whole object, while the receiving solid angle is small, i.e., $\Omega_s \gg \Omega_{rec}$, $\varphi_s(\vec{r}/z_0) = 1$ (wide-narrow or system 2).

3. The solid angles of the source and receiver are of the same order, and small as compared to the system vision field (narrow-narrow or system 3). For example, a laser is a source; an image converter, camera tube, etc. are used as photodetectors.

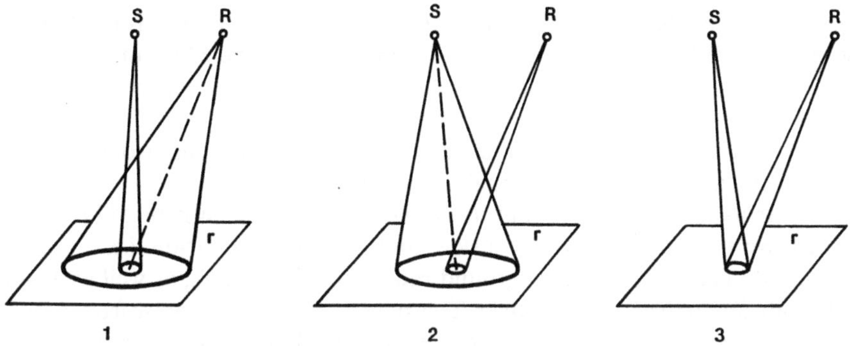


**Fig. 5.6.** Three types of observation systems: *S*, source; *R*, receiver; $\vec{r}$, object plane

The vision systems of types 1 and 3 are active, whereas system 2 may be both active and passive. The sun, a diverging laser beam, etc. may be radiation sources. Such systems may observe self-luminous objects as well. In these systems, photodetectors are, generally, multi-cell units (the eye, a camera, a camera tube, an infrared imager and so on). Thus, vision, photography and other very important and frequently encountered imaging situations can be assigned to the wide-narrow systems.

### 5.3.3 Comparison of Vision Systems

As seen from (5.3.2, 5), the PSF and MTF of the narrow-wide and wide-narrow systems are equal if the directional diagrams of the source in the former system and of the receiver in the latter are the same, and vice verse, i.e., $\varphi_s^{(1)}(\vec{r}/z_0) = \varphi_{rec}^{(2)}(\vec{r}/z_0)$ and $\varphi_{rec}^{(1)}(\vec{r}/z_0) = \varphi_s^{(2)}(\vec{r}/z_0)$.

Moreover, when $\varphi_s^{(1)}(\vec{r}/z_0)/z_0^2\Omega_s^{(1)} = \delta(\vec{r})$ and $\varphi_{rec}^{(1)}(\vec{r}/z_0) = 1$,

$$T^{(1)}(\vec{v}) = T^{(2)}(\vec{v}) = T(\vec{v}), \tag{5.3.8}$$

i.e., in the ideal case, the MTFs of systems 1 and 2 coincide with the MTF of the medium. As seen from (5.3.2), in the general case,

$$T^{(1)}(\vec{v}) = \frac{\varphi_s^{(1)}(\vec{v})}{z_0^2\Omega_s^{(1)}}T(\vec{v}), \qquad T^{(2)}(\vec{v}) = \frac{\varphi_{rec}^{(2)}(\vec{v})}{z_0^2\Omega_{rec}^{(2)}}T(\vec{v}), \tag{5.3.9}$$

where $\varphi_s^{(1)}(\vec{v})$ and $\varphi_{rec}^{(2)}(\vec{v})$ are the Fourier spectra of the corresponding directional diagrams. For example, if $\varphi_s^{(1)}(\vec{r}/z_0) = \exp(-\pi r^2/z_0^2\Omega_s)$, then $\varphi_s^{(1)}(\vec{v})/z_0^2\Omega_s = \exp(-v^2 z_0^2\Omega_s/4\pi)$.

For the narrow-narrow system MTF, from (5.3.7). we have

$$T^{(3)}(\vec{v}) = \frac{\displaystyle\iint_{-\infty}^{\infty} T^{(1)}(\vec{v}')T^{(2)}(\vec{v}-\vec{v}')\,d\vec{v}'}{\displaystyle\iint_{-\infty}^{\infty} T^{(1)}(\vec{v}')T^{(2)}(\vec{v}')\,d\vec{v}'}. \tag{5.3.10}$$

Here, $T^{(1)}(\vec{v})$ is the MTF of system 1 at $\Omega_s^{(1)} = \Omega_s^{(3)}$, and $T^{(2)}(\vec{v})$ is the MTF of system 2 at $\Omega_{rec}^{(2)} = \Omega_{rec}^{(3)}$. Formula (5.3.10) implies that the MTF of system 3 is always better than that of systems 1 and 2, i.e, the system with narrow directional diagrams of the source and receiver ensures higher contrast of te transferred image due to the fact that simultaneous selection of the signal from the given object element uses both directional diagrams.

As far as image contrast and its connection with the system MTF is concerned, we have confined our consideration to useful radiation, i.e., radiation from an object. However, background radiation of power $P_b$, as well as a valid signal, always arrives at the receiver. This may be background noise from natural sources (the sun, moon, night sky light, etc.), or medium backscattered radiation from an artificial source in active systems, i.e., backscattering noise (BSN). Background noises reduce the visible contrast in the object image, which becomes [5.6]

$$k_v = \frac{k}{1 + P_b/P_{VS}} . \tag{5.3.11}$$

Next, following [5.9], we compare the power characteristics (transfer coefficients of the valid signal and BSN) and the signal-to-noise ratio in active vision systems 1 and 2. Assume that the mean power $P_0$ of the light source, the area $\sum_{rec}$ of the receiver entrance pupil, the number $m$ of the elements in a frame and the recording time $t_f$ of the frame are equal in both cases. Also,

$$\varphi_s^{(1)} = \varphi_{rec}^{(2)} , \qquad \varphi_{rec}^{(1)} = \varphi_s^{(2)} , \qquad m = \frac{\Omega_{rec}^{(1)}}{\Omega_s^{(1)}} = \frac{\Omega_s^{(2)}}{\Omega_{rec}^{(2)}} . \tag{5.3.12}$$

As seen from (5.3.6), the above assumptions give

$$\eta_{VS}^{(2)} = \frac{\Omega_{rec}^{(2)}}{\Omega_{rec}^{(1)}} \eta_{VS}^{(1)} = \frac{1}{m} \eta_{VS}^{(1)} . \tag{5.3.13}$$

The BSN transfer coefficient obeys the relation

$$\eta_{BSN}^{(2)} = \frac{\Omega_{rec}^{(2)}}{\Omega_{rec}^{(1)}} \eta_{BSN}^{(1)} = \frac{1}{m} \eta_{BSN}^{(1)} . \tag{5.3.14}$$

In Table 5.3, the expressions in the small-angle approximation for the transfer coefficients $\eta_{VS}$ which determine the mean valid signal power on an image cell in

**Table 5.3.** Transfer coefficient $\eta_{VS}$ for different viewing systems.

| System | Narrow-wide (1) | Wide-narrow (2) | | Narrow-narrow (3) |
|---|---|---|---|---|
| | | Point source | Natural illumination | |
| $\eta_{VS}$ | $\frac{\Sigma_{rec}}{\pi z_0^2} S^2(z_0)$ | $\frac{\Sigma_{rec}}{\pi z_0^2} \frac{\Omega_{rec}}{\Omega_s} S^2(z_0)$ | $\frac{\Sigma_{rec} \Omega_{rec}}{\pi} S^2(z_0)$ | $\frac{4}{\pi} \Sigma_{rec} \Omega_{rec} \times E(\vec{r} = 0, z = 2z_0, \Omega = \Omega_s + \Omega_{rec})$ |

different systems are given. They result immediately from (5.3.6, 2). Here, the transfer coefficient is presented for observing in sun light, the irradiance $E_0 = 1$ being generated at the medium boundary $z = 0$. The quantity $E(\vec{r}, z, \Omega)$ is the cross-sectional irradiance distribution at a depth $z$ in the beam with an initial unit power and initial divergence solid angle $\Omega$.

We look to (5.1.5) for the signal-to-noise ratio in photoelectronic recording and, for the case in question, write it as

$$\delta = k_v \sqrt{\frac{S_\lambda P_0(\eta_{VS} + \eta_{BSN})t_{el}}{eB}} = k \sqrt{\frac{S_\lambda P_0 \eta_{VS} t_{el}}{eB(1 + \eta_{BSN}/\eta_{VS})}} . \tag{5.3.15}$$

Here the contrast $k$ of an image cell is independent of the background and can be calculated in terms of the reflection coefficient distribution $A(\vec{r})$ over the object surface and of the system MTF. The quantity $k_v$ is the visible contrast of the object image (taking into account the BSN), and is connected with $k$ by (5.3.11). In the particular case of a test object with a sinusoidal albedo distribution (5.2.12) and initial contrast $k_0$, the image contrast quality depends on the harmonic test object frequency $\vec{v}$:

$$k = k_0 T(\vec{v}), \qquad k_v = \frac{T(\vec{v})}{1 + \eta_{BSN}/\eta_{VS}} . \tag{5.3.16}$$

As seen from (5.3.8, 13–16), if for systems 1 and 2 $t_{el}^{(2)} = t_{el}^{(1)}$ (for example, a multiplier phototube being used in both systems as a receiver), then

$$\delta^{(2)} = \delta^{(1)}/\sqrt{m} , \tag{5.3.17}$$

i.e., system 2 is inferior to system 1 as regards the BSN value, due to less effective use of the source energy.

The application of the multielement photodetector (for example, an image converter, camera tube, etc.) with simultaneous recording of the entire frame in system 2, as is generally the case, changes the situation. In this instance, $t_{el}^{(2)} = t_f = mt_{el}^{(1)}$, and the signal-to-noise ratio in systems 1 and 2 appears to be equal, i.e.,

$$\delta^{(2)} = \delta^{(1)} . \tag{5.3.18}$$

### 5.3.4 Systems with Scattered Light Suppression

In modern vision systems, different techniques are used to reduce the harmful effect of scattered light and improve image quality. Some of them, as applied to vision and location systems in the atomsphere and in sea water, are described in Chap. 9. A great number of such methods have been suggested of late in the field of coherent optics. They will be discussed in Chap. 7. Nevertheless, it is worth emphasizing that, given the source and receiver parameters, we can estimate the quality characteristics for any particular system using the approach under consideration . To do so requires a knowledge of only some basic system characteristics such as the MTF, valid signal and BSN transfer coefficients which can, in turn, be calculated in terms of the radiative transfer theory. (Chaps. 3, 4). However, some qualitative comparison of the different systems can be made without resorting to special calculations. It is obvious that systems with backscattered noise suppression using polarization discrimination [5.35], pulse gating [5.36–38], etc., must possess a higher signal-to-noise ratio. It is seen from (5.3.15) that the SNR in the effective BSN suppression system is $(1 + \eta_{BSN}/\eta_{VS})^{1/2}$ times larger than in the system without suppression, given that their conditions are equal (for example, equal mean source power) [5.9].

The case is more involved when both the BSN and that portion of the scattered radiation which interacts with the object are cut off [5.39, 40] (in the limited case, the photodetector records only the unscattered component). In this instance, the use of two image quality criteria (the contrast $k_v$ and the magnitude of the SNR allowing only for shot system noises) makes it more difficult to reach a definite conclusion on the efficiency of system suppression. Then, in fact, the observation system MTF is improved (in the limit $T(\vec{v}) \to 1$), thus enhancing the image contrast $k_v$. At the same time, however, the valid signal transfer coefficient decreases, reducing the SNR (5.3.15). Therefore, the SNR can both increase and decrease, depending on which of the above factors prevails. The point is that the scattered light which transfers the image of the large details of object (compared with the spread function half-width) carries some useful information.

It is more convenient to assess the efficiency of vision systems with scattered light suppression by considering the quantity $\delta$ determined by (5.1.10), which includes both shot and transport noises. This quantity, when increased or decreased, provides information regarding higher or lower image quality. As seen from (5.1.10), for a sufficiently large photon number of an image cell ($\langle N \rangle \gg 1$), contrast enhancement is a key factor in the use of the scattered light suppression technique.

## 5.4 Visual Perception. Real Object Detection and Discrimination Range

We can calculate the medium transfer characteristics as well as the image pattern and quality characteristics of any object using the information presented in previous chapters. However, this proves to be insufficient to solve particular problems. The difficulty is that the eye and brain of the viewer are the final elements in vision. In practice, therefore, an estimation of the potential of a vision system and the ways in which it can be optimized requires us to consider at least some distinctions in the visual perception of real objects.

### 5.4.1 The Johnson Criteria

When we say that we see an object, we do not convey any information about the quality of the vision, which may be rather diverse: it ranges from discerning some

**Table 5.4.** Vision levels.

| Vision level | Meaning |
|---|---|
| Detection | Object appears in the field of vision |
| Determining orientation | Object shape and orientation discerned |
| Discrimination | Object is classified, e.g., as a building, vehicle, man, etc. |
| Identification | Object is identified, e.g., the vehicle is a pickup truck |

**Table 5.5.** Johnson criteria of vision quality.

| Vision level | Number of resolved half-periods of a stripped test object for a minimum size of a viewed object |
|---|---|
| Detection | $2 \pm 0.5$ |
| Determining orientation | $2.8 \pm 0.8$ |
| Discrimination | $8 \pm 1.6$ |
| Identification | $12.8 \pm 3$ |

fine details to a vague perception of a blurred contour of the object. It is clear that image quality specifies the level of vision, i.e., the possibility of solving one problem or another. Following [5.41], we shall define the levels of vision as detection, determining orientation, discrimination and identification (Table 5.4).

To make vision theory practical requires us to establish a connection between these levels of vision and the calculated and/or measured characteristics of vision systems. Different test objects, those with intermittent light and dark strips being among them, are used to measure the parameters of vision systems. As noted in Sect. 5.1, the condition $\delta > \delta^*$ can be regarded as a resolution criterion for such test objects. However, perception of a real object and its possible classification are dependent not only on image quality characteristics, but also on a number of individual features of the observer such as training, degree of attention, availability of preliminary information as well as on the picture content. Nevertheless, statistical processing of a great number of experimental results shows a correlation between the characteristics of the image quality of standard test objects and of real objects.

Johnson was the first to find such a correlation. Using military transport facilities, he compared an observer's ability to resolve a test object image and to detect, discriminate and identify real objects. As a result, Johnson assigned the number of resolved half-periods of an equivalent stripped test object within a minimum (critical) object size to each level of vision (Table 5.5). It is understood that an equivalent stripped test object is a rectangular test object having $n$ alternating longitudinal light and dark strips $l_{cr}/n$ wide ($n$ test object half-periods). Its total width is equal to the critical object size $l_{cr}$, whereas the length $l$ corresponds to the object size in the direction normal to the critical one (Fig. 5.7). For example, from the tables we know that a tank can be discriminated from

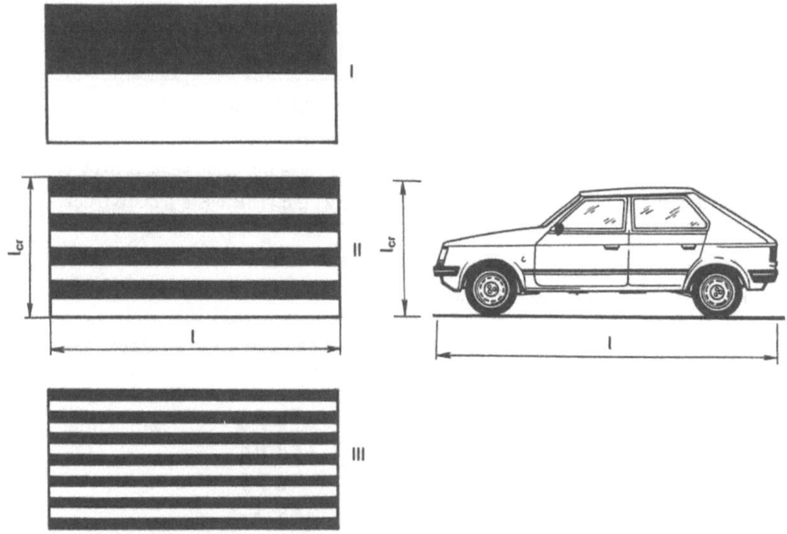


**Fig. 5.7.** Johnson criteria: I, detection; II, discrimination; III, identification

a truck, provided eight half-periods are resolved for a minimum object size. More detailed information is required to identify the type of object, i.e. it is necessary to resolve thirteen test object half-periods. These criteria have been verified in [5.16].

Thus, the Johnson criteria relate the visual perception of real objects in the given vision system to the parameters of the test object image quality, which may be predicted. In particular, one of the most important problems in estimating the range of real object detection reduces to one of detecting a rectangular test object twice as narrow as the critical size of a real object. The test object albedo is equal to the mean real object albedo, while the background albedo equals that of the real background.

### 5.4.2 Object Detection Range

We now consider in more detail the problem of the range of television or visual detection of an object in a scattering medium. We take it that in this case the observer compares the signals which arrive at two image cells, one for the object center and the other for the background level. Then, in (5.1.10)

$$\Delta N = C\Delta W, \qquad \langle N \rangle = CW_b, \tag{5.4.1}$$

where $\Delta W$ is the energy of a valid signal, i.e., the difference in the energies arriving at two compared image cells, and $W_b$ the background energy per image cell for the frame recording time. The proportionality factor $C$ is determined by the efficiency of the photodetector. Substituting (5.4.1) into (5.1.10) gives

$$\delta = \frac{\Delta W}{\sqrt{W_b/C + \alpha_2 W_b^2}} . \tag{5.4.2}$$

As noted in Sect. 5.1, the signal-to-noise ratio is a function of the image cell area $\Sigma_{el}$ over which the signal is integrated. The quantity $\Sigma_{el}$ is related to the solid receiving angle $\Omega_{rec}$. In the following analysis, we shall consider it in the object plane, i.e., set $\Sigma_{el} = \Delta_{rec} z_0^2$.

A minimum cell $\Sigma_{el}$ is determined by the parameters of the vision system. For a TV system, it is equal to the frame area divided by the number of elements it contains. However, the cell size can be increased by spatial integration of the image formed. Therefore, the question arises of the choice of an optimum image cell when the quantity $\delta$ and, correspondingly, the detection range are maximum.

Let us examine this point more closely. At small $\Sigma_{el}$, presumably, $\Delta W \sim \Sigma_{el}$, $W_b \sim \Sigma_{el}$ and, as seen from (5.4.2), $\delta \sim \Sigma_{el}^{1/2}$. In the other extreme case, where the cell area is greater than the total image area, $\Delta W$ no longer chnages with increasing $\Sigma_{el}$, while $\delta$ decreases. Therefore, we can state that $\Sigma_{el}$ dependence of the signal-to-noise ratio $\delta$ becomes maximum at some optimum $\Sigma_{el}^{opt}$. It is clear that for $\Sigma_{el} = \Sigma_{el}^{opt}$, the decision on the presence of an object in the medium turns out be most significant. It might be supposed that such image processing, i.e., the choice of an optimum element $\Sigma_{el}^{opt}$ and appropriate signal integration, is performed by the human vision system (eye + brain), both in direct viewing and in analysing a television image. This point has been verified experimentally in [5.42].

The quantity $\Sigma_{el}^{opt}$ depends on the distance $z_0$ to an object. Therefore, for a limited detection range $z_0^*$, two conditions $\Sigma_{el} = \Sigma_{el}^{opt}$ ($\Omega_{rec} = \Omega_{rec}^{opt}$) and $\delta = \delta^*$ must be obeyed simultaneously. Thus, estimation of $z_0^*$ reduces to the solution of the system of equations

$$\frac{d\delta(z_0, \Sigma_{el})}{d\Sigma_{el}} = 0$$

$$\delta(z_0, \Sigma_{el}) = \delta^* . \tag{5.4.3}$$

To solve the equation system (5.4.3), taking into account (5.4.2), requires us to know the $\Sigma_{el}$ dependence of $W_b$ and $\Delta W$. In the case of a homogeneous object with albedo $A_{ob}$ and area $\Sigma_{ob}$, we may write

$$\Delta W = W_0 \eta_{vs} |A_{ob} - A_b| \Psi(\Sigma_{ob}, \Sigma_{el}) , \tag{5.4.4}$$

where $W_0$ is the radiation source energy, and $A_b$ the albedo of the medium layer behind the object. The function

$$\Psi(\Sigma_{ob}, \Sigma_{el}) = \frac{1}{\Sigma_{el}} \iint_{-\infty}^{\infty} U(\vec{r}, \Sigma_{el}) \, d\vec{r} \iint_{-\infty}^{\infty} U(\vec{r}', \Sigma_{ob}) S'(\vec{r} - \vec{r}') \, d\vec{r}' \tag{5.4.5}$$

allows for the $\Delta W$ dependence on the size of the object and integration element. Here, $S'(\vec{r})$ is the normalized medium PSF for which $\iint_{-\infty}^{\infty} S'(\vec{r}) \, d\vec{r} = 1$, and the function $U(\vec{r}, \Sigma)$ is determined by (5.1.30).

Assuming that the vision angle of the receiver is not too large, $W_b \sim \Omega_{rec}$ and $\eta_{VS} \sim \Omega_{rec}$, i.e.,

$$W_b = W_b' \Sigma_{el}, \qquad \eta_{VS} = \eta_{VS}' \Sigma_{el}, \tag{5.4.6}$$

where $W_b'$ and $\eta_{VS}'$ are independent of $\Sigma_{el}$. Then, from the first equation in (5.4.3) and taking into account (5.4.2, 4, 6), we have

$$\frac{d \ln \Psi(\Sigma_{ob}, \Sigma_{el})}{d\Sigma_{el}} + \frac{1}{2\Sigma_{el}(1 + \alpha_2 C W_b)} = 0 . \tag{5.4.7}$$

The solution of this equation gives an optimum integration element $\Sigma_{el}^{opt}$ (or an optimum solid receiving angle $\Omega_{rec}^{opt} = \Sigma_{el}^{opt}/z_0^2$) at the definite distance $z_0$ from the object.

If we use the Johnson criteria and replace the real object by a rectangular test object with the length $y_{ob}$ equal to the length of the real object and the width $x_{ob}/2$ equal to its critical half-size, then $\Sigma_{el} = x_{el} y_{el}$ and, instead of (5.4.3), we obtain

$$\frac{\partial \ln \delta(z_0, x_{el}, y_{el})}{\partial x_{el}} = \frac{\partial \ln \Psi_1}{\partial x_{el}} + \frac{1}{2x_{el}(1 + \alpha_2 C W_b)},$$

$$\frac{\partial \ln \delta(z_0, x_{el}, y_{el})}{\partial y_{el}} = \frac{\partial \ln \Psi_1}{\partial y_{el}} + \frac{1}{2y_{el}(1 + \alpha_2 C W_b)},$$

$$\delta(z_0, x_{el}, y_{el}) = \delta^* \tag{5.4.8}$$

Here, $\Psi_1 = \Psi(x_{ob}/2, y_{ob}; x_{el}, y_{el})$ due to the rectangular Johnson test object which replaces the real one. An approximate expression for the function $\Psi$ is given is Sect. 6.6. The solution of the system (5.4.8), allowing for (5.4.2), gives the sizes $x_{el}^{opt}$ and $y_{el}^{opt}$ of the optimum spatial integration element with respect to two axes and the limited object detection range $z_0^*$. The applications of this detection concept will be discussed in Sects 8.3, 9.2, 3.

### 5.4.3 Object Discrimination Range

In this case, in conformity with the Johnson criteria, eight half-periods of the stripped test object, whose width is equal to the critical object size $x_{ob}$, must be resolved (Table 5.5 and Fig. 5.7). This restricts the size of the element of integration over the $x$ axis by the condition $x_{el} \leqslant x_{ob}/8$. Therefore, when considering the problem of object discrimination, we may take

$$\Sigma_{el}^{opt} = x_{ob} y_{ob}/8 . \tag{5.4.9}$$

Thus, the limited range of object discrimination $z_0^*$ is found from the equation

$$\delta(z_0, \Sigma_{el}^{opt}) = \delta^* \tag{5.4.10}$$

taking into account (5.4.2, 9). The quantity $\Delta W$ in (5.4.2) is the difference between the energy reaching the light image cell of the area $\Sigma_{el}^{opt}$, or the mean energy $W_b$ over the same area. In contrast to the object detection problem, in this case, both the valid signal energy $\Delta W$ and the background energy $W_b$ change. In fact, the constant radiation component from the observation object together with ambient noises and the BSN contribute to the background energy. It is equal to

$$W_b^{ob} = W_0 \eta_{VS} \frac{A_{ob} - A_b}{2} \Psi(x_{ob}, y_{ob}; x_{el}^{opt}, y_{el}^{opt}) . \tag{5.4.11}$$

The valid signal energy $\Delta W$ in the problem of Johnson test object observation is

$$\Delta W = \langle W_{VS} \rangle k_{st} , \tag{5.4.12}$$

where

$$\langle W_{VS} \rangle = W_0 \eta_{VS} \frac{A_{ob} + A_b}{2} \Psi(x_{ob}, y_{ob}; x_{el}^{opt}, y_{el}^{opt}) \tag{5.4.13}$$

is the constant component of the radiation energy from the object, and $k_{st}$ the contrast between adjacent cells at the center of the Johnson test object image. The formulas for $k_{st}$ are given in Sect. 6.5.2.

## 5.5 Television and Location Target Detection Systems

We have so far dealt with television and visual object detection. However, at present, pulse location is one of the ways of detecting an object. It is known that location systems, can also form an image. It is, therefore, natural to raise the question of the relationship between the potential of television and location target detection systems. For this purpose, we first consider a comparatively simple case of location where the direction to the object is known, and it is only required to confirm its existance and determine the distance to it. In fact, laser echo-ranging pulse location of the sea bed, for example, is implied here. It is clear that the detection range is greatest in this instance and can be taken as the upper limit of the pulse location range.

### 5.5.1 Location in a Given Direction (Laser Echo-Ranging)

Let a short light pulse be directed to an center of an object. A photodetector, whose axis is also oriented to the center of the object, is set near a source to register the time-dependent power of an echo signal (Fig. 5.8). At this instance, a location "splash" against a background of backscattering noise indicates the presence of an object and allows one to estimate the distance to it. The signal-to-noise ratio is (5.1.10),

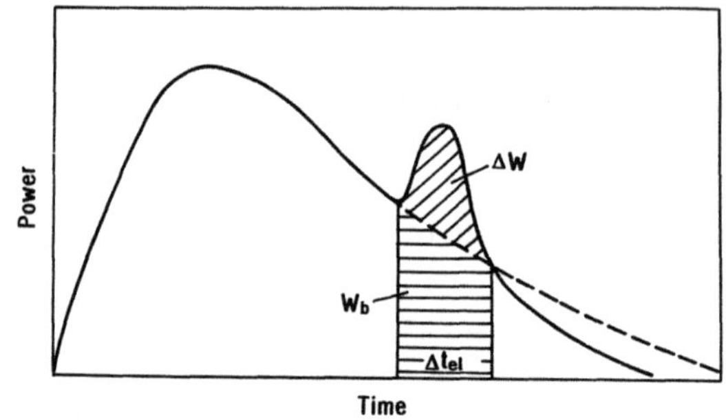


**Fig. 5.8.** Signal power in a location system

$$\delta = \frac{\Delta W}{W_b}\sqrt{\frac{CW_b}{1+\alpha_2 CW_b}}. \qquad (5.5.1)$$

Here, $W_b$ is the backscattering noise energy reaching the photodetector in the absence of an object for the time $\Delta t_{el}$ specified by the temporal resolution of the detector; $\Delta W$ if the energy of the location or valid signal for the time $\Delta t_{el}$ in the presence of an object.

We find the expressions for $\Delta W$ and $W_b$, and discuss the possibility of receiver angle optimization. The analysis will be all the more instructive if several simplifying assumptions are made.

We neglect the temporal spread of the location signal propagating in the medium and assume that the time $\Delta t_{el}$ is greater than the initial pulse time. The optimum quantity $\Delta t_{el}$, taking into account the temporal valid signal spread, is discussed in Sect. 9.5. Then, the valid signal $\Delta W$ can be estimated using (5.3.5). Assume, for simplicity, that the functions $A(\vec{r})$, $E_s(\vec{r})$ and $E_{rec}(\vec{r})$ have azimuthal symmetry and can be approximated by the normal distribution. Following (5.3.5), at $\vec{r}_1 = 0$, we obtain

$$\Delta W = \frac{W_0 \Sigma_{rec}\langle A\rangle S^2(z_0)}{\pi z_0^2}\,\frac{V_{rec}V_{ob}}{(V_{ob}+V_a)(V_s+V_{rec}+2V_r)}, \qquad (5.5.2)$$

$$V_a = \frac{(V_s+V_r)(V_{rec}+V_r)}{V_s+V_{rec}+2V_r}, \qquad (5.5.3)$$

where $W_0$ is the source energy; $\langle A\rangle$ is the average albedo of the object; $V_{ob} = \Sigma_{ob}/2\pi$, $\Sigma_{ob}$ is the area of the object; $V_s = \Omega_s z_0^2/2\pi$, $V_{rec} = \Omega_{rec} z_0^2/2\pi$, $V_r$ is the variance of the PSF; $V_a$ is the variance of the function $a(\vec{r}) = E_s(\vec{r})E_{rec}(\vec{r})$ (Sect. 5.3.1); $S(z_0)$ is the zeroth moment of the function $S(\vec{r}, z_0)$.

At $\langle A\rangle = 1$, $V_{ob} \to \infty$ and $V_{rec} \to \infty$ or $V_s \to \infty$. Formulas (5.5.2, 3) yield the relations for the transfer coefficients both in the narrow-wide and wide-narrow vision systems which are presented in Table 5.3.

In calculating $W_b$, we take it that at the time instant $t = 2z_0/v$ ($v$ being the light velocity in the medium) corresponding to the instant when the location signal arrives, the BSN is due to backscattered radiation of the $v\Delta t_{el}/2$ thick layer at depth $z_0$. Therefore, the expression for $W_b$ can be obtained immediately from (5.5.2) by substituting $\sigma x(\pi) v\Delta t_{el}/2$ for $\langle A\rangle$ ($\sigma$ being the scattering coefficient, $x(\pi)$ the backscattering indicatrix) and assuming that $V_{ob} \to \infty$:

$$W_b = \frac{W_0 \Sigma_{rec} S^2(z_0)}{\pi z_0^2} \frac{\sigma x(\pi) v\Delta t_{el}}{2} \frac{V_{rec}}{V_s + V_{rec} + 2V_r}. \tag{5.5.4}$$

The relations (5.5.1–4) are simple and convenient to use when analysing the potential of the location system. The limiting location range is determined from (5.5.1–4) at the condition $\delta \geqslant \delta^*$. We shall consider that the solid angle of the source is small and discuss the problem of the optimum choice of the solid angle of the receiver. The angle $\Omega_{rec}^{opt}$, at which the SNR is maximum, is regarded to be optimum. Therefore, the quantity $V_{rec}^{opt}$ and, correspondingly, $\Omega_{rec}^{opt}$ can be estimated from the condition

$$\frac{\partial \delta}{\partial V_{rec}} = 0. \tag{5.5.5}$$

In the general case, $V_{rec}^{opt}$ can easily be calculated from (5.5.1, 5) provided that $W_b$ is obtained with regard for all optical and geometric as well as energy parameters of the system. Here, we shall restrict our consideration to two important particular cases.

1. $\alpha_2 C W_b \gg 1$. This case occurs at very high source energy. The detection range is limited by system transport noises. As seen from (5.5.1–4), no optimum $V_{rec}$ exists here and the SNR is higher, the lower $V_{rec}$. We note that $V_{rec}$ may reduce to a certain limit since at $W_{rec} \to 0$, $W_b \to 0$ and the initial condition ($\alpha_2 C W_b \gg 1$) is violated.

2. Another situation is generally observed that $\alpha_2 C W_b \ll 1$ and the location range is restricted due to shot noises of the receiver (Sect. 9.5). In this instance, from (5.5.1–5), we have

$$V_{rec}^{opt} = \frac{V_{ob}(V_s + 2V_r) + V_r(V_s + V_r)}{\dfrac{V_s(V_s + V_r)}{V_s + 2V_r} - V_{ob}}. \tag{5.5.6}$$

As seen, an optimum $V_{rec}$ exists provided htat

$$V_{ob} < \frac{V_s(V_s + V_r)}{V_s + 2V_r}. \tag{5.5.7}$$

It should be noted that (5.5.7) yields a necessary, but not sufficient condition,

$$V_{ob} < V_s. \tag{5.5.8}$$

If the condition (5.5.7) is violated, then no optimum receiving angle exists, and the SNR is higher, the larger $V_{rec}$.

### 5.5.2 Image Forming Location

We now consider the more general problem of locating an object whose direction is unknown. To determine the position of the object requires scanning by a series of short pulses in some search area. The implementation of such a system is discussed, for example, in [5.43, 44]. The signal received from each source pulse is subdivided into a set of $\Delta t_{el}$ intervals and then stored. The $i$-th interval corresponds to the arrival of the backscattered light from the layer $z_i = vt_i/2$ distant from the locator. Narrow-beam scanning in the search area forms a set of images of $v\Delta t_{el}/2$ thick layers at different distances $z_i$ from the locator. Such a system, which we shall refer to as the image forming locator, is seen to be the pulsed narrow-wide vision system 1 (Sect. 5.3.2). Therefore, using the general analysis of vision systems given in Sect. 5.3.3, we can compare its characteristics with the appropriate parameters of the time-gating wide-narrow television system 2. All parameters of the location system have the subscript "1", whereas those of the television system are marked by "2".

Let us consider that the solid angle $\Omega_{rec}^{(1)} = \Omega_s^{(2)}$ is sufficiently large (much larger than the angular dimension of the PSF variance $V_r/z_0^2$), and that the solid angle $\Omega_s^{(1)} = \Omega_{rec}^{(2)}$ is sufficiently small. Moreover, assume that the source energy required to image one and the same frame is equal for both systems. This means that, if the locator sends pulses to inspect the frame of area $\Omega_s^{(2)} z_0^2$, then, in the television system 2, the image is formed by $m = \Omega_s^{(2)}/\Omega_s^{(1)}$ pulses. We also consider that the gating interval $\Delta t_{el}^{(2)} = \Delta t_{el}^{(1)}$, photodetector sensitivity, characteristics of transport noises and other similar parameters are equal for both systems. Thus, from the analysis given in Sect. 5.3.3, we come to the important conclusion that the SNR and target detection range are identical in the systems being compared.

As was assumed, the solid angle $\Omega_s^{(1)}$ of the source is small ($V_s \ll V_r$) in the location system. Therefore, in image processing, for example, where an operator observes an image on a screen, an optimum spatial integration element $\Sigma_{el}^{opt}$ can be chosen (Sect. 5.4.2) such that the SNR $\delta$ and the object detection range are maximum. However, scanning over a large search area with a small solid angle $\Omega_s^{(1)}$ requires using a laser with a high pulse repetition frequency. In fact, $m = \Omega_s^{(2)}/\Omega_s^{(1)}$ pulses must be sent during the frame recording time. In this connection, the television system 2 in which negligible $\Omega_{rec}^{(2)}$ can be achieved simply is preferable to the location system "1". In view of this, the question arises as to how much the solid angle $\Omega_s^{(1)}$ can be increased and, subsequently, the pulse repetition frequency reduced in the location system, without an essential decrease in the detection range. It is clear that such an increase can be up to $\Omega_s^{(1)} \approx \Sigma_{rec}^{opt}/z_0^2$.

Thus, we have come to the physically evident but important conclusion that the potentials of the television and location systems are generally equal if the requirements regarding the information derived from the search area are equal.

However, in some cases, where a priori information about the object sizes is available, for example, the location range can probably be increased by choosing the optimum scanning strategy. Indeed, there is then no need to scan at a small solid angle $\Omega_s^{(1)}$ over the entire frame. Using some frame points about some critical size $x_{ob}$ apart, we can reduce the number of frame elements to $m_1 = \Omega_{rec}^{(1)} z_0^2 / x_{ob}^2 < m$ and, thus, increase the energy spent per image cell during the frame recording time. It is clear that the price one pays for such an increase in the detection range is a reduction in the accuracy of determining the position of the object.

What are the limits of increasing the detection range in such a way? In the general case, this question is far from simple. However, it is not difficult to indicate the upper limit. It is the location range in the given direction (Sect. 5.5.1) with source energy intended for the entire frame imaging. It some instances, it may be 2–3 times as large as the television or visual detection range (Sect. 9.5).

## 5.6 Basic Characteristics of the Eye and Other Photodetectors

As seen from previous sections, image quality and the target detection range are specified in many respects by such photodetector characteristics as the threshold contrast, efficiency, spectral sensitivity, frame recording time, angular resolution, etc. These problems have been discussed extensively in specialized literature. We shall confine our consideration to some of the most important parameters of the most common photodetectors.

### 5.6.1 The Human Eye as a Radiation Receiver

The eye is the most ancient and important photodetector created by nature. In the course of evolution, the human vision system has been perfected. Suffice it to say that the eye, operating as a photon counter, can fix about ten quanta. The eye is a wide-angle optical system. Its correction for aberrations is surprising, its resolution under optimum viewing conditions being restricted to the diffraction limit.

A great many studies are devoted to the optics of the eye, the physiological and psychological aspects of visual perception [5.45–48]. We shall consider briefly some of the characteristics of the eye which specify image quality within the image transfer theory and allow the eye to be compared to other photodetectors (camera tube and camera).

It should be noted that the eye is a receiver which can see in sunlight. Therefore, its spectral sensitivity corresponds to the spectral distribution of solar radiation. As far as visual perception is concerned, light fields are generally characterized by photometric concepts and units which include the spectral sensitivity of the eye. The correlation between the radiant and photometric terms

**Table 5.6.** Summarized radiant and luminous system parameters.

| Radiant system | | Luminous system | |
|---|---|---|---|
| Name | MKS units | Name | MKS units |
| Radiant energy | Joule | Luminous energy | Talbot |
| Radiant flux | Watt | Luminous flux | Lumen |
| Radiant emittance | Watt/m$^2$ | Luminous emittance | Lumen/m$^2$ = Lux |
| Radiant intensity | Watt/sr | Luminous intensity | Candela |
| Radiance | Watt/m$^2$sr | Luminance | Meterlambert (cd/m$^2$) |
| Irradiance | Watt/m$^2$ | Illumination | Lumen/m$^2$ (Lux) |

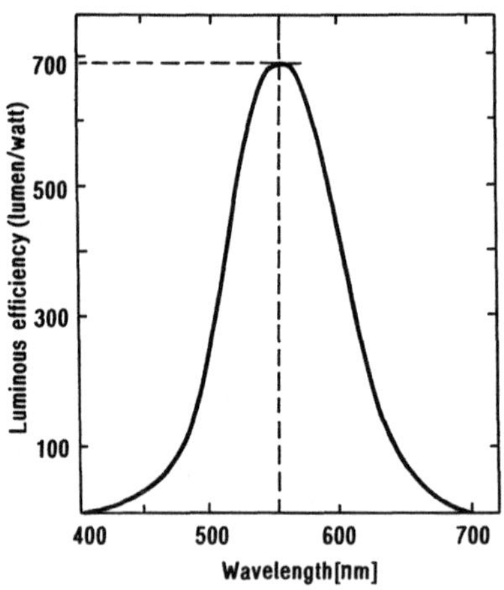


**Fig. 5.9.** Standard observer; conversion from watts to lumens

and MKS units is shown in Table 5.6 [5.49]. The transfer from radiometry to photometry only requires one to define the lumen, the luminous flux unit. Figure 5.9 shows the standard curve [5.50] for an "average observer", showing the effectiveness of each wavelength evoking the brightness sensation, i.e., the spectral sensitivity of an average eye. This curve is the function which converts watts into lumens. Below, in describing the eye, we shall use a luminous system for light characteristics.

It is very important that the human vision system is adaptive. Its parameters range wide, dependent on viewing conditions. Thus, the eye pupil diameter, which is generally 3 mm, varies from 2 mm at high luminance to 8 mm in the dark. For the eye focal distance of 17 mm, this means a variation in the relative aperture from 1 : 8 to 1 : 2. The exposure time changes similarly, from 0.05 s at high luminance ($10^3$–$10^4$ cd/m$^2$) to 0.2 s at threshold luminance ($\sim 10^{-6}$ cd/m$^2$). The mean eye persistence time is $\Delta t \approx 0.1$ s. It is worth noting that this agrees well with the response time of other human systems and, in particular, with the time of pulse propagation along the nerve fibers. Such an agreement, far from being mere chance, is biologically expedient and has developed through long evolution.

One more important characteristic of the eye is that its quantum efficiency $\eta$, which is the ratio of the number of photons used to the number of photons incident on the entrance pupil, changes with illumination. For green wavelength, the quantity $\eta$ is increased 10-fold (from 1 to 10%) as the irradiance is decreased by $10^8$-fold. It is about 5% for an average level of illumination [5.10].

The ability of the eye to adapt favours its operation in an extremely wide luminance range covering 10 orders, from $3 \times 10^{-6}$ to $3 \times 10^4$ cd/m$^2$ [5.11]. A peculiar biochemical amplifier [5.10] with a variable gain of about $10^6$ is also essential here. It is important that such an amplifier, as well as an eleectronic one in photoelectric recording, equally enhances both a valid signal and noise, without changing their proportion. Rose [5.10], for example, established a connection between this phenomenon and the vision noise generated during dark adaptation. In response to a random irritant, an unexpected knock on the door, for example, the gain grows drastically and the system transport noises are observed as splashes.

The eye is a multielement receiver. Its light sensitive layer (retina) contains about $10^8$ elements. Most of them (about 95%) are the rods whose sensitivity essentially depends on the preceding illumination. This, for example, is responsible for the increase in the quantum efficiency of the eye in dark adaptation. The rods afford vision at low irradiance. The cones constitute only about 5% of the retina light sensitive elements. They operate at relatively high illumination ($> 10^{-2}$ Lux), their sensitivity being a weak function of adaptation. They are, however, extremely important. The cones alone are responsible for colour vision. At illuminance above $10^2$ Lux, vision is almost completely due to the cones, while twilight vision at $10^{-2}$ to 10 Lux illuminance is afforded both by the cones and the rods. The highest vision quality is also due to the cones, which are located mainly around the fovea. Their density here is most prominent. It achieves $2 \times 10^5$ mm$^{-2}$ and controls the limiting angular eye resolution of 1–2′ within 1–2° consisting of the fovea diameter. It should be noted that this resolution is close to the diffraction limit for the minimum pupil diameter of 2 mm at high illuminance [5.10].

The sufficient vision quality, although at half as much angular resolution, is preserved within the yellow spot of the retina, i.e., some 8° along the vertical and 10° along the horizontal. The maximum field of vision of a fixed eye is 130 × 160° [5.51], a satisfactory image quality being afforded within the range 30 × 40° [5.11]. The decrease in the eye resolution from the center to the periphery is due, not to the optices of the eye, but to the increase in the resolution elements: the signals from tens of cones and hundreds of rods concentrate around a single nerve fiber [5.51, 52]. Such a grouping, together with a dominant rod vision, conforms to the requirements of eye operation at low illuminance levels, when increased sensitivity is the main requirement, and the resolution becomes less important. We note that as the luminous intensity of objects decreases, the SNR also reduce, thus increasing the limiting resolution cell and, hence, the angular resolution limit due to noise fluctuations. One can say, therefore, that the cooperation of many photoreceptors appears to be fully expedient not only on biological grounds (the decreasing number of nerve channels required for information transfer to the cerebrum), but also on physical grounds (increasing sensitivity, agreement between geometric and fluctuation resolution thresholds). In view of this cooperation, one can consider that the number of the resolution cells of the eye as a photodetector is some $10^5$, i.e., three orders less than the number of photoreceptors (about $10^8$).

One more characteristic of the eye is worth noting. Under optimum illumination conditions, the threshold contrast for adjacent image cells is independent of their luminous intensity and the angular size of the elements (if they are large enough) amounts to $k^* \approx 0.5$–$2\%$. In [5.13, 53, 54], some attempts have been made to account for the threshold origin using the theory of statistical decisions. To be specific, it is assumed in [5.13] that the Weber-Fechner law, from which the threshold contrast stems, is related to the transport noises of the neuron system. The system of receptors is taken to be similar to the number of parallel chains (pulse transfer lines), each being connected to one of the receptors, and containing transport noise sources. The Weber law results from the strong correlation of adjacent channels that transfer pulses from the receptors to the neuron system. This assumption requires experimental verification and an additional theoretical basis. In Sect. 5.1, such a connection between the threshold contrast and transport noises is expressed by (5.1.7).

### 5.6.2 Photographic and Photoelectric Recording

A photographic film is a facility widely used for image recording. Since the quantum efficiency of individual grains is 10 to 20%, the efficiency of a silver halogenide film amounts to some 1% [5.10]. Such a reduction in the sensitivity is due to the grain size dispersion required for the photographic latitude to be wide enough (i.e., to afford the dynamic irradiance range). Thus, the photographic film can be compared with the eye in its sensitivity at high irradiance, although it is much inferior at low irradiances. Nevertheless, the fact that the film is able to integrate incident energy (the frame recording time being limited only by the object stability) account for its wide use in recording very faint (but immovable) images, in astronomy for example. Similar information on the properties of photographic materials can be found in [5.27, 55].

The greatest progress has been achieved in improving the parameters of electronic imagers. In the last fifty years, the sensitivity of camera tubes has become $10^6$ higher. The modern superorticon tubes operate at a quantum efficiency of about 10% [5.10], i.e., approximately equal to the human eye efficiency at poor illumination and exceeding it in the daytime. The limiting quantum efficiency of $\sim 10\%$ is specified by vacuum elements with outer photoemissive effect. The use of the photoconductivity phenomenon (in vidicons) increases the efficiency by up to 100%. In recent years, solid semiconductor image detectors have shown promise [5.56]. The very compact detectors such as these ensure quantum conversion efficiency to 100% [5.57]. Their transfer function is linear in the range over many orders of the signal power. The detectors are superior to conventional camera tubes in their homogeneous sensitivity within the field of vision. The impact of inhomogeneous sensitivity on the observer's eye competes with that of noise [5.12, 58], and decreases the number of discernible irradiance ranges in imaging.

Finally, it is worth noting that modern electron systems surpass the human eye both on account of their quantum efficiency and ability to operate at very small object irradiance [5.10].

### 5.6.3 Notes on Infrared Imaging

We can understand the specific properties of vision systems operated in the infrared by reverting to Planck's formula for the spectral emissivity distribution

$$\varepsilon_f = \frac{2\pi h}{c^2}\frac{f^3}{e^{hf/kT}-1}\,. \tag{5.6.1}$$

Here, $\varepsilon_f$ is the spectral density of radiant emittance, $f$ the radiation frequency, $h$ the Planck constant, $k$ the Boltzmann constant, $c$ the light velocity and $T$ the absolute surface temperature.

Differentiating (5.6.1) with respect to $T$ gives the emittance contrast of the surface area, whose temperature is $\Delta T$ different from the mean temperature of the ambient background

$$k = \frac{\Delta\varepsilon_f}{\varepsilon_f} = \frac{hf\Delta T}{kT^2}(1-e^{-hf/kT})^{-1}\,. \tag{5.6.2}$$

It is seen from (5.6.2) that contrast degenerates with decreasing radiation frequency $f$. At the temperature difference $\Delta T = 0.2$ K against the background $T = 290$ K, the contrast $k = 0.8\%$ for $\lambda \approx 4$ µm and $k \approx 0.4 - 0.3\%$ within the atmospheric transparent spectral window of 8 to 14 µm. The IR contrast ranges from one to several percent for almost all real objects. It is the low contrast range which is responsible for the main problems in IR-image detection [5.59, 60].

As shot noise is proportional to the rms signal energy (5.1.2), the SNR improvement at low image contrast requires the use of a strong light-gathering optical system. The noises due to background radiation from the photodetector itself can be reduced by cooling the lenses, the receiver and its case [5.10]. Finally, the low image contrast range imposes one more essential condition on IR vision photodetectors. They must possess extremely homogeneous sensitivity within the field of vision.

# 6 Optical Transfer Function of a Scattering Medium

In the previous chapter, we showed that the optical transfer function (OTF) is one of the basic characteristics responsible for image transfer in viewing systems. When there is no feedback between subsystems (elements), the OTF of the entire system is represented as a product of the OTFs of its elements (5.2.21). Since the scattering medium is considered as one of the elements of the viewing system, such an approach allows one to study the OTF independently. In the following sections, we shall deal with the formulas commonly used to calculate the OTF of strongly scattering and essentially absorbing media, and consider the specific features of the OTFs of inhomogeneous, including stochastic media, as well as the OTF and viewing properties along oblique paths. In Sect. 6.4, we shall discuss the possibility of using the apparatus of transfer functions to describe image transfer through a thick scattering layer. In conclusion, we shall consider the connection between the medium OTF and the contrast observed in the image of real objects, and derive simple and convenient relations for the function $\Psi$ in the target detection theory (Sect. 5.4).

## 6.1 OTF of a Homogeneous Layer

In order to be specific, we assume that the scattering plane layer of optical thickness $\tau_0 = \varepsilon z_0$, through which an object is seen, is horizontally infinite. Therefore, we shall refer to the observation path for which the axis of sight is normal to the layer as vertical. Such a convention does not, of course, limit the generality of our discussion.

The methodologies and approaches used to calculate the OTF of the scattering media with different optical parameters were described in Chap. 4. We shall use these earlier results here. It is worth adding that, for the moment, efficient enough algorithms are available for OTF calculation using the Monte Carlo method [6.1–4] which, if necessary, can be applied to estimate the accuracy of some analytical formulas and straightforward calculations.

In vertical viewing, which is dealt with in Sect. 6.1.2, the medium point spread function possesses azimuthal symmetry. The OTF is then real, dependent only on the frequency modulus $|\vec{\nu}| = \nu = \sqrt{\nu_x^2 + \nu_y^2}$ and related to the PSF in terms of the Hankel transform (5.2.7). In this case, the medium MTF is equal to the normalized OTF, i.e.,

$$T(\nu) = \frac{S(\nu)}{S(\nu = 0)}, \tag{6.1.1}$$

no phase shifts being observed.

### 6.1.1 The Small Angle Approximation

For scattering media with moderate optical thickness and an extended scattering indicatrix, the OTF $S(\nu, z_0)$ can be estimated by the SAA (4.4.35). In Sect. 5.2.4, this expression resulted from the system aspect invariance. At $\nu = 0$, (4.4.35) implies that

$$S(\nu = 0, z_0) = \exp[-(\varepsilon - \sigma)z_0], \tag{6.1.2}$$

i.e., at a frequency $\nu = 0$, the OTF (total flux through the plane $z = z_0$) vanishes with the depth only by virtue of absorption.

At very high spatial frequencies, the OTF is determined only by unscattered light,

$$\lim_{\nu \to \infty} S(\nu, z_0) = \exp(-\varepsilon z_0). \tag{6.1.3}$$

Expressions (4.4.35) and (6.1.1) give the following MTF formula:

$$T(\nu) = \exp\left[-\sigma z_0 + \sigma \int_0^{z_0} x(\nu\xi)\, d\xi\right]. \tag{6.1.4}$$

If we introduce dimensionless angular frequency,

$$\nu^* = \nu z_0 = \omega^* = \omega\tau_0. \tag{6.1.5}$$

The analysis of the scattering medium transfer characteristics as functions of the frequency $\nu^*$ is more instructive. This is not due to chance. Really, in the radiative transfer theory, the similarity principle suggests equality of the different angular dimensions (Sect. 2.4). Therefore, the inclusion of angular frequency complies with the similarity principle.

We rewrite (6.1.4) as

$$T(\nu^*) = \exp[-\sigma z_0 \varphi_x(\nu^*)], \tag{6.1.6}$$

where

$$\varphi_x(\nu^*) = 1 - \int_0^1 x(\nu^* y)\, dy \tag{6.1.7}$$

is determined only by the scattering indicatrix. Equation (6.1.6) shows that in this case, the medium MTF depends only on the scattering thickness $\sigma z_0$ and the scattering indicatrix. For some of the more straightforward approximations of the scattering indicatrix, the function $\varphi_x(\nu^*)$ is simple. The appropriate expres-

**Table 6.1.** Functions $x(\beta)$, $x(\rho)$ and $\varphi_x(\nu^*)$.

| $x(\beta)$ | $x(\rho)$ | $\varphi_x(\nu^*)$ |
|---|---|---|
| $\dfrac{2ae^{-a\beta}}{\beta}$ | $\dfrac{a}{\sqrt{a^2+p^2}}$ | $1-\dfrac{a}{\nu^*}\ln\left[\dfrac{\nu^*}{a}+\sqrt{1+\left(\dfrac{\nu^*}{a}\right)^2}\right]$ |
| $2a^2e^{-a\beta}$ | $\dfrac{a^3}{(a^2+p^2)^{3/2}}$ | $1-\dfrac{1}{\sqrt{1+(\nu^*/a)^2}}$ |
| $\sum_{n=1}^{N} 2A_n a_n^2 e^{-a_n\beta}$ $\left(\sum_{n=1}^{N} A_n = 1\right)$ | $\sum_{n=1}^{N}\dfrac{A_n a_n^3}{(a_n^2+p^2)^{3/2}}$ | $1-\sum_{n=1}^{N}\dfrac{A_n}{\sqrt{1+(\nu^*/a_n)^2}}$ |
| $\dfrac{2}{b^2}e^{-\beta^2/2b^2}$ | $e^{-b^2p^2/2}$ | $1-\sqrt{\dfrac{\pi}{2}}(b\nu^*)^{-1}\operatorname{erf}\left(\dfrac{b\nu^*}{\sqrt{2}}\right)$ |
| $\sum_{n=1}^{N}\dfrac{2B_n}{b_n^2}e^{-\beta^2/2b_n^2}$ $\left(\sum_{n=1}^{N} B_n = 1\right)$ | $\sum_{n=1}^{N} B_n e^{-b_n^2p^2/2}$ | $1-\sqrt{\dfrac{\pi}{2}}\sum_{n=1}^{N} B_n(b_n\nu^*)^{-1}\operatorname{erf}\left(\dfrac{b_n\nu^*}{\sqrt{2}}\right)$ <br> $\operatorname{erf} x = \dfrac{2}{\sqrt{\pi}}\int_0^x e^{-t^2}\,dt$ |

sions are given in Table 6.1. We emphasize the indicatrix

$$x(\beta) = \frac{2a}{\beta}\mathrm{e}^{-a\beta}, \tag{6.1.8}$$

which provides a reasonable description of small-angle scattering in sea water at $a \approx 5-7$ [6.5]. The MTF equation is then

$$T(\nu^*) = \mathrm{e}^{-\sigma z_0}\left(\nu^*/a + \sqrt{1+(\nu^*/a)^2}\right)^{\sigma z_0/(\nu^*/a)}. \tag{6.1.9}$$

Formula (6.1.9), first derived in [6.5], has featured in various studies. It has the advantage of simplicity and of being a one-parameter representation.

Simple expressions for the function $\varphi_x(\nu^*)$ and, hence, for the MTF can also be derived by approximating the scattering indicatrix by means of the exponential or Gaussian function (Table 6.1). However, as is the case with the relation (6.1.8), one-parameter representations such as these do not make sufficient allowance for the real scattering indicatrix, for cloud scattering, for example [6.6]). In using this approach, we may represent the indicatrix as the series,

$$x(\beta) = \sum_{n=1}^{N} 2A_n a_n^2 \mathrm{e}^{-a_n\beta}, \tag{6.1.10}$$

or

$$x(\beta) = \sum_{n=1}^{N} (2B_n/b_n^2)\mathrm{e}^{-\beta^2/2b_n^2}. \tag{6.1.10'}$$

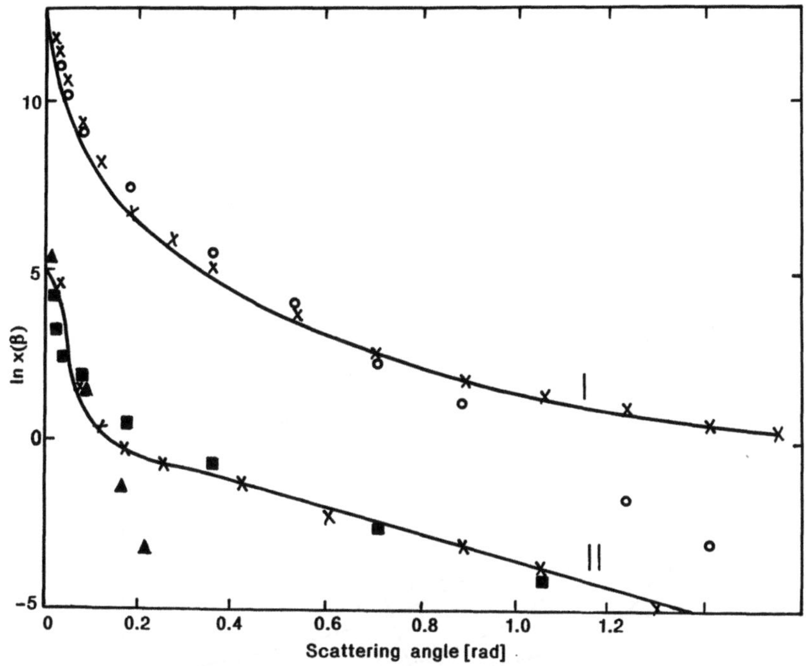


**Fig. 6.1.** Scattering indicatrices. Curve I shows the sea water indicatrix compared to the approximation of (6.1.10) (crosses) at $A_1 = 0.66, a_1 = 42$; $A_2 = 0.29, a_2 = 9.2$; $A_3 = 0.05, a_3 = 2.2$. Curve II gives the Cloud C.1 indicatrix approximated by (6.1.10) (crosses) at $A_1 = 0.22, a_1 = 86$; $A_2 = 0.23, a_2 = 48$, $A_3 = 0.55$, $a_3 = 4$. Circles, triangles, and squares calculated using (6.1.8) at $a = 7$, 30 and 3.1, respectively

An algorithm which allows one to expand functions as an exponential series was developed and discussed in [6.7]. Two or three terms of the indicatrix approximation (6.1.10, 10′) are sufficient for MTF calculations [6.6]. As seen from Fig. 6.1, (6.1.10) with $N = 3$, ensures good accuracy for $x(\beta)$ in ocean water and cloud scattering over a wide range of angles. Expression (6.1.8), which provides a reasonable description of small-angle scattering in sea water, cannot reproduce the $x(\beta)$ of clouds.

In contrast to the one-parameter approximation, the scattering indicatrix formulas (6.1.10, 10′) make more accurate allowance for scattering over all angles, and especially at small $\beta$. This ensures a more complete representation of the function $S(\nu)$ at large spatial frequencies $\nu$, where it is due to $x(\beta)$ behavior at small $\beta$. For example, if $x(\beta)$ is given as (6.1.10), equation (6.1.4) yields

$$T(\nu^*) = \exp\left[-\sigma z_0 + \sigma z_0 \sum_{n=1}^{N} \frac{A_n}{\sqrt{1 + (\nu^*/a_n)^2}}\right]. \tag{6.1.11}$$

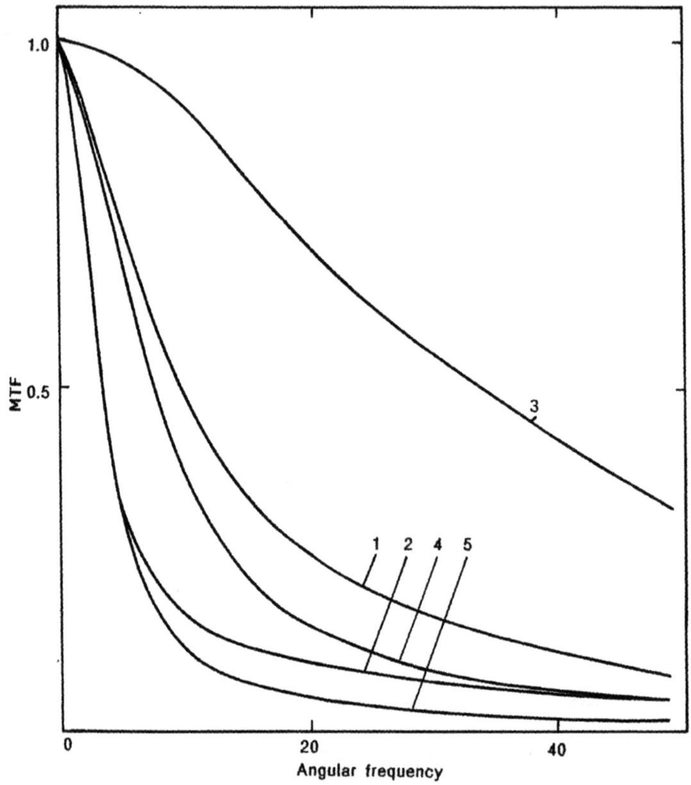


Fig. 6.2. MTF of sea water (curves 1, 4) and Cloud C.1 (curves 2, 3, 5) at $\sigma z_0 = 5$. Curves 1 and 2 give the MTF for $x(\beta)$ approximated by (6.1.10); curves 3–5 use the approximation (6.1.8) at $a = 30$, 7, and 3.1, respectively

The medium MTFs calculated using (6.1.9, 11) for the indicatrices of Fig. 6.1 are plotted in Fig. 6.2. As is seen, the one-parameter approximation of the indicatrix fails to satisfactorily represent the MTFs over a wide range of angular frequencies. At $\nu^* \to \infty$, or, to be specific at $\nu^*/a_n \gg 1$, (6.1.11) implies that

$$T(\nu^*) = e^{-\sigma z_0}\left(1 + \frac{\sigma z_0}{\nu^*} \sum_{n=1}^{N} A_n a_n\right). \tag{6.1.12}$$

This formula, for example, estimates the range of spatial frequencies where the medium MTF is determined by the unscattered light. In fact,

$$\frac{T(\nu^*) - T(\nu^* \to \infty)}{T(\nu^* \to \infty)} = \frac{\sigma z_0}{\nu^*} \sum_{n=1}^{N} A_n a_n .$$

We shall use the parameters $A_n$ and $a_n$ for the scattering indicatrix of sea water, as given in Fig. 6.1. We obtain that only at $v^* > 150\sigma z_0$ is the sea water MTF less than 20% different from its asymptotic value at $v^* \to \infty$.

Sometimes, the medium MTF expression in which the medium indicatrix is given by a set of angular moments $\beta_{2k}$ calculated from (4.4.31) can be useful. In this case, the function $\varphi_x(v^*)$ is of the form [6.6]

$$\varphi_x(v^*) = \sum_{k=1}^{\infty} (-1)^{k+1} \frac{\beta_{2k}}{(k!)^2(2k+1)} \left(\frac{v^*}{2}\right)^{2k} . \tag{6.1.13}$$

As follows from (6.1.6, 13), for low $v^*$, the MTF is defined by the Gaussian function

$$T(v^*) = \exp\left[-\frac{(v^*)^2}{2V_{v*}}\right], \tag{6.1.14}$$

where the value of

$$V_{v*} = \frac{6}{\sigma z_0 \beta_2} \tag{6.1.15}$$

vanishes in inverse proportion to the scattering thickness $\sigma z_0$.

In all the above SAA formulas for the OTF and MTF, one detail should be emphasized. Without renormalizing (4.4.6–8), the OTF values at small spatial frequencies, at $v = 0$ for example, prove to be overestimated because the losses due to backscattering have been neglected. At the same time, for high $v$, the OTF due to small-angle scattering is correctly described. Hence, the MTF values are underestimated at high spatial frequencies.

### 6.1.2 The Small-Angle Diffusion Approximation

In many cases, simple OTF and MTF expressions in the SADA are convenient and useful. As the spread function is SADA approximated using the Gaussian distribution (4.5.20), the medium OTF is also represented by the Gaussian distribution,

$$S(v, \tau_0) = S(\tau_0) \exp\left[-\frac{V_r(\tau_0) v^2}{2}\right]. \tag{6.1.16}$$

It is clear that (6.1.16), as well as (6.1.14), can be applied only at small spatial frequencies.

The modified SADA expressions are more commonly used (Sect. 4.5.7). On dividing the scattering indicatrix into two components as in (4.5.54), the medium OTF and MTF become

$$S(v, \tau_0) = \sum_{i=1}^{2} S_i(\tau_0) e^{-V_{r,i}(\tau_0) v^2/2} + e^{-\tau_0} , \tag{6.1.17}$$

$$T(\nu, \tau_0) = \frac{S(\nu, \tau_0)}{S(\nu = 0, \tau_0)} . \tag{6.1.18}$$

The components of the scattered radiation flux $S_i(\tau_0)$ and of the quantity $\varepsilon^2 V_{r,i}(\tau_0)$ are described by formulas (4.5.64, 70, 65, 71). The total flux through the plane $\tau = \tau_0$ is $S(\nu = 0, \tau_0) = \exp(-\tau_0) + S_1(\tau_0) + S_2(\tau_0)$. Given the parameters $\eta$, $\beta_{2,1}$ and $\beta_{2,2}$ in the indicatrix representation (4.5.54), we may employ formulas (6.1.17, 18). The values of these parameters typical for sea water are cited in Sect. 9.1.2, while those for cloud C.1 are presented in Sect. 4.5.7.

As compared with the SAA formulas, the expressions (6.1.16–18) are particularly convenient for calculating a radiant flux through a finite aperture. For this reason, they will be used in Sect. 6.6.

### 6.1.3 The Diffusion Approximation

The OTF of a strongly scattering thick layer is usually calculated using the data of the diffusion approximation. Hence, at $\tau_0 \gtrsim 4q$, the OTF can be estimated as in [6.8–10]

$$S(\nu, \tau_0) = \frac{\Lambda[(1 - 2mq)e^{-4mq} - (1 + 2mq)]}{8m[1 - e^{-2m(\tau_0 + 4q)}]} \times \left[\left(\frac{3}{1+m} - x_1\right)(1 - e^{-(1+m)\tau_0})e^{-m(\tau_0+4q)} - \left(\frac{3}{1-m} - x_1\right)(e^{-m\tau_0} - e^{-\tau_0})\right] + \left(1 - \frac{\Lambda q x_1}{4}\right)e^{-\tau_0} . \tag{6.1.19}$$

Here, $m = \sqrt{\gamma^2 + (\nu/\varepsilon)^2}$, $q = 1/(3 - x_1)$. We recall that the asymptotic attenuation coefficient $\gamma$ is approximately expressed in terms of the parameters $g$ and $\Lambda$ using (3.2.17, 18). Formula (6.1.19) requires us to solve the diffusion equation (4.3.9) where the source function is defined by (4.3.23) and we pass over from the radiation density to the irradiance using (4.3.16). In fact, the diffusion approximation gives a correct representation of the OTF for small spatial frequencies. In order that the asymptotic OTF value be correctly described at $\nu \to \infty$, in deriving (6.1.19) the diffusion approximation solution for the OTF was artificially supplemented by the term $-\Lambda q x_1 \exp(-\tau_0)/4$. At large $\tau_0$, this term does not alter the result for small frequencies; on the contrary, it gives correct OTF at $\nu \to \infty$.

As seen from (6.1.19), the effect of the scattering indicatrix on the OTF is taken into account only by the integral parameter $x_1$. Such a dependence is clearly valid only for optically thick layers and low spatial frequencies, i.e., within the applicability range of the DA.

For small frequencies (at $2mq < 1$), considering the term of order $m^2$, (6.1.19) suggests a more compact formula for the medium OTF,

$$S(\nu, \tau_0) = \frac{\sinh(5mq)}{\sinh[m(\tau_0 + 4q)]} - \frac{\sinh[m(\tau_0 + 5q)]}{\sinh[m(\tau_0 + 4q)]} e^{-\tau_0} + e^{-\tau_0} . \tag{6.1.20}$$

**Table 6.2.** MTF of thick strongly scattering layers.

| Applicability range | | $T(\nu)$ |
|---|---|---|
| $\gamma\tau_0 \gg 1$ | $\nu^*/\gamma\tau_0 < 1$ | $\exp[-(\nu^*)^2/2\gamma\tau_0]$ |
| | $\nu^*/\gamma\tau_0 > 1$ | $\frac{\nu^*}{\gamma\tau_0}\exp(-\nu^*)$ |
| $0 \leqslant \gamma\tau_0 \leqslant 1$ | $\nu^* < 1$ | $\exp[-(\nu^*)^2/6]$ |
| | $\nu^* \gg 1$ | $\nu^* \exp(-\nu^*)$ |

For very thick layers, (6.1.19) gives a very simple asymptotic formula,

$$S(\nu, \tau_0) = \frac{\sinh(5mq)}{\sinh[m(\tau_0 + 4q)]}. \tag{6.1.21}$$

At $\nu = 0$, formula (6.1.21) becomes the known asymptotic solution of the transfer equation (3.5.5, 6), thus demonstrating sufficient DA accuracy in representing the OTF at small spatial frequencies. It should be remembered that for different scattering indicatrices, the quantity $u_0$ ($\mu_0 = 1$) entering (3.5.6) ranges between 1.25 and 1.3.

In transfer theory, two extreme cases are often distinguished: (i) strong absorption in the layer when the portion of energy absorbed throughout the layer is large ($\gamma\tau_0 \gg 1$) and (ii) weak absorption in the layer ($0 \leqslant \gamma\tau_0 \ll 1$). For these two situations, Table 6.2 offers simple expressions which result from (6.1.21) and represent the medium MTF for small and comparatively large spatial frequencies.

When comparing the transfer properties of different scattering media, it is convenient to characterize the medium MTF by a single parameter. To do so, the MTF half-width, i.e., the $\nu$ value at which $T(\nu) = 0.5$, is sometimes used [6.11, 12]. However, we shall use another method. Let us characterize the function $T(\nu)$ by

$$V_\nu = \left[-\lim_{\nu \to 0} \frac{d^2 T(\nu)}{d\nu^2}\right]^{-1} = \frac{1}{V_r}. \tag{6.1.22}$$

The quantity $V_\nu$ is intimately connected with the MTF half–width and specifies the range of spatial frequencies transmitted through the scattering medium. If the PSF allows a Gaussian approximation, the MTF is also described by the Gaussian distribution (6.1.14) having the variance $V_{\nu*} = V_\nu z_0^2$. For a thick layer, $V_{\nu*} = \gamma\tau_0$. This implies that the MTF of the layer improves as its optical thickness $\tau_0$ increases. In the analysis which follows, we shall again discuss this interesting observation. For nonabsorbing or weakly absorbing layers ($\gamma\tau_0 \ll 1$), the quantity $V_{\nu*}$ stabilizes with increasing $\tau_0$, i.e., it is no longer a function of the optical thickness $\tau_0$. This results from the square-law variation of $V_r$ with the thickness

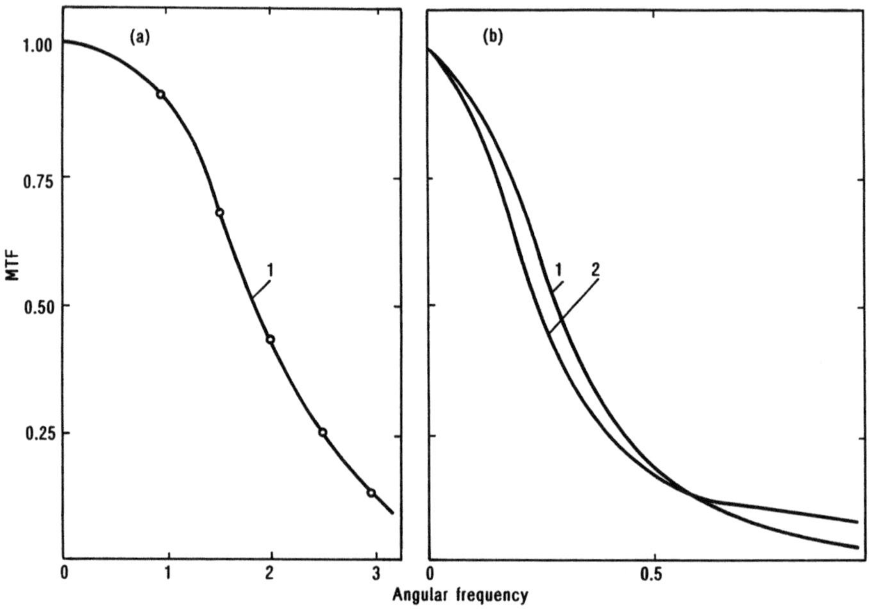


**Fig. 6.3.** Comparison of the MTF determined by (6.1.21) (curve 1) with rigorous calculation (*circles*) at $x(\beta) = 1$, $\Lambda = 0.9$, $\tau_0 = 4$ (a) and experimental data (curve 2) at $4q\gamma = 0.09$ and $\tau_0/4q = 5$ (b)

of a nonabsorbing thick layer (Sect. 4.3.3). Figure 6.3 provides additional information on the accuracy of MTF calculation within the diffusion approximation.

The approximate formulas discussed in Sects. 6.1–3 are sufficiently accurate for calculating the OTF for any optical parameters. The SAA, SADA and DA formulas complement one another.

### 6.1.4 MTF Dependence on Optical Medium Parameters

We first mention some general properties of the modulation transfer function. For the case of a symmetric spread function, the MTF decreases monotonically, $T(\nu = 0) = 1$, $T(\nu \to \infty) = \exp(-\tau_0)/S(\nu = 0)$. As $S(\nu \to \infty, \tau_0) = \exp(-\tau_0)$, $\iint_{-\infty}^{\infty} S(\vec{\nu}, \tau_0)\, d\vec{\nu}$ diverges. This divergency stems from the $\delta$-singularity of the PSF due to the unscattered radiation component. The local PSF properties, owing to the first and second orders of scattering (Sect. 4.2), cause the divergence of the integrals $\iint_{-\infty}^{\infty} S^{(i)}(\vec{\nu}, \tau_0)\, d\vec{\nu}$ at $i = 1, 2$, where $S^{(i)}(\vec{\nu}, \tau_0)$ is the $i$-th scattering order contribution to the medium OTF. This is seen, for example, from (6.1.11, 12).

The MTF dependence on the parameters of the optical medium is plotted in Figs. 6.4–6. For not very thick layers at $\tau_0 \leqslant 5$, the data have been obtained from the small-angle approximation, using (6.1.11). The DA formula (6.1.19) provided the data at $\tau_0 > 10$. The indicatrix parameters $a$ (in (6.1.8, 9)) and $x_1$ are related

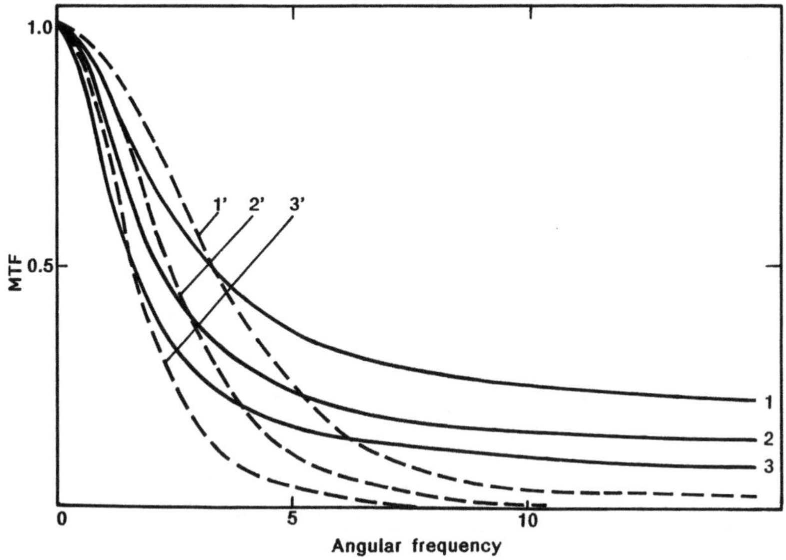


**Fig. 6.4.** Medium MTF at $g = 0.72$, $\tau_0 = 3$ (curves 1–3), 10 (curves 1′–3′) and $\Lambda = 0.7$ (curves 1, 1′), 0.9 (2, 2′) and 1 (curves 3, 3′)

as $x_1 = 3\arctan(2/a)/\arctan[2a/(a^2 - 1)]$, which at $a \gtrsim 5$ may be replaced by the more simple expression $x_1 = 3(1 - 1/a^2)$. The increase in the single scattering albedo results in an increase in the multiple scattering contribution and a broadening of the spread function, i.e., MTF degeneration. As mentioned above, within the scope of the SAA, $T(v^*)$ is a function of the scattering thickness $\sigma z_0$ alone, and decreases exponentially with increasing $\sigma z_0$ (6.1.6).

Figure 6.5 demonstrates the influence of the scattering indicatrix extension on the MTF. For the layers which are not very thick and strongly anisotropically scattering, the MTF is explicitly related to the scattering indicatrix by (6.1.6, 7, 9) and by the expressions in Table 6.1. At this point, (6.1.15) implies that the quantity $V_{v*}$ grows in proportion to $\beta_2^{-1}$ with increasing indicatrix extension.

We now consider the MTF dependence on the optical layer thickness.

For weakly absorbing media ($\gamma\tau_0 \ll 1$), $V_{v*}$ decreases monotonically with increasing $\tau_0$ and tends to an asymptotic value of 3 (Table 6.2). In the scattering medium with essential absorption, $T(v^*)$ changes nonmonotonically with $\tau_0$. As seen from Fig. 6.6, as long as the optical thickness increases, the half-width of the function $T(v^*)$ first reduces and the MTF then degenerates due to spread function broadening in the course of multiple scattering. At large $\tau_0$, the absorption of photons, which deviate noticeably from the beam axis and travel along a longer optical path, becomes effective. The spread function is seen to be narrower, and the MTF half-width grows. For an absorbing medium with strong anisotropic scattering (within the SADA applicability, Sect. 4.5), the process can

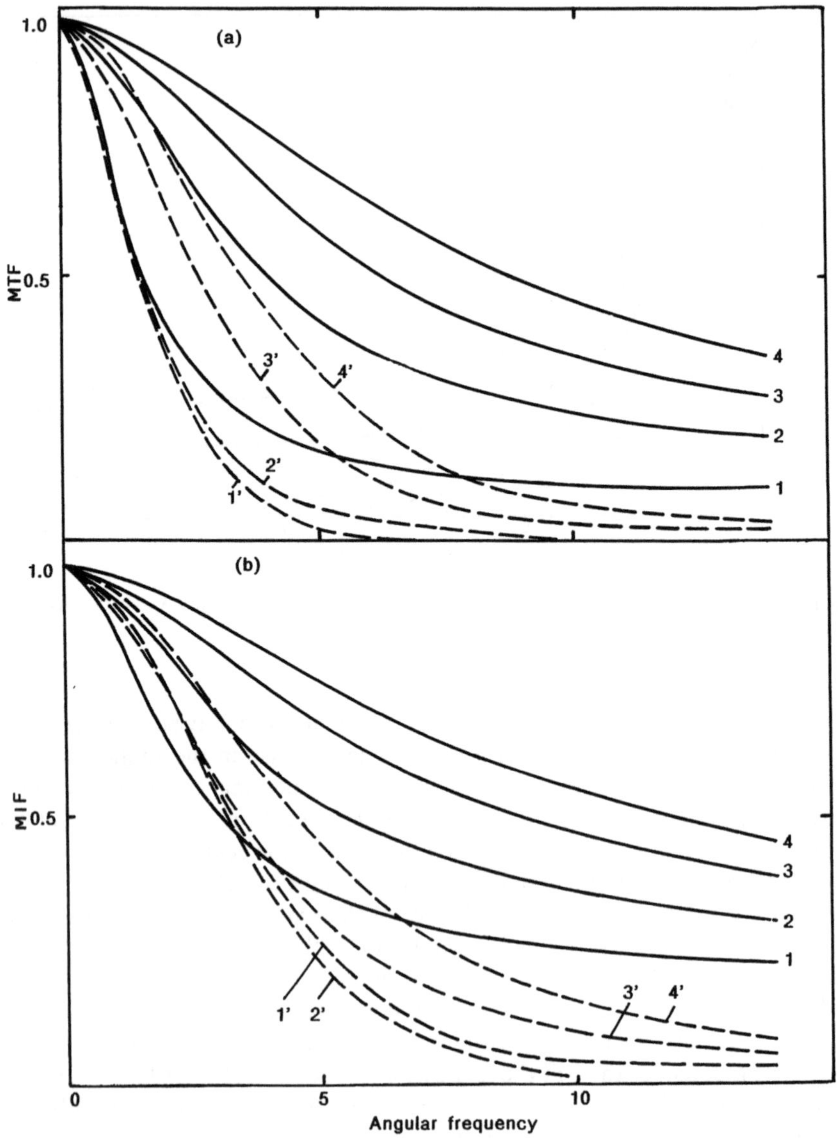


**Fig. 6.5.** Medium MTF at $\tau_0 = 3$ (curves 1–4), 10 (curves 1′–4′). (a) $\Lambda = 1$, (b) $\Lambda = 0.7$. Curves are given for different indicatrices at $g = 0.72$ (curves 1, 1′), 0.88 (curves 2, 2′), 0.95 (curves 3, 3′) and 0.97 (curves 4, 4′)

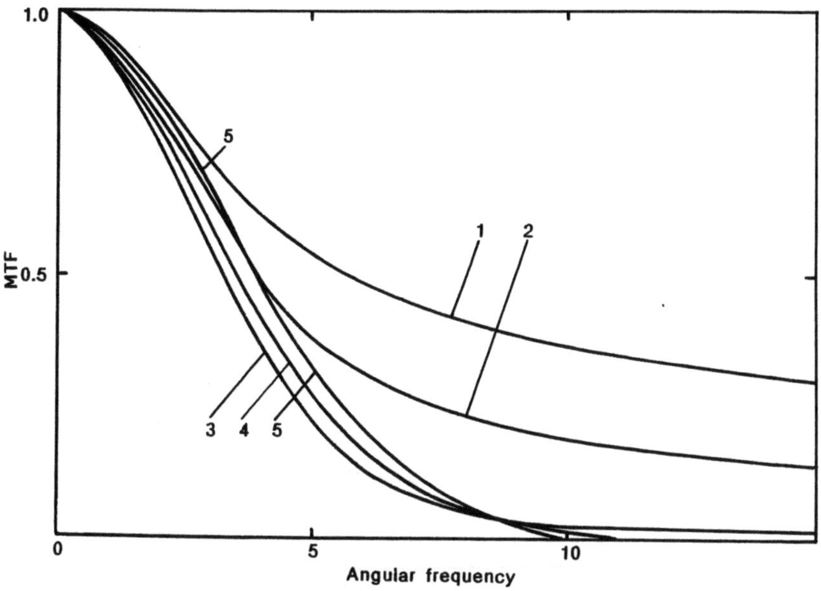


**Fig. 6.6.** Medium MTF at $\Lambda = 0.7$, $g = 0.72$. Curves correspond to different layer thicknesses $\tau_0 = 3$ (curve 1), 5 (curve 2), 10 (curve 3), 15 (curve 4) and 20 (curve 5)

conveniently be traced using the following formula

$$V_{\nu *} = \sqrt{\frac{2(1-\Lambda)}{\Lambda\beta_2}}\, f(s\tau_0)\,,$$

$$f(s\tau_0) = s\tau_0\left[1 - \frac{\tanh s\tau_0}{s\tau_0}\right]^{-1}. \qquad (6.1.23)$$

The relations (6.1.23) stem from (4.5.20, 23) for the SADA point spread function. It should be noted that $s = [\Lambda(1-\Lambda)\beta_2/2]^{1/2} = [\Lambda(1-\Lambda)(1-g)]^{1/2}$ (4.5.10). The function $f(s\tau_0)$ plotted in Fig. 6.7 depends only on the parameter $s\tau_0$ and is thus universal for media with different optical properties.

At small $\tau_0$, the quantity $V_{\nu *} \approx 6/\beta_2\Lambda\tau_0$, i.e., it decreases with increasing depth (thickness), as $\tau_0^{-1}$. At $\tau_0^{\min} = 1.6/s = 2.1/[\Lambda(1-\Lambda)\beta_2]^{1/2}$, it achieves its minimum and then increases. This means that at $\tau_0 = 1.6/s$, the half-width of the spread function expressed in angular measure reaches its maximum. At $s \approx 0.1$, typical of sea water, the quantity $\tau_0^{\min} \approx 16$. For $s\tau_0 \gg 1$, $V_{\nu *} = (1-\Lambda)\tau_0$, i.e. it increases linearly with $\tau_0$. We note that for strongly scattering and absorbing media, $V_{\nu *} = \gamma\tau_0$ in the diffusion approximation. At $1 - g \ll 1$, for $s \ll 1 - \Lambda$ and $\gamma \approx 1 - \Lambda$, the asymptotic, at $\tau_0 \gg 1$, values of $V_{\nu *}$ are equal in the DA and SADA.

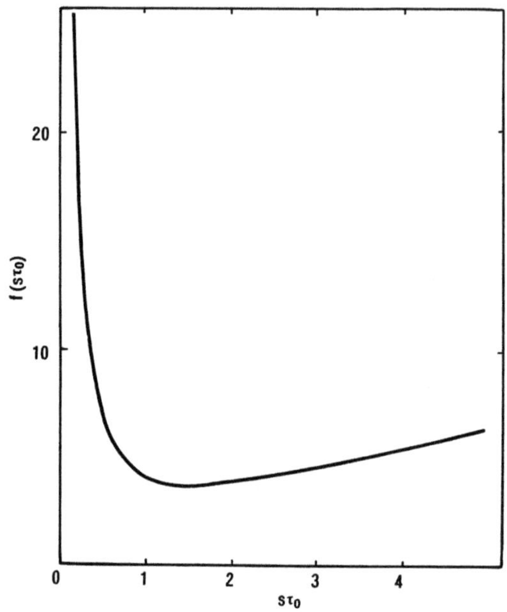


**Fig. 6.7.** Function $f(s\tau_0)$ from (6.1.23)

### 6.1.5 Scattering Layer MTF under Pulsed Source Illumination

We have so far dealt with the MTF as applied to medium illumination with a stationary source. In modern vision systems, pulsed lasers are commonly used. If, in this case, the receiver accumulates the radiation within the time interval $0 \leqslant t < \infty$, then the system MTF equals that for a relevant stationary source. Now, let the receiver accumulate signal photons only over a short time $\Delta t_{\text{rec}} = t_1 - t_0$ ($t_0$ being the time of detection onset) commensurate with the incident pulse duration, and $\Delta t_{\text{rec}} \ll \sqrt{2V_t}$. Here, $V_t$ is the variance of the time pulse spread on the receiver. Then, the system MTF depends both on $t_0$ and on $\Delta t_{\text{rec}}$, and may essentially differ from the MTF of the stationary source system.

In the analysis of the pulsed MTF, we revert to the PSF time dependence considered in Sects. 4.8.4, 5.

As seen from Fig. 4.18, when the pulse signal arrives at the plane $z = z_0$ ($t \gtrsim z_0/v$), the PSF has a maximum at $r = 0$, the variance of photon deviation from the axis being much less than in the stationary case. This means that in a viewing system which accumulates a signal from the object over a short time $\Delta t_{\text{rec}}$ after its arrival at the detector, the medium MTF is essentially better than in the similar stationary source system. The medium MTF approaches its stationary value with increasing $\Delta t_{\text{rec}}$.

The time dependent medium MTF can be expected to display some interesting specific features. For the time $t$, photons approach only a limiting region $r \leqslant \sqrt{(vt)^2 - z^2}$. The Fourier transform properties suggest that, in this case, the MTF oscillates at high frequencies. Another feature of the nonstationary MTF can be related to the PSF maximum shift from the beam axis (Fig. 4.18). For the time instants at which such a shift is observed, the nonstationary medium MTF is nonmonotonic. In real systems, these effects are not generally displayed by virtue of the averaging effect on the photodetector, since $\Delta t_{\mathrm{rec}} \gtrsim \sqrt{2V_t}$.

## 6.2 OTF of an Inhomogeneous Layer

The scattering medium through which an object is viewed is often inhomogeneous. This applies essentially to all natural media. We are aware of the stratification of the optical characteristics of sea water and the inhomogeneity of clouds. In atmosphere, higher and lower turbidity layers often alternate. Therefore, a discussion of the various aspects of viewing through inhomogeneous scattering media would seem to be very important.

Setting the dependence of the extinction and scattering coefficients, as well as of the scattering indicatrices given by the functions $\varepsilon(z)$, $\sigma(z)$ and $x(\beta, z)$, we consider first a comparatively thin layer with an essentially extended scattering indicatrix.

### 6.2.1 The Small-Angle Approximation

The stratification of optical characteristics is simple to account for in the SAA. Hence, for the MTF from (4.4.33), we have

$$T(v) = \exp\left[-\int_0^{z_0} \sigma(z)[1 - x(v(z_0 - z), z)]\,dz\right]. \tag{6.2.1}$$

It should be remembered that here as before, $z = z_0$ is the object plane.

Formula (6.2.1) was derived in [6.13]. It implies that the MTF is determined only by the distribution of the scattering characteristics along the viewing path, rather than being dependent on the stratification of the medium absorption coefficient. Let us compare the MTFs of homogeneous and inhomogeneous layers at equal scattering depths $\tau_\sigma = \int_0^{z_0} \sigma(z)\,dz$. In this instance, the scattering coefficient of the homogeneous layer $\sigma_{\mathrm{hom}} = \langle\sigma\rangle = (1/z_0)\int_0^{z_0} \sigma(z)\,dz$. Letting the MTF of such a layer be $T^{\mathrm{hom}}(v)$, we take it that the scattering indicatrix along the path is constant in both cases. We now have

$$\frac{T(\nu)}{T^{\mathrm{hom}}(\nu)} = \exp\left[\int_0^{z_0} [\sigma(z) - \langle\sigma\rangle] x(\nu(z_0 - z))\, dz\right]. \tag{6.2.2}$$

Consider the simplest case of an inhomogeneous optical path where the $z$ dependence of the scattering coefficient may be approximated by the linear function

$$\sigma(z) = \sigma_0 + \sigma_1 z\,. \tag{6.2.3}$$

Then $\langle\sigma\rangle = \sigma_0 + \sigma_1 z_0/2$ and, from (6.2.2), we have

$$\frac{T(\nu)}{T^{\mathrm{hom}}(\nu)} = \exp\left[\sigma_1 \int_0^{z_0} (z - z_0/2) x(\nu(z_0 - z))\, dz\right]. \tag{6.2.4}$$

As the function $x(\nu z)$ decreases monotonically,

$$\int_0^{z_0} (z - z_0/2) x(\nu(z_0 - z))\, dz \geqslant 0\,. \tag{6.2.5}$$

Therefore, if $\sigma_1 > 0$, i.e., the layers of higher turbidity are close to the object, then $T(\nu)/T^{\mathrm{hom}}(\nu) \geqslant 1$, and at all frequencies the MTF of the inhomogeneous layer is better than that of a homogeneous one. When $\sigma_1 < 0$, i.e., the layers of higher turbidity are remote from the object, the MTF values of the inhomogeneous layer are less than those of a homogeneous one at all frequencies.

For the inhomogeneous scattering layer, the relations (6.2.1, 6.1.22) give

$$V_{\nu *} = \frac{2z_0^2}{\int_0^{z_0} \sigma(z)\beta_2(z)(z_0 - z)^2\, dz}\,. \tag{6.2.6}$$

In the general case, (6.2.6) provides the same physical implications as those obtained for a particular inhomogeneity (6.2.3). As long as we set, for simplicity, that the indicatrix be independent of $z$, we have, instead of (6.2.6),

$$V_{\nu *} = b V_{\nu *}^{\mathrm{hom}} \tag{6.2.7}$$

Here, $V_{\nu *}^{\mathrm{hom}} = 6/\beta_2 \tau_\sigma$ is the value of $V_{\nu *}$ for the homogeneous layer with the same scattering thickness, and the parameter

$$b = \frac{z_0^2 \int_0^{z_0} \sigma(z)\, dz}{3 \int_0^{z_0} \sigma(z)(z_0 - z)^2\, dz} \tag{6.2.8}$$

defines the inhomogeneity effect on the MTF. If $\sigma(z)$ is a monotonically increasing function, then $b > 1$; when it decreases monotonically, $b < 1$. Thus, (6.2.7) yields a simple analytical estimate for the medium MTF variation with the concentration of scatterers in the layer.

### 6.2.2 OTF of an Inhomogeneous Strongly Scattering Layer

The OTFs of inhomogeneous strongly scattering layers have been discussed in [6.14, 15]. In those works, initial equations are formulated and the Green's function is studied for some partial dependences of optical medium parameters within the diffusion approximation. We shall confine our consideration to the exponential dependence of the extinction coefficient on the coordinate

$$\varepsilon = \varepsilon_0 \exp(-z/l_{\text{in}}) . \tag{6.2.9}$$

At $l_{\text{in}} > 0$, optically dense layers are located at the upper layer boundary; if $l_{\text{in}} < 0$, then they are concentrated near the object. Let us introduce a dimensionless inhomogeneity parameter $\chi = \varepsilon_0 l_{\text{in}}$. The greater $\chi$, the lower the layer inhomogeneity. For $\chi \to \infty$, $\varepsilon(z) \to \varepsilon_0$, i.e., the medium becomes homogeneous. For this case, the layer OTF is [6.14, 15]

$$S(\omega, \tau_0) = Q(\eta, \eta_0, \eta_1) \left[ -3 \int_0^{\tau_0} \sqrt{|\xi|}\, \Psi(\xi, \eta_0) e^{-\tau'} d\tau' + x_1 \sqrt{|\gamma\chi|}\, \Psi(\gamma\chi, \eta_0) - x_1 \sqrt{\eta}\, \Psi(\eta, \eta_0) e^{-\tau_0} \right], \tag{6.2.10}$$

where

$$Q(\eta, \eta_0, \eta_1) = \frac{A\varepsilon_0^2 \sqrt{\eta}}{8\pi\gamma} \frac{\Psi(\eta, \eta_1) - 2q\gamma[((1-2p)/2\eta)\Psi(\eta, \eta_1) + L(\eta, \eta_1)]}{\Psi(\eta_0, \eta_1)}, \tag{6.2.11}$$

$$\eta = \gamma(\chi - \tau_0), \quad \eta_1 = \gamma(\chi - \tau_0 - 2q), \quad p = \sqrt{1/4 + \omega^2\chi^2},$$
$$\omega = \nu/\varepsilon_0, \quad \eta_0 = \gamma(\chi + 2q), \quad \xi = \gamma(\chi - \tau'). \tag{6.2.12}$$

The functions $\Psi(\eta, \eta_1)$ and $L(\eta, \eta_1)$ are expressed in terms of the cylindrical functions of an imaginary argument

$$\Psi(\eta, \eta_1) = I_p(\eta) K_p(\eta_1) - K_p(\eta) I_p(\eta_1),$$
$$L(\eta, \eta_1) = I_{p-1}(\eta) K_p(\eta_1) - K_{p-1}(\eta) I_p(\eta_1). \tag{6.2.13}$$

If $\chi \to \infty$, then $\varepsilon(z) \to \varepsilon_0$ and (6.2.10) becomes the solution for the MTF of a homogeneous layer of optical thickness $\varepsilon_0 z_0$. It can be shown that for not so large gradients of the extinction coefficient ($|\chi| \gg 1$) and not so small spatial frequencies ($p \gg 1$), inhomogeneity exerts the same effect on the MTF as the layer thickness variation

$$\Delta\tau_0 = \frac{\tau_0 \omega^2}{2\chi(\omega^2 + \gamma^2)} . \tag{6.2.14}$$

If $\chi < 0$ (turbidity is higher near the object), then $\Delta\tau_0 < 0$, i.e., the MTF impoves. At $\chi > 0$, $\Delta\tau_0 > 0$ and the MTF of the inhomogeneous layer is more degenerated than that of the homogeneous layer of the same optical thickness. For $\omega \gg \gamma$, we have $\Delta\tau_0/\tau_0 = -1/2\chi$, i.e. the greater the inhomogeneity gradient, the more drastic the change in the MTF. In Fig. 6.8, the MTFs of inhomogeneous layers are compared with the MTFs of a homogeneous layer of the same optical thickness. As is seen, the $\varepsilon$ variation with the depth heavily distorts the MTF.

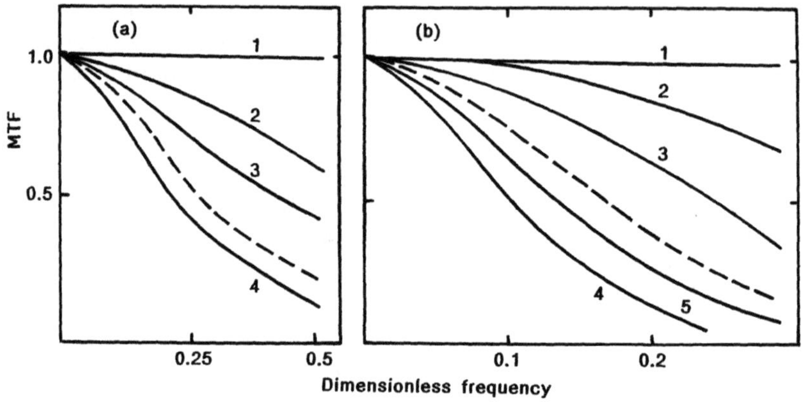


**Fig. 6.8.** MTF of homogeneous (*dashed lines*) and inhomogeneous (*solid lines*) layers of optical thickness $\tau_0 = 5$ (a) and 10 (b) at $q = 2$. Curves correspond to the inhomogeneity parameter values $\chi = -1$ (curve 1), $-3$ (curve 2), $-10$ (curve 3), 10 (curve 4) and 20 (curve 5)

However weak the inhomogeneity, it noticeably alters the spectrum of the spatial frequencies transmitted by the layer.

Thus, the qualitative effect of the inhomogeneity of optical characteristics on the MTF is always similar or the MTF increases at all frequencies as the turbidity moves to the lower layer boundary, i.e., to the observation object.

### 6.2.3 MTF Dependence on the Scattering Layer Position along the Observation Path

We consider in more detail the medium MTF in the particular case, dealt with in many works [6.16–21], of a diffusely luminous object viewed through a scattering layer $z_0'$ thick and a distance $H$ away from the object (Fig. 6.9). The

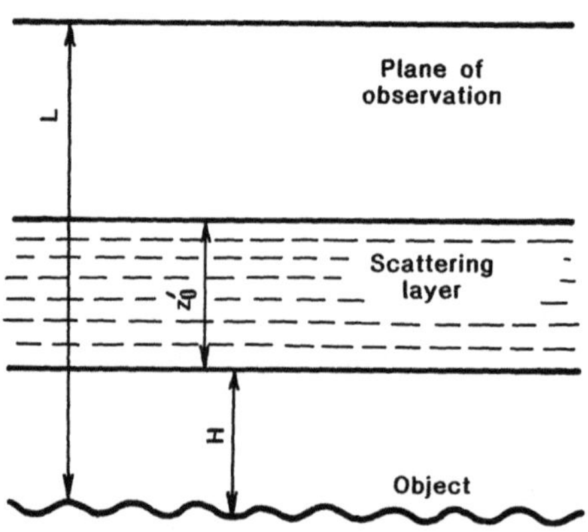


**Fig. 6.9.** Observation through a scattering layer

distance from the observation plane to the object $L = z_0$. A model such as this can be used, for example, to interpret observations of the Earth from space through cloud and a weakly scattering layer below cloud. These problems arise when in viewing through natural and artificial aerosol layers (smoke, smog, etc.).

The influence of the increase in the layer-object distance can easily be shown from very general considerations. In fact, the higher $H$ is, the larger the scattering area exposed to each element of the diffusely luminous object, the more essential the overlapping of the light beams arriving from different object elements and, hence, the more degenerated the MTF. The same phenomena can be described within the inverted scheme. The medium OTF is the Fourier transform of the irradiance distribution over the object plane due to a monodirectional point source (Sect. 5.2.5). The radiation from such a source transmitted through a scattering layer has a certain angular divergence. Therefore, as the layer boundary moves further away from the object plane (increasing $H$), the point spread function broadens, i.e., the MTF deteriorates.

It is also obvious from the inverted scheme that the medium PSF and OTF are independent of the distance $L - (z_0' + H)$ between the observation plane and the upper layer boundary.

We now proceed to a quantitative description of the phenomenon [6.16, 17].

Within the small-angle approximation, the layer-gap OTF is represented by (6.2.1), whose integration interval covers the entire path of observation. Let the scattering $z_0'$ thick layer be homogeneous and characterized by the scattering coefficient $\sigma$ and indicatrix $x(\beta)$. The layer inhomogeneity can simply be included, if required, and does not alter the situation. We rewrite (6.2.1) as

$$T(v) = \exp\left[-\sigma z_0' + \sigma \int_H^{H+z_0'} x(vz)\,dz\right]. \tag{6.2.15}$$

As is seen from a comparison of (6.2.15) and (6.1.4), the layer-gap OTF can be expressed in terms of the OTF of the homogeneous scattering $H$ and $H + z_0'$ thick layers [6.16],

$$T(\vec{v}) = \frac{T^{\text{hom}}(\vec{v}, H + z_0')}{T^{\text{hom}}(\vec{v}, H)}. \tag{6.2.16}$$

For example, for a medium with the scattering indicatrix (6.1.8), we have

$$T(v) = e^{-\sigma z_0'}\left[\frac{\sqrt{1 + (v(H + z_0')/a)^2} + v(H + z_0')/a}{\sqrt{1 + (vH/a)^2} + vH/a}\right]^{a\sigma/v}. \tag{6.2.17}$$

The relation (6.2.16) can immediately be written from the aspect invariance condition, using (5.2.21).

We now proceed to the dimensionless frequencies $v^* = vL$. On introducing a new integration variable $\xi = (z - H)/z_0'$ into (6.2.15), we obtain the following convenient expression for the layer-gap OTF [6.17],

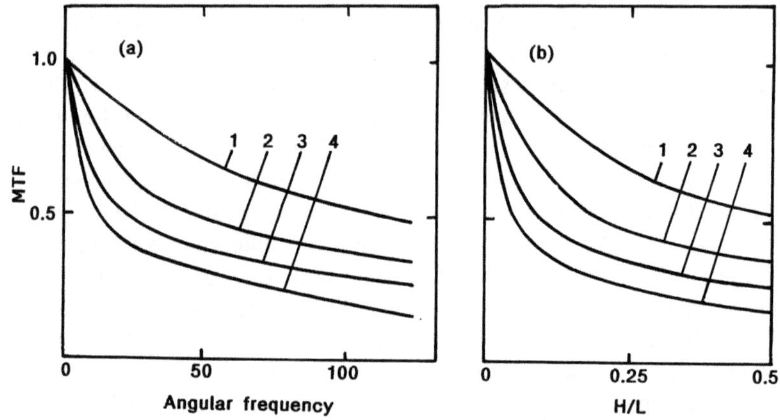


**Fig. 6.10.** Layer-gap MTF versus angular frequency (a) and parameter $t$ (b) for the Cloud C.1 indicatrix at $\sigma z_0' = 1$, $L = z_0' + H$. In (a) $t = 0.05$ (curve 1), 0.125 (curve 2), 0.25 (curve 3), 0.5 (curve 4). In (b) $v^* = 10$ (curve 1), 25 (curve 2), 50 (curve 3) and 100 (curve 4)

$$T(v^*) = \exp\left[-\sigma z_0'\left[1 - \int_0^1 x[v^*(l\xi + t)]\,d\xi\right]\right], \tag{6.2.18}$$

where $l = z_0'/L$, $t = H/L$.

If the scattering layer is thin or remote from the object, then $t \gg l$ and the relation (6.2.18) gives

$$T(v^*) = \exp[-\sigma z_0'(1 - x(v^* t))]\,. \tag{6.2.19}$$

We recall that $x(v^*)$ is the monotonically decreasing function. Therefore, $T(v^*)$ diminishes monotonically at all spatial frequencies $v^*$ with increasing $t$ (Fig. 6.10). In the figure, data are given at $\sigma z_0' = 1$. We can easily pass over to another $\sigma z_0'$ value. Following (6.2.18), we have

$$T(v^*, \sigma z_0') = [T(v^*, \sigma z_0' = 1)]^{\sigma z_0'}\,. \tag{6.2.20}$$

We now include one more useful estimate in our consideration. Using (6.2.6) for $V_{v*}$ of the inhomogeneous layer yields

$$V_{v*} = \frac{6L^2}{\beta_2 \sigma (z_0')^3}\left[1 + \frac{3H}{z_0'} + 3\left(\frac{H}{z_0'}\right)^2\right]^{-1}. \tag{6.2.21}$$

It should be noted that for the homogeneous $z_0'$ thick layer $V_{v*}^{\text{hom}}(z_0') = 6/\beta_2 \sigma z_0'$. Then, (6.2.21) becomes

$$V_{v*} = V_{v*}^{\text{hom}}(z_0')\left[1 + \frac{3H}{z_0'} + 3\left(\frac{H}{z_0'}\right)^2\right]^{-1} \frac{L^2}{(z_0')^2}\,. \tag{6.2.22}$$

As is seen from (6.2.22), the MTF deteriorates ($H$ decreases monotonically) with an increase in the distance between the layer and the object, or as $V_{v*}$ grows. At $H/z_0' \gg 1$,

$$V_{v*} = V_{v*}(z_0')\frac{(z_0'/H)^2}{3}\frac{L^2}{(z_0')^2} = V_{v*}(z_0')\frac{L^2}{3H^2}, \tag{6.2.23}$$

i.e., it decreases as $H^{-2}$ does. These relations were verified experimentally [6.17].

In summarizing the OTF data for inhomogeneous layers, we wish to point out that in all observations through inhomogeneous scattering medium, the MTF is better, i.e., at any spatial frequency, the contrast of an infinite harmonic test object is higher if the higher turbidity layers are located near the object. It is seen to degenerate monotonically as the turbid layer shifts in the direction of the detector. Such behavior is responsible for the well-known phenomenon referred to as the "shower curtain effect", in which the image of a person close to the shower curtain can be seen clearly, whereas a person farther from the shower curtain is difficult to see.

This result contradicts the data presented in [6.18–21], where it is stated that $T(v^*)$ is a nonmonotonic function of the scattering layer location along the path, or of the parameter $H/L$. This implication relies both on measurement and on the Monte Carlo simulations. Such an experimental result is most likely associated with an imperfect MTF estimation based on measuring the angular radiance distribution from a diffuse source (Chap. 5). Typical errors appearing in this case, and attributable to the truncation of the PSF wings, are analysed in [6.22]. The agreement between the experimental and Monte-Carlo simulation data in [6.18–21] is not surprizing since the theoretical Monte-Carlo method adequately represents the experimental situation. We shall consider the image contrast of an object of a given size as a function of the turbid layer position along the observation path in Sect. 6.5.3.

### 6.2.4 Stochastic Medium MTF

We now consider that the optical parameters of the medium vary at random along three spatial coordinates. The OTF expression for such a stochastic medium was derived in Sect. 4.4.7 to give the MTF [6.23]

$$\langle T(v)\rangle = \exp\bigg[-(\langle\sigma\rangle - 2\alpha_{\varepsilon\sigma} + \alpha_{\sigma\sigma})z_0 + (\langle\sigma\rangle - 2\alpha_{\varepsilon\sigma})\int_0^{z_0} x(v\xi)\,d\xi + \alpha_{\sigma\sigma}\int_0^{z_0} x^2(v\xi)\,d\xi\bigg]. \tag{6.2.24}$$

Figure 6.11 illustrates the MTF of a stochastic medium with $\Lambda = \langle\Lambda\rangle$ and a scattering indicatrix given by (6.1.8). The stochastic medium MTF is seen to be consistently better than that of the homogeneous medium. For slightly fluctuating optical parameters of the medium, the difference between $\langle T(v)\rangle$ and $T^{\mathrm{hom}}(v)$ is, of course, insignificant and, in fact, independent of $\Lambda$. However, with an increase in the extinction coefficient fluctuations and optical correlation radius $\langle\varepsilon\rangle R_c$, the MTF of the stochastic medium may be much higher than that of the homogeneous one. The difference is maximum at $v \to \infty$. In this case,

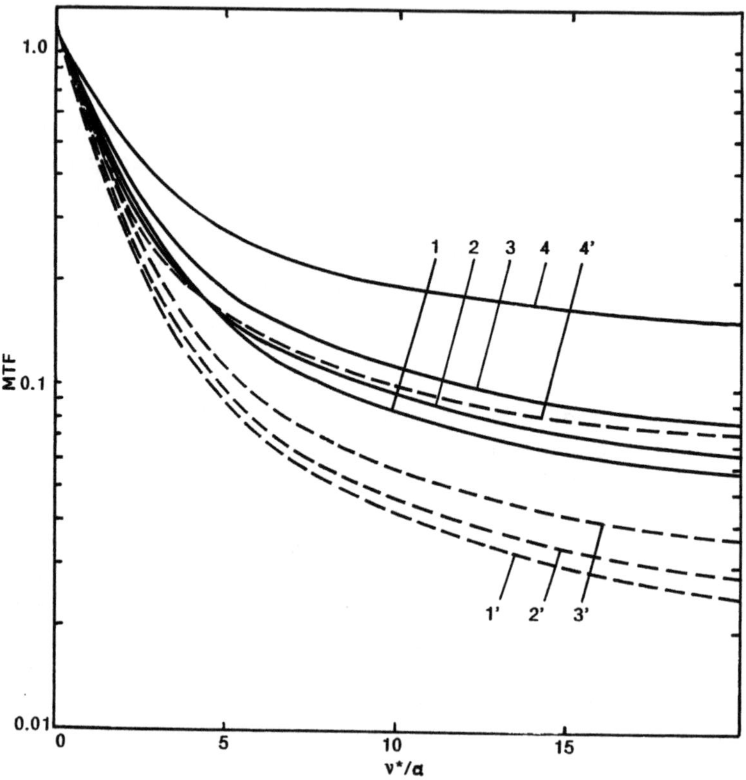


**Fig. 6.11.** Average MTF of a stochastic medium at $\langle\varepsilon\rangle z_0 = 5$, $\Lambda = 0.7$ (curves 1–4), 0.9 (curves 1′–4′) and $M = 0$ (curves 1, 1′), 0.03 (curves 2, 2′), 0.1 (curves 3, 3′) and 0.3 (curves 4, 4′)

$$\frac{\langle T(\nu)\rangle}{T^{\mathrm{hom}}(\nu)} = \exp[(2\alpha_{\sigma\kappa} + \alpha_{\sigma\sigma})z_0] \,. \tag{6.2.25}$$

At $\Lambda = \langle\Lambda\rangle$

$$\frac{\langle T(\nu)\rangle}{T^{\mathrm{hom}}(\nu)} = \exp[(2-\Lambda)\Lambda M \langle\varepsilon\rangle z_0] \,, \tag{6.2.26}$$

where $M = \sqrt{\pi/2}\langle\varepsilon\rangle R_c V_\varepsilon/\langle\varepsilon^2\rangle$. It is seen from (6.2.26) that the greatest deviations in the MTF for the stochastic and relevant homogeneous medium are encountered at $\Lambda = 1$, i.e., in a nonabsorbing medium.

Expression (6.2.24) at $\nu \to \infty$ and the obvious condition $\langle T(\nu)\rangle \leqslant 1$ provide a necessary but insufficient restriction for the fluctuations of medium parameters which may be considered within the approximation used,

$$2\alpha_{\varepsilon\sigma} - \alpha_{\sigma\sigma} < \langle\sigma\rangle \qquad \text{or } M < \frac{1}{2-\Lambda} \,. \tag{6.2.27}$$

## 6.3 Scattering Layer OTF along an Oblique Path. Phase Transfer Function

We have so far considered medium transfer characteristics as applied to observations along a vertical path. The case becomes complicated when observing along an inclined path. The point spread function is then the irradiance distribution from the beam which makes an angle $\arccos \mu_0$ with the normal to the layer (the angle which specifies the viewing direction). In this case, the PSF is dependent both on $r$ and on the azimuth $\varphi$, and the OTF is complex. Medium transfer characteristics for this case have not yet been adequately studied. However, some experimental work [6.24] and computer calculations [6.25–27] are worth noting. The small-angle approximation widely used in many vision problems is then not successful. Since the SAA disregards the light reflected by the medium, the medium boundary has no bearing on the solution. The only direction distinguished is the beam axis. Therefore, the light field characteristics in the plane normal to the axis are the result.

This problem may be solved using the small-angle diffusion and diffusion approximations. In the SADA, the point spread function for oblique medium irradiation is expressed by (4.5.26). Its simple Fourier transform gives the OTF of the scattering medium. Following [6.28], the OTF of a strongly scattering layer will be found here within the diffusion approximation, its properties being considered as applied to oblique path observations. We take it that a monodirectional unit power point source radiates at an angle $\arccos \mu_0$ to the normal to the layer. The source axis is regarded to be in the plane passing through the $x$ axis and the normal to the layer, and is directed to the point $x = 0$ in the $z = z_0$ plane, and intersects the $x$ axis in the $z = 0$ plane at the point having the coordinate $x = -z_0\sqrt{1-\mu_0^2}/\mu_0$ (Fig. 6.12a). Then, in the diffusional approximation and as in (6.1.9), for the medium OTF, we obtain

$$S(\vec{v}, \tau_0, \mu_0) = Q\left[\left(\frac{3}{m\mu_0 - \alpha} + x_1\right)(\mathrm{e}^{-\tau_0/\mu_0} - \mathrm{e}^{-m\tau_0 + \mathrm{i}\xi}) + \left(\frac{3}{m\mu_0 + \alpha} - x_1\right)(\mathrm{e}^{-(m+1/\mu_0)\tau_0} - \mathrm{e}^{-\mathrm{i}\xi})\mathrm{e}^{-m(\tau_0+4q)}\right] + \mu_0\left(1 - \frac{\Lambda q x_1}{4}\right)\mathrm{e}^{-\tau_0/\mu_0}, \tag{6.3.1}$$

where

$$Q = \frac{\Lambda\mu_0}{8m}\,\frac{1 + 2mq - (1 - 2mq)\mathrm{e}^{-4mq}}{1 - \mathrm{e}^{-2m(\tau_0+4q)}}, \tag{6.3.2}$$

$$\xi = v_x z_0\sqrt{1-\mu_0^2}/\mu_0, \qquad m = \sqrt{\gamma^2 + v^2/\varepsilon^2}, \tag{6.3.3}$$

$$\alpha = 1 + \mathrm{i}(v_x/\varepsilon)\sqrt{1-\mu_0^2}. \tag{6.3.4}$$

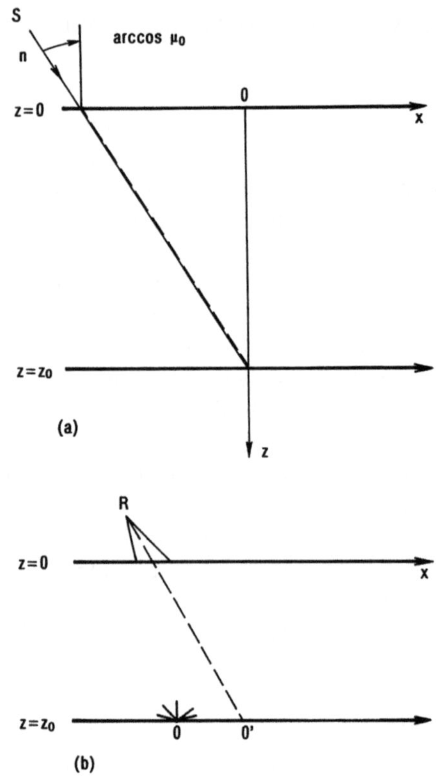


**Fig. 6.12.** (a) Oblique beam incidence on a scattering layer and (b) image shift along an oblique path of observation through a scattering layer: $S$, directional source; $R$, receiver; $O$, diffuse point source; $O'$, apparent source location

The function $S(\vec{v}, \tau_0, \mu_0)$ is complex and may be written as

$$S(\vec{v}, \tau_0, \mu_0) = |S(\vec{v}, \tau_0, \mu_0)| e^{-i\varphi(\vec{v}, \tau_0, \mu_0)} , \tag{6.3.5}$$

where

$$|S(\vec{v}, \tau_0, \mu_0)| = \sqrt{S_c^2(\vec{v}, \tau_0, \mu_0) + S_s^2(\vec{v}, \tau_0, \mu_0)} ,$$

$$\varphi(\vec{v}, \tau_0, \mu_0) = \arctan \frac{S_c(\vec{v}, \tau_0, \mu_0)}{S_s(\vec{v}, \tau_0, \mu_0)} \tag{6.3.6}$$

are the modulus and the phase of the OTF; $S_c(\vec{v}, \tau_0, \mu_0)$ and $S_s(\vec{v}, \tau_0, \mu_0)$ are the cosine and sine in the Fourier transform of the line spread function. The complex nature of the OTF for viewing along the oblique path results from the physically obvious azimuthal asymmetry of the point spread function. As before, the normalized OTF modulus $T(\vec{v})$ defines contrast changes in viewing a harmonic test object.

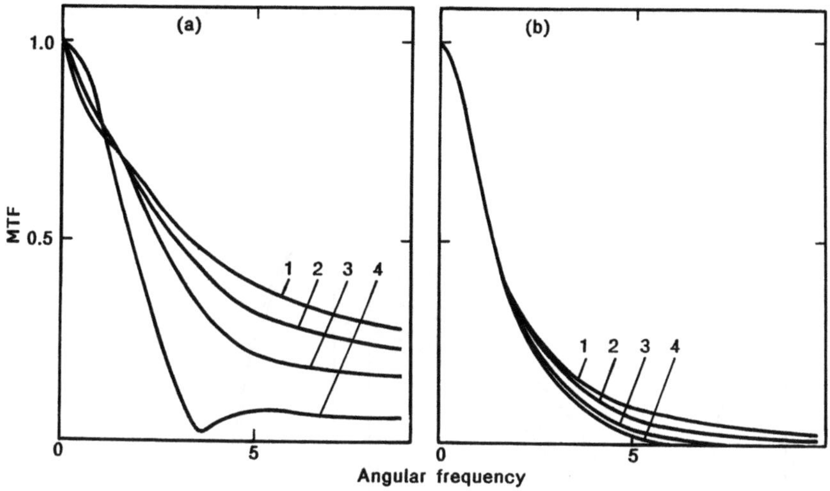


**Fig. 6.13.** MTF for oblique observation at $g = 0.863$, $\Lambda = 1$, (a) $\tau_0 = 2$, (b) $\tau_0 = 5$ and $\mu_0 = 1$ (curves 1), 0.9 (curves 2), 0.77 (curves 3), 0.5 (curves 4)

The difference of the phase $\varphi$ from zero and its dependence on the spatial frequency $\vec{v}$ has some consequences. For example, in viewing a harmonic test object, the strips are seen to shift along the $x$ axis, the magnitude of this shift being a function of the test object frequency. In the case of a more complex object, its image undergoes distortions. Therefore, for example, the image of a symmetric object appears to be asymmetric. As appears from the optical reciprocity theorem, the spread function asymmetry and the displacement of its maximum result, in particular, in the following. Let a point diffusely irradiating source be viewed through a scattering layer along an oblique path. Then, the maximum radiance of its image is shifted about the true source position (Fig. 6.12b).

We shall analyse the OTF dependence on the optical medium characteristics and the observation angle $\arccos \mu_0$. Set $v_y = 0$. As seen from Fig. 6.13, if the MTF $T(v^*)$ is presented as calculated by (6.3.1–5), the system MTF deteriorates with an increase in the observation angle and optical thickness $\tau_0$ of the layer. The higher $\tau_0$, i.e., the less significant the unscattered light effect, the weaker the MTF dependence on the scattering angle. At small $\tau_0$ and large enough angles $\arccos \mu_0$, the MTF minimum is attractive (Fig. 6.13a). Such a minimum implies that some spatial frequencies almost fail to be transmitted through the scattering layer. The explanation is simple. The spread function can be regarded as a superposition of the transmitted and scattered radiation. At large enough $\arccos \mu_0$, their maxima are markedly shifted about each other. For some spatial frequencies, their Fourier components prove to be close in the modulus and opposite in the phase. The total Fourier component is, therefore, about zero. This means that, in viewing a harmonic test object of a certain frequency, the

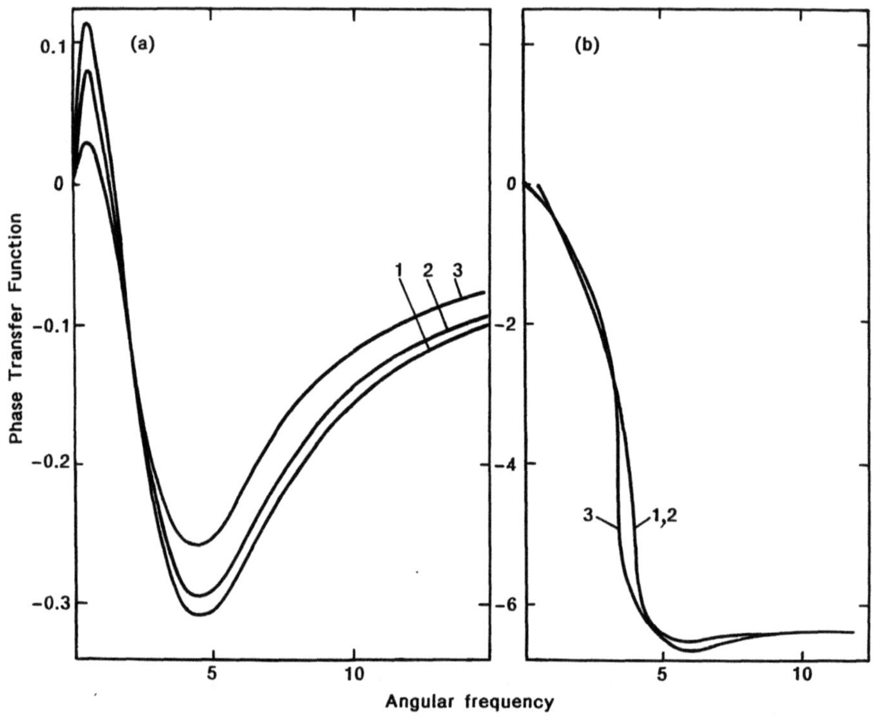


**Fig. 6.14.** Scattering layer PTF for oblique observation at $\tau_0 = 2$, $g = 0.863$ for (a) $\mu_0 = 0.9$, (b) $\mu_0 = 0.5$ and $\Lambda = 1$ (curve 1), 0.9 (curve 2), 0.7 (curve 3)

maximum scattered light radiance is equal to the minimum radiance of the unscattered component, and vice versa. As a result, the contrast of such a harmonic test object image turns out to be about zero.

Let us now analyse the phase transfer function. Figure 6.14 shows that the larger the observation angle, i.e., the larger the spread function shift about the point of the unscattered light reaching the $z = z_0$ plane, the higher $\varphi$. The phase shift is also seen to increase with $\tau_0$ and $\Lambda$, i.e., as the contribution of multiple scattering grows.

Actually, for all frequencies, the phase shift $\varphi < 0$ and only at very small $\nu^*$, $\varphi > 0$. We shall show and try to explain that, at small $\tau_0$, such a relationship really holds.

The Fourier transform properties imply that the quantity $x_c = \lim_{\nu \to 0} d\varphi/d\nu$ determines the center of gravity coordinate of the spread function. At small $\tau_0$ and given an oblique incident beam, the spread function of the medium with an anisotropic scattering indicatrix is drastically asymmetric. It rapidly drops when $x < 0$ and has a long tail which falls off slowly at $x > 0$. Therefore, the coordinate of its center of gravity tends to lie within $x > 0$ and $\lim_{\nu \to 0} d\varphi/d\nu > 0$. For large $\tau_0$, the spread function is less asymmetric and is shifted to $x < 0$. Then,

$\lim_{\nu \to 0} d\varphi/d\nu < 0$. This explanation is also supported by the fact that, for weakly anisotropic indicatrices (small $x_1$), $\varphi < 0$ at all frequencies.

At high $\tau_0$ and small $\nu^*$, where the Fourier component is dependent on the scattered light, the phase shift $\varphi$ is a linear function of the frequency $\nu^*$. This also indicates a weak spread function asymmetry. Similar PTF and MTF behavior along an oblique path can be observed in numerical calculations [6.25, 29] and experiments [6.24].

## 6.4 Nonlinear Distortions in Thick Scattering Layers

We have so far studied the OTF and MTF of a scattering medium, assuming that in the observation system, the principal conditions of the applicability of the apparatus of transfer functions such as system isoplanarity and the linear dependence of its response on the object albedo are obviously valid. These conditions may, however, be violated in observations through a strongly scattering layer [6.10].

To begin with, we recall that the system is known to be isoplanatic if it obeys the aspect invariance requirements (Sects. 5.2.2–4). Negligible backscattering is among them. Since the diffuse reflection coefficient of a thick scattering layer is high and may be of unity at $\Lambda = 1$ and $\tau_0 \to \infty$, in such layers the aspect invariance is, naturally, violated. In this connection, the use of the apparatus of transfer functions to describe the viewing through such media is reasonable from a practical point of view only in observation system 2 (Sect. 5.2.3). It should be noted that an image is formed here by recording the radiance in one and the same direction at different points of the horizontal $\vec{r}$ plane. This observation scheme applies to different situations.

Backscattering is responsible for both the violation of the aspect invariance and for the mutual reradiation, i.e., the interaction of the subsystems which compose an unified observation system. The radiation diffusely reflected by an object or emitted by a self-luminous object is backscattered by a medium layer, to be reflected from the object, and so on. Multiple reflections between the object and scattering medium violate the system linearity about the input signal spectrum $A(\vec{\nu})$. We consider each individual reflection from the medium or object as an impact of the relevant system element. Then (5.2.3) can be used at each stage. For the spectrum of the image transferred through the entire system, we have

$$E(\vec{\nu}) = E_1(\vec{\nu}) + E_2(\vec{\nu}) + E_3(\vec{\nu}) + \cdots, \qquad (6.4.1)$$

where

$$E_1(\vec{\nu}) = A(\vec{\nu})S(\vec{\nu}),$$

$$E_2(\vec{\nu}) = S(\vec{\nu}) \iint_{-\infty}^{\infty} A(\vec{\nu}')S_{\text{ref}}(\vec{\nu}')A(\vec{\nu} - \vec{\nu}')\,d\vec{\nu}',$$

$$E_3(\vec{v}) = S(\vec{v}) \iint_{-\infty}^{\infty} S_{ref}(\vec{v}')A(\vec{v} - \vec{v}')\,d\vec{v}'$$

$$\times \iint_{-\infty}^{\infty} A(\vec{v}'')S_{ref}(\vec{v}'')A(\vec{v}' - \vec{v}'')\,d\vec{v}'' ,$$

$$\ldots \tag{6.4.2}$$

Here, $S_{ref}(\vec{v})$ is the Fourier transform of the spatial irradiance distribution in the $z = z_0$ object plane. This irradiance is induced by the light reflected from the scattering medium illuminated with a diffuse unit power source at the same $z = z_0$ plane. As seen from (6.4.1, 2), nonlinear distortions appear in image transfer. Thus, the image spectrum $E(\vec{v})$ may have $\vec{v}$ frequencies that are absent in the $A(\vec{v})$ object spectrum. The possibility of such distortions appearing was mentioned in [6.30–32]. However, they may be manifested when a number of conditions are obeyed. Specifically, the layer reflection coefficient $r = S_{ref}(\vec{v} = 0)$, mean albedo $\langle A \rangle$ must be high. In practice, therefore, nonlinear distortions may frequently be neglected. As an example, we consider two specific cases.

1. The object of observation has a low contrast. It should be noted that a number of natural objects are characterized by low contrast, their mean albedo being sufficiently high. At this point, in calculating the integrals, we may approximately set $A(\vec{v}) = \langle A \rangle \delta(\vec{v})$. Hence,

$$E_2(\vec{v}) = A(\vec{v})S(\vec{v})S_{ref}(\vec{v})\langle A \rangle ,$$

$$E_3(\vec{v}) = A(\vec{v})S(\vec{v})S_{ref}^2(\vec{v})\langle A \rangle^2 ,$$

$$\ldots$$

$$E_{n+1}(\vec{v}) = A(\vec{v})S(\vec{v})S_{ref}^n(\vec{v})\langle A \rangle^n .$$

In this approximation, instead of (6.4.1), we obtain

$$E(\vec{v}) = \frac{A(\vec{v})S(\vec{v})}{1 - \langle A \rangle S_{ref}(\vec{v})} . \tag{6.4.3}$$

No nonlinear distortions are seen here, whereas the resultant OTF

$$S_{res}(\vec{v}) = \frac{E(\vec{v})}{A(\vec{v})} = \frac{S(\vec{v})}{1 - \langle A \rangle S_{ref}(\vec{v})} \tag{6.4.4}$$

depends on the layer characteristics $S(\vec{v})$ and $S_{ref}(\vec{v})$, and on the average albedo of the object. It is expressed in terms of $S(\vec{v})$, in the same way as the transmission coefficient of the layer with a reflecting lower boundary is expressed by the transmission and reflection of the layer with nonreflecting boundaries (Sect. 3.5.2). It is most simple to find $S_{res}(\vec{v})$ in the diffusion approximation. As follows from (4.3.14) and data presented in Sect. 3.5.2, it suffices to substitute $q(1 + \langle A \rangle)/(1 - \langle A \rangle)$ for the parameter $q$ in the $S(\vec{v})$ expression in order that $S_{res}(\vec{v})$ be found within the diffusion approximation.

2. The Fourier spectrum $S_{ref}(\vec{v})$ is much narrower than the object spectrum $A(\vec{v})$. It is encountered, for example, when an observation object is remote from a scattering layer (viewing the Earth's surface through thin clouds). Then, radiation from an isotropic point source reflected by the layer is spread over a large surface, and the spectrum $S_{ref}(\vec{v})$ can be regarded as concentrated in a small vicinity $\vec{v} \approx 0$. In this approximation, we have

$$E_{n+1}(\vec{v}) \approx S(\vec{v})A(\vec{v})\mathscr{L}^n , \tag{6.4.5}$$

where the quantity $\mathscr{L} = \iint_{-\infty}^{\infty} A(\vec{v}')S_{ref}(\vec{v}')\,d\vec{v}'$ is independent of the frequency $\vec{v}$.

Substituting (6.4.5) into (6.4.1) gives

$$E(\vec{v}) = \frac{A(\vec{v})S(\vec{v})}{1-\mathscr{L}} . \tag{6.4.6}$$

In this case, nonlinear distortions are not observed either, and the resultant OTF is

$$S_{res}(\vec{v}) = \frac{S(\vec{v})}{1-\mathscr{L}} . \tag{6.4.7}$$

The quantity $\mathscr{L}$ is easily found from the Parseval equality [6.33]

$$\mathscr{L} = \iint_{-\infty}^{\infty} A(\vec{v})S_{ref}(\vec{v})\,d\vec{v} = \iint_{-\infty}^{\infty} A(\vec{r})S_{ref}(\vec{r})\,d\vec{r}$$

$$\approx \langle A \rangle \iint_{-\infty}^{\infty} S_{ref}(\vec{r})\,d\vec{r} .$$

## 6.5 Object Image Contrast

We have so far discussed medium transfer characteristics such as the OTF and MTF which specify the contrast of individual elements in the object image. However, the connection between these characteristics and the contrast in the real object image is often far from trivial. For example, the decrease in the OTF at all spatial frequencies does not indicate that the real object contrast necessarily deteriorates. We shall examine the images of simple objects to ascertain this.

### 6.5.1 Small Object Contrast

Let us represent the medium PSF $S(\vec{r}, L)$ as a sum of the scattered and unscattered radiation components, i.e.,

$$S(\vec{r}, L) = e^{-\tau_0}\delta(\vec{r}) + S^s(\vec{r}, L) . \tag{6.5.1}$$

Here, $L$ is the distance between the object and the plane of observation. In conformity with (5.2.2) and (6.5.1), the irradiance distribution $E(\vec{r}', L)$ in the image of a diffuse self-luminous object with a radiant emittance $E_0(\vec{r})$ is

$$E(\vec{r}', L) = E_0(\vec{r}')\mathrm{e}^{-\tau_0} + E^s(\vec{r}', L)\,, \tag{6.5.2}$$

where

$$E^s(\vec{r}', L) = \iint_{-\infty}^{\infty} E_0(\vec{r}) S^s(\vec{r}' - \vec{r}, L)\, d\vec{r}\,. \tag{6.5.3}$$

If the object size is small compared with the $S^s(\vec{r}, L)$ half-width, then, from (6.5.3), we have

$$E^s(\vec{r}', L) = S^s(\vec{r}', L)\langle E_0\rangle \Sigma_{ob}\,, \tag{6.5.4}$$

where $\Sigma_{ob}$ is the object surface area, and $\langle E_0\rangle$ the average radiant emittance of the object. As seen from (6.5.4), in this case the scattered component $E^s(\vec{r}', L)$ in the distribution (6.5.2) does not include information about the radiant emittance distribution $E_0(\vec{r})$ of the object. Rather, it characterizes the structure of the inhomogeneous background against which the information signal $E_0(\vec{r}')\exp(-\tau_0)$ is observed. We take it, for the sake of certainty, that $E_0(\vec{r}) =$ const at $\vec{r} \in \Sigma_{ob}$, or the small object surface has uniform emittance, and determine the observed contrast as the ratio of the valid unscattered signal component to the background noise around the object. We shall also use this definition in Sect. 8.2 to evaluate the limiting detection range of self-luminous objects. Now,

$$k = \frac{\mathrm{e}^{-\tau_0}}{S^s(\vec{r}' = 0, L)\Sigma_{ob}}\,. \tag{6.5.5}$$

Thus, the contrast of the small object image degenerates with increasing object area and optical distance [6.34, 35].

### 6.5.2 Contrast in the Johnson Striped Test Object

As far as the range of object vision is concerned, we have used the Johnson criteria (Sect. 5.4.1). The reader is reminded that according to these criteria, the observer can discriminate a real object if 8 alternating light and dark strips (half-periods) of a striped test object are resolved per critical size (Fig. 5.7). The valid signal energy and the signal-to-noise ratio in its image depend on the contrast $k_{st}$ between the adjacent light and dark strips in the image center (Sect. 5.4.3). It is not difficult to find an expression for $k_{st}$. One can see that for the striped test object consisting of $m$ half-periods, the observed contrast

$$\begin{aligned} k_{st} &= \frac{E_1 - E_2}{E_1 + E_2} \\ &= \frac{4(A_1 - A_2)\sum_{i=0}^{m/2-1}(-1)^i \Psi_{2i+1} + 2A_1(\Psi_{m+1} - \Psi_{m-1})}{(A_1 + A_2)(\Psi_{m+1} + \Psi_{m-1})} - (-1)^{m/2} \\ &\quad \text{at even } m\,, \end{aligned} \tag{6.5.6}$$

while

$$k_{st} = 4\frac{(A_1 - A_2)\sum_{i=0}^{(m-1)/2}(-1)^i \Psi_{2i+1} - A_2\Psi_m}{A_1(\Psi_{m+2} + \Psi_m) + A_2(\Psi_m + \Psi_{m-2})}, \text{ at odd } m\,. \tag{6.5.7}$$

Here, $A_1$ and $A_2$ are the albedos of light and dark test object strips, respectively; $E_1$ and $E_2$ are the irradiances in the center of the light or dark strip image; the function $\Psi_i = \Psi(ix_{ob}/m, y_{ob}; x_{el}, y_{el})$ is determined by the relation (6.6.2); $x_{ob}$, $y_{ob}$ and $x_{el}$, $y_{el}$ are, respectively, the linear sizes of the object and the element in the space integration over the $x$, $y$ axes. It is assumed in (6.5.7) that the marginal strips have the albedo $A_1$. Formulas (6.5.6, 7) are exact. It should be noted that the exact expression (6.6.2), rather than an approximate relation of the form of (6.6.3), is required to calculate $k_{st}$. Otherwise, the calculation of the small difference of in (6.5.6, 7) may lead to a large error.

An approximate $k_{st}$ may be found in another fashion. Replacing a striped test object with a sinusoidal one having a spatial frequency $v = m\pi/x_{ob}$ and taking $x_{el} \to 0$, $y_{el} \to 0$ gives for $k_{st}$

$$k_{st} = \frac{k_0 T(v)}{\Psi(x_{ob}, y_{ob}; 0, 0)}, \tag{6.5.8}$$

which describes the contrast in the image of a sinusoidal finite object [6.36]. Here, $k_0 = (A_1 - A_2)/(A_1 + A_2)$. The accuracy of (6.5.8) is illustrated by the data in Table 6.3, giving $k_{st}$ at $m = 8$ and $k_0 = 1$ $(A_2 = 0)$ for observations in sea water, whose optical model is presented in Sect. 9.1.2. An exact $k_{st}$ has been calculated by (6.5.6, 7; 6.6.2), its approximate values by (6.5.8). As seen from the table, the accuracy of (6.5.8) is satisfactory.

### 6.5.3 Finite Object Contrast as a Function of the Scattering Layer Position along the Observation Path

We have shown in Sect. 6.2 that, as the distance between the scattering layer and the object increases, the medium MTF, i.e. the contrast in the image of an infinite

**Table 6.3.** Contrast of a self-luminous stripped test object in sea water.

| | $\sigma x_{ob} = \sigma y_{ob}$ | | | | | |
|---|---|---|---|---|---|---|
| | 0.1 | | 0.2 | | 0.4 | |
| $\sigma z_0$ | Exact | Approx. | Exact | Approx. | Exact | Approx. |
| 1 | 0.595 | 0.587 | 0.591 | 0.574 | 0.641 | 0.611 |
| 3 | 0.354 | 0.352 | 0.241 | 0.238 | 0.196 | 0.190 |
| 5 | 0.258 | 0.255 | 0.132 | 0.133 | 0.074 | 0.076 |
| 7 | 0.176 | 0.177 | 0.072 | 0.075 | 0.029 | 0.037 |

sinusoidal test object, decreases monotonically at all spatial frequencies. The question now is to know how the contrast in the image of real objects changes. The observed contrast of a small object is described by the relation (6.5.5). As appears from (5.2.6),

$$S^s(\vec{r}\,' = 0, L) = \frac{1}{4\pi^2} \iint_{-\infty}^{\infty} S^s(\vec{v}, L)\, d\vec{v}\,, \tag{6.5.9}$$

where $S^s(\vec{v}, L)$ is the scattered component of the medium OTF. Since for all $\vec{v}$, $S^s(\vec{v}, L)$ decreases with increasing distance between the layer and the object (Sect. 6.2), the image contrast enhances monotonically, as seen from (6.5.5, 9).

At first glance, the increase in the small object image contrast might seem to contradict the MTF decrease, i.e. the degeneration of the contrast of a sinusoidal infinite test object at any frequency. In fact, the above inferences show that the results are not contradictory at all. On the contrary, one of them may follow from the other.

We now consider a somewhat more intricate case. Let a striped test object consisting of two light and one dark strips of width $b$ be viewed in the observation scheme of Fig. 6.9. Its image contrast $k_{st}$ is described by (6.5.7) at $m = 3$. In Fig. 6.15, $k_{st}$ is plotted as a function of $H/z_0'$ for different $b$. This dependence is different for different $b$ and is attributed to the broadening of the spread function with $H/z_0'$ (Sect. 6.2.3).

If $b$ is less than the PSF half-width, even at $H/z_0' = 0$, then the object may be regarded as small, the contrast enhancing monotonically with $H$. When $b$ is much greater than the PSF half-width within the entire $H/z_0'$ range, the object may be

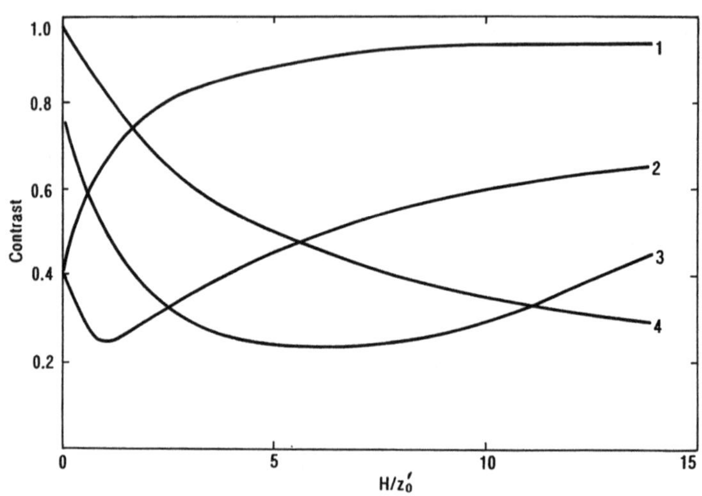


**Fig. 6.15.** Test object image contrast $\kappa_{st}$ as a function of scattering layer placement on the object-observer path at $\sigma z_0 = 2$ and $\sigma b = 0.01$ (curve 1), 0.1 (curve 2), 0.5 (curve 3), 2 (curve 4)

considered infinite, the monotonic decrease in its contrast being an immediate result of the monotonic MTF reduction with growing $H/z_0'$.

For intermediate $b$ values, $k_{st}$ is a nonmonotonic function of $H/z_0'$, or of the parameter $t = H/L$ ($L$ being the fixed distance between the object and the plane of observation). At some distance between the scattering layer and the object plane, the contrast $k_{st}$ is minimum. In [6.18–21], such a contrast dependence on the parameter $t$ is referred to as the $t$-effect. Its physical explanation is simple. At small $H/z_0'$, the strip width $b$ is large compared with the PSF half-width, and the contrast degenerates with increasing $H/z_0'$. At large $H/z_0'$, the PSF half-width exceeds the object size, and the observed contrast enhances with $H/z_0'$ in the same way as the small object contrast. A minimum $k_{st}$ occurs when $H/z_0'$ is such that $b$ is commensurate with the PSF half-width. If the medium PSF is distinctly two-scale, for example in cloud with the indicatrix C.1, and can be described by the modified SADA relation (4.5.7), then the $H/z_0'$ dependence of $k_{st}$ may have two minima.

Finally, the contrast $k_{st}$ as a nonmonotonic function of the distance to the scattering layer may occur only when observing light strips against a dark background. On the contrary, when dark strips are viewed against an infinite light background, the image contrast degenerates monotonically with increasing $H/z_0'$, which is clear from the physical explanation of the phenomenon.

## 6.6 The Function $\Psi$ in Object Detection and Discrimination

In Sect. 5.4, we dealt with the function $\Psi(\Sigma_{ob}, \Sigma_{el})$ determined by the relation (5.4.5) and its specific case $\Psi(x_{ob}, y_{ob}; x_{el}, y_{el})$, which are required to estimate the ranges of object detection and discrimination in a scattering medium. As seen in Sect. 6.5.2, the function $\Psi$ can describe simply the contrast of a striped test object. However, analytical expressions for $\Psi$ are desirable in these problems. Therefore, we shall find its approximate formulas.

It follows from (5.4.5) that

$$\Psi(\Sigma_{ob}, \Sigma_{el}) = \frac{1}{4\pi^2 \Sigma_{el}} \iint_{-\infty}^{\infty} U(\vec{v}, \Sigma_{ob}) U(\vec{v}, \Sigma_{el}) T(\vec{v})\, d\vec{v}\,, \tag{6.6.1}$$

where $U(\vec{v}, \Sigma)$ is the Fourier transform of the function $U(\vec{r}, \Sigma)$ determined by the relation (5.1.30). For a homogeneous rectangular object,

$$\begin{aligned}\Psi(\Sigma_{ob}, \Sigma_{el}) &\equiv \Psi(x_{ob}, y_{ob}; x_{el}, y_{el}) \\ &= \frac{1}{4\pi^2} \iint_{-\infty}^{\infty} \frac{\sin(v_x x_{ob}/2)}{v_x/2} \frac{\sin(v_y y_{ob}/2)}{v_y/2} T(v_x, v_y) \\ &\quad \times \frac{\sin(v_x x_{el}/2)}{v_x x_{el}/2} \frac{\sin(v_y y_{el}/2)}{v_y y_{el}/2} dv_x\, dv_y\,.\end{aligned} \tag{6.6.2}$$

We take it that the medium PSF and $U(\vec{r}, \Sigma)$ have a symmetry center and approximate the integrand in (6.6.1) by a normal distribution. Then, in the modified SADA where the medium MTF $T(\vec{v})$ is described by (6.1.17, 18), we have, from (6.6.1),

$$\Psi(\Sigma_{\rm ob}, \Sigma_{\rm el}) = \frac{1}{S(\tau_0)} \left\{ e^{-\tau_0} \left[ \frac{V_{\rm ob}^x V_{\rm ob}^y}{(V_{\rm ob}^x + V_{\rm el}^x)(V_{\rm ob}^y + V_{\rm el}^y)} \right]^{1/2} + \sum_{i=1}^{2} S_i(\tau_0) \left[ \frac{V_{\rm ob}^x V_{\rm ob}^y}{(V_{r,i} + V_{\rm ob}^x + V_{\rm el}^x)(V_{r,i} + V_{\rm ob}^y + V_{\rm el}^y)} \right]^{1/2} \right\}, \quad (6.6.3)$$

where $S(\tau_0) = S(v = 0, \tau_0)$ is the total flux through the plane $\tau = \tau_0$; $V_{\rm ob}^x$, $V_{\rm ob}^y$ and $V_{\rm el}^x$, $V_{\rm el}^y$ are the variances of $U(\vec{r}, \Sigma_{\rm ob})$ and $U(\vec{r}, \Sigma_{\rm el})$ along the $x$ and $y$ axes. We may consider that

$$(V_{\rm ob}^x V_{\rm ob}^y)^{1/2} = \Sigma_{\rm ob}/2\pi\,, \qquad (V_{\rm el}^x V_{\rm el}^y)^{1/2} = \Sigma_{\rm el}/2\pi\,. \quad (6.6.4)$$

For a rectangular object,

$$V_{\rm ob}^x = x_{\rm ob}^2/2\pi\,, \qquad V_{\rm ob}^y = y_{\rm ob}^2/2\pi\,,$$
$$V_{\rm el}^x = x_{\rm el}^2/2\pi\,, \qquad V_{\rm el}^y = y_{\rm el}^2/2\pi\,. \quad (6.6.5)$$

Let us note the physical implication of the function $\Psi$. First, for simplicity, we shall set $\Sigma_{\rm el} = 0$. In this case, as seen from (5.4.5), the function $\Psi$ describes the ratio of the irradiance in the center of a homogeneous object image (neglecting the noise component) to the irradiance of an infinite object. Therefore, at $\Sigma_{\rm ob} \to \infty$, $\Psi \to 1$.

At $\Sigma_{\rm el} = 0$, the function $\Psi$ has one more instructive physical interpretation. It is the ratio of the flux through the area $\Sigma_{\rm ob}$ to the total flux in the $\tau = \tau_0$ plane when the medium is illuminated by a monodirectional point source. If $\Sigma_{\rm el} \neq 0$, then the first interpretation of the function $\Psi$ must take into account the $\Sigma_{\rm el}$-averaged quantities, while in the second one, the source with the directional diagram $U(\vec{r}, \Sigma_{\rm el})/z_0^2$ changes for the monodirectional one.

# 7 Image Transfer in Coherent Light

The last twenty years are notable for the rapid development of coherent optics and, specifically, for the great strides made in holography. This might have been discovered as early as the last century since the theoretical principles of optical holography are based on classical physics. Nevertheless, the numerous applications of holography would have been impossible without lasers, which ensure high temporal and spatial coherence. At the present time, the methods of coherent-holography are widely used not only in optics and related fields, but also in computer art, biology etc. However, they find their most natural application in problems related to information transfer and processing, and to image transfer, in particular. In this chapter, we shall touch upon some aspects of the problem concerned with the transfer of the image of remote diffusely reflecting (rough) objects through a scattering medium, relating it to the problems of imaging in incoherent systems. New problems will be considered in the light of the principles of vision theory presented in Chap. 5.

## 7.1 Coherent-Holography Imaging Through a Scattering Medium

The impressive advances achieved thanks to Leith and Upatnieks [7.1–3], as well as to Denisyuk [7.4], seemed to promise that holography, which provides a complete recording of information, would also stimulate progress in the problem of vision in scattering media by improving the image quality and, essentially, increasing the observation range. However, it soon became clear that on account of the very completeness of the information recorded, holography, as it is today, fails to solve these problems. In fact, a perfect image reconstructed by a hologram provides a complete illusion of a real object. At the same time the hologram, being a substitute for an observation object, also substitutes for a scatterer. Therefore, the object details in a normal hologram-reconstructed image cannot be resolved any better than by direct viewing.

Special coherent-holography approaches have been successful in improving image quality, among them: time-average holography [7.5–11], limited temporal coherence (LTC) [7.12–14] and limited spatial coherence [7.15] combined with reference-free image plane holography (RFIPH) [7.16–18]. They may be used to enhance the visibility of remote objects and do not require such unrealistic conditions as, for instance, the presence of a reference point source near an object, as required in the method of equal object and reference wave distortions [7.19]. The main idea of these methods is exposure time averaging of the wavefront. By virtue of the scatterer motion (in time-average holography), or owing to the different times-of-arrival of the reference and object waves at the hologram (in the LTC technique), the radiation scattered by the medium proves to be incoherent with respect to the reference wave, does not contribute to the interference pattern and is cut off at the reconstruction stage. This essentially increases the contrast of the reconstructed image.

We shall consider these techniques in more detail.

### 7.1.1 Time-Averaged Holography

The basic concept behind this technique is described in [7.5, 6]. Its typical setup is shown in Fig. 7.1a. Some of the radiation from a laser S is allowed to pass

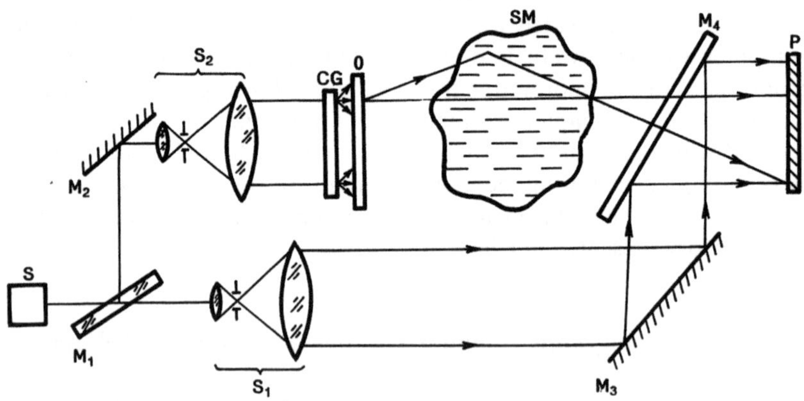


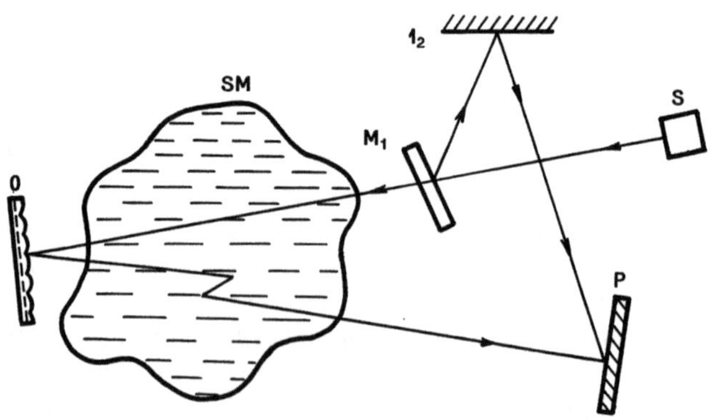


**Fig. 7.1.** (a) Time-averaged holography and (b) limited time coherence technique

through a beam splitter and travels to a reference beam which forms the system $S_1$. The other portion arrives at an object O via the object beam forming system $S_2$. Upon passing through the object and scattering medium SM, the radiation arrives at a photographic plate P where a hologram is recorded. In this scheme, the object is a transparency, its diffusivity being simulated by superposing clouded glass CG. Such an experimental geometry, which may be referred to as the scheme of information transfer through a scattering medium, has provided the majority of experimental results. We can easily imagine how this technique is realized in the active vision system variant where a source and a photodetector are set adjacent on one side of a diffusely reflecting object.

As already noted, this method implies that the radiation scattered by a time-variable medium proves to be incoherent with respect to a reference beam, and is recorded on a hologram as uniform background. A stable interference pattern, which forms an image at the stage of reconstruction, is mainly due to the radiation arriving directly from an immovable object without being scattered.

The available results of the theoretical analysis of the method apply to two somewhat different modifications. It is assumed in one case [7.7] that scattering particles move in a certain direction, or at random. The frequency of the radiation scattered over moving particles is Doppler-frequency shifted with respect to the reference radiation. In such a situation, the velocity of the scattering particles and the exposure time of the hologram, at which the scattered radiation practically, incoherent with respect to the reference beam, have been related in a single scattering approximation. In the other case [7.8–10], it is assumed that a large number of holograms of one and the same object are recorded on a photographic plate for the recording time at different (uncorrelated) realizations of the scattered field.

For the medium with moving scatterers, we can roughly assume that $n \approx t_e/\Delta t_c$, where $t_e$ is the exposure time and $\Delta t_c$ the time correlation radius of the scattered field. The probability theory may be used to relate some characteristics of the reconstructed image quality to the number of superimposed holograms. For example, as shown in [7.9], the contrast of fine image details transferred by the unscattered radiation is $k = k_0/(1 + 1/nm)$, where $k_0$ is the initial object contrast and $m$ the ratio of the unscattered-to-scattered light-induced irradiances on a hologram. We note, by way of comparison, that in ordinary photography (radiance time-averaging), the contrast is $k = k_0/(1 + 1/m)$ [7.9].

Thus, time-averaging or the superposition of a large number of holograms essentially enhances the image contrast due to the elimination of scattered radiation. Experiments [7.10, 11] have proved the efficiency of this technique and the reliability of the theoretical inferences. It is possible, in the second case, to fall outside the frame of the single-scattering approximation. However, it is also assumed that useful information about the object is transferred only by unscattered radiation and that all of the scattered light ruins the image. It is simply regarded as a factor which decreases the contrast and produces noise in the image. As noted in Sect. 5.3.4, this is true only when fine object details are observed. Moreover, the approximation of the complete field correlation within the time correlation radius $\Delta t_c$ and its complete incorrelation beyond this range for media with moving scatterers may turn out to be rather crude. The question of choosing $\Delta t_c$ for real conditions also remains unresolved.

### 7.1.2 Limited Time Coherence (LTC) Method

This method was first suggested in [7.12]. The schematic diagram (Fig. 7.1b) differs from normal holographic recording only in that it uses a source of limited coherence length $\Delta l_{coh}$ and in the possibility it offers for changing the delay time in the reference beam. In this case, only that portion of the object beam whose path length in the medium is less than $\Delta l_{coh}$ shorter as against the reference arm length is recorded on a hologram. By varying the delay time of the reference beam with respect to the object beam, we may select the scattered radiation which arrives from various depths. Thus, the LTC method as well as the laser sounding technique can be used to explore the medium structure [7.13, 14, 20]. In such a method, the depth resolution is specified by the coherence length $\Delta l_{coh} \sim \lambda^2/\Delta\lambda$, where $\lambda$ is the radiation wavelength and $\Delta\lambda$ the spectral range of the source radiation. For example, at $\lambda = 500$ nm and $\Delta\lambda = 2.5$ nm, $\Delta l_{coh} \sim 100$ $\mu$m. This is the resolution achieved in the interference experiment [7.20]. We should like to mention that such a high spatial resolution in a pulsed laser sounding system can be attained if the source pulse time and the receiver time resolution range from $10^{-12}$ s to $10^{-13}$ s [7.21, 22].

By setting the delay time in the reference beam a double distance from the observation object, we may eliminate the radiation scattered by the medium, backscattering noise, for example, thus enhancing the image contrast. Some of the potentials of LTC are discussed in [7.12, 14]. The BSN suppression efficiency is shown to be rather high. Thus, the LTC technique provides for satisfactory imaging through a scattering medium under conditions where a photographic picture would be blurred. Convincing illustrations are given in [7.14].

### 7.1.3 Reference-Free Image Plane Holography (RFIPH)

The techniques mentioned above require the use of a reference wave. This leads to considerable technological difficulties when viewing remote objects. However, the situation is different in the case of reference-free holography [7.15, 16–18, 23, 24] which makes use of a speckled pattern which appears in recording the waves reflected from a rough object surface illuminated by spatially coherent laser radiation. One of the earliest techniques of this method [7.23], known as the intensity hologram method has been used to view through a turbulent medium [7.24]. In this case, the image autocorrelation, rather than the image itself is reconstructed and, therefore, only inversion symmetry objects can be viewed.

The RFIPH does not suffer from this shortcoming. Such a hologram recording (Fig. 7.2a) does not, in fact, differ from ordinary photography. However, the

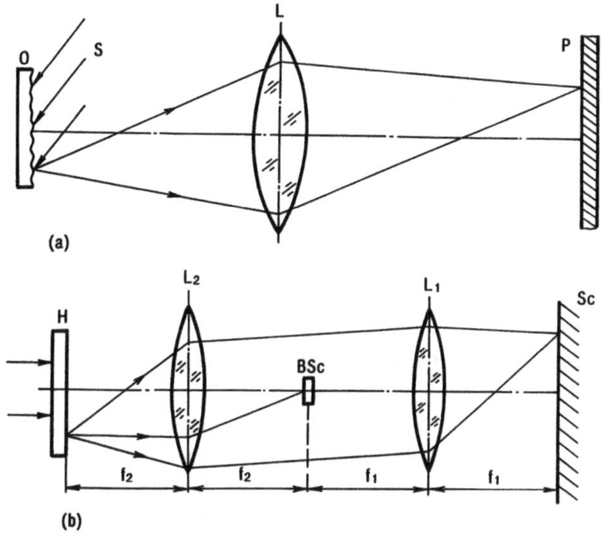


**Fig. 7.2.** (a) Recording and (b) reconstruction of an image using reference-free image plane holography: O, object; S, radiation source (laser); L, $L_1$, $L_2$, lenses; P, photographic plate; H, hologram; BSc, blocking screen; Sc, image screen; $f_1$ and $f_2$, focal distances of lenses

object of observation is illuminated, by spatially coherent laser radiation. The photographic image appears to be coated with speckles. Such a speckled pattern occurs due to the interference of waves with a random phase distribution which arrive at the given image point from different points of a diffusely reflecting object. The speckled pattern visibility depends on the proportion between the spatial coherence radius of the light beam and the spread function half-width, whereas the amplitude is, in addition, determined by the reflection coefficient of the given object area.

In illuminating the reference-free image plane hologram (RFIPH) with an arbitrary source, the radiation is diffracted on the speckled pattern to reconstruct a positive image. As is the case with ordinary photography, the brightest positive areas correspond to the darkest negative fragments. The blocking screen BSc set on the axis (Fig. 7.2b) filters the zeroth order diffraction and the radiation diffracted at low spatial frequencies of a negative photographic image. The elementary RFIPH theory can be found in [7.16, 17, 25]. The RFIPH has the advantage of having low sensitivity to the object shift as well as not requiring a reference wave [7.18].

We shall highlight some of the ways in which the RFIPH promises to improve the quality of imaging through a scattering medium. In [7.15], the use of limited spatial coherence radiation at the stage of hologram recording is proposed. The result is that some of the scattered radiation that falls at the wings of the turbid medium PSF arrives at the given hologram point from the object areas outside the region of spatial coherence of the incident beam and, hence, does not contribute to the speckled pattern recording. Thus, the effective point spread function appears to be narrower, and the contrast of the reconstructed image is enhanced. The modulation transfer function of the viewing system combining the RFIPH with limited spatial coherence radiation was studied in [7.26].

When an image is transferred through the medium with moving scatterers, the RFIPH and time-averaged holography allow for both cut-off of the scattered radiation and an improvement in the image contrast by increasing the exposure time [7.27].

## 7.2 Comparison of Holographic and Incoherent Vision Systems

In a theoretical analysis of holographic imaging with interference pattern recording, the amplitude-phase characteristics of the radiation field should be taken into account. However, a calculation of the radiation fields in a random-inhomogeneous medium involves major mathematical difficulties. For this reason, the amplitude-phase approach fails to relate the image quality characteristics to the optical properties of the scattering medium. This, in turn, prevents one from estimating coherent-holographic observation systems for natural scattering media. On the other hand, in incoherent vision systems, the image quality can now be effectively estimated using the radiative transfer technique (Chaps. 3–6) which allows the contrast and signal-to-noise ratio of an object image to be calculated, in each particular situation, depending on the optical medium parameters.

The question arises as to whether it is possible to find the conformity between the above two systems of vision in scattering media and to apply the methods developed for the incoherent technique to evaluate the image quality characteristics in coherent and, specifically, in holographic recording. Indeed, such a possibility exists in a number of situations thanks to the connection between the correlation function of the wave field in the medium and the photometric parameters given by the radiative transfer equation.

### 7.2.1 Mutual Coherence Function as Related to Radiance

Let us consider a scalar (disregarding a state of polarization) stationary wave field $u(\vec{R}, t)$. The second-order coherence manifested in interference effects can be characterized by the two-point cross-correlation function for the electric field correlation at two space-time points $(\vec{R}_1, t_1)$ and $(\vec{R}_2, t_2)$ [7.28]

$$\Gamma(\vec{R}_1, t_1; \vec{R}_2, t_2) = \langle u(\vec{R}_1, t_1) u^*(\vec{R}_2, t_2) \rangle . \tag{7.2.1}$$

Here, averaging is taken over the time much in excess of the light wave oscillation period. For a stationary field, the cross-correlation function depends on the difference $t_d = t_2 - t_1$:

$$\Gamma(\vec{R}_1, \vec{R}_2, t_d) = \langle u(\vec{R}_1, t) u^*(\vec{R}_2, t + t_d) \rangle . \tag{7.2.2}$$

The function $\Gamma(\vec{R}_1, \vec{R}_2, t_d)$ is the *mutual coherence function*. As seen from (7.2.2),

$$\Gamma(\vec{R}, \vec{R}, 0) = \langle |u(\vec{R})|^2 \rangle = w(\vec{R}) \tag{7.2.3}$$

is the *energy density*. In the coherence theory, $w(\vec{R})$ is termed the field intensity. This should not be confused with the photometric radiation intensity.

The normalized mutual coherence function

$$\gamma(\vec{R}_1, \vec{R}_2, t_d) = \frac{\Gamma(\vec{R}_1, \vec{R}_2, t_d)}{[w(\vec{R}_1) w(\vec{R}_2)]^{1/2}} \tag{7.2.4}$$

is frequently called the *complex degree of coherence*.

The time and spatial field coherences are frequently, although approximately, separated. The time coherence is characterized by the autocorrelation function $\Gamma(\vec{R}, \vec{R}, t_d)$ which is related to the *spectral field density* in terms of the Fourier transform

$$w(\vec{R}, \Omega) = \int_{-\infty}^{\infty} \Gamma(\vec{R}, \vec{R}, t_d) e^{-i2\pi\Omega t_d} dt_d , \tag{7.2.5}$$

$$\Gamma(\vec{R}, \vec{R}, t_d) = \int_{-\infty}^{\infty} w(\vec{R}, \Omega) e^{i2\pi\Omega t_d} d\Omega . \tag{7.2.6}$$

The half-width of the function $\Gamma(\vec{R}, \vec{R}, t_d)$ is responsible for the coherence time $\Delta t_{coh}$ which is connected to the half-width $\Delta\Omega$ of the spectral field density as

$$\Delta t_{coh} \sim 1/\Delta\Omega . \tag{7.2.7}$$

As seen from (7.2.6) and (7.2.3),

$$\int_{-\infty}^{\infty} w(\vec{R}, \Omega) d\Omega = \Gamma(\vec{R}, \vec{R}, 0) = w(\vec{R}) . \tag{7.2.8}$$

*The mutual spectral density* $w(\vec{R}_1, \vec{R}_2, \Omega)$, defined as the Fourier transform of the mutual coherence function $\Gamma(\vec{R}_1, \vec{R}_2, t_d)$, is the generalization of the spectral field density concept to the case of different $\vec{R}_1$ and $\vec{R}_2$ points.

The spatial coherence is characterized by the two-point correlation function $\Gamma(\vec{R}_1, \vec{R}_2, 0) \equiv \Gamma(\vec{R}_1, \vec{R}_2)$.

Introducing the coordinates

$$\vec{R} = \frac{\vec{R}_1 + \vec{R}_2}{2} , \qquad \vec{R}_d = \vec{R}_2 - \vec{R}_1 \tag{7.2.9}$$

gives $\Gamma(\vec{R}_1, \vec{R}_2) = \Gamma(\vec{R}, \vec{R}_d) = \langle u(\vec{R} - \vec{R}_d/2, t) u^*(\vec{R} + \vec{R}_d/2, t) \rangle$.

If the function $\Gamma(\vec{R}, \vec{R}_d)$ varies slightly as $\vec{R}$ is changed by the magnitude comparable with the wavelength, then the function $\Gamma(\vec{R}, \vec{R}_d)$ beyond the radiation source region may be represented as [7.29–31]

$$\Gamma(\vec{R},\vec{R}_{\rm d}) \approx \iint_{4\pi} j(\vec{R},\vec{n})\,{\rm e}^{-{\rm i}(2\pi/\lambda)\vec{n}\vec{R}_{\rm d}}\,d\vec{n}\,, \tag{7.2.10}$$

where the function $j(\vec{R},\vec{n})$ obeys the radiative transfer equation.

For the relatively narrow (at least, within the small angle approximation domain) angular coherence function spectrum $j(\vec{R},\vec{n})$, instead of (7.2.10), we have [7.30, 32]

$$\Gamma(\vec{r},z,\vec{r}_{\rm d}) = \iint_{-\infty}^{\infty} j(\vec{r},z,\vec{n}_\perp)\,{\rm e}^{-{\rm i}(2\pi/\lambda)\vec{n}_\perp\vec{r}_{\rm d}}\,d\vec{n}_\perp\,. \tag{7.2.11}$$

In this case, the angular spectrum $j(\vec{r},z,\vec{n}_\perp)$ can be expressed in terms of $\Gamma(\vec{r},z,\vec{r}_{\rm d})$ as

$$j(\vec{r},z,\vec{n}_\perp) = \frac{1}{\lambda^2}\iint_{-\infty}^{\infty} \Gamma(\vec{r},z,\vec{r}_{\rm d})\,{\rm e}^{{\rm i}(2\pi/\lambda)\vec{n}_\perp\vec{r}_{\rm d}}\,d\vec{r}_{\rm d}\,. \tag{7.2.12}$$

Again, $(\vec{r},z)$ are the coordinates of the point $\vec{R}$, $\vec{r}$ and $\vec{n}_\perp$ the projections of the vectors $\vec{R}$ and $\vec{n}$ onto the plane normal to the axis of propagation, i.e., $z = \text{const}$ (Fig. 4.1).

We note that in the general case, the quantity $j(\vec{r},z,\vec{n}_\perp)$ is not equivalent to the radiance [7.30]. For example, it may assume negative values. However, if the coherence function $\Gamma(\vec{r},z,\vec{r}_{\rm d})$ varies slowly with $\vec{r}$, as compared to its variation with $\vec{r}_{\rm d}$, then in averaging over $\vec{r}$ segments much in excess of the coherence area, the angular spectrum $j(\vec{r},z,\vec{n}_\perp)$ is purely positive and can be interpreted as the radiance. Since it is exactly these averages which will concern us in the following analysis, we shall not differentiate between the angular coherence function spectrum and the radiance. For future reference, we shall also use $I(\vec{r},z,\vec{n}_\perp)$ rather than $j(\vec{r},z,\vec{n}_\perp)$.

Thus, the angular radiance distribution at the point $(\vec{r},z)$ is related to the coherence function $\Gamma(\vec{r},z,\vec{r}_{\rm d})$ in terms of the Fourier transform (7.2.11, 12). This means that the solution of the transfer equation provides not only photometric, but also (through the coherence function) some wave field characteristics. The electrodynamic grounds for the transfer equation and the correlation between the radiative transfer and multiple scattering wave theories are given in detail in [7.29, 31, 33–36].

The relation (7.2.11) is written for the scatterers at rest. If they are moving, then the radiation field is a function of time, and we must consider the mutual coherence function $\Gamma(\vec{R},\vec{R}_{\rm d},t_{\rm d})$. In this case, the relation

$$\Gamma(\vec{r},z,r_{\rm d},t_{\rm d}) = \iint_{-\infty}^{\infty} I(\vec{r},z,\vec{n}_\perp,t_{\rm d})\,{\rm e}^{-{\rm i}(2\pi/\lambda)\vec{n}_\perp\vec{r}_{\rm d}}\,d\vec{n}_\perp \tag{7.2.13}$$

generalizes (7.2.11) [7.37]. Here, $I(\vec{r},z,\vec{n}_\perp,t_{\rm d})$ is the solution to the equation

$$\left(\frac{\partial}{\partial z} + \vec{n}_{\perp}\nabla_{\vec{r}} + \varepsilon\right) I(\vec{r}, z, \vec{n}_{\perp}, t_{\mathrm{d}})$$
$$= \frac{\sigma}{4\pi} \iint_{-\infty}^{\infty} I(\vec{r}, z, \vec{n}'_{\perp}, t_{\mathrm{d}}) x(|\vec{n}_{\perp} - \vec{n}'_{\perp}|, t_{\mathrm{d}})\, d\vec{n}'_{\perp} \,, \quad (7.2.14)$$

while the function $x(|\vec{n}_{\perp} - \vec{n}'_{\perp}|, t_{\mathrm{d}})$ has the physical implication of the scattering indicatrix extended to the moving scatterers. It is the modulus of the autocorrelation function for the total field generated in the far region by the radiation scattered over all moving scatterers of a single scattering volume [7.38]. The conventional scattering indicatrix is

$$x(|\vec{n}_{\perp} - \vec{n}'_{\perp}|) = x(|\vec{n}_{\perp} - \vec{n}'_{\perp}|, t_{\mathrm{d}} = 0) \,. \quad (7.2.15)$$

The Fourier transform of the function $I(\vec{r}, z, \vec{n}_{\perp}, t_{\mathrm{d}})$

$$I(\vec{r}, z, \vec{n}_{\perp}, \Omega) = \int_{-\infty}^{\infty} I(\vec{r}, z, \vec{n}_{\perp}, t_{\mathrm{d}}) \mathrm{e}^{-\mathrm{i}2\pi\Omega t_{\mathrm{d}}}\, dt_{\mathrm{d}} \quad (7.2.16)$$

is the frequency spectrum (spectral density) of the radiance for moving scatterers. Applying the $t_{\mathrm{d}}$ Fourier transform to equation (7.2.14) gives that the function $I(\vec{r}, z, \vec{n}_{\perp}, \Omega)$ obeys the transfer equation

$$\left(\frac{\partial}{\partial z} + \vec{n}_{\perp}\nabla_{\vec{r}} + \varepsilon\right) I(\vec{r}, z, \vec{n}_{\perp}, \Omega)$$
$$= \frac{\sigma}{4\pi} \int_{-\infty}^{\infty} d\Omega' \iint_{-\infty}^{\infty} I(\vec{r}, z, \vec{n}'_{\perp}, \Omega') x(|\vec{n}_{\perp} - \vec{n}'_{\perp}|, \Omega - \Omega')\, d\vec{n}'_{\perp} \quad (7.2.17)$$

and the function

$$x(\vec{n}_{\perp}, \Omega) = \int_{-\infty}^{\infty} x(\vec{n}_{\perp}, t_{\mathrm{d}}) \mathrm{e}^{-\mathrm{i}2\pi\Omega t_{\mathrm{d}}}\, dt_{\mathrm{d}} \quad (7.2.18)$$

describes spectrum spreading due to the Doppler effect in the event of single scattering on moving particles. Equation (7.2.17) has an obvious physical implication and might have been formulated immediately on phenomenological grounds.

### 7.2.2 Quality Characteristics of Rough Object Images in Reference Wave Holography

Let us return to the analysis of the holographic image quality. It seems natural to use the contrast $k$ and signal-to-noise ratio $\delta$ in this description, as is the case for incoherent systems. Holographic imaging involves hologram recording and reconstruction stages. It is obvious that both characteristics play a role in the final result of the process, i.e., in the reconstructed image. Bearing this in mind, we shall consider holographic imaging in more detail.

We take it that the field of the object wave propagating from the object in the scattering medium is given by the function $u_{ob}(\vec{r}, t)$, where $\vec{r}$ is the hologram point coordinate. The function $u_r(t)$ prescribes the reference wave field. For the recording time $t_e$, the exposure distribution on the hologram is

$$H(\vec{r}) = \int_0^{t_e} |u_r(t) + u_{ob}(\vec{r}, t)|^2 \, dt = H_r + H_{ob} + H_{r,ob} + H^*_{r,ob} , \tag{7.2.19}$$

where

$$H_r = \int_0^{t_e} u_r^2(t) \, dt , \qquad H_{ob} = \int_0^{t_e} u_{ob}^2(\vec{r}, t) \, dt , \tag{7.2.20}$$

$$H_{r,ob} = \int_0^{t_e} u_r^*(t) u_{ob}(\vec{r}, t) \, dt . \tag{7.2.21}$$

We generally assume that the amplitude transmission coefficient of the developed hologram is of the form

$$t_a(\vec{r}) = \langle t_a \rangle - \chi[H(\vec{r}) - \langle H \rangle] , \tag{7.2.22}$$

where $\langle t_a \rangle$ is the average amplitude transmission coefficient determined by the average exposure $\langle H \rangle$, and $\chi$ the proportionality constant dependent on $\langle H \rangle$ and the characteristic film curve.

In reconstructing the hologram with a plane wave at the amplitude $u_{rec}$, the radiation field just behind the hologram is $u_i(\vec{r}) = u_{rec} t_a(\vec{r})$. The angular coherence function spectrum which, for example, forms a real object image, is [7.39]

$$j_i(\vec{n}_\perp) = \frac{\chi^2 u_{rec}^2 t_e H_r}{\lambda_2^2} \iint_{-\infty}^{\infty} \Gamma(\vec{r}_d) e^{i(2\pi/\lambda_2)\vec{n}_\perp \vec{r}_d} \, d\vec{r}_d , \tag{7.2.23}$$

where

$$\Gamma(\vec{r}_d) = \frac{1}{t_e H_r \Sigma_h} \iint_{\Sigma_h} H_{r,ob}(\vec{r}) H^*_{r,ob}(\vec{r} + \vec{r}_d) \, d\vec{r} , \tag{7.2.24}$$

and $\lambda_2$ is the reconstructing radiation wavelength. In (7.2.24), integration is over the hologram surface area $\Sigma_h$. If the hologram is large as compared to the function $\Gamma(\vec{r}_d)$ half-width, i.e. with the coherence area of the field $u_i(\vec{r}_d)$, then the function $j_i(\vec{n}_\perp)/\Sigma_h$, as mentioned in Sect. 7.2.1, signifies the radiance $I_i(\vec{n}_\perp)$ of the reconstructed image.

The radiant flux through an image resolution cell is

$$P_i(\vec{n}_\perp) = \Sigma_h \iint_{\Delta\vec{\Omega}_2} I_i(\vec{n}_\perp) \, d\vec{n}_\perp . \tag{7.2.25}$$

Here, the resolution cell is governed by the hologram resolution diffraction angle $|\Delta\vec{\Omega}_2| \sim \lambda_2^2/\Sigma_h$ and the hologram-to-reconstructed image distance.

We now arrive at the contrast of two adjacent image cells

$$k = \frac{P_i(\vec{n}_{\perp,1}) - P_i(\vec{n}_{\perp,2})}{P_i(\vec{n}_{\perp,1}) + P_i(\vec{n}_{\perp,2})} = \frac{\Delta P_i(\vec{n}_\perp)}{2\langle P_i(\vec{n}_\perp)\rangle}, \tag{7.2.26}$$

where $\langle P_i(\vec{n}_\perp)\rangle = [P_i(\vec{n}_{\perp,1}) + P_i(\vec{n}_{\perp,2})]/2$ is the mean radiant flux, and $\Delta P_i(\vec{n}_\perp)$ the difference of the fluxes through adjacent cells which provides useful information about the object of observation.

Due to the discrete granular structure of the photographic material, the reconstructed object image contains a noise component in addition to the valid signal. In view of the coherent field interference, the variance of the fluctuations of the resultant flux $P$ through an image cell is [7.39]

$$V_P = \langle P^2\rangle - \langle P\rangle^2 = \langle P_{\text{noi}}\rangle^2 - 2\langle P_i\rangle\langle P_{\text{noi}}\rangle, \tag{7.2.27}$$

where the noise flux $P_{\text{noi}}(\vec{n}_\perp) \approx I_{\text{noi}}(\vec{n}_\perp)|\Delta\vec{\Omega}_2|\Sigma_h$ and $I_{\text{noi}}(\vec{n}_\perp)$ is the noise irradiance.

Considering (7.2.26, 27), for the signal-to-noise ratio in the object image, we obtain

$$\delta = \frac{\Delta P_i}{\sqrt{V_P}} = k\frac{2\langle P_i\rangle/\langle P_{\text{noi}}\rangle}{\sqrt{1 + 2\langle P_i\rangle/\langle P_{\text{noi}}\rangle}}, \tag{7.2.28}$$

i.e., the quantity $\delta$ is determined by the image contrast and the ratio $\langle P_i\rangle/\langle P_{\text{noi}}\rangle$. The ratio $\langle P_i\rangle/\langle P_{\text{noi}}\rangle$ depends on the number, $\langle N\rangle$, of the object wave photons arriving at the hologram in the angular spectrum range $\Delta\vec{\Omega}_1$ and contributing to the interference pattern. The quantity $|\Delta\vec{\Omega}_1| \sim \lambda_1^2/\Sigma_h$ and conforms to the resolution diffraction angle in the course of hologram recording. Here, $\lambda_1$ is the radiation wavelength at the recording stage. The careful analysis made in [7.39] taking into account the random distribution of the photoemulsion grains has demonstrated that, under optimum conditions, when $H_r \gg H_{ob}$ and the exposure is chosen within the linear fragment of the characteristic curve,

$$\frac{\langle P_i\rangle}{\langle P_{\text{noi}}\rangle} \approx \frac{2}{\pi}\eta\langle N\rangle. \tag{7.2.29}$$

Here, $\eta$ is the quantum sensitivity of the photoemulsion.

Substituting (7.2.29) into (7.2.28), we get

$$\delta = k\frac{(4/\pi)\eta\langle N\rangle}{\sqrt{1 + (4/\pi)\eta\langle N\rangle}}. \tag{7.2.30}$$

For the number of photons used $\eta\langle N\rangle \gg 1$, when the film-grain noise is less than the signal (7.2.29), from (7.2.30), we have

$$\delta = k\sqrt{(4/\pi)\eta\langle N\rangle}. \tag{7.2.31}$$

We consider the relation (7.2.24) to determine the photon number $\langle N\rangle$. The quantity $(1/\lambda_1^2)\int\int_{-\infty}^{\infty} \Gamma(\vec{r}_d)\exp[i(2\pi/\lambda_1)\vec{n}_\perp\vec{r}_d]\,d\vec{r}_d$ can be interpreted as the hologram area $\Sigma_h$ and the exposure time $t_e$ average of the angular distribution of the power of the object radiation forming an interference pattern on the hologram. Therefore,

$$\langle N \rangle = \frac{t_e \Sigma_h}{\lambda_1^2 hf} \iint_{\Delta\Omega_1} d\vec{n}_\perp \iint_{-\infty}^{\infty} \Gamma(\vec{r}_d) e^{i(2\pi/\lambda_1)\vec{n}_\perp \vec{r}_d} d\vec{r}_d \, . \tag{7.2.32}$$

Here, $hf$ is the energy of the photon used for hologram recording.

Consequently, (7.2.26, 30), taking into account (7.2.25, 32), are the solution to the problem stated and describe the characteristics of the image quality in holographic observation systems.

We now compare the signal-to-noise ratio for a hologram reconstructed image with that in the incoherent vision system. Assume, for the sake of certainty, that photomultiplier recording is used for imaging in the incoherent system. A $D$-dia objective with focal distance $F$ and a $d$-dia diaphragm located in the focal objective plane are used. First, compare (7.2.31) with (5.1.3, 5). In both cases, the signal-to-noise ratio in the object image is found from one and the same expression, the difference in the coefficient being insignificant. The quantity $\delta$ is always proportional to the image contrast and to $(\eta\langle N \rangle)^{1/2}$. The photon number $\eta\langle N \rangle$, in turn, is in proportion to the exposure time $t_e$, receiving aperture (hologram $\Sigma_h$ or objective $\pi D^2/4$ areas, respectively) and angular system resolution ($|\Delta\vec{\Omega}_1| \sim \lambda_1^2/\Sigma_h$ or $|\Delta\vec{\Omega}_{rec}| = \pi d^2/4F^2$). It should be noted that $\langle N \rangle$ does not increase with the hologram area since, in this case, the solid angle $|\Delta\vec{\Omega}_1|$ diminishes (the spatial frequency transmission band is broadened). This will, in fact, be the situation for the incoherent system where the objective diameter $D$ grows, but the aperture ratio $D^2/F^2$ is fixed. In this case, $\langle N \rangle$ can be increased in spatial integration (Sect. 5.1) by increasing the receiver diaphragm diameter $d$. The spatial resolution of the system is, naturally, degenerated here. In the coherent system, the same effect may be achieved by reducing the spatial coherence degree of the reconstructing beam.

The following three situations are worth noting.

In both systems, the expressions for the signal-to-noise ratio are identical under optimum hologram recording conditions, at $\eta\langle N \rangle \gg 1$, for example, when the film-grain noises can be disregarded. If these conditions are violated, the quantity $\delta$ then proves to be lower in the holographic system. Such a situation can be regarded as natural because, in contrast to the incoherent vision system, the holographic system includes an additional radiation recording stage which also introduces noises. In fact, (7.2.30) can be written as

$$\delta = \frac{k}{\mu_{gr}} \sqrt{\frac{4}{\pi} \eta \langle N \rangle} \, , \tag{7.2.30$'$}$$

where $\mu_{gr} = \sqrt{1 + \pi/4\eta\langle N \rangle}$ determines the gain due to the film-grain noises. For a larger hologram area $\Sigma_h$, $\eta\langle N \rangle$ is increased and $\mu_{gr}$ reduced.

In reference wave holography, the information regarding each object element is recorded simultaneously over the entire hologram, which is much larger than the area of each emulsion grain. This is probably why, in holographic reference wave systems, the film-grain noise is essentially lower than in conventional photography [7.40] where the object element information is concentrated over an appropriate film area. For this reason, for example, reference-free image plane holograms with localized information recording compare unfavourably with the reference wave technique due to film-grain noises. This is a kind of compensation for the advantages of reference-free recording.

The signal-to-noise ratio is independent of the power of the reconstructing wave. In effect, if the power $u_{\text{rec}}^2$ is small, then appreciable additional noises may interfere at the stage of reconstructed image recording. However, the signal-to-noise ratio cannot be improved by increasing $u_{\text{rec}}^2$. The situation is similar to that in electronic amplifiers: the input signal and noise are, at least, increased equally, which does not change the signal-to-noise ratio. Of course, special methods of noise suppression and signal detection through noise are not implied here.

### 7.2.3 Contrast and Signal-to-Noise Ratio in Time-Averaged Holography and the Limited Time Coherence Technique

We shall use the results of the previous section to analyze in more detail image construction and transfer in time-average holography and the LTC technique [7.41].

Let the radiation field from a source to an object of observation be characterized by the complex amplitude $u_0(t)$. Taking into account medium scattering and object reflection, we can represent the object wave field at the hologram point $\vec{r}$ as

$$u_{\text{ob}}(\vec{r}, t) = \int_{-\infty}^{\infty} u_0(t')G(\vec{r}, t, t')\,dt' , \tag{7.2.33}$$

where $G(\vec{r}, t, t')$ is the Green function of the system defining transfer to a point $\vec{r}$ at an instant $t$ of the amplitude of the field emitted at a time $t'$.

For each given time instant $t$, we assume that the scattering medium is frozen and that the function $G(\vec{r}, t, t')$ depends on the difference argument $t - t'$ alone. The function $G(\vec{r}, t, t')$ for a nonstationary medium is, however, time $t$ dependent. We emphasize this dependence by writing

$$G(\vec{r}, t, t') = G(\vec{r}, t, t - t') . \tag{7.2.34}$$

Here, the second argument represents comparatively small variations in the Green's function.

The field of the reference wave delayed for the time $t_0$ is governed by the function

$$u_r(t) = Cu_0(t - t_0) . \tag{7.2.35}$$

The quantity $C^2$ defines the reference-to-object radiation power ratio. Substituting (7.2.33–35) into (7.2.21), we obtain the time average of the integrand over $\Delta t$, which exceeds the coherence time of the source $u_0(t)$ but is less than the characteristic time of the function $G(\vec{r}, t, t')$ variation with the scatterer velocity. Then, instead of (7.2.21), we have

$$H_{\mathrm{r,ob}} = C\langle u_0^2\rangle \int_0^{t_e} dt \int_{-\infty}^{\infty} \gamma_0(t_0 - t')G(\vec{r}, t, t')\,dt' \,. \tag{7.2.36}$$

In (7.2.36), $\gamma_0(t') = \langle u_0(t)u_0^*(t + t')\rangle/\langle u_0^2(t)\rangle$ is the time coherence degree of the source.

Formulas (7.2.20) and (7.2.35) yield

$$H_{\mathrm{r}} = C^2 t_e \langle u_0^2(t)\rangle \,. \tag{7.2.37}$$

Substituting (7.2.36, 37) into (7.2.24), we arrive at

$$\begin{aligned}\Gamma(\vec{r}_{\mathrm{d}}) = \frac{\langle u_0^2\rangle}{t_e^2\Sigma_{\mathrm{h}}} \iint_{\Sigma_{\mathrm{h}}} d^2\vec{r} \int_0^{t_e} dt \int_{-\infty}^{\infty} \gamma_0(t_0 - t'')G(\vec{r}, t, t'')\,dt'' \\ \times \int_0^{t_e} dt' \int_{-\infty}^{\infty} \gamma_0^*(t_0 - t''')G^*(\vec{r} + \vec{r}_{\mathrm{d}}, t', t''')\,dt''' \,.\end{aligned} \tag{7.2.38}$$

If the mean distance between scattering particles is much less than the source coherence length, then, for the random distribution of the scatterers over the medium, the Green's function $G(\vec{r}, t, t')$ in (7.2.38) can be regarded as $\delta$ correlated in time $t'$. In this approximation, from (7.2.38), we have

$$\begin{aligned}\Gamma(\vec{r}_{\mathrm{d}}) = \frac{\langle u_0^2\rangle}{t_e^2\Sigma_{\mathrm{h}}} \iint_{\Sigma_{\mathrm{h}}} d^2\vec{r} \int_0^{t_e} dt \int_{-t}^{t_e - t} dt_{\mathrm{d}} \\ \times \int_{-\infty}^{\infty} \gamma_0^2(t_0 - t'')G(\vec{r}, t, t'')G^*(\vec{r} + \vec{r}_{\mathrm{d}}, t + t_{\mathrm{d}}, t'')\,dt'' \,.\end{aligned} \tag{7.2.39}$$

We introduce the correlation function

$$\Gamma_G(\vec{r}_{\mathrm{d}}, t_{\mathrm{d}}, t'') = \langle G(\vec{r}, t, t'')G^*(\vec{r} + \vec{r}_{\mathrm{d}}, t + t_{\mathrm{d}}, t'')\rangle_{\vec{r},t} \,. \tag{7.2.40}$$

Here, averaging is made over the hologram area and exposure time $t_e$. When the hologram size is sufficiently large as compared to the correlation radius of the object field $u_{\mathrm{ob}}(\vec{r}, t)$, the surface averaging is equivalent to averaging over the spatial ensemble realizations of the scatterer distribution. Assuming an ergodic system in (7.2.40), we may restrict to averaging over $\vec{r}$. This means that

$$\Gamma_G(\vec{r}_{\mathrm{d}}, t_{\mathrm{d}}, t'') = \frac{1}{\Sigma_{\mathrm{h}}} \iint_{\Sigma_{\mathrm{h}}} G(\vec{r}, t, t'')G^*(\vec{r} + \vec{r}_{\mathrm{d}}, t + t_{\mathrm{d}}, t'')\,d\vec{r} \tag{7.2.41}$$

and is independent of $t$.

Taking into account (7.2.41) and the $t_{\mathrm{d}}$ evenness of the function $\Gamma_G(\vec{r}_{\mathrm{d}}, t_{\mathrm{d}}, t'')$, from (7.2.39), we have

$$\Gamma(\vec{r}_{\mathrm{d}}) = \frac{2\langle u_0^2\rangle}{t_e^2} \int_0^{t_e} dt \int_0^{t} dt_{\mathrm{d}} \int_{-\infty}^{\infty} \gamma_0^2(t_0 - t')\Gamma_G(\vec{r}_{\mathrm{d}}, t_{\mathrm{d}}, t')\,dt' \,. \tag{7.2.42}$$

Replacing the function $\Gamma_G(\vec{r}_d, t_d, t')$ by its spectral representation

$$\Gamma_G(\vec{r}_d, t_d, t') = \int_{-\infty}^{\infty} \Gamma_G(\vec{r}_d, \Omega, t') e^{i2\pi\Omega t_d} d\Omega \qquad (7.2.43)$$

and taking into account that $\Gamma_G(\vec{r}_d, \Omega, t')$ is the even function of $\Omega$, equation (7.2.42) gives

$$\Gamma(\vec{r}_d) = \langle u_0^2 \rangle \int_{-\infty}^{\infty} \left( \frac{\sin(\pi\Omega t_e)}{\pi\Omega t_e} \right)^2 d\Omega \int_{-\infty}^{\infty} \gamma_0^2(t_0 - t) \Gamma_G(\vec{r}_d, \Omega, t)\, dt\,. \qquad (7.2.44)$$

Substituting (7.2.44) into (7.2.32), for the photon number $\langle N \rangle$ which determines the signal-to-noise ratio, we have

$$\langle N \rangle = \frac{t_e \Sigma_h |\Delta\vec{\Omega}_1|}{hf} I_{eff}(\vec{n}_\perp)\,, \qquad (7.2.45)$$

where

$$I_{eff}(\vec{n}_\perp) = \langle u_0^2 \rangle \int_{-\infty}^{\infty} \left( \frac{\sin(\pi\Omega t_e)}{\pi\Omega t_e} \right)^2 d\Omega \int_{-\infty}^{\infty} \gamma_0^2(t_0 - t) I_n(\vec{n}_\perp, \Omega, t)\, dt\,, \qquad (7.2.46)$$

$$I_n(\vec{n}_\perp, \Omega, t) = \frac{1}{\lambda_1^2} \iint_{-\infty}^{\infty} \Gamma_G(\vec{r}_d, \Omega, t) e^{i(2\pi/\lambda_1)\vec{n}_\perp \vec{r}_d} d\vec{r}_d\,. \qquad (7.2.47)$$

The image contrast is also expressed in terms of the function $I_{eff}(\vec{n}_\perp)$:

$$k = \frac{\Delta I_{eff}(\vec{n}_\perp)}{2 I_{eff}(\vec{n}_\perp)}\,. \qquad (7.2.48)$$

For the "frozen" medium whose scatterers are at rest, $I_n(\vec{n}_\perp, \Omega, t) = I_n(\vec{n}_\perp, t)\delta(\Omega)$. Then, in the LTC technique, from (7.2.46), we obtain

$$I_{eff}(\vec{n}_\perp) = \langle u_0^2 \rangle \int_{-\infty}^{\infty} \gamma_0^2(t_0 - t) I_n(\vec{n}_\perp, t)\, dt\,. \qquad (7.2.49)$$

The function

$$\begin{aligned} I_n(\vec{n}_\perp, t) &= \int_{-\infty}^{\infty} I_n(\vec{n}_\perp, \Omega, t)\, d\Omega \\ &= \frac{1}{\lambda_1^2} \iint_{-\infty}^{\infty} \Gamma_G(\vec{r}_d, t_d = 0, t) e^{i(2\pi/\lambda_1)\vec{n}_\perp \vec{r}_d} d\vec{r}_d \end{aligned} \qquad (7.2.50)$$

obeys the usual nonstationary transfer equation (4.8.1) and is frequently termed the photon path (time-of-arrival) distribution function. It defines the angular and time distributions of the radiance at a hologram point when the medium is illuminated by a pulsed unit power source.

In time-average holography, assuming $\gamma_0(t) \approx 1$, i.e., the source coherence time is much greater than the half-width of the photon time-of-arrival distribution function, from (7.2.46), we arrive at

$$I_{\text{eff}}(\vec{n}_\perp) = \langle u_0^2 \rangle \int_{-\infty}^{\infty} I_n(\vec{n}_\perp, \Omega) \left( \frac{\sin \pi \Omega t_e}{\pi \Omega t_e} \right)^2 d\Omega \,. \qquad (7.2.51)$$

Here, the function

$$I_n(\vec{n}_\perp, \Omega) = \int_{-\infty}^{\infty} I_n(\vec{n}_\perp, \Omega, t)\, dt \qquad (7.2.52)$$

is the angular distribution of the spectral radiance of the object radiation which obeys the transfer equation (7.2.17). It is assumed that the medium is illuminated by a monochromatic unit power source ($\Omega = f - f_0$, $f_0$ being the source radiation frequency).

As seen from (7.2.49, 51), $I_{\text{eff}}(\vec{n}_\perp)$ is the effective, i.e., recorded in coherence, radiance of the object radiation.

The relations (7.2.31, 45, 48), taking into account (7.2.49, 51) determine the quality characteristics of the holographic image in terms of the photometric quantities $I_n(\vec{n}_\perp, t)$ and $I_n(\vec{n}_\perp, \Omega)$ in the LTC and time-average holography. They also allow the characteristics of the image so constructed to be compared with the appropriate characteristics in the cooresponding incoherent vision systems.

It is clear from the underlying physics that the LTC technique application could be similar to the application of signal time gating on the photodetector in the incoherent pulsed-source vision system. Such systems, with pulsed cut-off of the backscattered noise, were considered in Chap. 5. The relations (7.2.45, 46, 49) point to a rather profound analogy. When the frame recording energies and the gating function $\gamma_0^2(t_0 - t)$ are equal in the pulsed and LTC techniques, the contrast and signal-to-noise ratio are the same for both systems.

This means that the pulse time gating technique is analogous to the holographic LTC technique in incoherent optics.

As seen from (7.2.51), the application of an ordinary spectral filter with the transmission function $(\sin(\pi\Omega t_e)^2/(\pi\Omega t_e)^2$ in incoherent optics conforms to the use of time-average holography with a monochromatic source. For equal exposure times $t_e$ and mean source power, the contrast and signal-to-noise ratio must be the same in these two systems. If the source is other than monochromatic, then the time-average holography has major advantages. It is not only because one cannot provide an optical filter with a transmission band of fractions of a hertz (the exposure time $t_e$ being some seconds). The application of a spectral filter with a very small half-width $\Delta\Omega \sim t_e^{-1}$ in the incoherent systems causes high energy losses. At the same time, the coherent systems use almost all the signal energy (7.2.45, 46). Here, the radiation from each harmonic signal component, broadened due to moving scatterers, is recorded in the spectral range $\Delta\Omega$. Indeed, (7.2.46) integrates over the difference frequency $\Omega = f - f_0$, regardless of the harmonic component $f_0$.

Such a comparison might be pursued further. However, we shall only make general remarks since no comprehensive studies have been carried out. It seems that for the majority of coherent vision systems for scattering media, an analogue can be found in incoherent optics, and vice versa. Thus, the image quality characteristics of both systems will be identical. The possibility of such a conformity is by no means surprising. When observing rough objects, the mean light field characteristics in holographic systems can be determined within the framework of incoherent optics [7.42].

Such analogues, as well as the possibility of describing the image quality in coherent systems in terms of photometric quantities, allow the well-developed radiative transfer theory and estimations of incoherent vision systems to be used for theoretical studies of coherent systems (Chaps. 3–6). This technique can readily be used to analyse the contrast and signal-to-noise ratio as functions of the optical parameters of the scattering medium such as the optical layer thickness, scattering indicatrix, etc., as well as to estimate the limiting vision range. These possibilities are illustrated below for time-average holography.

### 7.2.4 Contrast and Signal-to-Noise Ratio as Functions of the Averaging Time and Optical Parameters of a Scattering Medium

We shall consider two viewing schemes. The model of self-luminous object viewing in vision theory is the same as in Fig. 7.1a. We shall refer to it as the information transfer scheme. The other (active observation) does not differ from the experimental geometry seen in Fig. 7.1b. Here, the radiation from the source passes through the scattering media and, at the same time, illuminates all diffusely reflecting objects. The reflected radiation is again allowed to pass through the scattering medium and arrives at the photographic plate. A reference wave is produced by the system of mirrors $M_1$ and $M_2$.

For future reference and for convenience, instead of (7.2.51), we shall write its equivalent:

$$I_{\mathrm{eff}}(\vec{n}_\perp) = \frac{2\langle u_0^2\rangle}{t_e^2}\int_0^{t_e} dt \int_0^t I_n(\vec{n}_\perp, t_d)\, dt_d\,, \tag{7.2.53}$$

where $I_n(\vec{n}_\perp, t_d)$ is the Fourier transform of $I_n(\vec{n}_\perp, \Omega)$. Expressions (7.2.53) and (7.2.51) are related in the same fashion as (7.2.42) and (7.2.44).

For the information transfer scheme, considering the photometric implication of $I_n(\vec{n}_\perp, t_d)$ and taking as our basis the optical reciprocity theorem, we can write, by analogy with (5.3.4),

$$\langle u_0^2\rangle I_n(\vec{n}_\perp, t_d) = \iint_{-\infty}^{\infty} I_o(\vec{r}\,')S(\vec{r} - \vec{r}\,', z, t_d)\, d\vec{r}\,'\,. \tag{7.2.54}$$

Here, $z$ is the object-to-hologram distance, $\vec{r}$ the coordinate of the object plane point to which the vector $-\vec{n}$ is directed, and $I_o(\vec{r})$ the radiance in the object plane. The function

$$S(\vec{r}, z, t_d) = \Gamma(\vec{r}, z, \vec{r}_d = 0, t_d) = \iint_{-\infty}^{\infty} I_n(\vec{r}, z, \vec{n}_\perp, t_d)\, d\vec{n}_\perp \tag{7.2.55}$$

generalizes the point spread function for a nonstationary medium; $I_n(\vec{r}, z, \vec{n}_\perp, t_d)$ obeys the transfer equation (7.2.14) at the boundary condition when the medium is illuminated by a monochromatic unit power point source.

For the system MTF which is dependent on the exposure time, the expressions (7.2.54, 53) provide

$$T(\vec{v}, t_e) = \frac{a(\vec{v}, t_e)}{a(\vec{v} = 0, t_e)}, \tag{7.2.56}$$

where

$$a(\vec{v}, t_e) = \frac{2}{t_e^2} \int_0^{t_e} dt \int_0^{t} S(\vec{v}, z, t_d)\, dt_d\,, \tag{7.2.57}$$

$$S(\vec{v}, z, t_d) = \iint_{-\infty}^{\infty} S(\vec{r}, z, t_d) e^{-i\vec{v}\cdot\vec{r}} d\vec{r}\,. \tag{7.2.58}$$

The function $a(\vec{v}, t_e)$ operates as the optical transfer function of the medium.

For the image-average photon number, from (7.2.45, 53, 54), we have

$$\langle N \rangle = \frac{\langle I_0 \rangle \Sigma_h t_e |\Delta\vec{\Omega}_1|}{hf} a(\vec{v} = 0, t_e)\,. \tag{7.2.59}$$

In considering the active viewing system, we shall disregard backscattering noise. In this case, the function $I_n(\vec{n}_\perp, t_d)$ is also described by the relation (7.2.54) in which, however, the object radiance distribution $I_0(\vec{r})$ is replaced by

$$I_0(\vec{r}, t_d) = \frac{\langle u_0^2 \rangle}{\pi} A(\vec{r}) \iint_{-\infty}^{\infty} S(\vec{r}, z, t_d)\, d\vec{r}$$
$$= \frac{\langle u_0^2 \rangle}{\pi} A(\vec{r}) S(\vec{v} = 0, z, t_d)\,. \tag{7.2.60}$$

Here, $A(\vec{r})$ is the albedo of the diffusely reflecting object, $u_0$ the amplitude of the field of the illuminating beam on the object without a scatterer; the function $S(\vec{v} = 0, z, t_d)$ includes medium transmission and time coherence changes due to motion of the scatterer. In this case, the expressions for the number $\langle N \rangle$ and the function $a(\vec{v}, t_e)$ are, correspondingly,

$$\langle N \rangle = \frac{\langle u_0^2 \rangle \langle A \rangle \Sigma_h t_e |\Delta\vec{\Omega}_1|}{\pi hf} a(\vec{v} = 0, t_e)\,, \tag{7.2.61}$$

$$a(\vec{v}, t_e) = \frac{2}{t_e^2} \int_0^{t_e} dt \int_0^{t} S(\vec{v} = 0, z, t_d) S(\vec{v}, z, t_d)\, dt_d\,. \tag{7.2.62}$$

An interesting and important result of (7.2.62) deserves mention. In time-average holography, the MTF of the active vision system is not equal to the MTF in the information transfer scheme (the analogue for self-luminous object observation). On the contrary (Sect. 5.3), the MTF of the incoherent vision system where the directional diagram of the source is wide and that of the receiver narrow is the same as the MTF in the self-luminous object viewing system.

We consider to be specific, that scatterers of radius $a$ are in Brownian motion. In this case, the function $x(\vec{n}_\perp, t_d)$ is [7.38]

$$x(\vec{n}_\perp, t_d) = x(\vec{n}_\perp)e^{-(2\pi/\lambda)^2 n_\perp^2 \mathscr{D} t_d}, \tag{7.2.63}$$

where $\mathscr{D} = kT/6\pi\eta a$ is the translational diffusion coefficient, $k$ the Boltzmann constant, $T$ the temperature and $\eta$ the viscosity. For a polydisperse medium with the particle size distribution function $f(a)$, we have

$$x(\vec{n}_\perp, t_d) = \int_0^\infty f(a) x(\vec{n}_\perp) e^{-(2\pi/\lambda)^2 n_\perp^2 \mathscr{D} t_d} da. \tag{7.2.64}$$

It should be remembered that here the quantity $\mathscr{D}$ and the scattering indicatrix $x(\vec{n}_\perp)$ are the functions of a particle radius. For simplicity, we regard the medium to be monodisperse and use the relation (7.2.63).

Making allowance for (7.2.55, 58) and using the known solution to equation (7.2.14) (Sect. 4.4.5), we write the expression for the function $S(\vec{v}, z, t_d)$ as

$$S(\vec{v}, z, t_d) = \exp\left[-\varepsilon z + \frac{\sigma z}{vz}\int_0^{vz} x(p, t_d)\, dp\right], \tag{7.2.65}$$

where

$$\begin{aligned} x(p, t_d) &= \frac{1}{4\pi}\iint_{-\infty}^{\infty} x(\vec{n}_\perp, t_d) e^{-i\vec{n}\cdot\vec{p}} d\vec{n}_\perp \\ &\approx \frac{1}{2}\int_0^\infty x(\beta) e^{-(2\pi/\lambda)^2 \beta^2 \mathscr{D} t_d} J_0(p\beta)\beta\, d\beta. \end{aligned} \tag{7.2.66}$$

Here, $J_0(p\beta)$ is the Bessel function, and $|\vec{n}_\perp| \approx \beta$ is set at small scattering angles.

For arbitrary $t_e$ and scattering medium parameters, the solution (7.2.65), including (7.2.57, 62), provides estimates for the function $a(\vec{v}, t_e)$. We then find the system MTF $T(\vec{v}, t_e)$, the photon number $\langle N \rangle$ and the signal-to-noise ratio in the object image.

We shall now consider two limiting cases for which analytical solutions have been derived.

1. The exposure time $t_e$ is small. Assuming $u = (2\pi/\lambda)^2 \mathscr{D}\beta_2 t_e < 1$ ($\beta_2$ being, as before, the mean-square scattering angle) and restricting to two terms in the exponential series in (7.2.66), from (7.2.65), we obtain

$$S(\vec{v}, z, t_d) = \exp[-\varepsilon z + \sigma z x'(vz) - \sigma z (2\pi/\lambda)^2 \mathscr{D} x_2'(vz) t_d]. \tag{7.2.67}$$

Here,

$$x'(vz) = \frac{1}{2vz}\int_0^{vz} dp \int_0^\infty x(\beta) J_0(p\beta)\beta\, d\beta, \tag{7.2.68}$$

$$x_2'(vz) = \frac{1}{2vz}\int_0^{vz} dp \int_0^{\infty} x(\beta)J_0(p\beta)\beta^3\, d\beta\,. \tag{7.2.69}$$

If, by the normalization condition, $x'(vz = 0) = 1$, then $x_2'(vz = 0) = \beta_2$, i.e., the mean-square scattering angle.

Substituting (7.2.67) into (7.2.57, 62), we obtain for the information transfer scheme

$$a(\vec{v}, t_e) = \psi(\xi_1 t_e)e^{-\varepsilon z + \sigma z x'(vz)} \tag{7.2.70}$$

and for the active viewing scheme

$$a(\vec{v}, t_e) = \psi(\xi_2 t_e)e^{-2\varepsilon z + \sigma z[1 + x'(vz)]}\,. \tag{7.2.71}$$

Here,

$$\psi(\xi_i t_e) = \frac{2(\xi_i t_e + e^{-\xi_i t_e} - 1)}{(\xi_i t_e)^2}\,, \qquad i = 1, 2\,, \tag{7.2.72}$$

$$\xi_1 = (2\pi/\lambda)^2\sigma z\mathscr{D}x_2'(vz)\,,$$

$$\xi_2 = (2\pi/\lambda)^2\sigma z\mathscr{D}[x_2'(vz) + x_2'(vz = 0)]\,. \tag{7.2.73}$$

2. $t_e \to \infty$. In this case, the function $a(\vec{v}, t_e)$ is determined only by unscattered radiation. In fact, the spectral filter $(\sin(\pi\Omega t_e)^2/(\pi\Omega t_e)^2$ in (7.2.52) becomes ultimately narrow and, as follows from (7.2.57, 62), the contribution of the scattered component vanishes in proportion to $t_e^{-1}$. Therefore, for the two schemes under consideration, we have

$$a(\vec{v}, t_e \to \infty) = e^{-\varepsilon z}\,, \tag{7.2.74}$$

$$a(\vec{v}, t_e \to \infty) = e^{-2\varepsilon z}\,. \tag{7.2.75}$$

We compare the image quality characteristics in the time-average holography at different averaging times $t_e$. We shall rely on $t_e = 0$, which corresponds to observation when the scatterers are at rest. In order to obtain a correct comparison and expose a pure effect of scatterer motion, the condition of constant exposure, i.e., $\langle I_0 \rangle t_e = \text{const}$ or $\langle u_0^2 \rangle t_e = \text{const}$ must be obeyed. For the above two viewing schemes, Fig. 7.3 shows the modulation transfer function $T(\vec{v}, t_e)/T(\vec{v}, 0)$ and $\delta(\vec{v}, t_e)/\delta(\vec{v}, 0)$ ratios in imaging the same harmonic test object. Taking into account (7.2.59, 61, 31), we have

$$\frac{\delta(\vec{v}, t_e)}{\delta(\vec{v}, 0)} = \frac{T(\vec{v}, t_e)}{T(\vec{v}, 0)}\sqrt{\frac{a(\vec{v} = 0, t_e)}{a(\vec{v} = 0, 0)}}\,. \tag{7.2.76}$$

The calculations have been made for the medium with the scattering indicatrix $x(\beta) = 2a\exp(-a\beta)/\beta$. In this case,

$$x'(vz) = \frac{\ln(vz/a + \sqrt{1 + (vz/a)^2})}{vz/a}\,,$$

$$x_2'(vz) = \frac{2 + (vz/a)^2}{a^2[1 + (vz/a)^2]^{3/2}}\,. \tag{7.2.77}$$

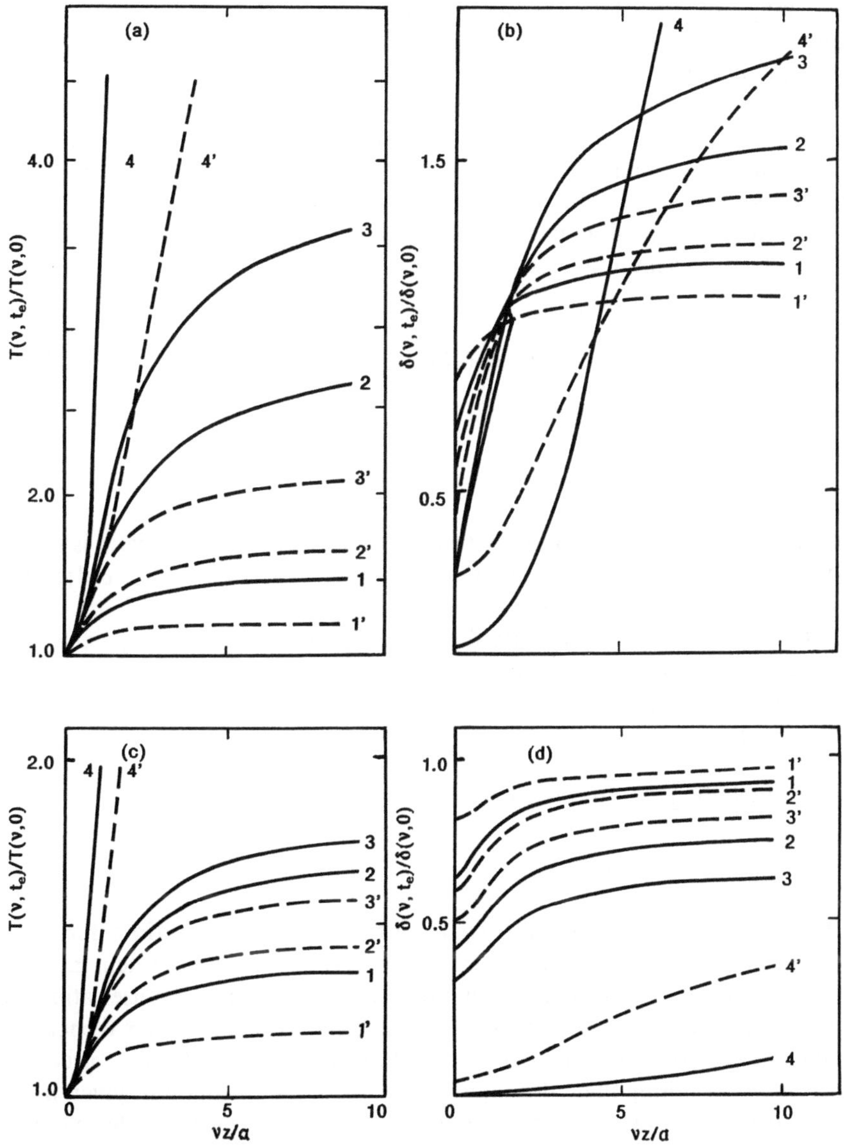


**Fig. 7.3.** MTF $(a, c)$ and signal-to-noise ratio $(b, d)$ in the information transfer $(a, b)$ and active observation $(c, d)$ systems of time-averaged holography at $\sigma z = 7$ (curves 1–4), 3 (curves 1′–4′); $u = 0.2$ (curves 1, 1′), 0.6 (curves 2, 2′), 1 (curves 3, 3′) and $\infty$ (curves 4, 4′)

As seen from the figure, the system MTF grows with the averaging time $t_e$ at all space frequencies, i.e., the image contrast is improved. This is due, as already noted, to the cut-off of the radiation portion which is scattered on the path from the object to the hologram by moving scatterers, and is incoherently recorded on the hologram. However, this causes signal energy losses, i.e., photon number decrease as well as contrast enhancement. Therefore, the signal-to-noise ratio is more intricately dependent on $t_e$. It will be recalled that the quantity $\delta$ is a function of both the contrast and the photon number $\langle N \rangle$ (7.2.31). In the information transfer system, the effect of the contrast enhancement at large spatial frequencies is more pronounced and, therefore, the signal-to-noise ratio grows with $t_e$. At small $vz$, the scattered light contains rather a large amount of information about the object. So, in spite of the contrast increase, the energy losses prove to be predominant and cause a reduction in $\delta$. This is most evident at $t_e \to \infty$. In this limiting case,

$$T(\vec{v}, \infty) = 1 , \qquad T(\vec{v}, 0) = \mathrm{e}^{-\sigma z[1-x'(vz)]} , \tag{7.2.78}$$

$$a(\vec{v} = 0, \infty) = \mathrm{e}^{-\varepsilon z} , \qquad a(\vec{v} = 0, 0) = \mathrm{e}^{-(\varepsilon-\sigma)z} , \tag{7.2.79}$$

and from (7.2.76), we have

$$\frac{\delta(\vec{v}, \infty)}{\delta(\vec{v}, 0)} = \mathrm{e}^{\sigma z[1/2 - x'(vz)]} . \tag{7.2.80}$$

At spatial frequency $v$, for which $x'(vz) < 0.5$, the signal-to-noise ratio is seen to increase with the averaging time. On the contrary, at $x'(vz) > 0.5$, $\delta$ decreases with increasing averaging time.

In the general case, as seen from Fig. 7.3ab, for each spatial frequency, there is an optimum averaging time $t_e^{\mathrm{opt}}$ at which $\delta(\vec{v}, t_e^{\mathrm{opt}})$ is maximum.

In the active viewing system,

$$a(\vec{v} = 0, \infty) = \mathrm{e}^{-2\varepsilon z} , \qquad a(\vec{v} = 0, 0) = \mathrm{e}^{-2(\varepsilon-\sigma)z} , \tag{7.2.81}$$

and, by including (7.2.76, 78), we have

$$\frac{\delta(\vec{v}, \infty)}{\delta(\vec{v}, 0)} = \mathrm{e}^{-\sigma z x'(vz)} . \tag{7.2.82}$$

It follows from (7.2.78, 82) and Fig. 7.3cd that, in contrast to the information transfer scheme, time-average holography in the active viewing system improves the image contrast, while it causes a reduction in the signal-to-noise ratio at all spatial frequencies. This is attributed to additional losses in the signal energy by virtue of radiation scattering along the path from the source to the object.

It is worth noting that, although the results of the calculations given in Fig. 7.3 apply to the particular case of the Brownian motion of particles, the inferences made are valid for any nonstationary medium. This is, at least, supported by the fact that (7.2.79, 80, 82) for the limiting case $t_e \to \infty$ can be used for any scatterer motion.

Thus, if the radiation energy arriving from the object of observation to the hologram is sufficiently high so that the signal-to-noise ratio $\delta$ essentially exceeds the threshold $\delta^*$, then the time-average holography always seems to be reliable. It increases the image contrast and, hence, improves its quality and vision range. In the case of a relatively weak valid signal, when the photon number $\langle N \rangle$ is small and the signal-to-noise ratio $\delta \approx \delta^*$, the time-average holography may both improve and degenerate the image quality. Then, the above relations provide estimates for the optimum averaging time $t_e$, depending on the viewing scheme, optical and dynamic characteristics of the medium, and the spatial frequency spectrum of the object.

Furthermore, in analysing the active vision system, we have disregarded the backscattered noise (BSN). At large BSN, the cut-off is always efficient (Sect. 5.3.4). The required averaging time $t_e'$ is far below the time necessary for an essential improvement in the system MTF to be achieved. The quantity $t_e'$ can easily be evaluated in the single backscattering approximation. For example, in the case of Brownian motion, from the condition $4(2\pi/\lambda)^2\mathscr{D}t_e' \sin^2(\beta/2) \sim 1$ [7.38], at $\beta = \pi$, we have $t_e' \sim [4(2\pi/\lambda)^2\mathscr{D}]^{-1}$. Thus, for scattering in water, at $\lambda = 0.5\ \mu$m and particle radius $a \approx 1\ \mu$m, $t_e' = 10^{-2}$ s, whereas the characteristic averaging time at which the contrast can essentially be improved is $t_e \approx 0.5$ s ($u \approx 1$).

# 8 Viewing in Atmosphere

In this chapter, which introduces the third part of the book, aspects of image transfer theory are applied to study the characteristic properties of viewing in some important, albeit specific, scattering media.

There is no need to emphasize the importance of the role played by viewing in atmosphere in everyday life. In relatively pure atmosphere, we can see objects tens of kilometers away. Scattering is insignificant and it is atmospheric turbulence that is mainly responsible for the deterioration in image quality. This problem has been elaborated in the literature and will not feature in our discussion. In atmospheric aerosols, as well as in conditions of fog and cloud, the range of vision can be reckoned in kilometers, or hundreds, and even tens of meters. Multiple light scattering by atmospheric aerosols and water droplets plays a key role. The specific features of object viewing under such conditions are dealt with in this chapter. The relevant information on the optical parameters of atmospheric aerosols and clouds will be given in Sect. 8.1.

From the scientific point of view, we shall single out three problems. The first is the visibility of self-luminous objects, or radiation sources, for example, the lights of an airport runway, laser beacons and landing searchlights, vehicle headlights and so on. The second problem concerns object visibility in sunlight. This is also the classical problem of determining the meteorological visibility range (MVR) measured daily in all the world's airports. In these two problems, we are essentially concerned with the detection (generally visual) of an object, rather than with the identification of object details. A more precise degree of viewing in atmosphere, i.e., where the identification of object detail is involved, constitutes the third problem.

## 8.1 Optical Parameters of the Atmosphere

Our current knowledge of the optical properties of the atmosphere relies upon vast experimental data derived from consistent studies over many years, using ground-based stations and observatories, balloons and satellites, lidar sounding, etc. This information is difficult to review. Fortunately, it is classified and accumulated in different atmospheric models, viz. climatological and optico-location models [8.1–10], which highlight the optical properties of this complex medium which varies with time and space.

### 8.1.1 Cloudless Atmosphere

Current models of cloudless atmosphere include information on the altitude concentration profile and the microphysical and chemical properties of scattering and absorbing components, taking into account regional, meteorological and other specific features [8.1–10]. Using these data and assuming aerosol particles to be spherical in shape, the altitude and spectral dependence of optical characteristics, such as scattering and absorption coefficients and scattering indicatrices, were derived from the Mie formulas. We shall apply the most complete and up-to-date atmosphere models, viz. the climatological [8.9] and opticolocation [8.7, 8] models, to characterize the optical parameters of the atmosphere in the visible. Absorption in atmospheric gases may be neglected, with the exception of the ozone Chappuis band with a maximum around 0.6 μm. The ozone layer is encountered in the atmosphere at an altitude of about 18–20 km.

The scattering and absorbing properties of the atmosphere are formed as a set of molecular and aerosol component properties. For the simplest model of the atmosphere as a homogeneous layer, the main optical characteristics such as the total optical thickness $\tau$, single scattering albedo $\Lambda$ and scattering indicatrix are determined as

$$\tau = \tau_{\mathrm{mol}} + \tau_{\mathrm{aer}} + \tau_{\mathrm{oz}} ,$$

$$\Lambda = (\tau_{\mathrm{mol}} + \Lambda_{\mathrm{aer}}\tau_{\mathrm{aer}})/\tau ,$$

$$x(\cos\beta) = \frac{\tau_{\mathrm{mol}}x_{\mathrm{mol}}(\beta) + \Lambda_{\mathrm{aer}}\tau_{\mathrm{aer}}x_{\mathrm{aer}}(\beta)}{\tau} . \tag{8.1.1}$$

Here, $\tau_{\mathrm{mol}}$, $x_{\mathrm{mol}}(\beta)$ and $\tau_{\mathrm{aer}}$, $\Lambda_{\mathrm{aer}}$, $x_{\mathrm{aer}}(\beta)$ are the molecular and aerosol component characteristics averaged over the whole atmosphere thickness, and $\tau_{\mathrm{oz}}$ the optical ozone absorption thickness. The molecular component characteristics $\tau_{\mathrm{mol}}$ and $\tau_{\mathrm{oz}}$ are stable in comparison [8.1, 2, 10]. The molecular scattering properties were discussed briefly in Sect. 2.2, where the molecular scattering indicatrix (2.2.14) and spectral dependence $\tau_{\mathrm{mol}} \sim \lambda^{-4}$ were given. The quantities $\tau_{\mathrm{mol}}$ and $\tau_{\mathrm{oz}}$ are cited in Table 8.1 in accordance with the model [8.1]. Molecular scattering and ozone absorption may vary insignificantly according to latitude and season. This is taken into account in the updated models [8.2]. The aerosol component is very unstable. In visible, the $\tau_{\mathrm{aer}}$ range is sufficiently wide, being between 0.05 and

**Table 8.1.** Mean optical thickness $\tau_i = \int_0^\infty \varepsilon_i(z)\,dz$ of molecular and aerosol atmosphere components.

| $\lambda$ [μm] | $\tau_{\mathrm{mol}}$ [8.1] | $\tau_{\mathrm{oz}}$ [8.1] | $\tau_{\mathrm{aer}}$ [8.9] | $\tau_{\mathrm{full}}$ |
|---|---|---|---|---|
| 0.45 | 0.223 | 0.0012 | 0.28 | 0.504 |
| 0.55 | 0.098 | 0.031 | 0.225 | 0.354 |
| 0.8 | 0.0215 | 0.0034 | 0.145 | 0.150 |

1 [8.11]. The spectral $\tau_{aer}(\lambda)$ behavior is also variable, although in the majority of cases both $\tau_{aer}$ and $\tau_{mol}$ decrease with growing $\lambda$. Sometimes, the approximation $\tau_{aer}(\lambda) = \tau_{aer}(0.55)\,(\lambda/0.55)^{-\alpha}$ may be used, where $\lambda$ is measured in μm, and the Angström parameter $\alpha < 4$. Most frequently, $\alpha \approx 0$–$2$, but $\alpha \lesssim 0$ is also encountered. It should be noted that a similar approximation for $\tau_{mol}(\lambda)$ gives $\alpha \approx 4.08$ [8.7].

Some representative conditions are used in modeling an aerosol component [8.2, 4, 5, 9, 12]: the atmosphere above land and ocean, the atmosphere in different seasons, for example, in spring-summer and autumn-winter [8.4]; the atmosphere when it is unperturbed and under extreme conditions such as volcanic eruptions, dust storms, etc. The regional classification is more detailed: country, urban, arid, arctic, oceanic and coastal maritime, and so on. For example, the state of the unperturbed atmosphere, i.e., when there is no great aerosol formation, above the continent is characterized by $\tau_{aer}(\lambda = 0.55\ \mu m) \lesssim 0.2$ and the monotonic quasi-neutral function $\tau_{aer}(\lambda)$. The appropriate $\tau_{aer}$ values in the model of [8.9] are listed in Table 8.1. The total optical thickness of an unperturbed atmosphere is about 0.35 at $\lambda = 0.55$ μm. The generalization of the experimental data on the state of the atmosphere above the ocean [8.12, 13] illustrates the regional classification. Three types of states are distinguished, such as central ocean atmosphere with mean $\tau_{aer} = 0.07$ and $\alpha = 0.4$; coastal atmosphere, including internal seas, with $\tau_{aer} = 0.2$ and $\alpha = 0.9$, and the atmosphere above the ocean in the presence of a sand storm in the Sahara at $\tau_{aer} = 0.42$ and $\alpha = 0.45$.

However, the applicability of the model of a homogeneous atmospheric layer is restricted, and it is often necessary to take account of the optical inhomogeneity of the atmosphere. Such an inhomogeneity is associated with the gravity effect, and with genetic reasons such as the different sources and conditions of aerosol formation at different atmospheric levels. Since the horizontal variations of the optical characteristics in the atmosphere are much smoother than the vertical ones, only vertical stratification is considered in models. One more simplification of models arises from the assumption that, within a certain (more or less thick) atmospheric layer, the aerosol composition remains constant, i.e., the size distribution function $f_i(a)$, refractive index $m_i = n_i - in_i'$ and $N_i/N$ relative number concentration are valid for each $i$-th type of particles. The only parameter which varies with the altitude $h$ within each layer is the total concentration $N(h)$. Depending on the different nature of the external and internal sources which produce aerosol particles, the atmosphere is usually stratified to include (i) a boundary layer (up to $h_1 \approx 0.3$–$0.5$ km) whose properties are controlled by the underlying surface (ocean, ground surface); (ii) a turbulent mixing layer ($h_2 \approx 2$–$6$ km); (iii) the troposphere with the upper boundary at $h_3 \approx 10$–$12$ km and (iv) the stratosphere, ($h_4 \approx 20$–$30$ km). The strata above 20–30 km introduce only small changes in the total atmosphere characteristics. The strata altitudes $h_i$ depend on the season and region. In Table 8.2, the data for the type of aerosols in each stratum, the distribution of the optical thicknesses of the molecular and

**Table 8.2.** Vertical stratification of the atmosphere.

| | Layer, upper boundary altitude | Aerosol | $\tau_{mol}$ | Mean background $\tau_{aer}$ | Extreme aerosol optical thickness |
|---|---|---|---|---|---|
| 4 | Stratosphere 20 km | Stratospheric | 0.049 | 0.002–0.003 | 0.1–0.3 volcanic outbreaks |
| 3 | Troposphere 10–12 km | Continental or | 0.058 | 0.017–0.025 | |
| 2 | Turbulent mixing layer, 2–6 km | background | | | $\tau_{aer}^{max} \leqslant 3$ dust storm in a desert<br>$\tau_{aer}^{max} \lesssim 1$ idustrial regions |
| 1 | Boundary layer | Depending on underlying surface:<br>maritime<br>urban<br>arid<br>background | 0.022 | 0.2–0.3 | $\tau_{aer}^{max} \leqslant 3$ Sahara outbreaks above the ocean<br>$\tau_{aer}^{min} \approx 0.05$ above the ocean<br>$\tau_{aer}^{min} \approx 0.12$ above the continent |

**Table 8.3.** Spectral variation of the optical characteristics of basic atmospheric aerosols (*C*, continental; *U*, urban; *M*, maritime; *BS*, background stratospheric).

| | $\varepsilon/\varepsilon(0.55)$ | | | | $\Lambda$ | | | | $g$ | | | |
|---|---|---|---|---|---|---|---|---|---|---|---|---|
| $\lambda$ [μm] | *C* | *U* | *M* | *BS* | *C* | *U* | *M* | *BS* | *C* | *U* | *M* | *BS* |
| 0.4 | 1.4 | 1.48 | 1.08 | 1.38 | 0.901 | 0.66 | 0.987 | 1 | 0.646 | 0.6 | 0.744 | 0.739 |
| 0.55 | 1 | 1 | 1 | 1 | 0.891 | 0.647 | 0.989 | 1 | 0.637 | 0.591 | 0.743 | 0.726 |
| 0.86 | 0.577 | 0.542 | 0.91 | 0.469 | 0.841 | 0.588 | 0.986 | 1 | 0.633 | 0.583 | 0.754 | 0.65 |

aerosol components, and the $\tau_{aer}$ range are summarized. The main models used for different atmospheric aerosols are Continental, Urban, Maritime and Background Stratospheric. Their chemical and microphysical composition, as well as the optical properties, viz. $\Lambda$, $x(\beta)$, extinction per unit concentration calculated on their basis can be found in [8.5–9]. Table 8.3 points out the spectral variability of the main optical parameters of these aerosols. It should be noted that the functions $\varepsilon(\lambda)$ and $\Lambda(\lambda)$, as shown in [8.7], may have more distinct spectral selectivity than in global models. This is a peculiarity of aerosols with a high concentration of soot or hepatite ($Fe_2O_3$). The degree of asymmetry of the scattering indicatrix is illustrated by the data in Table 8.4. It gives the values of $\langle g \rangle$ for different atmospheric layers, as well as for the atmosphere as a whole (0–30 km). We have restricted our consideration to the profiles of three comparatively pure atmospheres viz. continental (excluding deserts), urban and maritime (where there is no great dust formation). The tabulated data demonstrate the main trend in the behavior of the scattering indicatrices: decreasing asymmetry

**Table 8.4.** Asymmetry parameter $\langle g \rangle$ of the scattering indicatrices averaged over the layers and across the atmosphere for the regions with different types of aerosols in the ground layer (*C*, continental; *U*, urban; *M*, maritime aerosols).

| | | | $h = 0$–$2$ km | | | $h = 0$–$30$ km | | |
|---|---|---|---|---|---|---|---|---|
| $\lambda$ [μm] | $h = 12$–$30$ km | $h = 2$–$12$ km | *C* | *U* | *M* | *C* | *U* | *M* |
| 0.4 | 0.05 | 0.083 | 0.494 | 0.555 | 0.303 | 0.29 | 0.44 | 0.14 |
| 0.55 | 0.115 | 0.176 | 0.569 | 0.572 | 0.52 | 0.42 | 0.52 | 0.31 |
| 0.86 | 0.23 | 0.356 | 0.611 | 0.577 | 0.7 | 0.556 | 0.56 | 0.57 |

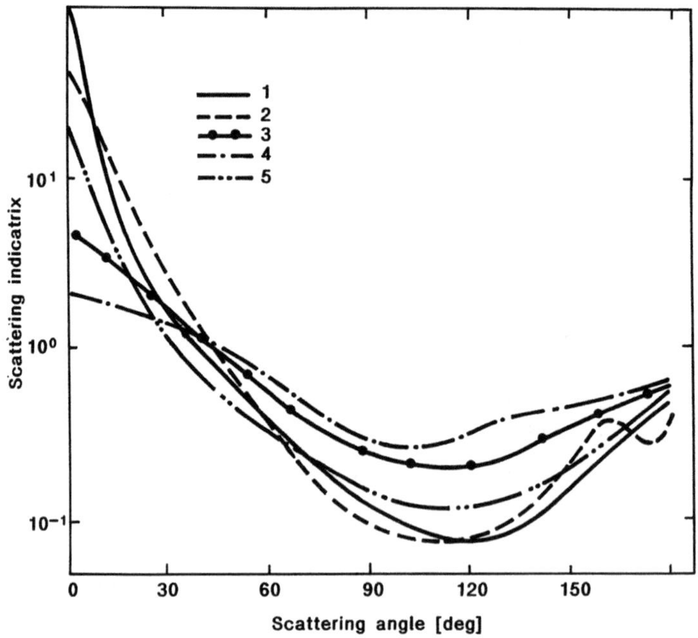


**Fig. 8.1.** Variation with altitude of scattering indicatrices for middle latitudes in summer. Curve 1 is the scattering indicatrix for a boundary layer model; curve 2, for a turbulent mixing layer; curve 3, for the upper troposphere; curve 4, for the regions above and below a stratospheric layer; curve 5, for a stratospheric layer

degree with increasing altitude $h$ and decreasing $\lambda$ which is, in the main, due to the increasing contribution of the molecular component to the total indicatrix. One more example of the variation of $x(\beta)$ with altitude is shown in Fig. 8.1. In this case (summer, middle latitudes [8.5, 6]), the appropriate $g(h)$ behavior is somewhat different: the asymmetry parameter $g$ grows with $h$ below the stratospheric layer.

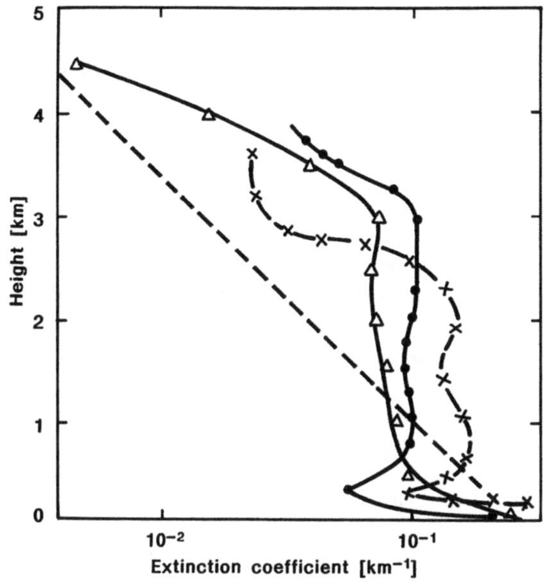


**Fig. 8.2.** Vertical variation of the extinction coefficient. Comparison of lidar sensing data [8.8] for middle latitudes under anticyclonic conditions with the *Elterman* model [8.1] at $\lambda = 0.69\ \mu m$

In Fig. 8.2., the experimental altitude extinction coefficient profile is compared with the *Elterman* model [8.1]. At ground level, aerosol extinction, even for pure atmosphere, is much in excess of the molecular one and completely determines the horizontal visibility range $S_m = 3.5/\sigma$ (Sect. 8.3). For the pure atmosphere above the continent, $S_m \approx 50$ km. Unfortunately, the anthropogenic effect on the atmosphere is such that urban inhabitants are deprived of such pure air.

The information presented highlights the properties of the atmosphere as a scattering medium necessary to model viewing along vertical, oblique and horizontal paths, including air- and spacecraft observations.

### 8.1.2 Cloud and Fog

For the cenral European region of the Soviet Union, the probability of cloudiness is about 60%. Depending on the season, the altitude of the lower cloud boundary ranges between 0.5 and 2 km, its thickness being, on average, from several hundreds of meters to some kilometers. Low cloudiness is also encountered. If the altitude $z_l$ of the lower cloud boundary is below 200 m, then the extinction coefficient $\varepsilon$ increases with decreasing altitude [8.14]. At the Earth's surface, the meteorological visibility range in cold weather is observed to be

$$S_m = 15.3z_l + 1280\ \mathrm{m}\,. \tag{8.1.2}$$

For clouds of Sc type with $z_l > 200$ m, the haze of the layer under a cloud does not reach the Earth and, hence, the $z_l$ variation does not affect the horizontal visibility range at ground level. In this case, the extinction coefficient of the lower cloud layer is almost independent of the altitude and is about 1 km$^{-1}$. In multiple layers of clouds, the extinction coefficient between the clouds is frequently the same as in a cloudless atmosphere. Data on the microstructure of clouds and the optical characteristics of the atmosphere can be found elsewhere [8.15].

Let us consider the dependence of optical cloud parameters on their microstructure.

Particle size distribution in cloud and fog is frequently approximated by the bimodal function [8.16]

$$\begin{aligned} f(a) &= f_l(a) + f_{sm}(a) \qquad \text{at } a_1 < a < a_l \\ &= f_l(a) \qquad \text{at } a > a_l \end{aligned} \tag{8.1.3}$$

where $f_l(a) = A_l a^{\mu} \exp(-\mu a/a_m)$, $f_{sm}(a) = A_{sm} a^{-\nu}$. The gamma distribution $f_l(a)$ gives a good description of the large-size fraction, whereas the Junge inverse power law is used to represent the size distribution of water-immersed condensation centers in clouds, and of particles in the atmospheric hazes. Here, $A_l$ and $A_{sm}$ are determined by the normalization of the function $f(a)$ and the fraction proportion, $a_1 \approx 0.5$ µm; the parameters $\nu$ and $a_l$, modal radius $a_m$ and quantity $\mu$, specify the gamma distribution half-width range as $3 \leqslant \nu \leqslant 4.5$; $3\ \mu\mathrm{m} \leqslant a_l \leqslant 10\ \mu\mathrm{m}$, $1\ \mu\mathrm{m} \leqslant a_m \leqslant 10\ \mu\mathrm{m}$, $2 \leqslant \mu \leqslant 15$ [8.16, 17].

For typical situations, the range of $a_m$ and $\mu$ characterizing the large-size fraction is essentially narrower. Following the data of [8.11] for the main types of clouds (Cumulus, Cu; Cumulo-nimbus, Cb; Stratus, St; Nimbo-stratus, Ns; Alto-stratus, As and Alto-cumulus, Ac), $a_m = 5$–$6$ µm, $\mu = 2$–$3$. This is the standard international classification for types of clouds [8.11]. The cloud liquid-water content $q$ (water weight per 1 m$^3$) is 0.1–0.2 g/m$^3$ and reaches 1.2 g/m$^3$ only in heavy cumulus clouds. For fogs $q \approx 0.1$ g/m$^3$, $a_m \approx 5$ µm; $\mu \approx 3$ for advective and $\mu \approx 6$ for radiative fogs, respectively.

The scattering coefficient is the main parameter which undergoes the greatest changes with variation in the microstructure and is, in the main, responsible for the characteristics specifying image transfer in clouds. It may be calculated using (2.2.13), but we shall write it in another form,

$$\sigma = \frac{3}{4}\frac{C_V \int_0^{\infty} Q_s(a,\lambda)a^2 f(a)\,da}{\int_0^{\infty} a^3 f(a)\,da}\,. \tag{8.1.4}$$

**Table 8.5.** Scattering coefficient of cumulative fraction $\sigma_{sm}[m^{-1}]$ ($C_V^{sm} = 10^{-6}$, $\lambda = 0.63$ μm).

| $a_e$ | $\nu$ = 3 | 3.5 | 3.8 | 4.1 | 4.5 |
|---|---|---|---|---|---|
| 3 | 1.38 | 1.65 | 1.86 | 2.04 | 2.34 |
| 5 | 0.93 | 1.23 | 1.47 | 1.74 | 2.10 |
| 10 | 0.57 | 0.87 | 1.11 | 1.41 | 1.87 |

Here, $Q_s$ is the efficiency factor for scattering, and $C_V$ the volumetric concentration of scatterers, i.e., the volume fraction occupied by particles. At $C_V = 10^{-6}$, the liquid-water content $q = 1$ g/m$^3$. In Table 8.5, the scattering coefficient of the cumulative (small particle) fraction $\sigma_{sm}[m^{-1}]$ as calculated by the Mie formulas is given at $q = 1$ g/m$^3$. The scattering indicatrices of the cumulative fraction are not very extended. Although the particle number $N_{sm}$ of the cumulative fraction may essentially be greater than the number of large particles, generally $C_V^{sm} \ll C_V^l$ [8.16]. It is regarded, as a rule, that the contribution of the cumulative fraction to the optical characteristics of the clouds is small. It is worth noting that the small-size fraction of the particle-size distribution has not been adequately studied. As shown by experimental data [8.16], at least for some cases $\sigma_{sm}/\sigma$ attains values of 0.1–0.2, but does not exceed 0.3.

Information on optical cloud parameters is presented in several studies [8.11, 17, 19]. The contribution of the cumulative fraction to the optical cloud characteristics is usually neglected. Moreover, the microstructure of the large-size fraction is described not only by the gamma-distribution, but also by the log-normal and modified gamma-distributions. Optical parameters are derived from the Mie formulas. In Fig. 8.3, the scattering indicatrices are given for different clouds at $\lambda \approx 0.7$ μm [8.8].

Detailed tables for $\sigma$, $\Lambda$ and $x(\beta)$ can be found, for example, in [8.19]. However, another approach is possible. To describe the optical properties of the large-size fraction, we may use the analytical solutions obtained assuming scattering by large spheres (Sect. 2.2 and Table 2.1). For the particle size distribution $f_1(a)$, we have [8.20]

$$\begin{aligned} \varepsilon &= 1.5\, C_V/a_{32}\,, \\ g &= g_0(n) - 1.8/\rho_{21}\,, \\ \Lambda &= 1 - f(n) 4\pi n' a_{32}/\lambda\,. \end{aligned} \tag{8.1.5}$$

It should be noted that $g_0(n)$ is the asymptotic asymmetry parameter at $\rho \to \infty$, its values for different $n$ being cited in Table 2.2; $f(n) = 0.666n^2[1 - (1 - n^{-2})^{3/2}]$. At $n = 1.33$, $g_0 = 0.884$ and $f(n) = 0.934$. By definition,

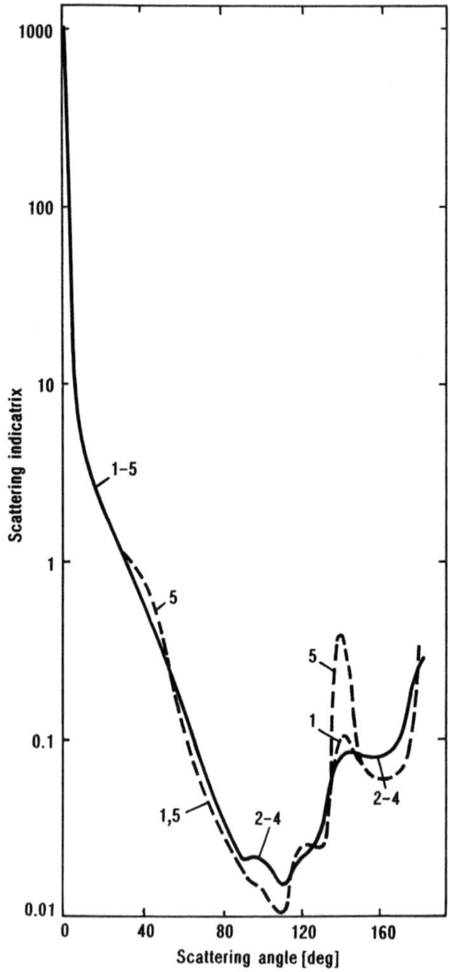


**Fig. 8.3.** Scattering indicatrices for the clouds of Cumulonimbus, Cb (curve 1); Cumulus, Cu (curve 2), Strato-cumulus, Sc (curve 3); Stratus, St (curve 4) and Cumulus congestus, Cu. cong (curve 5)

$$a_{32} = \frac{\int_0^\infty a^3 f(a)\,da}{\int_0^\infty a^2 f(a)\,da},$$

$$\rho_{21} = \frac{2\pi a_{21}}{\lambda} = \frac{2\pi}{\lambda}\frac{\int_0^\infty a^2 f(a)\,da}{\int_0^\infty a f(a)\,da}. \tag{8.1.6}$$

For the gamma-distribution, $a_{32} = a_m(1 + 3/\mu)$, $\rho_{21} = 2\pi a_m(1 + 2/\mu)/\lambda$. The error in $\varepsilon$ calculated using (8.1.5) is less than 10% at $\rho_{32} > 35$; that in the $\Lambda$ and $g$ calculation, less than 1.5% at $n > 1.3$, $\mu \in 2$–$10$ and $n' \leqslant 10^{-3}$ [8.20]. As seen from the above representative characteristics of the large-size cloud fraction, $a_{32} \in 10$–$25$, i.e., at $\lambda \approx 0.55$ μm $\rho_{32} \in 100$–$250$. Thus, for clouds in the visible ($n = 1.33$, $n' = 0$), we have

$$\Lambda = 1\,,$$

$$\varepsilon = 1.5\frac{C_V}{a_m(1 + 3/\mu)}\,,$$

$$g = 0.88 - \frac{1.8}{\rho_m(1 + 2/\mu)}\,. \tag{8.1.7}$$

These simple formulas provide an analysis of the microstructure effect on the optical cloud parameters without applying extensive Mie calculations.

For a typical range of microstructure parameters and liquid-water content of clouds and fog, in the visible range of the spectrum $\varepsilon$ variation is 6–30 km$^{-1}$. For heavy cumulus clouds, $\varepsilon \approx 150$ km$^{-1}$. In the visible and near IR, $\varepsilon$ is independent of the wave-length.

The above relations for $\varepsilon$, $\Lambda$ and $g$ offer a simple solution to many problems in cloud optics. We note that the integral multiple scattering characteristics depend only on two generalized scattering medium parameters, viz. $\sigma(1 - g)z$ and $(1 - \Lambda)/\Lambda(1 - g)$ (Sect. 4.7). These are exactly the parameters on which the optical cloud properties are dependent.

Let us calculate, for example, the transmission of a thick cloud in the visible ($\Lambda = 1$), using (3.5.3) and (8.1.7). We obtain

$$t(\mu_0) = \frac{u_0(\mu_0)}{1 + 0.2W(1 - 10/\rho_{21})/a_{32}}\,, \tag{8.1.8}$$

$u_0(\mu_0) = 3(1 + 2\mu_0)/7$. This useful relation connects, in a straightforward manner, the cloud transmission coefficient with its liquid-water content $W = C_V z_0$, and microstructure parameters $\rho_{21}$ and $a_{32}$.

For typical cloud models, $\Lambda = 1$ is set in the visible as there is no water absorption there. Measurements, however, show that even in extremely thick clouds, the albedo does not exceed 0.7–0.8. Using these values in

$$1 - \Lambda = 3(1 - g)\left[\frac{\ln r(\mu_0)}{4u_0(\mu_0)}\right]^2$$

which results from (3.3.13; 3.2.11), we obtain the estimate $1 - \Lambda \approx 0.002$–$0.004$. There is no general agreement on the origin of this absorption, and it is frequently related to aerosol absorption.

We have estimated cloud transmission and absorption using the asymptotic formulas given in Sect. 3.4.5 and the simple relation for $\varepsilon$, $g$, $\Lambda$ (formulas (8.1.7)).

The combination of these two approaches, viz. the transfer equation asymptotics and geometrooptical approximation for optical parameters, suggests an easy solution to a large number of other problems in cloud optics.

## 8.2 Light Source Visibility

Light from powerful radiation sources (self-luminous objects) can be detected some kilometers away, even in cloudy and foggy atmosphere. The direction to the light source will be determined when it is possible to distinguish its radiation (valid signal) from the background. Generally, the maximum light signal power $P_s$ registered in the direction of the light source and the background power $P_b$ are compared. The quantity

$$k = \frac{P_s - P_b}{P_b} = \frac{P_s}{P_b} - 1 \tag{8.2.1}$$

characterizes the observed object contrast. The natural background being negligible, $P_b$ is the power of the radiation scattered from the source itself.

We first consider the situation where the souce emits a relatively wide beam with small angular divergence. These may be the sun, spaceship laser, etc. We suppose that the source energy is sufficiently large so that the object visibility range is limited due to the contrast decrease to $k \leqslant k^*$, where $k^*$ is the threshold contrast of the receiving system (Sect. 5.1). Moreover, we assume, as in Sect. 6.5, that the valid signal $P_s - P_b$ is determined by the unscattered radiation. In this case, the source detection range may be extremely large at small angular divergence of the source. In fact, the scattered light radiance $I^s \sim I_0 \Omega_s$, where $I_0$ is the incident radiance and $\Omega_s$ the incident beam solid angle. Hence, by reducing $\Omega_s$, at any given optical depth $\tau$ we can attain a situation such that the unscattered light radiance $I^0 = I_0 \exp(-\tau)$ will exceed the that of the scattered light $I^s$.

To assess the ratio $P_s/P_b$ and the limiting detection depth for a source (self-luminous object) requires a knowledge of the angular distribution of the scattered radiation in the angular region of the unscattered light. Such an estimation, based on the experimental data for model scattering media, was made as far back as 1966 [8.18]. We shall, however, use the theoretical results (Chaps. 3 and 4) to compare the limiting detection depths $\tau^*$ of different objects.

Choosing the receiving solid angle $\Omega_{\rm rec}$ as less or equal to the solid angle $\Omega_s$ may give maximum $\tau^*$. Disregarding the light background from foreign radiation sources, for the light flux in the focal plane of the receiver objective, we obtain $P_s - P_b = I_0 \Omega_{\rm rec} \Sigma_{\rm rec} \exp(-\tau)$, $\Sigma_{\rm rec}$ being the object area, and $P_b = I_0 \Omega_s \Omega_{\rm rec} \Sigma_{\rm rec} I_n(\tau, \mu = 1)$. Here, $I_n(\tau, \mu = 1)$ is the radiance of the scattered light in the direction normal to the layer at the point of observation normalized to the radiant emittance $I_0 \Omega_s$ of an infinitely wide radiation source. In the asymptotic theory (Sect. 3.5) and within the diffusion approximation,

$$I_n(\tau, \mu = 1) = \frac{S(\tau)u_0(1)}{\pi} \approx \frac{81}{49\pi} \frac{\sinh 4q\gamma}{\sinh[\gamma(\tau + 4q)]} . \tag{8.2.2}$$

Here, we assume, for the sake of certainty, that both the source and the photo-dedector are located at the opposite plane layer boundaries. As seen from (4.5.7) and Table 4.4, within the SADA

$$I_n(\tau, \mu = 1) = \frac{S(\tau)}{2\pi V_\theta} = \frac{1 - \Lambda}{2\pi s} \frac{e^{-(1-\Lambda)\tau}}{(u_0 + \tanh s\tau)\cosh s\tau} , \tag{8.2.3}$$

where

$$u_0 = \frac{\Omega_s}{4\pi} \frac{1 - \Lambda}{\beta_2 \Lambda} = \frac{\Omega_s}{4\pi} \frac{1 - \Lambda}{\Lambda(1 - g)} ,$$

$$s = \sqrt{\Lambda(1 - \Lambda)\beta_2/2} = \sqrt{\Lambda(1 - \Lambda)(1 - g)} .$$

For large depths $\tau$, the radiance of the scattered light obeys the exponential law with the asymptotic attenuation coefficient $\gamma$. We recall that in the SADA $\gamma = 1 - \Lambda + s$. As the contrast

$$k = \frac{e^{-\tau}}{I_n(\tau, \mu = 1)\Omega_s} , \tag{8.2.4}$$

its dependence on the optical depth $\tau$ is approximately $k \sim \exp[-(1 - \gamma)\tau]$.

Given a certain value for the threshold contrast $k^*$ and using the above expressions for $I_n(\tau, \mu = 1)$, from (8.2.4), we obtain the limiting source detection range

$$\tau^* = -\frac{1}{1 - \gamma} \ln[k^* \Omega_s I_n(\tau^*, \mu = 1)e^{\gamma\tau^*}] . \tag{8.2.4'}$$

Equation (8.2.4') is transcendental. However, it may be regarded as a convenient expression for an iterative $\tau^*$ calculation. For example, within the SADA, at large $\gamma\tau$

$$I_n(\tau, \mu = 1)e^{\gamma\tau} \approx (u_0 + \tanh s\tau)^{-1} \approx (1 + u_0)^{-1} \approx 1 ,$$

and even a zero approximation

$$\tau^* \approx -\frac{\ln(k^* \Omega_s)}{\Lambda - s}$$

gives satisfactory accuracy.

The threshold contrast is set at $k^* = 0.03$. For this case, the values of $\tau^*$ are listed in Table 8.6 for two scattering indicatrices: $g = 0.867$ (the first value) and $g = 0.967$ (the second value). The unbracketed numbers are DA calculated, while the bracketed ones are obtained within the SADA approximation. For the atmosphere, in the visible $\Lambda \approx 1$. Therefore, we shall consider exactly this variant. As seen from the table, the 300-fold decrease in the angular divergence of the

**Table 8.6.** Values of $\tau^*$ at $k^* = 0.03$ for finite and infinite self-luminous sources.

| | Infinite source $2\gamma_s = 2\sqrt{\Omega_s/\pi}$, $\Omega_{rec} \leqslant \Omega_s$ | | | Finite source $2\gamma_{rec} = 2'$ | | |
|---|---|---|---|---|---|---|
| | $A$ | | | | | |
| $2\gamma_s$ | 1 | 0.9 | 0.7 | 1 | 0.9 | 0.7 |
| 2′ | 20.4<br>19.7 | 24.9(25.1)<br>22.8(22.6) | 35.1(36)<br>30.4(30.0) | 37.1<br>36.8 | 43.9(44.1)<br>40.4(39.6) | 60(61)<br>52.2(52.2) |
| 30′ | 14.8<br>14.2 | 17.9(18.3)<br>16.4(16.1) | 25.3(25.9)<br>21.8(21.2) | 31.6<br>31.3 | 37.1(37.3)<br>34.2(33.1) | 50.4(52.1)<br>43.9(43.4) |
| 1° | 13.3<br>12.7 | 16.1(16.5)<br>14.8(14.4) | 22.7(23.3)<br>19.7(18.9) | 30.2<br>29.9 | 35.4(35.6)<br>32.5(31.4) | 48.0(49.6)<br>41.8(41.2) |
| 10° | 8.5<br>8.1 | 10.2(10.6)<br>9.3(8.7) | 14.3(14.7)<br>12.4(11.5) | 25.5<br>25.3 | 29.7(29.8)<br>27.2(26.0) | 39.9(41.2)<br>34.7(34.0) |
| 180° | — | — | — | 20.6<br>20.3 | 23.6(23.3)<br>21.5(21.2) | 31.4(32.5)<br>27.2(27.8) |

source radiation results in a double increase in $\tau^*$. The variation of the scattering indicatrix within the range of interest does not actually have an impact on the limiting depth of the source detection.

We note that $\tau^*$ depends on the product $k^*\Omega_s$, rather than on $k^*$ and $\Omega_s$ separately (8.2.4′). Therefore, the data of Table 8.6 provide an estimation of the $k^*$ impact on the limiting range of detection of an infinitely wide self-luminous object.

We now proceed to the detection of finite sources (lamps, beacon lights, searchlights, lasers, etc.) used for navigation, communication, information transfer, and so on. Some experimental $\tau^*$ data for such objects with directional and diffuse radiance diagrams were obtained in [8.21–23]. However, as before, it is more convenient here to use the results of calculations.

At large distances $l$, the observed angular dimensions of a target are, generally, less than $\Omega_s$ and $\Omega_{rec}$. Moreover, because of the radiation divergence, the light spot of unscattered beams exceeds the area of the receiver objective. Then, the fluxes of the unscattered light $P^{(0)} = I_0 \exp(-\tau)\Sigma_s\Sigma_{rec}/l^2$ and scattered light $P^s = I_n(\rho = 0, \tau, \mu = 1)\Sigma_{rec}\Omega_{rec}\iint_{\Omega_s} I_0(\theta)\Sigma_s \cos\theta\, d\vec{n}$ are received. Here, $I_n(\rho = 0, \tau, \mu = 1)$ is the scattered light radiance in the normal direction at the point of observation $(\tau, \rho = 0)$ related to the flux emitted by the light source. In this case, setting $I_0 = \text{const}$, we have

$$k = \frac{e^{-\tau}}{\pi^2\gamma_{rec}^2 l^2 I_n(\rho = 0, \tau, \mu = 1)\sin^2\gamma_s} \tag{8.2.5}$$

Here, it is considered that for a small receiving angle, $\Omega_{rec} = \pi\gamma_{rec}^2$, $2\gamma_s$ and $2\gamma_{rec}$ being the aperture angles of the source and receiver.

Within the diffusion approximation and assuming the Gaussian form of the point spread function,

$$I_n(\rho = 0, \tau, \mu = 1) = \frac{\tau^2}{l^2} \frac{I_n(\tau, \mu = 1)}{2\pi V_\rho},$$

where the variance of the spread function (Table 4.1)

$$V_\rho = \frac{\tau + 4q}{\gamma} \operatorname{cotanh}[\gamma(\tau + 4q)] - \frac{5q}{\gamma} \operatorname{cotanh} 5q\gamma .$$

For the sake of simplicity, we disregard the $\gamma_s$ dependence of $I_n(\rho = 0, \tau, \mu = 1)$.

Within the SADA,

$$l^2 I_n(\rho = 0, \tau, \mu = 1) = \frac{S(\tau)}{2\pi V_\rho} \frac{1}{2\pi V_{\rho\theta}},$$

the functions $S(\tau)$, $V_\rho$ and $V_{\rho\theta}$ being determined by the formulas given in Table 4.4. In this case, the dependence of the scattered light radiance due to a finite source on the optical depth $\tau$ is also mainly specified by the factor $S(\tau)$. Therefore, taking into account (8.2.5), it is convenient to write the equation for $\tau^*$ as

$$\tau^* = -\frac{1}{1-\gamma} \ln[l^2 I_n(\rho = 0, \tau, \mu = 1) e^{\gamma\tau^*} k^* \pi^2 \gamma_{\text{rec}}^2 \sin^2 \gamma_s] . \tag{8.2.6}$$

The DA and SADA values of $\tau^*$ are cited in Table 8.6. The use of a finite light source is seen to increase $\tau^*$. This increase is especially pronounced for sources with wider angular diagrams. In fact, we may consider $\tau^* \approx 35$–$40$ to be the limiting values for atmospheric observations in the visible range of the spectrum.

## 8.3 Object Visibility in Sunlight

The visibility range of an object illuminated by an external source is smaller than that of a self-luminous object. The source radiation scattered by the medium generates additional background, thus deteriorating the object contrast.

Problems connected with object visibility in daylight have been studied in a number of monographs [8.24–28]. Among them, the classical problem of visibility along a horizontal path considered as far back as 1925 by *Koshmider* [8.29] is thought to be the most thoroughly investigated.

### 8.3.1 Meteorological Visibility Range

Let a small black object (albedo $A = 0$) be at an optical distance $\tau$ from the viewer. In the absence of an object, the radiance of the scattered sunlight on the photoreceiver (background radiance in the $\vec{n}_b$ direction) is

$$I(\vec{n}_b) = \int_0^\infty B(\tau', \vec{n}_b) e^{-\tau'} d\tau' . \tag{8.3.1}$$

Here, $B(\tau', \vec{n}_b)$ is the source function due to the multiple scattering of solar radiation in the atmosphere. It is assumed in (8.3.1) that the optical thickness of the atmosphere in the horizontal direction is infinite. If the photodetector axis is oriented to the black object, then the detected radiance is

$$I(\vec{n}_s) = \int_0^\tau B(\tau', \vec{n}_s) e^{-\tau'} d\tau' . \tag{8.3.2}$$

The observation path being homogeneous and the object quite small so that it does not distort the light field, we have

$$B(\tau', \vec{n}_b) = B(\tau', \vec{n}_s) = B , \tag{8.3.3}$$

and the observed object contrast

$$k = \frac{I(\vec{n}_s) - I(\vec{n}_b)}{I(\vec{n}_b)} = -e^{-\tau} . \tag{8.3.4}$$

The visibility range depends on the system threshold contrast $k^*$ and is specified by the obvious condition $k = k^*$. Taking into account (8.3.4), we get

$$\tau^* = -\ln|k^*| . \tag{8.3.5}$$

Setting $|k^*| \approx 0.03$ (threshold vision contrast), we have

$$\tau^* = \varepsilon S_m = 3.5 . \tag{8.3.6}$$

Such visibility of the black object is termed the *meteorological visibility range.* This concept is widely used in atmospheric optics.

In deriving (8.3.4), the object was constrained to be small. In the general case, where this condition may be violated, the horizontal visibility range $\tau^*$ is found from the relation

$$|k| = |k^*| = \iint_{\Sigma_{ob}} S(\vec{r}, \tau^*) d\vec{r} , \tag{8.3.7}$$

which is inferred from the general theory of object detection discussed in Sect. 5.4.2. Here, $S(\vec{r}, \tau^*)$ is the medium PSF at the distance $\tau^*$, the domain of integration being the object surface. The quantity $\tau^*$ as a function of the optical size $\varepsilon d$ of a square object having a side $d$ and albedo $A = 0$ is plotted in Fig. 8.4. It is obtained for viewing through cloud at $\Lambda = 1$ and a scattering diagram C.1 (Sect. 2.2) for different values of the threshold contrast $k^*$. The condition that the object be small is seen to hold at $\varepsilon d/\tau^* \lesssim 0.03$–$0.05$, i.e., for a visible angular object size less than 2°.

### 8.3.2 Visibility Range in Clouds

One more problem that can be rigorously solved is related to the visibility of an object hidden behind clouds and fog, where the asymptotic deep regime is realized.

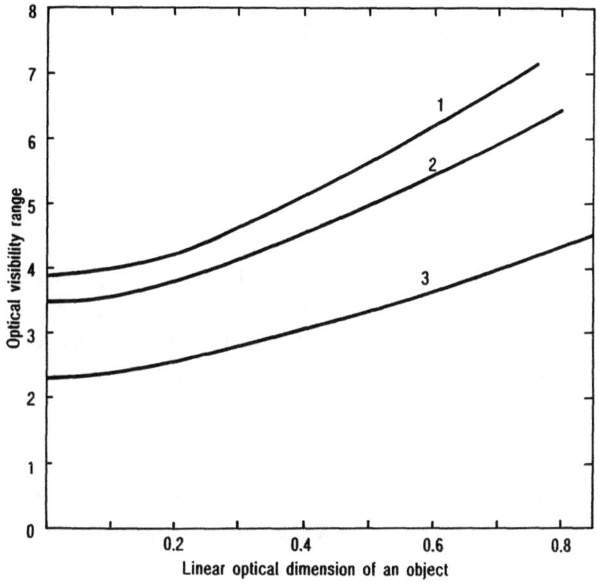


**Fig. 8.4.** Visibility range of a black object ($A_{ob} = 0$) as a function of size for threshold contrast values $k^* = 0.02$ (curve 1), 0.03 (curve 2) and 0.1 (curve 3)

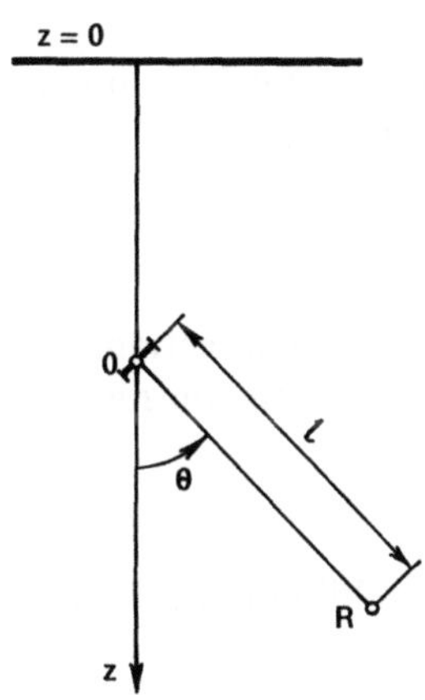


**Fig. 8.5.** Object (O) and receiver (R) positions in a scattering medium

Let an object be $z$ deep from the upper cloud boundary $l$ distant from the receiver, the object-receiver axis being normal to the object surface and forming an angle $\theta$ to the downward layer normal $z$ (Fig. 8.5). It is also assumed that the angular object sizes are larger than the angular receiver resolution. At the same time, the object is small enough not to distort the light field around it, its radiance (the light background is neglected) being proportional to $\exp(-\tau)$ ($\tau = \varepsilon l$). Given the above assumptions, the observed contrast is [8.24]

$$k = k_0 e^{-(1-\gamma\cos\theta)\tau} . \tag{8.3.8}$$

Here,

$$k_0 = \frac{AE(z, \theta + \pi)/\pi - I(z, \theta)}{I(z, \theta)} \tag{8.3.9}$$

is the true object contrast, $E(z, \theta + \pi)$ the object irradiance from the viewer's standpoint, $I(z, \theta)$ the radiance at the point O where there is no object, $A$ the object albedo, and $\gamma$ the dimensionless asymptotic attenuation coefficient. As seen from (8.3.9), the true object contrast $k_0$ is independent of $z$ since the angular radiance distribution does not depend on $z$. The quantity $k_0$ may be both positive (the object radiance is greater than the medium radiance) and negative. For a number of specific cases, expression (8.3.9) may be cast in another more convenient form in terms of the medium $r = E(z, \pi)/E(z, 0)$ and object $A$ albedos [8.24]. Given $I(z, \theta) \approx E(z, \theta)/\pi$, i.e., a weakly anisotropic angular radiance distribution, we have:

for viewing from above ($\theta = 180°$) $k_{0,\downarrow} = A/r - 1$ ,
for horizontal viewing ($\theta = 90°$) $k_{0,\rightarrow} = A - 1$ ,
for viewing from below ($\theta = 0°$) $k_{0,\uparrow} = Ar - 1$ .

The true contrast is seen to be determined by $A$ and $r$. In horizontal viewing, however, it is independent of the optical medium parameters and vanishes if the object albedo tends to unity. In very thick clouds, $r$ is close to unity. Therefore, for any viewing direction, the true contrast is

$$k_0 = A - 1 < 0 \tag{8.3.10}$$

and does not depend on the viewing direction. If in this case, $A = 1$, then $\gamma = 0$ and, as seen from (8.3.8), the observed contrast does not depend on the angle $\theta$ and is

$$k = k_0 e^{-\tau} . \tag{8.3.11}$$

Using (8.3.10, 11) the object visibility range in the cloud is found as

$$\tau^* = \ln \frac{1 - A}{|k^*|} . \tag{8.3.12}$$

In formula (8.3.12), it is assumed that $|k_0| = 1 - A \geqslant |k^*|$. Otherwise, $\tau^* = 0$. For $|k^*| = 0.03$ and at $A$ equal to 0; 0.3; 0.7; and 0.9 and 0.97, the limiting visibility range, correspondingly, amounts to 3.5; 3.1; 2.3; 1.9 and 0. We note that in conformity with (8.3.8, 10), for horizontal viewing paths ($\theta = 90°$), the relation (8.3.12) holds, irrespective of the medium properties. Setting $|k^*| = 0.03$ and assuming the object to be a black body ($A = 0$), we have $\tau^* = \varepsilon S_m = 3.5$. As expected, we have again arrived at (8.3.6).

## 8.4 Vision Characteristics in Cloud and Fog

We shall discuss some details concerning vision in cloud and fog. Their salient features are, first of all, bound up with the absence of absorption in the visible. Other important distinctions cover a wide range of medium microstructures, high spatial inhomogeneity and the time instability of the parameters. This restricts the amount and accuracy of the information available on the microphysical parameters of the medium. Such a situation rules out high accuracy in a calculation of vision characteristics and often makes it unnecessary, in contrast to a simple estimation. The latter would allow one, even with some loss in accuracy, to immediately relate the image quality characteristics to the microphysical cloud and fog parameters using simple expressions. Therefore, we shall consider possible ways of developing such an approach, as well as the general analysis [8.30].

### 8.4.1 OTF and Signal-to-Noise Ratio

It is most natural to use the DA formulas (6.1.19–21) (Table 6.2) to calculating the OTF of sufficiently thick nonabsorbing scattering layers such as, for example, clouds in the visible. In the case of relatively thin cloudy layers or haze, the SAA data appear to be effective [(6.1.4) and Table 6.1]. The SAA applicability for calculating the OTF of a layer with the single scattering albedo $\Lambda = 1$ and a scattering indicatrix less extended than that for sea water is not so obvious. This problem is discussed in [8.31–32], where the SAA is compared Monte Carlo simulations to show that at $\tau_0 \lesssim 3$, the error in the SAA calculation of the OTF for all spatial frequencies is no more than 20%. At large spatial frequencies, $\nu^* \gtrsim 15$, this approximation satisfactorily (within 20%) describes the OTF up to $\tau_0 \approx 7$. The normalized OTF as a function of the angular spatial frequency $\nu^* = \nu z_0$ for different $\tau_0$, $\Lambda$ and scattering indicatrices are plotted in Figs. 6.4–6.

Let us analyze the signal-to-noise ratio $\delta$. We emphasize the dependence of $\delta$ on the optical properties of the medium by rewriting (5.3.15) as

$$\delta = \xi \delta_{\mathrm{m}}, \tag{8.4.1}$$

where the factor $\xi$ is determined by the parameters of the source-receiver system, and $\delta_{\mathrm{m}}$ is dependent only on the scattering properties. For example, for a self-luminous object with the harmonic emittance distribution (5.2.12), we have

$$\xi = k_0 \sqrt{\frac{E_0 \Sigma_{\mathrm{rec}} \Omega_{\mathrm{rec}} S_\lambda t_{\mathrm{el}}}{2\pi e B}}, \tag{8.4.2}$$

$$\delta_{\mathrm{m}} = \frac{S(\nu)}{\sqrt{S(\nu = 0)}}. \tag{8.4.3}$$

If the object is illuminated by the sun in zenith and is observed through a cloud down the vertical, then, allowing for backscattering noise and double

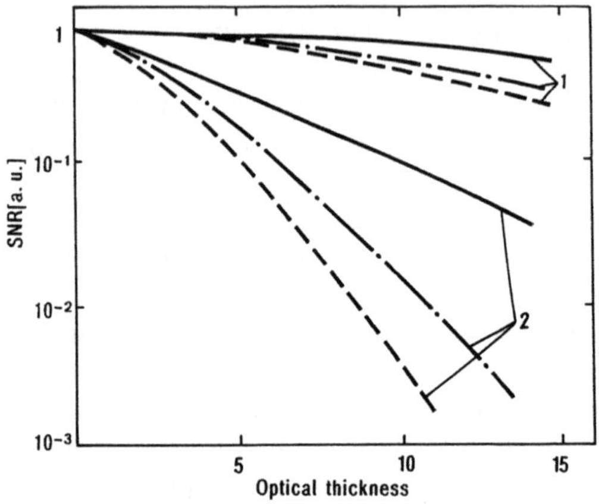


**Fig. 8.6.** Quantities $\delta_{1,m}$ ($\nu = 0$) (*solid line*), $\delta_{2,m}$ ($\nu = 0$) (*dashed lines*) and $\delta_{3,m}$ ($\nu = 0$) (*dash-dot lines*) at $g = 0.866$ and $\Lambda = 1$ (curves 1), 0.7 (curves 2)

sunlight transmission through the cloud, we obtain

$$\delta_\mathrm{m} = \frac{\langle A\rangle S(\nu)S(\nu = 0)}{\sqrt{\langle A\rangle S^2(\nu = 0) + r}}, \tag{8.4.4}$$

where $r$ is the diffuse reflection coefficient of the cloud. Equation (8.4.4) is written in the approximation of an isotropic angular radiance distribution of the reflected radiation. The accuracy of such an approximation is illustrated in Fig. 3.3. Formulas (8.4.3, 4) demonstrate the same $\delta_\mathrm{m}$ dependence on the spatial frequency $\nu$ as in the case of the medium OTF. Therefore, in view of the OTF data presented in Chap. 6, we shall restrict our consideration to $\delta_\mathrm{m}$ at $\nu = 0$.

For self-luminous and sunlighted objects, (8.4.3, 4) give

$$\delta_\mathrm{m}(\nu = 0) = \delta_{1,\mathrm{m}}(\nu = 0) = \sqrt{S(\nu = 0)}\,, \tag{8.4.5}$$

$$\delta_\mathrm{m}(\nu = 0) = \delta_{2,\mathrm{m}}(\nu = 0) = \frac{\sqrt{A}S(\nu = 0)}{\sqrt{1 + (r/\langle A\rangle S^2(\nu = 0))}}\,. \tag{8.4.6}$$

For the system with BSN suppression, from (8.4.6) at $r = 0$, we find

$$\delta_\mathrm{m}(\nu = 0) = \delta_{3,\mathrm{m}}(\nu = 0) = \sqrt{A}S(\nu = 0)\,. \tag{8.4.7}$$

The dependence of $\delta_{i,\mathrm{m}}(\nu = 0)$ ($i = 1, 2, 3$) on $\tau_0$ at $A = 1$ is shown in Fig. 8.6.

### 8.4.2 Cloud Microstructure Effect on the OTF and SNR

We considered the optical cloud parameters $\sigma$ and $x(\beta)$ as functions of the microstructure characteristics in Sect. 8.1.2. Using these data, we shall now study the microstructure effect on cloud transfer characteristics. Table 8.7 gives

**Table 8.7.** Value of $z_0 V_{\nu*}$ for large drop clouds [m] ($C_V^l = 10^{-6}$).

| $\mu$ | $a_m$ [μm] | | | |
|---|---|---|---|---|
| | 1 | 3 | 6 | 10 |
| 5 | 11.7 | 77.2 | 196 | 379 |
| 15 | 9.9 | 49.6 | 143 | 222 |

$z_0 V_{\nu*} = 6/\sigma_1 \beta_{2,1}$, which defines the cloud OTF half-width and depends only on the microstructure parameters. The data are obtained using (8.1.7) and assuming $\beta_2 = 1 - g$ (Sect. 4.7).

Let us consider in some detail the effect of cloud microstructure on the OTF. To do so, we introduce the function

$$Y(\nu^*) = -[\ln S(\nu^*)]/z_0 . \tag{8.4.8}$$

As seen from (4.4.35), in the SAA the function

$$Y(\nu^*) = \varepsilon - \sigma \int_0^1 x(\nu^* \xi)\, d\xi \tag{8.4.9}$$

for a homogeneous medium is determined only by its microstructure and does not depend on the layer thickness. The important property of $Y(\nu^*)$, viz. its additivity with respect to the contribution of different fractions and especially the contributions of individual scatterers, makes it particularly suited to analysis since it allows for the dependence on the microstructure parameters of each individual fraction. The function $Y(\nu^*)$ calculated by (8.4.9) is given in Fig. 8.7. As seen, the reduction in the distribution half-width ($\mu$ increase) results in higher $Y(\nu^*)$ values. This complies with the decrease of the OTF half-width $V_{\nu*}$ in Table 8.7.

The quantity $V_{\nu*}$ and function $Y(\nu^*)$ respond more rapidly to the changes in the modal radius $a_m$. The $a_m$ growth from 1 to 10 μm causes a 20 to 30 times reduction in $Y(\nu^*)$. Such a strong dependence is mainly due to the effect of the parameter $a_m$ on the scattering coefficient (8.1.7). However, the change in the scattering indicatrix with particle size variation is also a factor. For instance, the parameter $g$ increases as $a_m$ (8.1.7). For more extended indicatrices, the function $Y(\nu^*)$ tends more slowly to its asymptotic value at $\nu^* \to \infty$.

The influence of the scattering indicatrix on the OTF of the medium with changing microstructure is well illustrated by analyzing the function $[S(\nu^*)]^{1/\tau_0}$. As seen from (4.4.35), the OTF of the layer of some arbitrary thickness $\tau_0$ can be represented within the small-angle approximation as $S(\nu^*) = [S(\nu^*, \tau_0 = 1)]^{\tau_0}$. The function $[S(\nu^*)]^{1/\tau_0}$ is independent of $\tau_0$. At a constant single scattering albedo, it depends only on the scattering indicatrix and layer position on the observation path [8.33].

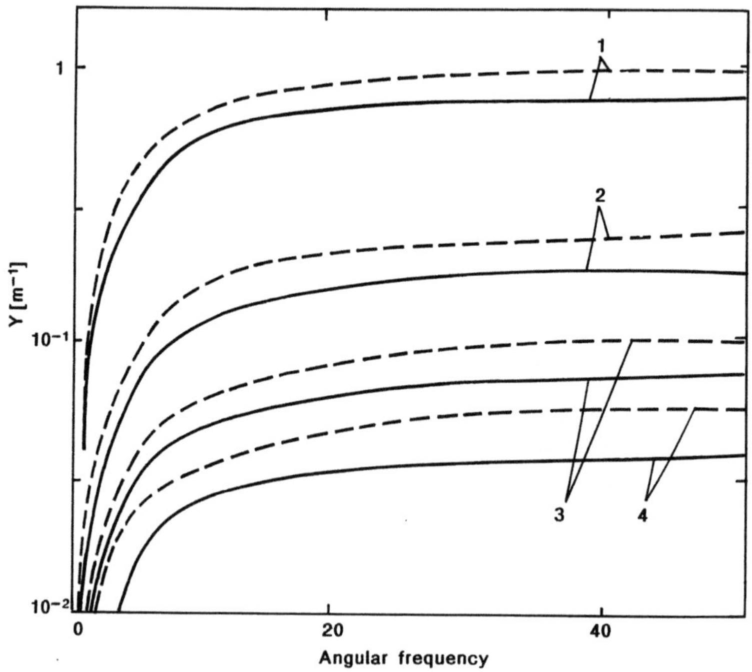


**Fig. 8.7.** Function $Y(\nu^*)$ for a large-drop fraction at $\lambda = 0.63$ µm, $C_V^l = 1$ µm$^3$cm$^{-3}$ and $\mu = 5$ (*solid lines*), 15 (*dashed lines*); $a_m = 1$ µm (curves 1), 3 (curves 2), 6 (curves 3), 10 (curves 4)

For a homogeneous path, from (4.4.35), we have

$$[S(\nu^*)]^{1/\tau_0} = \exp\left[-1 + \Lambda \int_0^1 x(\nu^*\xi)\,d\xi\right]. \tag{8.4.10}$$

Figure 8.8, taken from [8.34], gives $[S(\nu^*)]^{1/\tau_0}$ as calculated by (8.4.10) for water clouds and fogs of different microstructures at the wavelength $\lambda =$ 0.63 µm. As seen, with the modal radius $a_m \approx 10$ µm, for example, the OTF approaches the asymptote at $\nu^* \approx 10^3$, i.e., the contrast is independent of the image cell size when its angular sizes are less than 3′.

The OTF essentially depends on the distribution of the optical characteristics along the observation path. Thus, when observed from some height, objects on the ground are better viewed through ground fog than through cloud with the same light scattering characteristics, but close to the observer. The OTFs of inhomogeneous media have been thoroughly discussed in Chap. 6. In Fig. 8.9, experimental measurements [8.33] are shown for the OTFs of the stratus along the 424 m path. It also presents OTF calculations by (8.4.10), using the experimental data on the stratus microstructure in their middle and at the bottom [8.16]. Figure 8.9 illustrates an essential change in the cloud OTF with the microstructure and spatial inhomogeneity of the optical parameters. It should

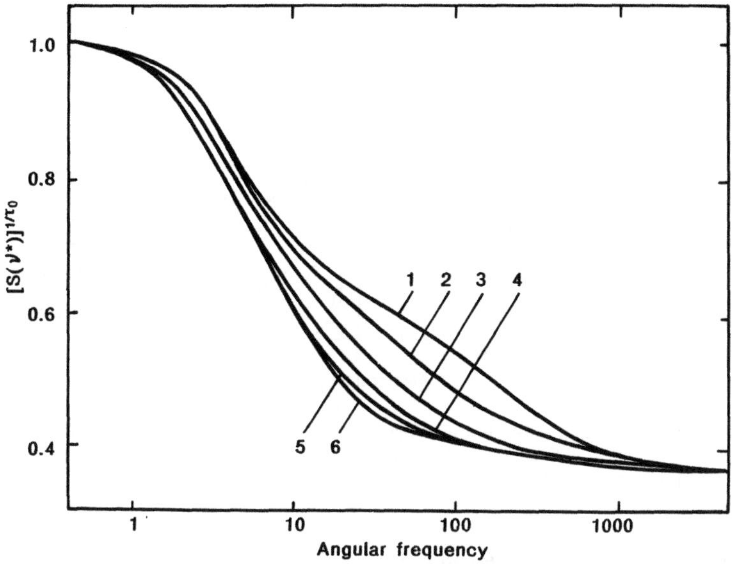


**Fig. 8.8.** OTF of a cloud of unit optical thickness for particle size distribution $f_1(a)$ at $\mu = 6$ and $a_m = 10$ µm (curve 1), 4 (curve 2), 2 (curve 3), 1 (curve 4) and for $f_{sm}(a)$ at $\nu = 3$ (curve 5); 4.5 (curve 6). From [8.34]

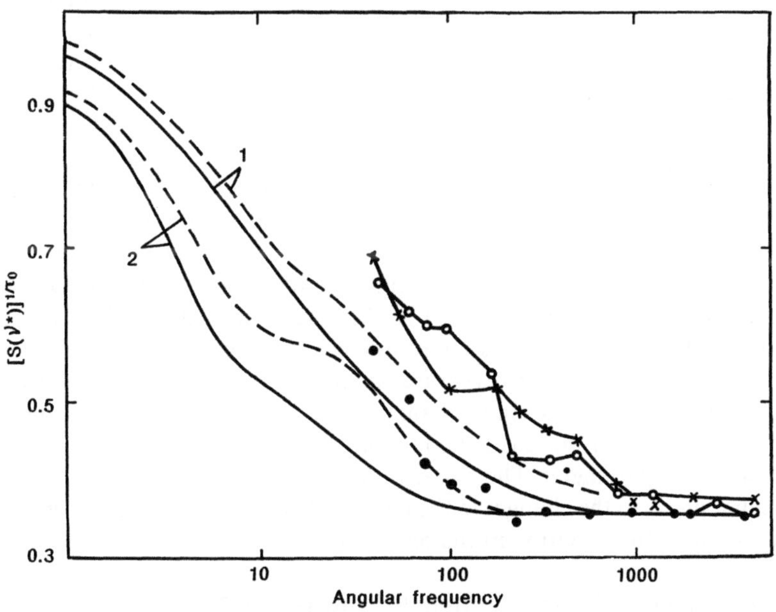


**Fig. 8.9.** Curves 1–3 give experimental results for the OTFs of clouds near a viewer (points), near a middle path (*circles, crosses*); curves 1, 2 show the calculated results using microstructure data in the middle (curves 1) and at the bottom (curves 2) of a cloud; *solid lines*, the cloud is near a viewer; *dashed lines*, homogeneous path

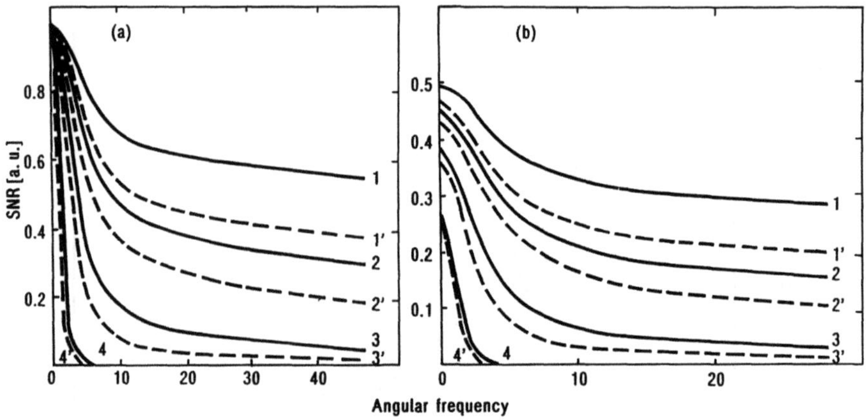


**Fig. 8.10.** SNR in viewing a self-luminous object (a) and in sunlight (b) through a large-drop cloud at $\lambda = 0.63\ \mu m$, $\tau_0 = 3$, $\mu = 5$ (curves 1–4), 15 (curves 1′–4′): $a_m = 10\ \mu m$ (curves 1, 1′), 6 (curves 2, 2′), 3 (curves 3, 3′), 1 (curves 4, 4′)

be noted that, at present, the experimental data on the OTF of atmospheric aerosols are not sufficient to bear comparison with the theoretical estimates which allowing for the microstructure of a disperse medium. Moreover, such a comparison requires synchronous measuring of the OTF and spatial distribution of the microstructure parameters. Unfortunately, no such data are as yet available.

We consider briefly the microstructure effect on the power of a valid signal and backscattering noise, and the signal-to-noise ratio. As the scattering coefficient decreases and the scattering indicatrix becomes more extended, the valid signal power grows, whereas the BSN degenerates. Such a change in the optical medium properties improves the OTF. As seen from (8.4.2, 3), the OTF is mainly responsible for the SNR changes with changing microstructure. Therefore, we may say that any change in the cloud microstructure, which leads to an improvement in the medium OTF, at the same time decreases the BSN in the observation system and increases the SNR. In Fig. 8.10, $\delta_m$ is plotted as a function of the angular spatial frequency for a cloud model which consists only of large drop fractions. The effect of microstructure changes on the SNR is stronger at high spatial frequencies, i.e., the changes are most pronounced in the image quality of fine object details. In this case, the modal particle radius $a_m$ is the determining microstructure parameter. The cloud water content being fixed, its effect is due, in the first place, to the changes in the medium OTF which are, for the most part, bound up with the variation in the scattering coefficient.

### 8.4.3 Estimation of Cloud OTF from Microstructure Data

The scattering coefficient is the basic optical characteristic which undergoes the greatest changes with microstructure and on which the medium OTF and SNR primarily depend. For a large drop fraction, it can be found from (8.1.7). Vision characteristics are less dependent on the scattering indicatrix. Its approximation, therefore, permits a comparatively crude approach.

As mentioned in Sect. 6.1, two representative regions of small (0–8°) and comparatively large scattering angles can be distinguished for the scattering indicatrices of clouds. In this connection, we may, in approximation (6.1.10), use the sum of two terms only

$$x(\beta) = B_1 \mathrm{e}^{-b_1\beta} + B_2 \mathrm{e}^{-b_2\beta} . \tag{8.4.11}$$

The fractions of radiation scattered within the regions of small and large angles are $B_1' = B_1/b_1^2$ and $B_2' = 1 - B_1' = B_2/b_2^2$, respectively. The analysis of the Mie calculations shows that the changes in the microstructure alter the proportion between these fractions, the fraction of light $B_1'$ scattered into the small angle region increasing with $a_m$. For the visible range within 1 μm < $a_m$ < 10 μm, we may approximately assume

$$B_1 \approx 0.4(1 + 2\mathrm{e}^{-0.04\rho_m}) , \qquad \rho_m = 2\pi a_m/\lambda , \tag{8.4.12}$$

and $b_1 = 4\ \mathrm{rad}^{-1}$, $b_2 = 30\ \mathrm{rad}^{-1}$ irrespective of the microstructure and, hence, of the wavelength $\lambda$.

Using (8.4.11) at $\Lambda = 1$, for the function $Y(v^*)$, from (8.4.9), we have

$$Y(v^*) = \sigma\left[1 - \frac{B_1'}{\sqrt{1 + (v^*/b_1)^2}} - \frac{B_2'}{\sqrt{1 + (v^*/b_2)^2}}\right] . \tag{8.4.13}$$

Figure 8.11 gives the function $Y(v^*)$ as calculated for different parameters of the large-drop fraction. As seen from the figure, the accuracy of the approximate approach is satisfactory. The error in the $Y(v^*)$ calculation does not exceed 20%. We note, as follows from (8.1.7, 4.12), that the same uncertainty is encountered for the error of 15–20% in the $a_m$ estimation.

We now estimate the contribution of the accumulative fraction. As noted in Sect. 8.1.2, in agreement with the experimental data, $\sigma_{sm}/\sigma_1$ does not exceed 0.3. This means that the effect of the accumulative fraction on the optical cloud parameters and the vision conditions in them is small by comparison. In view of this fact and the relatively low $\sigma_{sm}$ sensitivity to the parameters $v$ and $a_{sm}$, the dependence of the scattering coefficient $\sigma$ on the microstructure can be disregarded, and some mean value of $\sigma_{sm} \approx 1.5 \cdot 10^6 C_V^{sm}\ \mathrm{m}^{-1}$ set in the visible. Then, taking into consideration both fractions, for the cloud scattering coefficient, we finally obtain

$$\sigma \approx 1.5 \cdot 10^6 \left[\frac{C_V^1}{a_m(1 + 3/\mu)} + C_V^{sm}\right] . \tag{8.4.14}$$

Here, $a_m$ is in μm, $\sigma$ in $\mathrm{m}^{-1}$. For both fractions, the differences in the scattering indicatrix can also be neglected. The estimates show that even assumptions as crude as these on the accumulative fraction introduce an additional error of no more than 20% into the $Y(v^*)$ calculation.

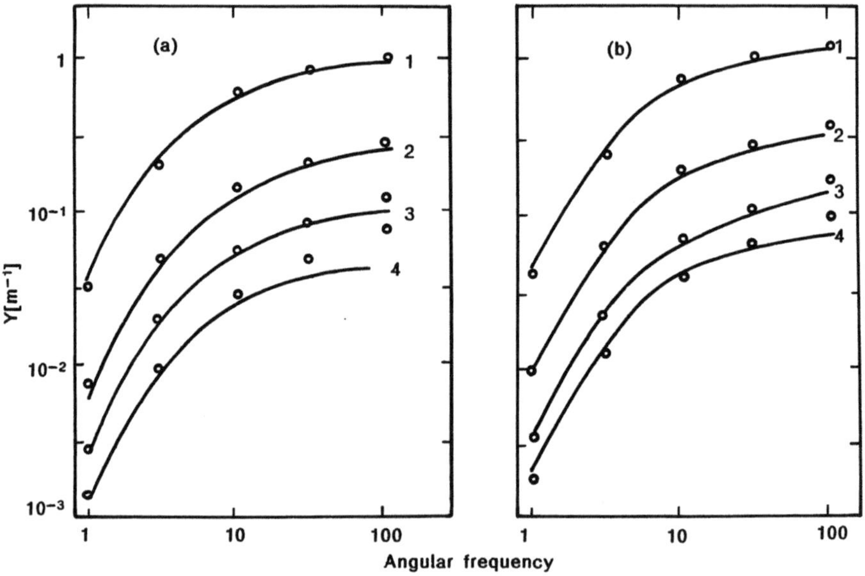


**Fig. 8.11.** Function $Y(v^*)$ at $\lambda = 0.63$ μm, $C_V^1 = 10^{-6}$, (a) $\mu = 5$, (b) $\mu = 15$ and $a_m = 1$ μm (curves 1), 3 (curves 2), 6 (curves 3), 10 (curves 4). *Lines* show the results of calculation by (8.4.9) with the scattering coefficient and indicatrix derived from the Mie formulas; *circles*, approximation of (8.4.12, 13)

Thus, an estimation of the cloud MTF from the known parameters $C_V^{sm}$, $C_V^1$, $a_m$ and $\mu$ reduces to (8.4.8, 12), taking into account (8.4.13, 14).

## 8.5 Viewing Through Stochastic Clouds

Many natural scattering media, particularly clouds, are characterized by a stochastic distribution of the optical parameters. The stochastic distribution of the medium inhomogeneities is, naturally, responsible for the stochasticity of the radiation field characteristics. In this case, an additional noise component is induced by the fluctuations of the optical medium parameters and deteriorates the contrast, thereby reducing the vision range. The general structural formulas for the signal-to-noise ratio which allow for this factor were given in Sect. 5.1.5. The optical transfer function of the stochastic medium was found in Sects. 4.4.7 and 6.2.4. We now consider the OTF of the system as a whole, and calculate the power of the fluctuational component of the flux arriving at the photodetector for different conditions of viewing through clouds. We shall also analyse the signal-to-noise ratio with respect to the two competing factors, viz. the arrival of additional noise and the simultaneous OTF improvement in the stochastic medium as compared with the homogeneous medium of the same mean optical depth (Sect. 6.2.4).

### 8.5.1 Viewing System OTF and Signal Power Fluctuations

Naturally, in the stochastic medium with fluctuating optical parameters, the medium PSF $S(\vec{r}, \vec{r}_1)$ and the system PSF $a(\vec{r}, \vec{r}_1)$ may not possess invariance, i.e., in the general case $a(\vec{r}, \vec{r}_1) \neq a(\vec{r} - \vec{r}_1)$. Here, $\vec{r}_1$ is the coordinate of the point to which the source and detector axes are directed. We represent the function $a(\vec{r}, \vec{r}_1)$ as a sum of the mean and fluctuating component. Assuming that the fluctuations of optical medium parameters are statistically uniform in space, we obtain

$$a(\vec{r}, \vec{r}_1) = \langle a(\vec{r}, \vec{r}_1)\rangle + \tilde{a}(\vec{r}, \vec{r}_1), \tag{8.5.1}$$

$$\langle a(\vec{r}, \vec{r}_1)\rangle = \langle a(\vec{r} - \vec{r}_1)\rangle . \tag{8.5.2}$$

Here, the average of $\vec{r}_1$ is taken. The function $\langle a(\vec{r} - \vec{r}_1)\rangle$ specifies the mean power of a valid signal

$$\langle P_{\mathrm{VS}}\rangle = P_0\langle A\rangle \iint_{-\infty}^{\infty} \langle a(\vec{r} - \vec{r}_1)\rangle \, d\vec{r}, \tag{8.5.3}$$

providing information on the object structure, and the variable power component

$$\Delta P(\vec{r}_1) \doteq P_0 \iint_{-\infty}^{\infty} [A(\vec{r}) - \langle A\rangle]\langle a(\vec{r} - \vec{r}_1)\rangle \, d\vec{r}, \tag{8.5.4}$$

where $P_0$ is the radiation source power, and $A(\vec{r})$ the albedo distribution in the object plane. The quantities

$$\langle a(\vec{v})\rangle = \iint_{-\infty}^{\infty} \langle a(\vec{r} - \vec{r}_1)\rangle \mathrm{e}^{-\mathrm{i}\vec{v}\cdot\vec{r}} \, d\vec{r} \tag{8.5.5}$$

and

$$\langle T(\vec{v})\rangle = \frac{\langle a(\vec{v})\rangle}{\langle a(\vec{v} = 0)\rangle} \tag{8.5.6}$$

may, correspondingly, be termed the mean OTF and mean MTF of the viewing system.

The function $\tilde{a}(\vec{r}, \vec{r}_1)$ defines the medium-inhomogeneity postulated fluctuation component of the power

$$\tilde{P}_{\mathrm{VS}}(\vec{r}_1) = P_0 \iint_{-\infty}^{\infty} A(\vec{r})\tilde{a}(\vec{r}, \vec{r}_1) \, d\vec{r}. \tag{8.5.7}$$

It will be recalled (Sect. 5.3.1) that

$$a(\vec{r}, \vec{r}_1) = \frac{\Sigma_{\mathrm{rec}}\Omega_{\mathrm{rec}}}{\pi} E_{\mathrm{s}}(\vec{r}, \vec{r}_1) E_{\mathrm{rec}}(\vec{r}, \vec{r}_1), \tag{8.5.8}$$

where $E_{\mathrm{s}}(\vec{r}, \vec{r}_1)$ is the irradiance distribution in the object plane induced by a real unit power source, and $E_{\mathrm{rec}}(\vec{r}, \vec{r}_1)$ the similar distribution due to some fictitious source whose directional diagram is identical to the responsivity diagram of the real photodetector. We separate the mean and fluctuation portions in the functions $E_s(\vec{r},\vec{r}_1)$ and $E_{rec}(\vec{r},\vec{r}_1)$, and rewrite (8.5.8) as

$$a(\vec{r},\vec{r}_1)=\frac{\Sigma_{rec}\Omega_{rec}}{\pi}[\langle E_s(\vec{r}-\vec{r}_1)\rangle+\tilde{E}_s(\vec{r},\vec{r}_1)][\langle E_{rec}(\vec{r}-\vec{r}_1)\rangle+\tilde{E}_{rec}(\vec{r},\vec{r}_1)]. \tag{8.5.9}$$

Hence, for the mean PSF of the system of viewing through a stochastic scattering medium, we have

$$\begin{aligned}\langle a(\vec{r}-\vec{r}_1)\rangle &= \langle a(\vec{r},\vec{r}_1)\rangle \\ &= \frac{\Sigma_{rec}\Omega_{rec}}{\pi}[\langle E_s(\vec{r}-\vec{r}_1)\rangle\langle E_{rec}(\vec{r}-\vec{r}_1)\rangle \\ &\quad + \langle \tilde{E}_s(\vec{r},\vec{r}_1)\tilde{E}_{rec}(\vec{r},\vec{r}_1)\rangle],\end{aligned} \tag{8.5.10}$$

i.e., the PSF $\langle a(\vec{r}-\vec{r}_1)\rangle$ is dependent on the correlation between the fluctuation components $\tilde{E}_s(\vec{r},\vec{r}_1)$ and $\tilde{E}_{rec}(\vec{r},\vec{r}_1)$. If no such correlation exists (for example, when the radiation source and receiver are far away from each other), then

$$\langle a(\vec{r}-\vec{r}_1)\rangle = \frac{\Sigma_{rec}\Omega_{rec}}{\pi}\langle E_s(\vec{r}-\vec{r}_1)\rangle\langle E_{rec}(\vec{r}-\vec{r}_1)\rangle, \tag{8.5.11}$$

and the OTF of the viewing system is determined, as for the homogeneous medium, by the convolution

$$\langle a(\vec{v})\rangle = \frac{\Sigma_{rec}\Omega_{rec}}{\pi}\iint_{-\infty}^{\infty}\langle E_s(\vec{v}')\rangle\langle E_{rec}(\vec{v}-\vec{v}')\rangle\, d\vec{v}'. \tag{8.5.12}$$

As seen from (8.5.9), the fluctuation component $\tilde{a}(\vec{r},\vec{r}_1)$ specifying the fluctuation $\tilde{P}_{VS}(\vec{r}_1)$ is

$$\begin{aligned}\tilde{a}(\vec{r},\vec{r}_1) &= \frac{\Sigma_{rec}\Omega_{rec}}{\pi}[\langle E_s(\vec{r}-\vec{r}_1)\rangle\tilde{E}_{rec}(\vec{r},\vec{r}_1)+\langle E_{rec}(\vec{r}-\vec{r}_1)\rangle\tilde{E}_s(\vec{r},\vec{r}_1) \\ &\quad + [\tilde{E}_s(\vec{r},\vec{r}_1)\tilde{E}_{rec}(\vec{r},\vec{r}_1)-\langle\tilde{E}_s(\vec{r},\vec{r}_1)\tilde{E}_{rec}(\vec{r},\vec{r}_1)\rangle]].\end{aligned} \tag{8.5.13}$$

Here, the first term is related to the fluctuations along the object-to-receiver path, and the second to the fluctuations of the object irradiance distribution.

Let an object be illuminated by an infinitely wide parallel beam (sunlight). Disregarding the third term in (8.5.13), we obtain

$$\tilde{P}_{VS}(\vec{r}_1) = \tilde{P}_1(\vec{r}_1) + \tilde{P}_2(\vec{r}_1), \tag{8.5.14}$$

where

$$\tilde{P}_1(\vec{r}_1) = \frac{E_{sun}\Sigma_{rec}\Omega_{rec}\langle S(z)\rangle}{\pi}\iint_{-\infty}^{\infty} A(\vec{r})\tilde{E}_{rec}(\vec{r},\vec{r}_1)\, d\vec{r}, \tag{8.5.15}$$

$$\tilde{P}_2(\vec{r}_1) = \frac{E_{\mathrm{sun}}\Sigma_{\mathrm{rec}}\Omega_{\mathrm{rec}}}{\pi} \iint_{-\infty}^{\infty} A(\vec{r})\langle E_{\mathrm{rec}}(\vec{r}-\vec{r}_1)\rangle \tilde{S}(\vec{r},z)\, d\vec{r}\,. \tag{8.5.16}$$

The quantities $\langle S(z)\rangle$ and

$$\tilde{S}(\vec{r},z) = \iint_{-\infty}^{\infty} \tilde{S}(\vec{r},\vec{r}_1,z)\, d\vec{r}_1 \tag{8.5.17}$$

are the mean and fluctuation components of the irradiance coefficient in the medium illuminated by an infinitely wide parallel beam, and $E_{\mathrm{sun}}$ the irradiance generated by the sunlight at the cloud boundary.

Consider, for example, observation of a low-contrast object ($\Delta A(\vec{r}) = A(\vec{r}) - \langle A\rangle \ll \langle A\rangle$) when a photodetector with a narrow directional diagram is used. Then $A(\vec{r}) \approx \langle A\rangle$, $E_{\mathrm{rec}}(\vec{r},\vec{r}_1) = S(\vec{r},\vec{r}_1,z)$ and

$$\tilde{P}_1(\vec{r}_1) = \frac{E_{\mathrm{sun}}\Sigma_{\mathrm{rec}}\Omega_{\mathrm{rec}}\langle A\rangle}{\pi} \langle S(z)\rangle \tilde{F}(\vec{r}_1,z)\,, \tag{8.5.18}$$

$$\tilde{P}_2(\vec{r}_1) = \frac{E_{\mathrm{sun}}\Sigma_{\mathrm{rec}}\Omega_{\mathrm{rec}}\langle A\rangle}{\pi} \iint_{-\infty}^{\infty} \langle S(\vec{r}-\vec{r}_1,z)\rangle \tilde{S}(\vec{r},z)\, d\vec{r}\,, \tag{8.5.19}$$

where

$$\tilde{F}(\vec{r}_1,z) = F(\vec{r}_1,z) - \langle F(z)\rangle\,, \tag{8.5.20}$$

$$F(\vec{r}_1,z) = \iint_{-\infty}^{\infty} S(\vec{r},\vec{r}_1,z)\, d\vec{r}\,. \tag{8.5.21}$$

The quantity $F(\vec{r}_1,z)$ is the total radiation flux through the $z$ = const plane when the medium is illuminated by a monodirectional unit power point source oriented to the point $\vec{r}_1$.

### 8.5.2 Irradiance and Radiation Flux Fluctuation Variances

We calculate the variances of the random functions $\tilde{S}(\vec{r},z)$ and $\tilde{F}(\vec{r}_1,z)$ which affect the variance of the power fluctuation component $\langle \tilde{P}^2_{\mathrm{VS}}\rangle$ and, hence, the signal-to-noise ratio. First, we write a general expression for the quantity $\langle|\tilde{I}(\vec{v},z,\vec{p})|^2\rangle$ which, at $\vec{p}=0$, determines the power spectrum of the irradiance fluctuations in the medium subject to different illumination conditions. It will be recalled that the Fourier spectrum of the fluctuation component of the irradiance $\tilde{I}(\vec{v},z,\vec{p})$ is expressed by (4.4.53–56). At $\vec{p}=0$, we then obtain [8.35]

$$\langle|I(\vec{v},z,\vec{p}=0)|^2\rangle = \frac{1}{4\pi^2}\int_0^z dz' \iint_{-\infty}^{\infty} \Phi(\vec{v},\vec{v}',z)\langle I(\vec{v}',z-z',\vec{v}z')\rangle^2 \times [I_{\mathrm{n}}^{\mathrm{hom}}(\vec{v},z',0)]^2\, d\vec{v}'\,, \tag{8.5.22}$$

where

$$\Phi(\vec{v},\vec{v}',z) = \Phi_{\varepsilon\varepsilon}(\vec{v}-\vec{v}') - 2\Phi_{\varepsilon\sigma}(\vec{v}-\vec{v}')x(\vec{p}=\vec{v}z) + \Phi_{\sigma\sigma}(\vec{v}-\vec{v}')x^2(\vec{p}=\vec{v}z)\,, \tag{8.5.23}$$

$$\Phi_{\substack{\varepsilon\varepsilon\\ \varepsilon\sigma\\ \sigma\sigma}}(\vec{v}) = \int_{-\infty}^{\infty} \Gamma_{\substack{\varepsilon\varepsilon\\ \varepsilon\sigma\\ \sigma\sigma}}(\vec{v},z')\,dz'\,, \tag{8.5.24}$$

and the functions $\Gamma_{\varepsilon\varepsilon}(\vec{v},z)$, $\Gamma_{\varepsilon\sigma}(\vec{v},z)$ and $\Gamma_{\sigma\sigma}(\vec{v},z)$ are determined by expressions such as (4.4.59, 60).

The relation (8.5.22), taking into account (4.4.64), provides the variances of the appropriate random functions. Assuming $I_0(\vec{v},\vec{p}) = 1$, for the variance of the fluctuations of the total radiation flux $F(\vec{r}_1,z)$ at depth $z$, we have

$$V_F = \langle \tilde{F}^2(\vec{r}_1,z)\rangle = \langle|\tilde{I}_n(\vec{v}=0,z,\vec{p}=0)|^2\rangle$$
$$= \frac{1}{4\pi^2}\int_0^z dz' \iint_{-\infty}^{\infty} \Phi_{\kappa\kappa}(\vec{v})\langle I_n(\vec{v},z-z',0)\rangle^2 [I_n^{\mathrm{hom}}(0,z',0)]^2\, d\vec{v}\,. \tag{8.5.25}$$

The function $\langle I_n(\vec{v},z,\vec{p})\rangle$ is expressed by (4.4.65).

Given the correlation functions of the optical medium parameter fluctuations as (4.4.68) and approximating the scattering indicatrix spectrum by the function

$$x(\vec{p}) = 1 - \beta_2 p^2/4\,, \tag{8.5.26}$$

we get [8.36]

$$V_F = \sqrt{2\pi}\,R_c\langle F(z)\rangle^2 V_\kappa \int_0^z \frac{dy}{1+\langle\sigma\rangle\beta_2 y^3/3R_c^2}\,. \tag{8.5.27}$$

It should be noted that the approximation (8.5.26), as seen from (4.4.35), conforms to the Gaussian description of the homogeneous medium OTF.

As appears from the optical reciprocity theorem, the coefficient of variation $\sqrt{V_F}/\langle F(z)\rangle$ is equal to the coefficient of variation $\sqrt{V_I}/\langle I\rangle$ of the fluctuations of the radiance $I(\vec{r}_1)$ in viewing a homogeneous Lambertian surface. Therefore, the quantity $V_F$ may characterize the power of fluctuation noises on the photodetector when a self-luminous homogeneous object is observed.

We assume that the absorption coefficient in the medium does not fluctuate, i.e., $V_\kappa = 0$. Such a situation occurs, for example, in clouds with no absorption in the visible. As seen from (8.5.27), $V_F = 0$. This means that there are no signal power fluctuations on the photodetector when a homogeneous isotropically luminous surface is being viewed. This attractive and important result becomes evident, taking into account the specific features of essentially anisotropic scattering. In fact, in this case, as in the SAA, we may take it that radiation is only scattered to the front semisphere and neglect the dispersion of photon paths. In a nonabsorbing medium, therefore, the total radiation flux through the $z$ = const plane is independent of the spatial scattering coefficient distribution in the medium and is constant at all depths.

In observing a self-luminous object in a nonabsorbing medium, image radiance fluctuations may occur in the presence of a variable component of the initial radiant emittance of the object. For example, if a harmonic test object of high spatial frequency $\nu \gg 2\pi/R_c$, $\nu \gg 2/z\sqrt{\beta_2}$ and initial contrast $k_0$ is observed, then the variance of the radiance fluctuations in the image of such a self-luminous object is described by the expression [8.36]

$$V_I = \sqrt{2\pi} R_c \langle I \rangle^2 \int_0^z \left[ \frac{V_\kappa}{1 + \langle\sigma\rangle \beta_2 y^3/3R_c^2} + \frac{k_0^2}{2} V_\varepsilon T^2(\nu) \right] dy \,, \tag{8.5.28}$$

which, subject to the condition

$$(\langle\sigma\rangle \beta_2 z^3/6)^{1/2} \gg R_c \,, \tag{8.5.29}$$

(the halfwidth of the spread function being large compared with the characteristic size of the inhomogeneities) takes on the form

$$V_I = \sqrt{\frac{\pi}{2}} \langle I \rangle^2 \left[ \frac{\pi}{\sqrt{3}} \left( \frac{3}{\langle\sigma\rangle \beta_2 R_c} \right)^{1/3} V_\kappa R_c^2 + k_0^2 V_\varepsilon R_c z T^2(\nu) \right] . \tag{8.5.30}$$

From (8.5.30), it follows that the inhomogeneities of the absorption coefficient are responsible for the highest noise level, although the radiation power fluctuations on the photodetector occur when any of the optical medium parameters fluctuate. In observing a self-luminous object with the emittance modulation depth $k_0 = 1$, the two percent fluctuations of the absorption coefficient in the medium with the single scattering albedo $\Lambda = 0.8$ have the same result as the fluctuations of the scattering coefficient (about 90%) [8.36].

We now consider the variance, $V_S$, of the irradiance fluctuations in the medium illuminated by an infinitely wide beam. In this case $I_0(\vec{\nu}, \vec{p}) = 4\pi^2 \delta(\vec{\nu})$. Therefore, making the same assumption as for (8.5.27), we have [8.35]

$$\begin{aligned} V_S = \langle \tilde{S}^2(\vec{r}, z) \rangle &= \frac{1}{4\pi^2} \int\!\!\int_{-\infty}^{\infty} \langle |\vec{I}_n(\vec{\nu}, z, 0)|^2 \rangle \, d\vec{\nu} \\ &= \sqrt{2\pi} R_c^3 \langle S(z) \rangle^2 \int_0^z \left[ \frac{V_\varepsilon}{\varphi(y)} - \frac{2\langle \tilde{\varepsilon}\tilde{\sigma} \rangle}{\varphi(y) + \beta_2 y^2/2} + \frac{V_\sigma}{\varphi(y) + \beta_2 y^2} \right] dy \,, \end{aligned} \tag{8.5.31}$$

where

$$\varphi(y) = R_c^2 + \langle\sigma\rangle \beta_2 y^2 (z - 2y/3) \,. \tag{8.5.32}$$

One more important situation should be emphasized. In a stochastic medium, as well as in a homogeneous layer, the average (over the observation point $\vec{r}$ coordinate, or over the medium realization ensemble) of the function $S(\vec{r}, z)$ is equal to the average (taken over the beam incidence point $\vec{r}_1$ coordinate or medium realization ensemble) of the total light flux $F(\vec{r}_1, z)$ at depth $z$, i.e., $\langle S(\vec{r}, z) \rangle = \langle F(\vec{r}_1, z) \rangle = \langle S(z) \rangle$. Nevertheless, a similar equality does not hold

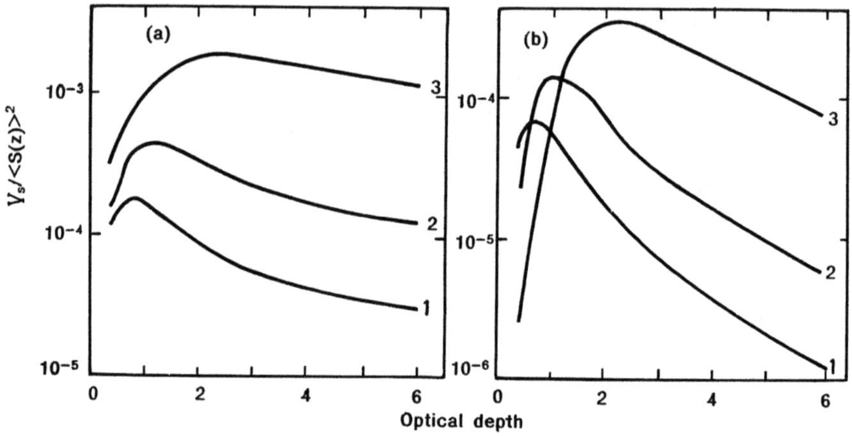


**Fig. 8.12.** Squared variation coefficient of irradiance fluctuations at $\sqrt{V_\varepsilon}/\langle\varepsilon\rangle = 0.1$, (a) $\Lambda = 0.7$, (b) $\Lambda = 0.99$ and $\langle\varepsilon\rangle R_c = 0.05$ (curves 1), 0.1 (curves 2), 0.5 (curves 3)

for the corresponding fluctuation parameters, i.e., $\tilde{S}(\vec{r}_1, z) \neq \tilde{F}(\vec{r}_1, z)$. Moreover, $V_S \neq V_F$ (8.5.27, 30). Specifically, for $\Lambda = 1$, i.e., at $\kappa = 0$, the variance $V_F = 0$, but $V_S \neq 0$.

The dependence of $V_S/\langle S(z)\rangle^2$ on the optical depth $\langle\varepsilon\rangle z$ calculated using (8.5.31) is illustrated in Fig. 8.12. The relative irradiance fluctuations are seen to increase with the characteristic inhomogeneity size $R_c$ and decreasing $\Lambda$. At some values of $\langle\varepsilon\rangle z$, $V_S/\langle S(z)\rangle^2$ achieves a maximum whose position depends mainly on the inhomogeneity size and, to a lesser extent, on $\Lambda$. As $\langle\varepsilon\rangle R_c$ grows and $\Lambda$ reduces, the maximum shifts to greater optical depths. Its position can approximately be estimated by [8.35]

$$(\langle\sigma\rangle z)_{\max} = 4.9(\langle\sigma\rangle R_c)^{2/3} . \tag{8.5.33}$$

### 8.5.3 Signal-to-Noise Ratio

We shall return to the question of signal power fluctuations on the photodetector in viewing through cloud under natural lighting in order to compare the contribution of the components $\tilde{P}_1(\vec{r}_1)$ and $\tilde{P}_2(\vec{r}_1)$.

From (8.5.18, 19), we have

$$\langle\tilde{P}_1^2\rangle = \left[\frac{E_{\text{sun}}\Sigma_{\text{rec}}\Omega_{\text{rec}}}{\pi}\langle A\rangle\langle S(z)\rangle\right]^2 V_F , \tag{8.5.34}$$

$$\langle\tilde{P}_2^2\rangle = \left[\frac{E_{\text{sun}}\Sigma_{\text{rec}}\Omega_{\text{rec}}\langle A\rangle}{\pi}\right]^2 \iint_{-\infty}^{\infty} \langle S(\vec{v}, z)\rangle^2 \langle|\tilde{S}(\vec{v}, z)|^2\rangle\, d\vec{v} . \tag{8.5.35}$$

Consider that $S(\vec{v}, z) = I_n(\vec{v}, z, \vec{p} = 0)$, whereas $|\tilde{S}(\vec{v}, z)|^2 = \langle|\tilde{I}(\vec{v}, z, \vec{p} = 0)|^2\rangle$ at $I_0(\vec{v}, \vec{p}) = 4\pi^2\delta(\vec{v})$ (8.5.22) (Sect. 4.4.7). Subject, as before, to the approximations

(4.4.68; 8.5.26) and the condition (8.5.29), for the medium at $\Lambda = 1$, we obtain

$$\langle \tilde{P}_2^2 \rangle = C\left[\frac{E_{\text{sun}}\Sigma_{\text{rec}}\Omega_{\text{rec}}\langle A\rangle}{\pi}\right]^2 \frac{(\langle\sigma\rangle R_c)^3}{(\langle\sigma\rangle z)^4\beta_2}\frac{V_\sigma}{\langle\sigma\rangle^2}, \tag{8.5.36}$$

where the numerical coefficient $C \approx 1.05$.

A comparison of the variances $\langle \tilde{P}_1^2 \rangle$ and $\langle \tilde{P}_2^2 \rangle$ shows [8.37] that $\langle \tilde{P}_1^2 \rangle \gg \langle \tilde{P}_2^2 \rangle$ for the medium with appreciable absorption ($\Lambda \lesssim 0.7$). In the case of a non-absorbing medium, $\langle \tilde{P}_1^2 \rangle \ll \langle \tilde{P}_2^2 \rangle$ even at 100% modulation of the initial object irradiance ($k_0 = 1$) and maximum $\langle \tilde{P}_1^2 \rangle$. It should be noted that (8.5.34, 35) were obtained for a low contrast object. We may show that at $k_0 = 1$, $\langle \tilde{P}_1^2 \rangle$ is also described by (8.5.34), the variance $V_F$ being found from the condition $V_F/\langle F\rangle^2 = V_I/\langle I\rangle^2$ and the relations (8.5.28, 30). For example, at $\langle\sigma\rangle z = 5$, $\langle\sigma\rangle R_c = 0.25$, $\beta_2 = 0.1$ and $k_0 = 1$, the ratio $\langle \tilde{P}_1^2 \rangle / \langle \tilde{P}_2^2 \rangle \approx 0.28$. Therefore, for viewing through cloud in the visible ($\Lambda = 1$), the quantities $\langle \tilde{P}_1^2 \rangle$ and $\langle \tilde{P}_1 \tilde{P}_2 \rangle$ may generally be neglected, and we may consider that additional noises in the viewing system due to the fluctuations of the optical medium parameters along the path of observation are completely determined by $\langle \tilde{P}_{\text{VS}}^2 \rangle = \langle \tilde{P}_2^2 \rangle$.

Given certain assumptions, we may use (8.5.36) to calculate the signal-to-noise ratio in the system of viewing through a stochastic cloud. For the sake of simplicity, we disregard the spatial correlation of fluctuation noises of the medium and transfer noises of the system. Setting that the medium and shot noises are statistically independent, we can write the signal-to-noise ratio as [8.38]

$$\delta = \frac{k\sqrt{\langle N\rangle}}{\sqrt{1 + \langle N\rangle(\tilde{\eta}/\langle\eta\rangle)^2}}, \tag{8.5.37}$$

where the observing image contrast

$$k = \frac{\Delta P}{\langle P\rangle}, \tag{8.5.38}$$

and the quantity $\langle N\rangle$, as before (Sect. 5.1), is a mean number of photons used per image cell. In the case of sunlight, the mean transfer coefficient $\langle\eta\rangle = \langle P\rangle/E_{\text{sun}}$, whereas the fluctuation component transfer coefficient $\tilde{\eta} = \sqrt{\langle \tilde{P}^2\rangle}/E_{\text{sun}}$.

The backscattering noise $P = P_{\text{VS}}$, $\eta = \eta_{\text{VS}}$ is neglected. If the backscattering noise arrives at the photodetector at the same time as a valid signal, then

$$P = P_{\text{VS}} + P_{\text{BSN}}, \qquad \eta = \frac{P_{\text{VS}} + P_{\text{BSN}}}{E_{\text{sun}}} = \eta_{\text{VS}} + \eta_{\text{BSN}}. \tag{8.5.39}$$

Setting the power fluctuations $\tilde{P}_{\text{VS}}$ and $\tilde{P}_{\text{BSN}}$ statistically independent, we have

$$\langle \tilde{P}^2\rangle = \langle \tilde{P}_{\text{VS}}^2\rangle + \langle \tilde{P}_{\text{BSN}}^2\rangle, \qquad \tilde{\eta}^2 = \tilde{\eta}_{\text{VS}}^2 + \tilde{\eta}_{\text{BSN}}^2. \tag{8.5.40}$$

The quantities $\tilde{P}_{\text{VS}}$ and $\tilde{P}_{\text{BSN}}$ seem to be statistically independent when the backscattering noise and signal are due to scattering from different medium inhomogeneities which are far apart (noncorrelated). Such a situation may occur, for example, in viewing through a rather thick medium layer when the quantity $P_{VS}(\vec{r}_1)$ is specified by the entire medium thickness, and the BSN is induced only by some small area near the source and the photodetector.

As the stochastic property of the medium has only a smal effect on the mean light field characteristics (Sect. 4.4.7), the quantity $\langle\eta\rangle$ in (8.5.37) is, in fact, equal to the transfer coefficient $\eta^{hom}$ for the homogeneous medium. In particular, the quantity $\langle\eta_{VS}\rangle$ can be calculated using the formula cited for an appropriate case in Table 5.3. If the medium does not absorb the equalities $\langle\eta_{VS}\rangle = \eta_{VS}^{hom}$ and $\langle N\rangle = N^{hom}$ are valid over the small angle approximation. At $\Lambda \neq 1$, these equalities are approximate (4.4.67).

Compare the signal-to-noise ratio for observations in stochastic and homogeneous media with equal mean values of the optical parameters. For the sake of simplicity, we disregard the backscattering noise. Taking a harmonic test object at frequency $\nu$ to be an object of observation, we rewrite formula (8.5.37) as

$$\delta = \delta^{hom} \frac{\langle T(\nu)\rangle / T^{hom}(\nu)}{\sqrt{1 + \langle N\rangle(\tilde{\eta}_{VS}/\langle\eta_{VS}\rangle)^2}}, \qquad (8.5.41)$$

where $\delta^{hom} = k_0 T^{hom}\sqrt{N^{hom}}$ is the signal-to-noise ratio in the homogeneous medium, and $T^{hom}(\nu)$ and $\langle T(\nu)\rangle$ are the MTF of the homogeneous and stochastic media, respectively.

We analyse (8.5.41) for three situations.

1. $\langle N\rangle(\tilde{\eta}_{VS}/\langle\eta_{VS}\rangle)^2 \ll 1\,, \qquad \langle N\rangle(\tilde{\eta}_{VS}/\langle\eta_{VS}\rangle)^2 < [\langle T(\nu)\rangle T^{hom}(\nu)]^2 - 1$

As $\langle T(\nu)\rangle \geqslant T^{hom}(\nu)$ (Sect. 6.2.4), these inequalities may hold, for example, due to the small number $\langle N\rangle$ of photons used. Then, as seen from (8.5.41), $\delta > \delta^{hom}$. The higher signal-to-noise ratio in the stochastic as opposed to the homogeneous medium is due to the improvement in the medium MTF. Additional fluctuation noises of the medium are less significant here. A more detailed analysis shows, however, that the $\delta$ increase is small and can be achieved only for very small $\delta^{hom}$ (at least, for $\delta^{hom} < 1$).

2. $[\langle T(\nu)\rangle / T^{hom}(\nu)]^2 - 1 < \langle N\rangle(\tilde{\eta}_{VS}/\langle\eta_{VS}\rangle)^2 \leqslant 1\,.$

This condition means that additional fluctuation noises, although less than shot photodetector noises, balance the medium MTF improvement effect and $\delta < \delta^{hom}$ results. The decrease in the signal-to-noise ratio is, however, insignificant here. Taking into account some enhancement of the image contrast, we may consider that, in this case, the medium stochasticity may both degenerate and improve the image. This situation can also be realized only for comparatively small $\delta$ ($\delta^{hom} \lesssim 1$).

3. $\langle N\rangle(\tilde{\eta}_{VS}/\langle\eta_{VS}\rangle)^2 \gg 1\,.$

This situation seems to be the most interesting. It may occur at sufficiently high medium inhomogeneity as well as at high power of the signal $\langle N \rangle$. The noises due to power fluctuations of the recorded radiation are greater than the shot photodetector noise, and prevail. At the same time, $\delta \ll \delta^{\text{hom}}$. It is important that in such a situation, the signal-to-noise ratio cannot be enhanced by increasing the signal power or photodetector sensitivity.

We shall estimate the quantity $\sqrt{V_\sigma}/\langle \sigma \rangle$, the scattering coefficient fluctuations in a nonabsorbing medium, at which the signal-to-noise ratio does not exceed $\delta^*$, even shot noises being neglected.

At $\langle N \rangle \, (\tilde{\eta}_{\text{VS}}/\langle \eta_{\text{VS}} \rangle)^2 \gg 1$, from (8.5.41) and taking into account (8.5.36), we obtain

$$\frac{\sqrt{V_\sigma}}{\langle \sigma \rangle} \geqslant \frac{k}{\delta^*} \frac{(\langle \sigma \rangle z)^2 \sqrt{\beta_2}}{\sqrt{C}(R_c \langle \sigma \rangle)^{3/2}} . \tag{8.5.42}$$

Setting, for example, $\delta^* = 5$, $k \approx 0.05$, $\langle \sigma \rangle z = 2$, $\beta_2 = 0.1$ and $\langle \sigma \rangle R_c \approx 0.3$, we find $\sqrt{V_\sigma}/\langle \sigma \rangle \gtrsim 0.08$.

Thus, even for scattering coefficient fluctuations of about 10%, in viewing through clouds in sunlight, the fluctuation noises due to the stochastic medium structure may restrict the visibility range.

# 9 Underwater Vision and Location in Sea Water

Underwater photography, filming and television have made it possible for the man in the street to admire the fantastic underwater world. Unfortunately, it is not only the thirst for knowledge and search for adventure which inspire human beings to deal with problems of vision in the sea. They are motivated by the quite pragmatic aims of developing and monitoring the ocean's resources, aims dictated by stark reality and the need to forecaste the future.

In this chpater, basing ourselves on image transfer theory and eleborate detection concepts, we shall consider a number of classical and current underwater vision problems, starting from the visibility of Sekky's disc which plays a crucial role in optical oceanography, and going on to laser location, vision and bathymetry systems.

In Sect. 9.2, the problem of detecting of underwater light sources, as well as visibility above water in transmitted sunlight, is considered together with white disc visibility. Underwater TV systems, their different types, a calculation of their main characteristics, an estimation of the visibility range and an analysis of the progress made are dealt with in Sect. 9.3. Laser location systems, laser bathymetry and sea-bed detection depth are discussed in Sect. 9.5.

Ocean exploration and observation through a rough sea surface are becoming more and more important. The specific features of image transfer through a rough surface, taking into account wind waves and swells, as well as some necessary data on surface models, are given in Sect. 9.4. This information, together with the theory of image transfer in a stochastic medium, including fluctuation noise correlation (Sect. 5.1.5), provides a key to solving this intricate problem. Some comments on taking account of the effect of the atmosphere on the characteristics of viewing from above through a sea surface supplement the discussion.

The chapter begins with relevant information on sea water as a scattering medium. We should like to draw the reader's attention to a simple model of optical sea water properties dealt with in Sect. 9.1.2, which is very convenient for research into methodology and forecasting. This model, as well as the sea light field calculation technique, may be used to solve a wide range of problems in hydrooptics, optical oceanography and remote sensing of the ocean. Solutions are given to illustrate such a procedure.

## 9.1 Optical Properties of Sea Water

### 9.1.1 Experimental Data

Electromagnetic radiation is completely absorbed in the surface water layer. The only radiation that penetrates fairly deeply into water lies within a small spectral interval corresponding to the visible. Light absorption by ocean water is determined by a complex combination of numerous factors. Among them, absorption by natural water containing dissolved salts as well as pigments of suspended particulate matter and soluble organic substances is predominant. The absorption ability of natural water does not actually change in different regions of the world's ocean. The concentration of suspended particles and soluble organic substances, including a yellow substance, fluctuates over a wide range. Light is scattered by small mineral and large organic suspended particles, as well as due to water density fluctuations. Optically active components of water arise due to phytoplankton generation in the ocean, rivers and glacial run-off, and the eolian contribution (from aerosol particles in the atmosphere).

As a rule, in oceanic waters away from coastal regions and areas of upwellings, particle concentration decreases with increasing depth, the rate of variation of large particle concentration being higher than that of smaller ones. However, as in atmosphere, layers of higher particle concentration may exist at certain depths $z$ (40–100 m). Starting from $z > 300$ m, the optical properties of water do not alter in the majority of cases.

The vertical distribution of particle concentration is frequently associated with the hydrological structure of water. With sharp changes in temperature and stratificiation which cause a water density gradient, large changes in particle concentration are observed. Turbidity layers may also be due to the accumulation of plankton which migrate, depending on the illumination, weather and amount of biogenic (nutrient) substances.

The concentration of particles is minimum in the tropical region of the world's ocean and near the poles. Such a latitudinal structure of water is, first and foremost, peculiar to surface layers.

These components of ocean water which determine its light scattering and absorption properties are responsible for the optical parameters of an elementary volume. Detailed data on the optical properties of the world's ocean waters can be found in monographs [9.1–5]. We shall restrict our consideration to the range and most probable values of the optical characteristics of ocean water.

The order of the scattering coefficient $\sigma$ of ocean water in the blue-green region ranges from $10^{-2}$ to $10^{0}$ m$^{-1}$. The highest values and greatest variations in the scattering coefficient are observed in coastal regions. In ocean waters away from coastal and upwelling regions, the representative $\sigma$ values range from 0.10 to 0.15 m$^{-1}$; for deep water, 0.05–0.10 m$^{-1}$ [9.5]. Above 50–100 m, $\sigma$ is no less than 0.05–0.06 m$^{-1}$ [9.3]. The lowest $\sigma$ values measured in the ocean are about

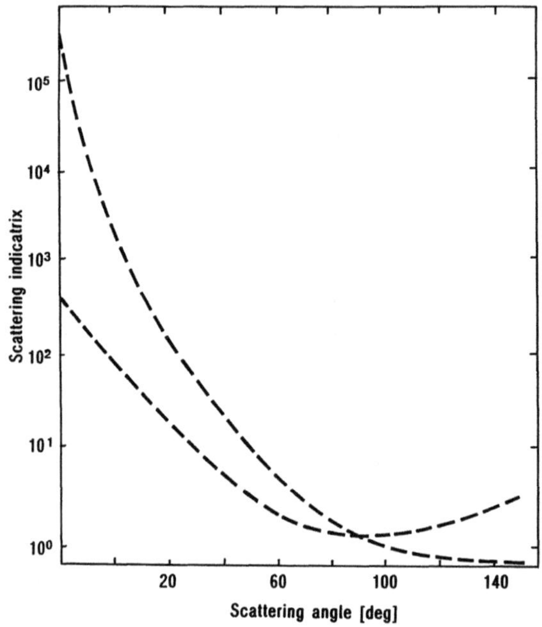


**Fig. 9.1.** Oceanic scattering indicatrices (normalized at 90°)

$0.02\ m^{-1}$. They are observed at a depth of 500 to 1000 m [9.3]. It should be noted that at $\lambda = 550$ nm, the scattering coefficient of pure sea water is less than $0.002\ m^{-1}$, i.e., even in the most transparent ocean waters, the contribution of molecular scattering to $\sigma$ does not exceed 10%. Suspended matter plays a key role in light scattering by sea water. Therefore, as the wavelength increases, the scattering coefficient $\sigma$ decreases according to the power law $\lambda^{-\alpha}$, where $\alpha$ ranges from 0.9 to 1.8.

An interpretation of the available experimental data on the scattering indicatrices shows that all of them are concentrated in the domain bounded by the curves in Fig. 9.1 [9.1]. The curves are normalized to unity at $\beta = 90°$. Note that the $x$-axis has been expended below 20°. The indicatrices are essentially forward extended. For sea water, the scattering angle corresponding to the solid angle which covers half the scattered light flux ranges from 2° to 10° [9.3]. The forward/back ratio $F/(1 - F)$ ranges between 8 and 80, whereas the asymmetry parameter $g$ is between 0.84 and 0.96. In some extreme situations (for example upper ocean layers with a high concentration of large biological particles), the forward/back ratio is above 100. The degree of asymmetry decreases with increasing depth because the fraction of large suspension particles reduces. The mean values of $\sigma(\beta)$ (at $\sigma = 0.11\ m^{-1}$), chosen from 41 indicatrices measured in the Black Sea, Atlantic and Pacific Ocean [9.3], are given in Table 9.1.

**Table 9.1.** Mean values of $\sigma(\beta)$.

| $\beta$ [deg] | $\langle\sigma(\beta)\rangle$ [$m^{-1}$] | $\beta$ [deg] | $\langle\sigma(\beta)\rangle$ [$m^{-1}$] |
|---|---|---|---|
| 0.5 | 255 | 10 | 1.38 |
| 1 | 103 | 15 | 0.503 |
| 2 | 40.2 | 45 | 0.035 |
| 4 | 11.3 | 90 | $7\cdot10^{-3}$ |
| 6 | 4.27 | 135 | $7\cdot10^{-3}$ |

The absorption coefficient $\kappa$ in the visible range of the spectrum is somewhat smaller, but of the same order as the scattering coefficient. All the absorption spectra have a minimum whose location is determined by the superposition of two spectral curves, one for absorption by soluble organic substances and suspended particulate matter which increases in the short wavelength region, and the other for absorption by water which increases with $\lambda$. In transparent waters with insignificant substance absorption, minimum absorption by sea water is observed at $\lambda \approx 470$–$490$ nm, being practically equal to pure water absorption (0.003–0.006 $m^{-1}$). For surface ocean waters away from coastal and upwelling regions, it is typical for the minimum to occur at $\lambda \approx 510$ nm and $\kappa \approx 0.01$–$0.02$ $m^{-1}$. In turbid waters, the minimum shifts to 550 nm. In contrast to the scattering coefficient, maximum $\kappa$ values are encountered 50–200 m deep, rather than near the surface (0–50 m) [9.3]. The position of the absorption maximum determines the spectral composition of the daylight at ocean depth. Absorption is most variable in the short wavelength range of the visible. In the red, $\kappa(\lambda)$ curves are universal, being dominated by the absorption wavelength of pure water.

Of all the experimental data available on the optical characteristics of the ocean, the information on the extinction coefficient $\varepsilon$ is most impressive. A number of monographs [9.2, 4] provide maps and plans of the $\varepsilon$ variation for various ocean regions. The ocean water extinction coefficient essentially depends on scattering, and ranges from 0.05–0.06 $m^{-1}$ to 2–3 $m^{-1}$ for different geographic regions. The mean values of $\varepsilon$ for the world's ocean are between 0.2 and 0.3 $m^{-1}$. The extinction coefficient is generally high in a surface layer, tending to decrease with increasing depth. As already mentioned, at some depth (around an abrupt density change) higher turbidity layers are observed. The spectral functions $\varepsilon(\lambda)$ are plotted in Fig. 9.2. As the transparency goes up, the minimum spectral distribution $\varepsilon(\lambda)$ is seen to shift from the yellow-green to the blue, whereas the $\varepsilon(\lambda)$ spectrum approaches that of pure water. Table 9.2 lists the mean $\varepsilon$ values and the variation coefficient $G$ at different $\lambda$. These results were obtained from the treatment of 274 samples of typical relatively pure water [9.3].

Comprehensive experimental data on the single-scattering albedo $\Lambda = \sigma/\varepsilon$ are available only for $\lambda = 550$ nm. For example, in the surface waters of the Indian Ocean ($\leqslant 100$ m), the mean $\Lambda$ is equal to 0.75 and the r.m.s. deviation is 0.05. For deeper waters (100 m), $\Lambda = 0.58$ and the r.m.s. deviation equals 0.10. The decrease in $\Lambda$ with increasing depth is due to decreasing scattering, whereas

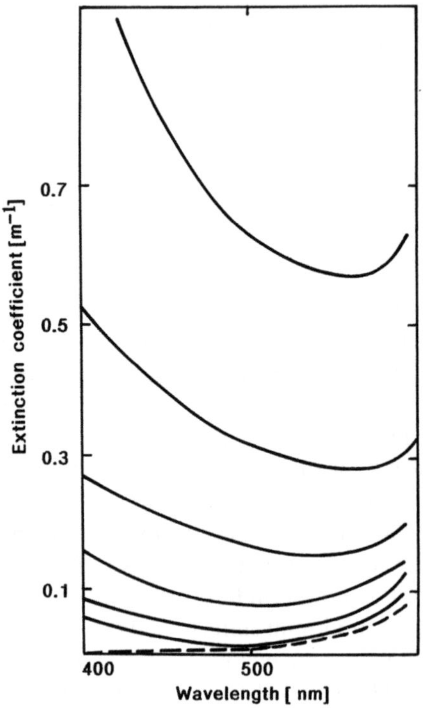


**Fig. 9.2.** Oceanic water extinction coefficients [9.3]. *Solid lines* show the data obtained in various regions of the ocean; *dashed lines*, that for pure sea water

**Table 9.2.** Mean sea water extinction $\langle \varepsilon \rangle$ and variation coefficient $G$ at different wavelengths.

| $\lambda$ [nm] | $\langle \varepsilon \rangle$ [m$^{-1}$] | $G$ [%] | $\lambda$ [nm] | $\langle \varepsilon \rangle$ [m$^{-1}$] | $G$ [%] |
|---|---|---|---|---|---|
| 410 | 0.169 | 36 | 510 | 0.107 | 42 |
| 430 | 0.148 | 37 | 530 | 0.113 | 39 |
| 450 | 0.131 | 40 | 550 | 0.116 | 37 |
| 470 | 0.117 | 42 | 570 | 0.123 | 33 |
| 490 | 0.107 | 44 | | | |

absorption at this wavelength only changes slightly and is close to pure water absorption.

In Sect. 9.1.2, we shall consider the correlation between $\Lambda$ and the extinction coefficient $\varepsilon$ for a number of regions in the world's ocean.

### 9.1.2 Simple Model of Optical Sea Water Characteristics

In order to carry out various methodological studies, diagnostic estimations of underwater vision and location systems, it is convenient to have a simple model of optical ocean properties. One of these can be built using the known physical model of sea water as a scattering and absorbing medium [9.6]. According to this model, sea water is a composition of large organic and small terrigenous (mineral) particles, pure sea water and soluble yellow substances. Optical constants of organic and mineral fractions, particle size distribution and other reference data are given in [9.6].

**Scattering Indicatrix.** According to this model, the scattering coefficient and scattering indicatrix of sea water are

$$\sigma = \sigma_w + C_l\sigma_l^0 + C_{sm}\sigma_{sm}^0 , \tag{9.1.1}$$

$$x(\beta) = \frac{\sigma_w x_w(\beta) + C_l\sigma_l^0 x_l(\beta) + C_{sm}\sigma_{sm}^0 x_{sm}(\beta)}{\sigma} , \tag{9.1.2}$$

where $\sigma_w$ and $x_w(\beta)$ are the scattering coefficient and scattering indicatrix of pure water; $\sigma_l^0$, $\sigma_{sm}^0$ and $x_l(\beta)$, $x_{sm}(\beta)$ the scattering coefficients (for unit concentration) and scattering indicatrices of large organic and small mineral particulate fractions of hydrosols; $C_l$ and $C_{sm}$ the concentrations of large and small particles. At $\beta < 2°$, the scattering indicatrix of sea water is completely determined by the large organic fraction; at $\beta \geqslant 15°$, by a small mineral one. Pure water scattering essentially contributes only at $\beta > 90°$. For $90° < \beta < 180°$, the scattering indicatrix only changes slightly and, in the following analysis, we consider it to be isotropic.

Bearing in mind the comments made above, the scattering indicatrix of sea water can be cast into the form

$$\begin{aligned} x(\beta) &= \eta x_l(\beta) + (1-\eta)x_{sm}(\beta) , \qquad 0 \leqslant \beta \leqslant \pi/2 \\ &= 2(1-F) , \qquad \pi/2 < \beta \leqslant \pi , \end{aligned} \tag{9.1.3}$$

where the quantity

$$\eta = \frac{C_l\sigma_l^0}{C_l\sigma_l^0 + C_{sm}\sigma_{sm}^0} = \frac{\sigma_l C_l/C_{sm}}{\sigma_{sm} + \sigma_l C_l/C_{sm}} \tag{9.1.4}$$

is the portion of light scattered by large organic particles, and $F$ is again the portion of light scattered into the forward semisphere. Processing a large number of scattering indicatrices of sea water has revealed a simple correlation between $1-F$ and the mean squared scattering angle $\beta_2$ (4.4.31), which is used in small-angle approaches (Sects. 4.4, 5)

$$1 - F \approx \beta_2/3 . \tag{9.1.5}$$

It will be recalled that in Sect. 4.7, we underlined the correlation between $1-F$ and the asymmetry parameter $g$ (Fig. 4.14). On the other hand, the parameter $\beta_2$ for the indicatrix (9.1.3) is

$$\beta_2 = \eta\beta_{2,l} + (1-\eta)\beta_{2,sm} , \tag{9.1.6}$$

where $\beta_{2,1}$ and $\beta_{2,\mathrm{sm}}$ are the values of the parameter $\beta_2$ for large organic and small mineral fractions. Processing the data given in [9.6] yields

$$\beta_{2,1} \approx 3.45 \times 10^{-3}, \qquad \beta_{2,\mathrm{sm}} \approx 0.27 . \tag{9.1.7}$$

Thus, taking into account (9.1.3–7), we obtained a single-parameter description of the scattering indicatrix. The quantities $\sigma_1^0$, $\sigma_{\mathrm{sm}}^0$ and the functions $x_1(\beta)$, $x_{\mathrm{sm}}(\beta)$ are known [9.6]. For the angles $0 < \beta < \pi/2$, the indicatrices $x_1(\beta)$ and $x_{\mathrm{sm}}(\beta)$ allow a simple approximation, which is convenient for analytical calculation (Sect. 6.1.1),

$$x_1(\beta) = 2b_1^2 e^{-b_1\beta}, \qquad b_1 = 41.7\ \mathrm{rad}^{-1} \tag{9.1.8}$$

$$x_{\mathrm{sm}}(\beta) = 2b_{\mathrm{sm}}^2 e^{-b_{\mathrm{sm}}\beta}, \qquad b_{\mathrm{sm}} = 4.6\ \mathrm{rad}^{-1} \tag{9.1.9}$$

The single scattering indicatrix parameter $\eta$ (the portion of light scattered by a large fraction) can be derived, for example, using the $C_1^0/C_{\mathrm{sm}}^0$ ratio (9.1.4). In terms of $\eta$ and applying (9.1.5, 6), the integral indicatrix parameters $\beta_2$ and $F$ are determined. The variation of $\eta$, i.e., the change in the concentration proportion $C_1/C_{\mathrm{sm}}$ provides a general description of the variability of scattering indicatrices in different regions of the world's ocean.

The analysis of experimental data shows that the characteristic indicatrices of pure ocean waters vary only slightly. A similar situation is also encountered for more turbid sea waters which contain a greater number of mineral particles. The values of $\eta \approx 0.69$ for ocean waters and $\eta \approx 0.62$ for sea waters are most typical. Therefore, for predictions and methodological calculations, we can suggest some "sea" and "ocean" models of representative indicatrices with the parameters listed in Table 9.3.

**Single-scattering Albedo.** It is essentially more difficult to measure the single scattering albedo $\Lambda$ than the extinction coefficient $\varepsilon$. The amount and accuracy of the experimental data available on $\Lambda$ are much less. Therefore, a simple approximate relationship between $\Lambda$ and $\varepsilon$, which results immediately from the physical model of optical sea water properties seems to be of particular significance.

The extinction coefficient for sea water is the sum of the extinction coefficients of all its optically active components

$$\varepsilon = \varepsilon_{\mathrm{part}} + \kappa_{\mathrm{ye}} + \varepsilon_{\mathrm{w}} , \tag{9.1.10}$$

**Table 9.3.** Parameters of representative indicatrices.

| Type | Parameters | | | Notes |
|---|---|---|---|---|
| | $a$ | $\beta_2$ | $1-F$ | |
| Ocean | 0.69 | 0.086 | 0.029 | $\beta_{2,1} = 3.45 \cdot 10^{-3}$ |
| Sea | 0.62 | 0.105 | 0.035 | $\beta_{2,\mathrm{sm}} = 0.27$ |

where $\varepsilon_{part}$ and $\varepsilon_w$ are the extinction coefficients of suspended particles and water; $\kappa_{ye}$ is the absorption coefficient of soluble yellow substances. In relatively pure waters, even at $\lambda = 500$–$540$ nm, the absorption of suspended particules is small as compared to pure water absorption and can be neglected ($\varepsilon_{part} \approx \sigma_{part}$). Then, taking into account that $\varepsilon_w = \kappa_w + \sigma_w$ and $\sigma_{part} + \sigma_w = \sigma$, we have

$$\varepsilon = \sigma + \kappa_{ye} + \kappa_w \,. \tag{9.1.11}$$

In the transparency window ($\lambda \approx 510$ nm), the absorption of yellow substances is small. According to [9.5], for different world ocean regions, $\kappa_{ye}/\varepsilon = 0.02$–$0.07$. Therefore, the model can approximately allow for yellow substance absorption in terms of the mean value $\kappa_{ye}/\varepsilon = 0.045$. Then, considering that at $\lambda \approx 510$ nm, $\kappa_w \approx 0.035\ \mathrm{m}^{-1}$ [9.5], from (9.1.11), we obtain

$$\Lambda = 0.955 - \frac{0.035}{\varepsilon}, \qquad [\varepsilon] = \mathrm{m}^{-1} \,. \tag{9.1.12}$$

The relation (9.1.12) was derived in [9.7], showing its high accuracy within $0.1\ \mathrm{m}^{-1} < \varepsilon < 0.5\ \mathrm{m}^{-1}$ for different world ocean regions.

Thus,the proposed one-parameter model of optical water properties requires that only the parameter $\eta$ or water type (sea, ocean), which determines the scattering indicatrix (Table 9.3), be given, in addition to the extinction coefficient $\varepsilon$, which defines the quantity $\Lambda$ by (9.1.12).

## 9.2 Object Visibility in Sea Water

### 9.2.1 Light Source Visibility

Let us consider a limiting visibility range of a radiation source in the sea. As in Sect. 8.2, we assume that the source energy is sufficiently large and that the visibility range limit is due to the generation of the visible contrast below the threshold system contrast. A general analysis of the behavior of the radiance contrast of a self-luminous object against the scattered radiation background was made in Sect. 8.2. Here, we use the data of Table 8.6 which covers a wide range of $\Lambda$ values, including those typical of sea water ($\Lambda = 0.7$–$0.9$). It is seen that $\tau^*$ essentially increases with absorption due to the lower scattered radiation background. The range of a point source detection is about twice as large as that of an infinitely wide source. The same proportion is valid if a diffuse source is substituted for one with a narrow directional diagram. The change in the scattering indicatrix does not produce any strong effect on $\tau^*$. Moreover, the real range of sea water scattering indicatrices is not very wide. In the case of an infinitely wide source of sufficiently large power, the visibility range in water may reach $\tau^* = 30$–$35$ at the aperture angle $2\gamma_s = 2'$ and $\tau^* = 18$–$25$ at $2\gamma_s = 30'$ (solar radiation divergence). For a finite source, at $2\gamma_s = 2'$, maximum $\tau^* = 50$–$60$. Clearly, depths as large as these may be considered possible only by convention. Real limitations to the visibility range are generally attributed to the negligible energy of the valid signal, which cannot be detected against noise, rather than to the low contrast.

### 9.2.2 Range of Visibility of a Sunlit Object at Ocean Depth

The analysis of the visibility of underwater objects illuminated by external sources may be based on (8.3.8, 9), taking into account sea water properties. As follows from (8.3.8), the visibility range at large depth is

$$\tau^* = \frac{1}{1 - \gamma \cos\theta} \ln \frac{|k_0|}{k^*} . \tag{9.2.1}$$

We recall that here, $k_0$ is the true object contrast, $k^*$ the threshold contrast of the receiver, $\theta$ the angle between the $z$ axis and the object-viewer direction (Fig. 8.5), and $\gamma$ the dimensionless asymptotic attenuation coefficient. Setting $k^* = 0.03$ (threshold vision contrast), we consider the limiting cases of viewing black and white objects. For a black object (albedo $A = 0$, contrast $k_0 = -1$), from (9.2.1), we have

$$\tau^* = \frac{3.5}{1 - \gamma \cos\theta} , \tag{9.2.2}$$

i.e. the visibility range depends on the spatial orientation of the viewer and the object. For three definite orientations (viewing from below, in the horizontal direction and from above), we have

$$\tau^*_{\downarrow} : \tau^*_{\rightarrow} : \tau^*_{\uparrow} = \frac{1}{1 + \gamma} : 1 : \frac{1}{1 - \gamma} . \tag{9.2.3}$$

The black object is most visible when observed from below, its visibility being most poor in the opposite direction. The differences in $\tau^*$ are the greater, the larger $\gamma$, i.e., the lower the single scattering albedo $\Lambda$ and the less extended the scattering indicatrix in the forward direction.

We consider one more extreme case where the object albedo $A \approx 1$. To calculate $\tau^*$ using (9.2.1) requires a knowledge of the contrast $k_0$ determined by (8.3.9). In Sect. 8.3.2, we assumed weak anisotropy of the angular radiance distribution $I(z, \theta)$. In water, the angular radiance distribution $I(z, \theta)$ is essentially extended in the forward direction, such an assumption being only valid for viewing from above [$I(z, \theta) \approx$ const at $\theta > \pi/2$]. Here $k_{0,\downarrow} = A/r - 1$. This assumption is not required when viewing from below. Because the sea albedo $r$ is negligible, $k_{0,\uparrow} \approx -1$ may be set. Therefore, at $k^* = 0.03$, $A \approx 1$ and $r \approx 0.03$, and from (9.2.1), we get

$$\tau^*_{\downarrow} = \frac{1}{1 + \gamma} \ln \frac{A}{0.03r} \approx \frac{7}{1 + \gamma} , \qquad \tau^*_{\uparrow} \approx \frac{3.5}{1 - \gamma} , \tag{9.2.4}$$

or

$$\frac{\tau_{\downarrow}^{*}}{\tau_{\uparrow}^{*}} = \frac{2(1-\gamma)}{1+\gamma} . \tag{9.2.5}$$

### 9.2.3 Sekky's Disc Depth of Visibility

In hydrooptics, the problem of the depth of visibility of a white Sekky's disc observed through water surface is of particular interest. This is connected with the estimation of water extinction from the visibility depth of the Sekky's disc lowered from a ship long ago and adopted in hydrooptical practice. The simplicity of such observations is the reason why they are so widely used, resulting in a large number of measurements of the visibility depth $z_d$ of a standard white 30 cm diameter disc in different regions of the world's ocean [9.1, 8].

As noted in some studies,

$$z_d = \beta/\varepsilon , \tag{9.2.6}$$

where $\beta$ is independent of the optical water properties. This essentially means that the optical white disc detection depth $\varepsilon z_d$ is constant. In practice, this simple relation is used to proceed from the measured depth $z_d$ to the extinction coefficient $\varepsilon$. However, more careful analysis shows that for different regions of the world's ocean, the quantity $\beta$ has a wide range, while the white disc visibility depth does not depend on $\varepsilon$ alone.

It is shown in [9.1] that, approximately,

$$\beta = \frac{1}{1+\alpha/\varepsilon} \ln \frac{A-r}{rk^*} , \tag{9.2.7}$$

where $\alpha$ is the vertical attenuation coefficient for the irradiance in sea water. Taking a real range of $\alpha/\varepsilon$ and $r$, we may show that, at $A \approx 0.9$, maximum $\beta \approx 7.5$. At comparatively large sea albedo $r$, $A \approx 0.6$ and low sea irradiance with oblique sun rays, the quantity $\beta \approx 4$. The most probable $\beta$, in agreement with (9.2.7), lies between 4.7 and 5.

The problem of white disc visibility has been elaborated in [9.9] where a number of factors which affect $z_d$, such as the dependence of the contrast sensitivity of sight on the spreading width of the disc/background boundary, the impact of sea roughness, possible screening of sun rays by a viewer ship are taken into account. As shown, for example, viewing from the illuminated, rather than from the shady side of a ship may decrease $z_d$ by 20%, and small sea roughness, by 30–35%.

It is worth considering the problem of the visibility of Sekky's disc from point of view of theory of object detection developed in Sect. 5.4.2. In fact, the quantity $z_d$ may be considered to be the limiting range of detection of the object, with given size and albedo, illuminated by the sun and viewed through the water

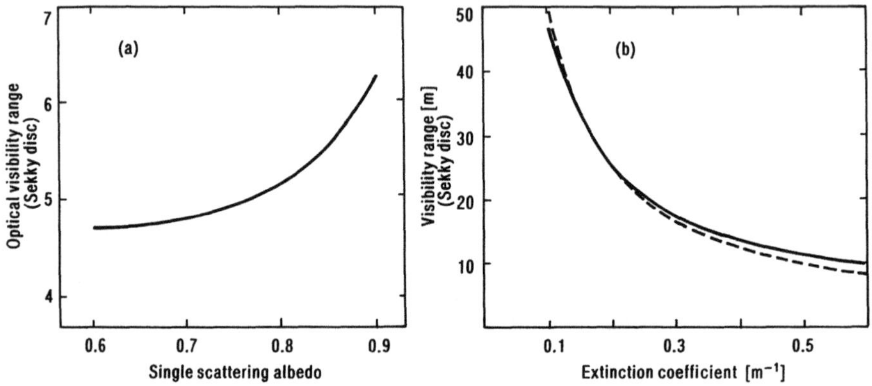


**Fig. 9.3.** Sekky's disc range of visibility $\beta$ vs. scattering albedo $\Lambda$ (a) and $z_d$ vs. $\varepsilon$ (b). *Dashed curve* shows $z_d = 5/\varepsilon$

surface. In Fig. 9.3, $\beta = \varepsilon z_d$ is plotted against $\Lambda$ and the depth $z_d$ is shown as related to $\varepsilon$ [for the suggested model (Sect. 9.1.2), $\Lambda = 0.955 - 0.035/\varepsilon$] in agreement with the detection theory developed in Sect. 5.4.2 for a 30 cm diameter disc with $A = 0.8$ at a threshold vision contrast of $k^* = 0.03$. As seen, $\beta = \varepsilon z_d$ increases with $\Lambda$. The data of the calculation within the detection theory are in good agreement with the results of [9.9]. In Fig. 9.3b, the dashed line represents $z_d = 5/\varepsilon$. At $\beta = 5$, (9.2.6) is seen to give a good description of $z_d$ in ocean waters, for which the simple model suggested in Sect. 9.1.2 can be applied.

## 9.3 Underwater Television

Problems concerned with the development of the world's ocean have required advanced facilities for viewing in water and, in particular, the design of underwater television systems. Nowadays, underwater television systems are commonly used to search for and develop natural ocean resources, view the sea-bed, underwater construction, fishery, search for and inspect sunken ships, etc. For example, the authors of review [9.10] remind us that after a prolonged search for atomic bombs sunk in Palomar (Spain) in 1966, specialists in the US Naval Forces clearly understood the importance of improvements, however modest, in underwater television.

In the USSR, underwater television systems were first developed in the fifties at the Institute of Oceanology, USSR Academy of Sciences [9.11]. The design and improvement of lasers for the blue-green have offered ample scope for progress in this field [9.10, 12, 13]. Pioneering experiments using lasers in underwater television date back to the early seventies and owe much to optical frequency doublers. By now, different television systems have been developed and employed, including portable cameras for divers (at 30–40 m depths), stationary underwater equipment placed on the sea-bed, underwater TV searchers attached to towing vehicles and underwater self-propelled TV cameras [9.14]. Television systems can also be set on board ship, helicopter or plane. In this case, observation is through a rough sea surface.

The main shortcoming of underwater television systems is the small limiting distance between the receiving-transmitting system and the object of observation. The operating range of existing systems is no more than 100 m, even in pure ocean waters (at $\varepsilon \approx 0.1\ \mathrm{m}^{-1}$). For many systems, it does not exceed the range of visual observations. Nevertheless, as compared with visual observation, underwater television has the advantage of offering the possibility of operating at any depth, unlimited time of operating under water, documentary reliability and independence. The implementation of wireless communication lines will ensure operation when the object of observation and the viewer who might, for example, be on board ship or on shore, are far apart.

### 9.3.1 Underwater TV Systems

Viewing systems of all the types (narrow-wide (1), wide-narrow (2) and narrow-narrow (3)) described in Sect. 5.3.2 are commonly used in underwater TV techniques. Each of these systems may employ both continuous and pulsed light sources.

As already stated in Chap. 5, in any system the light reflected by illuminated sea water (BSN) and different background noises reach the receiver, along with the valid signal reflected by the object. In daytime, at small and moderate observation depths, the most essential background is generated by scattered sunlight. The design features and efficiency of underwater TV systems are mainly specified by the type of system, as well as by the BSN and background suppression techniques.

In continuous source systems, the BSN can be reduced by separating the source and the receiver [9.14], as well as by applying polarization cut-off [9.15–17]. A certain increase in the source-to-receiver distance may essentially suppress the BSN and enhance the visible object contrast. A large source-to-receiver distance, however, is inconvenient for design reasons.

Polarization cut-off consists of the following. The BSN is mainly induced by quasi-single scattered light which preserves the polarization of an initial beam [9.18]. If the object of observation has a large albedo $A$, then its multiple scattering causes almost complete depolarization of the reflected radiation [9.19]. A polarizer, which transmits only a cross-polarization component relative to an incident beam, is placed in front of the receiver. In this case, the BSN will be drastically suppressed, while the valid signal will be only twice as weak. The observed contrast will be noticeably enhanced. As the albedo $A$ decreases, the reflected light partially preserves the incident beam polarization. Therefore, in viewing objects with a small albedo, the polarizational discrimination of the BSN is less effective than the base increase [9.14, 17].

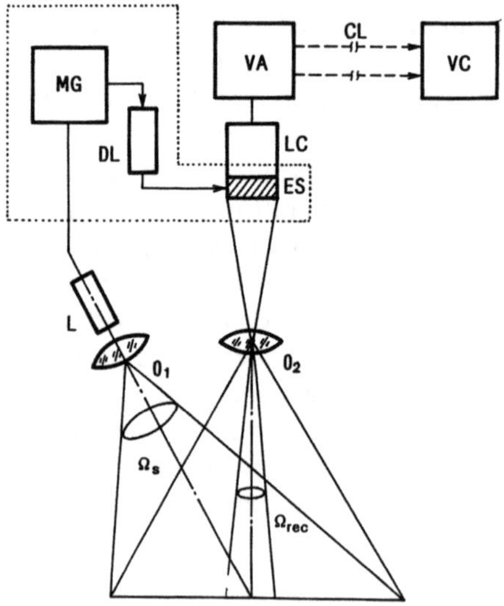


**Fig. 9.4.** Pulsed laser system 2 [9.14]: L, laser; $O_1$ and $O_2$, source and receiver lenses; ES, electronic shutter; LC, light converter (image converter and camera tube); MG, master generator; DL, delay line; VA, video amplifier; CL, communication line; VC, video control; *dotted line*, BSN cut-off unit

When used in vision systems, pulsed sources provide time gating. In this instance, a light converter (photomultiplier, image converter, etc.) equipped with an electronic shutter which "opens" only for the time-of-arrival of the signal reflected by the object is used. In this way, a greater BSN portion is cut-off. At the same time, time gating essentially suppresses the light background induced at the photodetector entrance by different sources (for example, sunlight). In pulsed systems, as well as in continuous source systems, noise is also quenched by spectral narrow-band light filters.

Specific schematic diagrams and a description of underwater TV systems can be found in [9.10–14, 20, 21]. We shall confine ourselves to information on the most commonly used systems, as appears in reviews [9.14, 21].

Underwater television employs sytem 2, i.e., the system which involves illumination of the whole field of vision and a narrow receiving diagram (Fig. 5.4). In spite of the progress made in developing pulsed lasers, systems with continuous wave sources are used more commonly. A video signal arrives at an amplifier and then, at a video control device along the communication line.

In the pulsed system 2 (Fig. 9.4), the pulsed laser is started by the master oscillator which, at that instant, sends an electric pulse to a shutter of the light converter via the delay line. The image converter, which is connected to the camera tube is generally used as a receiver in such systems. It is required that at least one pulse occur at the time when the frame is received. Making allowance for the standard TV frame frequency, this agrees with the repetition frequency not less than 50 Hz.

A unit, consisting of a master oscillator, delay line and electronic shutter (the dotted line in Fig. 9.4) realizes BSN cut-off by opening the light converter for the time-of-arrival of an object reflected signal.

The block-diagram of the continuous wave source system 2, compared to that in Fig. 9.4, has no BSN time cut-off device. The implementation of systems 1 and 3 requires, in a pulsed regime, a pulse repetition frequency of between $10^4$ and $10^7$ Hz, depending on the resolution and frame recording time. On account of technological problems, systems 1 and 3 are currently implemented in underwater television, based as they are on continuous wave sources which are essentially different in field scanning devices.

A search TV system of type 3 placed on a vehicle is described in [9.12]. In this system, synchronous scanning by the source and receiver diagrams is realized along one coordinate; scanning is achieved by moving the TV-system vehicle along the other. For the continuous source system 3, the scanner provides complete frame image scanning [9.13].

We have so far been concerned with classical underwater TV systems. Recent advances in developing holographic and interference systems for this purpose have been outlined (Chap. 7). Limited temporal and spatial coherence systems, as well as time-average holography, may be used for underwater vision problems. The main estimates given in Chap. 7 were made exactly for scattering media of sea-water type.

### 9.3.2 MTF and Valid Signal and Noise Energy in Underwater Vision Systems

Contrast and signal-to-noise ratio were the main quantities introduced in Chap. 5 to characterize the quality of the image formed by one system or another. In the majority of systems which use high-sensitive TV tubes, a photomultiplier or image converter connected to a camera tube, shot noises essentially exceed dark noises. Dark noises exceed shot noises in the tube only in those systems based on the vidicon and its modifications, and intended to operate in bright daylight [9.14]. Therefore, for the majority of systems, the signal-to-noise ratio is calculated from (5.3.15) or, when taking into account the system transport noises, from (5.4.2). They imply that an estimation of the image quality in the vision system requires a knowledge of the MTF $T(\nu)$, which determines the contrast $k$, as well as the valid signal and backscattering noise energy. If the noise from external sources reaches the photodetector, then its energy $W_{ext}$ per picture element for a recording time $t_{el}$ must be known.

The procedure and formulas for calculating vision system characteristics were presented in previous chapters. Below, we shall highlight those which are recommended for estimating and comparing underwater TV systems. In so doing, a number of specific factors must be considered.

Design calculations of vision systems are most often associated with estimating the efficiency of one system or another, or seeking ways to optimize it. Since the range of hydrooptical parameters for different ocean regions is sufficiently wide, comparatively low accuracy is required in calculating light field characteristics. In such cases, different approximate procedures are appropriate. Sea water is characterized by an essentially extended scattering indicatrix ($g = 0.8$–$0.97$) and appreciable absorption ($\Lambda = 0.5$–$0.9$). The small-angle (Sect. 4.4) and small-angle diffusion (Sect. 4.5) approximations may, therefore, be successful in underwater vision calculations, whereas the background energy may be determined using the quasi-single scattering approximation (Sects. 3.6, 4.6) or asymptotic solutions at $(1-\Lambda)/(1-g) \gg 1$ (Sect. 3.3.3). In particular, sea water OTF and MTF may be calculated using (4.4.35; 6.1.9, 11).

**Medium and System MTF.** In [9.22–24], the sea water indicatrix is approximated as $x(\beta) = 2a\exp(-a\beta)/\beta$ at $a = 5$–$7$, which in the SAA gives a simple MTF formula (6.1.9),

$$T(\nu) = \mathrm{e}^{-\sigma z_0}\left[\nu z_0/a + \sqrt{1+(\nu z_0/a)^2}\right]^{a\sigma/\nu}. \tag{9.3.1}$$

The modified SADA provides good results (Sects. 4.5.7; 6.1.2). In this approximation, the medium MTF is

$$T(\nu) = \mathrm{e}^{-\sigma z_0} + \sum_{i=1}^{2} \xi_i \mathrm{e}^{-V_{r,i}\nu^2/2}, \tag{9.3.2}$$

where

$$\xi_1 = \mathrm{e}^{-(1-\eta)\sigma z_0} - \mathrm{e}^{-\sigma z_0}, \qquad V_{r,1} = \frac{\eta\sigma\beta_{2,1}z_0^3}{6(1-\mathrm{e}^{-\eta\sigma z_0})},$$

$$\xi_2 = \mathrm{e}^{-(1-\eta)\sigma z_0}, \qquad V_{r,2} = \frac{(1-\eta)\sigma\beta_{2,2}z_0^3}{6[1-\mathrm{e}^{-(1-\eta)\sigma z_0}]}. \tag{9.3.3}$$

Representative sea water parameters $\eta$, $\beta_{2,1}$ and $\beta_{2,2}$ were given in Sect. 9.1.2. For the proposed model, $\beta_{2,1} = \beta_{2,1}$ and $\beta_{2,2} = \beta_{2,\mathrm{sm}}$. Formula (9.3.2) is more convenient than (9.3.1) to calculate the radiant flux through a finite aperture, or to determine the function $\Psi$ in the problem of object detection range (Sects. 5.4.2; 6.6). Moreover, it can be applied to the entire range of depths, including large depths where the SAA error is essential.

In practical calculations, it is advisable to estimate the range of spatial frequencies, from which one may use the asymptotic MTF $T(\nu) = \exp(-\sigma z_0)$ due to unscattered radiation. The estimates derived using (6.1.11) show that the sea water MTF differs from its asymptotic value at $\nu \to \infty$ by less than by 20% for $\nu/\sigma \gtrsim 100$–$150$.

Equations (9.3.1, 2) describe the medium MTF. We now consider the MTF of the whole viewing system. First, we estimate the frequency range, where the system MTF is determined by the medium MTF and is independent of the source (receiver) characteristics. Let a source with the directional diagram $\varphi_s^{(1)} = \exp(-\theta^2/\gamma_s^2)$ be used in system 1, or a receiver with the sensitivity diagram $\varphi_{\mathrm{rec}}^{(2)} = \exp(-\theta^2/\varphi_{\mathrm{rec}}^{(2)})$ be used in system 2. Then, for the MTFs of systems 1 and 2, we have

$$T^{(1)}(\nu) = T(\nu)\mathrm{e}^{-(\nu z_0\gamma_s)^2/4},$$

$$T^{(2)}(\nu) = T(\nu)\mathrm{e}^{-(\nu z_0\gamma_{\mathrm{rec}})^2/4}. \tag{9.3.4}$$

Hence, it follows that, at spatial frequencies $\nu z_0 < \gamma_s^{-1}$ or $\nu z_0 < \gamma_{rec}^{-1}$, one may consider that $T^{(1)}(\nu) = T^{(2)}(\nu) \approx T(\nu)$. For higher frequencies, the source (receiver) deteriorates the system MTF.

In sea viewing systems, the ultimate resolution is determined by turbulent fluctuations of the sea water refractive index. Indeed, one may assume, for example, that

$$T^{(1)}(\nu) = T(\nu)e^{-(\nu z_0 \gamma_s)^2/4} T_{fl}(\nu)\,, \tag{9.3.5}$$

where $T_{fl}(\nu)$ is the medium MTF determined by turbulent fluctuations. The calculations made in [9.25–27] and measurements performed in [9.28] show that $T_{fl}(\nu)$ noticeably differs from unity only at $\nu z_0 > 10^3$ rad$^{-1}$. Such a frequency band is transmitted by the system (without water scattering) at $\gamma_{s(rec)} < 10^{-3}$ rad. In underwater television problems, implying lower angular resolution, turbulent fluctuations can be disregarded.

We now consider the narrow-narrow system MTF $T^{(3)}(\nu)$ expressed by (5.3.10). Figure 9.5 shows $T^{(3)}(\nu)$ as calculated by (5.3.10) for the indicatrix (6.1.8) when $T(\nu)$ is given by (9.3.1). The data are taken from [9.29]. For comparison, the dashed curve depicts the MTF of system 2 at $\sigma z_0 = 5$. The MTF of system 3 is seen to be much better, i.e., the resolved frequency band is much wider than in systems 1 and 2.

**Valid Signal and BSN Energy.** We consider the energy of the valid signal and noises in underwater viewing systems. General formulas for the valid signal $\eta_{vs}$ power transfer coefficient are given in Table 5.3. They contain $S(z_0)$ which, in the

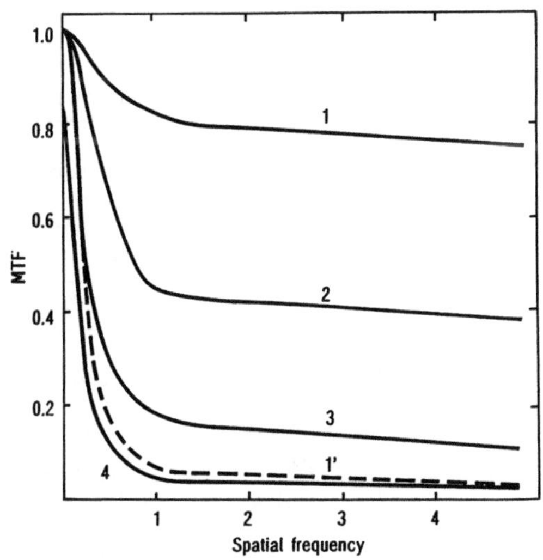


**Fig. 9.5.** MTF versus $(\nu/\sigma)\sqrt{\beta_2/2}$ for system 3 at $\gamma_s = \gamma_{rec} = 5'$ and $\sigma z_0 = 5$ (curve 1), 6 (curve 2), 7 (curve 3), 8 curve 4). For comparison, the *dashed line* (curve 1') shows the MTF of system 2 at $\gamma_{rec} = 5'$ and $\sigma z_0 = 5$ for a medium with indicatrix (6.1.8) at $a = 7$

SAA, is $S(z_0) = \exp[-(\varepsilon - \sigma F)z_0]$. The expressions for $\eta_{VS}$ in different viewing systems can be found in a straightforward manner from (5.3.6), by substituting appropriate solutions for the irradiance distributions $E_s(\vec{r})$ and $E_{rec}(\vec{r})$. For this reason, we may, for example, use analytical solutions for the spread function in the SADA and modified SADA (Sect. 4.5).

The main procedures for calculating the BSN were discussed in Sects. 3.6; 4.6, 8. The BSN from a stationary source at a sufficiently large receiving aperture angle ($\gamma_{rec} > 10°$) may be calculated using (4.6.1). In pulsed viewing systems, the time dependence of the BSN may be estimated for an arbitrary time instant $t$, using the appropriate methods described in 4.8.3; for moderate $t$, at a large receiving angle $\gamma_{rec}$, (4.8.30, 31) may be employed. The BSN time dependence is illustrated in Figs. 4.15, 17.

The above combination of formulas and procedures may be used to calculate the characteristics of the systems of viewing in sea water under any conditions. Here, we shall consider a specific case of underwater object vision using system 2. To solve the object detection and discrimination problems (Sects. 5.4.2, 3) requires one to calculate the energy of the background $W_b$, as well as $W_{VS} = W_0 \eta_{VS}$ and the function $\Psi(\Sigma_{ob}, \Sigma_{el})$, which determine the energy of the valid signal $\Delta W$ in formula (5.4.2) and the background component $W_b^{ob}$ (See (5.4.4, 11)). The quantity $W_{VS}$ signifies the valid signal energy in observing a homogeneous object with albedo $A(\vec{r}) = 1$. The function $\Psi(\Sigma_{ob}, \Sigma_{el})$ was discussed in Sect. 6.6. The quantity $W_b$ may be represented as a sum of the BSN energy and external (sunlight) background $W_{ext}$ energies, i.e., $W_b = W_{BSN} + W_{ext}$.

Assuming that the viewing depths are not very large, we write the formulas for $W_{VS}$, $W_{BSN}$ and $W_{ext}$ (Table 9.4) using the quasi-single scattering approxima-

**Table 9.4.** Formulas for calculating signal and noise energies in underwater TV systems.

| | | |
|---|---|---|
| Passive system | Valid signal energy | $W_{VS} = \dfrac{W_s \mu_s t_{atm} t_{sur}}{\pi} \Sigma_{rec} \Omega_{rec} e^{-(\varepsilon - \sigma F)[z_0/\mu_s' + (z_0 - z_{rec})/\mu_0']}$ |
| | Sunlight background energy | $W_{ext} = \dfrac{W_s \mu_s t_{atm} t_{sur}}{\pi} \Sigma_{rec} \Omega_{rec} \dfrac{\Lambda(1-F)}{2(1-\Lambda F)(\mu_s' + \mu_0')} e^{-(\varepsilon - \sigma F) z_{rec}/\mu_s'}$ |
| Active system | Valid signal energy | $W_{VS} = \dfrac{W_0 \Sigma_{rec} \Omega_{rec}}{\pi \Omega_s [(z_0 - z_{rec})/\mu_0']^2} e^{-2(\varepsilon - \sigma F)(z_0 - z_{rec})/\mu_0'}$ |
| | BSN energy | $W_{BSN} = \dfrac{W_0 \Sigma_{rec} \Omega_{rec}}{\pi \Omega_s [(z_0 - z_{rec})/\mu_0']^2} \dfrac{\Lambda(1-F)}{4(1-\Lambda F)} [e^{-2(\varepsilon - \sigma F) z_1/\mu_0'} - e^{-2(\varepsilon - \sigma F) z_2/\mu_0'}]$ |
| | Sunlight background energy | $W_{ext} = \dfrac{W_s \mu_s t_{atm} t_{sur}}{\pi} \Sigma_{rec} \Omega_{rec} \dfrac{\Lambda(1-F)}{2(1-\Lambda F)(\mu_s' + \mu_0')} e^{-(\varepsilon - \sigma F) z_{rec}/\mu_s'}$ |

tion which provides sufficiently high accuracy in representing light field characteristics in underwater viewing systems.

Let us comment on Table 9.4. As assumed, the object of observation is at a depth $z_0$, the photodetector at $z_{rec} < z_0$ under the water surface; arccos $\mu'_0$ and arccos $\mu'_s$ are the angles of observation direction and unscattered light (upon refracting in water), with the vertical to the water surface; arccos $\mu_s$ is the angle of unscattered sunlight radiation in the atmosphere. The quantity $W_{VS}$ in the passive observation system and $W_{ext}$ depend on the energy $W_s = E_s \Delta\lambda t_{el}$ of the solar radiation incident on the normal unit plane at the upper boundary of the atmosphere within the spectral interval $\Delta l$ for the picture element recording time $t_{el}$. For $\lambda = 550$ nm, the solar constant $E_s = 0.2$ J/m$^2$nm.

In the pulsed time gating observation system, the quantity $W_{BSN}$ depends on the depth $z_1$ and $z_2$ which correspond to the onset and end of a time gate. In this case, the picture element recording time (for a single pulse) $\Delta t_{el} = 2(z_2 - z_1)/v\mu'_0$; $v$ is the light velocity in water. In the formulas of Table 9.4, $t_{sur} \approx 0.98$ is the sea surface transmission, $t_{atm}$ the transmission coefficient of sunlight in the atmosphere on its path to the sea surface. In approximate estimations, we may set $t_{atm} \approx 1$. The quantity $F$ of sea water is presented in Table 9.3.

If the TV system is at some altitude $H$ above the sea surface, then the image transfer problem becomes much more difficult. The atmosphere and rough surface are two additional sources of image distortion and noise. Some basically new aspects associated with image transfer through a rough sea surface will be considered briefly in the next section. We shall, therefore, make some comments on atmosphere effects on the MTF, valid signal and background, assuming a smooth sea surfaces.

1. If the observation altitude is small, or of any magnitude under high-visibility conditions, then the atmospherical optical thickness is small. In this case, the influence of the atmosphere on the normalized PSF and MTF of the ocean-atmosphere system can be neglected.
2. At a fixed value of $\Omega_{rec}$, in the active vision system, the energy $W_{VS}$ decreases with increasing altitude mainly because of the angular spread of the beam along its path from the source to the object, i.e., the reduced object irradiance. Therefore, in the formula for $W_{VS}$ (Table 9.4), the factor $\mathscr{L}^{-2} = [H/\mu_0 + z_0/m\mu'_0]^{-2}$ will appear in place of the factor $[(z_0 - z_{rec})/\mu'_0]^{-2}$. In the passive observation system, the value of $W_{VS}$ is independent of $H$.

   When the image cell area $\Sigma_{el}$ is fixed, an additional energy decrease with increasing altitude $H$ is observed both in active and passive observation systems, and is due to the reduction in the solid receiving angle $\Omega_{rec} = \Sigma_{el}/\mathscr{L}^2$. It should be noted that, in this case, the background energy $W_b$ decreases, correspondingly.

   Some additional losses in the valid signal energy are connected with the extinction in the atmosphere and radiation transmission through the sea surface. These losses are dependent on the radiation incidence and observation angles, meteorological visibility range or altitude and do not usually exceed 40–50%.

3. In daytime observations, additional background noise is induced due to the reflection of transmitted solar radiation from the sea surface, as well as to the reflection of solar radiation from an atmospheric layer of thickness $H$. The role of the latter component becomes more important as $H$ increases.
4. In active laser systems, an additional BSN component appears in connection with the backscattering in the atmosphere. It is, however, much smaller than sea water BSN, growing essentially only at large optical depths of an object. Therefore, it is hardly necessary to allow for this noise when analyzing television systems, but may be important in laser bathymetry problems.

These observations, together with the approaches described in Sects. 3.4, 6, 7, provide a key to allow for the effect of the atmosphere on image quality in viewing underwater objects through the sea surface.

### 9.3.3 Limiting Ranges of Underwater Vision

In viewing with a continuous source system 2 (wide-narrow), the underwater range of vision is dependent on the water type, object reflection coefficient and characteristic sizes of its details, which are $\varepsilon z_0 = 3–9$ [9.14] and can be compared with the visual observation range. This value may be higher in pulsed systems with BSN suppression. Figure 9.6 illustrates the ranges of square object detection and discrimination (according to Johnson, Sect. 5.4.1) in sea water when an underwater TV system is used. The calculations have been made according to

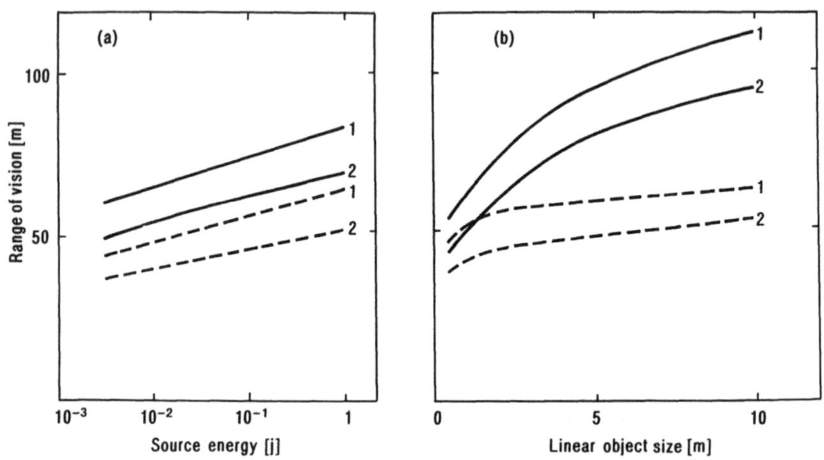


**Fig. 9.6.** Detection (*solid lines*) and discrimination (*dashed lines*) ranges of a d-side square in ocean water as functions of (a) source energy $W_0$ at $d = 2$m, and of (b) $d$ at $W_0 = 0$ 1 J for an albedo $A = 0.8$ (curves 1), 0.2 (curves 2) at $\varepsilon = 0.1$ m$^{-1}$, $\Lambda = 0.6$

the detection and discrimination concept given in Sects. 5.4.2, 3 by (9.3.2, 3), those cited in Table 9.4, and using the model of optical sea water characteristics (Sect. 9.1.2).

Improvement of the image quality and range of vision is a key problem in underwater television. The analysis of different vision systems and the estimation of possible SNR amendment by BSN suppression made in Chap. 5, as well as the comparison of coherent and incoherent vision systems in Chap. 7, apply completely to underwater vision systems. In summarizing the results, we can state the following: all the improvements are intended to suppress the BSN, or to simultaneously solve two problems, such as BSN suppression and MTF improvement, by cutting off some portion of the scattered light as it interacts with the observation object. In the latter case, the valid signal transfer coefficient inevitably decreases. Such systems are, therefore, preferable when the mean source radiation power is high and the vision threshold limited due to low contrast.

We shall now consider the problem of increasing the visibility range, and its ultimate potential. A simple analysis will help us to visualize the situation. Formula (5.3.15) is used for the signal-to-noise ratio $\delta$. For the sake of certainty, we consider systems 1 and 2 at $\Omega_s^{(1)} = \Omega_{rec}^{(2)}$. It is assumed that the BSN is completely suppressed in the system, i.e., $\eta_{BSN} = 0$. Moreover, for simplicity, we assume that a harmonic test object is being observed. Now, $\delta$ depends only on the observation range $z_0$ due to $T(\nu)$ and $\eta_{VS}$. We emphasize this dependence by writing $T(\nu, z_0)$, $\eta_{VS}(z_0)$ and $\delta(\nu, z_0)$.

The formulas of Table 5.3 give

$$\eta_{VS}(z_0) = \frac{\Sigma_{rec}}{\pi z_0^2}\exp(-2\kappa z_0)\,.$$

It should be noted that at the initial object contrast $k_0 = 1$, $T(\nu, z_0) \approx 1$ at $\nu \ll 1$ and $T(\nu, z_0) = \exp(-\sigma z_0)$ at $\nu \gg 1$. In view of this, we rewrite (5.3.15) as

$$\delta(\nu, z_0) = C\frac{e^{-\kappa z_0}}{\kappa z_0}\sqrt{\kappa^2 \Sigma_{rec} P_0 t_{el}}\,, \qquad \nu \ll 1\,, \tag{9.3.6}$$

$$\delta(\nu, z_0) = C\frac{e^{-\varepsilon z_0}}{\varepsilon z_0}\sqrt{\varepsilon^2 \Sigma_{rec} P_0 t_{el}}\,, \qquad \nu \gg 1\,. \tag{9.3.7}$$

Here, $C$ is the receiver constant. Given some threshold SNR $\delta^*$, which is determined by the required image transfer quality (Sect. 5.1.4), from (9.3.6, 7), we find the limiting vision depth

$$\kappa z_0^* + \ln(\kappa z_0^*) = -\ln\delta^* + \frac{1}{2}[ln(\kappa^2\Sigma_{rec}) + \ln(P_0 t_{el})] + \ln C\,, \qquad \nu \ll 1 \tag{9.3.8}$$

$$\varepsilon z_0^* + \ln(\varepsilon z_0^*) = -\ln\delta^* + \frac{1}{2}[ln(\varepsilon^2\Sigma_{rec}) + \ln(P_0 t_{el})] + \ln C\,, \qquad \nu \gg 1 \tag{9.3.9}$$

For the optical range $\varepsilon z_0^* \gg 1$, the second terms on the l.h.s. in (9.3.8, 9) may be neglected, suggesting obvious but rather unfavorable inferences. If some limiting observation depth $z_0^*$ is achieved at the given $\delta^*$, then an order-of-magnitude increase in the mean source power $P_0$ results in increase an $\varepsilon z_0^*$ about unity. This is also seen in Fig. 9.6. A similar dependence occurs when the receiving aperture $\Sigma_{rec}$ and frame recording time $t_{el}$ are increased. Generally speaking, light extinction in water is the key factor responsible for the range of vision. The range of vision is inversely proportional to the extinction coefficient. It would seem useful, therefore, to employ a frequency-tuned laser to adjust the source wavelength to the maximum spectral transparency of water in the observation region.

It should be noted that for the fixed visibility range $z_0$ and SNR $\delta \gtrsim \delta^*$, the image contrast may be improved by increasing $P_0$, $t_{el}$, $\Sigma_{rec}$, etc. The difficulties in improving the limiting range of vision are due to the logarithmic dependence of $z_0^*$ on all system parameters.

The limiting range of vision in active systems seems to be less than the visual range of a self-luminous object of sufficiently high radiance. According to prediction and experiment, $\varepsilon z_0 = 15$–$20$ depending, on the object parameters and water characteristics [9.14, 28, 30].

The prospects for underwater television coincide, in many respects, with those for television in general, and the application of colour and stereo television. However, the specific features of sea must be taken into account. For example, as noted in [9.17], even when the distance from the camera tube to the object of observation is small, the spectral extinction of sea water essentially distorts colour reproduction.

The potential for a wider application of underwater television is connected, to a large extent, with the implementation of robotics, wireless communication lines, blue-green frequency-tuned lasers of high mean power and large light pulse repetition frequency.

## 9.4 Image Transfer Through a Rough Sea Surface

We have so far considered water surface to be smooth. However, a dead calm is a very rare phenomenon. As a rule, sea surface is rough because of wind waves and swells, which introduce additional distortions to the image of underwater objects observed from above-water vehicles.

### 9.4.1 Rough Sea Surface Model

The Pierson-Moskowitz spectrum is most frequently used to model a rough wind sea surface [9.31]. According to the modification suggested in [9.32, 33] for a one-dimensional fully developed wind wave, the elevation energy spectrum is

$$G_w^2(\vec{v}) = \frac{a_1}{v^4} e^{-y_1/v^2 - b_1 v^2} \delta(\varphi - \varphi_0) , \tag{9.4.1}$$

where $a_1 = 4.05 \times 10^{-3}$, $y_1 = 0.74\, g^2/v^4$, $g$ is the acceleration of gravity, $v$ the wind velocity, and $\varphi_0$ the azimuth of the wave propagation direction. In contrast to the ordinary Pierson-Moskowitz spectrum, this modification involves introducing the parameter $b_1$ in (9.4.1). This cuts off the elevation spectrum at high frequencies and is chosen such that the dependence of the variance of the wind-wave slopes on the wind velocity complies with the *Cox–Munk* data [9.34].

$$\sigma_{\theta,\mathrm{w}} = \iint_{-\infty}^{\infty} G_{\mathrm{w}}^2(\vec{v}) v^2 \, d\vec{v} = 3.16 \times 10^{-3} v \, . \tag{9.4.2}$$

According to (9.4.1), the variance of the wind-wave elevations $\sigma_{\mathrm{h,w}}^2$ is described by the expression

$$\sigma_{\mathrm{h,w}}^2 = \iint_{-\infty}^{\infty} G_{\mathrm{w}}^2(\vec{v}) \, d\vec{v} \approx \frac{a_1}{2y_1} = \frac{a_1 v^4}{1.48 g^2} \, . \tag{9.4.3}$$

The sea surface is never smooth, even when it is calm. When the wind stops blowing, the sea surface is covered with long-period swells. The swell spectrum is well described by a linear hydrodynamic sea surface model. Therefore, for the one-dimensional case, it may be written as

$$G_{\mathrm{sw}}^2 = \frac{a_2}{v^4} e^{-y_2/v^2 - b_2 v^2} \delta(\varphi - \varphi_0) \, , \tag{9.4.4}$$

where

$$a_2 = 217\, h_{\mathrm{sw}}^2/(g T_{\mathrm{sw}}^2)^2 \, , \qquad y_2 = 683/(g T_{\mathrm{sw}}^2)^2 \, ,$$

$b_2 = 1.208 \times 10^{-4} (g T_{\mathrm{sw}}^2)^2$, and $h_{\mathrm{sw}}$ and $T_{\mathrm{sw}}$ are the mean height and mean period of swells. The spectral density of swells and of fully developed windy sea calculated from (9.4.1, 4) are presented in Fig. 9.7. For the variances of swell slopes and elevations, from (9.4.4), we have

$$\sigma_{\theta,\mathrm{sw}}^2 \approx 0.81 a_2^2 \, , \qquad \sigma_{\mathrm{h,sw}}^2 \approx a_2/2y_2 \approx h_{\mathrm{sw}}^2/2\pi \, . \tag{9.4.5}$$

### 9.4.2 Image Transfer Characteristics

The modern theory of image transfer through a rough sea surface is based on the general statements and results given in Chaps. 5 and 6. It has been most elaborated in [9.29, 35–42]. Its basic concepts are set out concisely. We shall discuss the main results of these works.

As in (5.3.5), the photodetector power input per picture element in viewing through a rough sea surface is

$$P_{\mathrm{VS}}(\vec{r}_1') = \frac{P_0 \Sigma_{\mathrm{rec}} \Omega_{\mathrm{rec}}}{\pi m^2} \iint_{-\infty}^{\infty} A(\vec{r}) E_{\mathrm{s}}(\vec{r}, \vec{r}_1) E_{\mathrm{rec}}(\vec{r}, \vec{r}_1) \, d\vec{r} \, . \tag{9.4.6}$$

Here, $m$ is the water refractive index, $\vec{r}_1$ the object plane point where the source and receiver axes intersect, and $\vec{r}_1'$ the conjugate point in the image plane. The functions $E_{\mathrm{s}}(\vec{r}, \vec{r}_1)$ and $E_{\mathrm{rec}}(\vec{r}, \vec{r}_1)$ describe the irradiance distribution in the

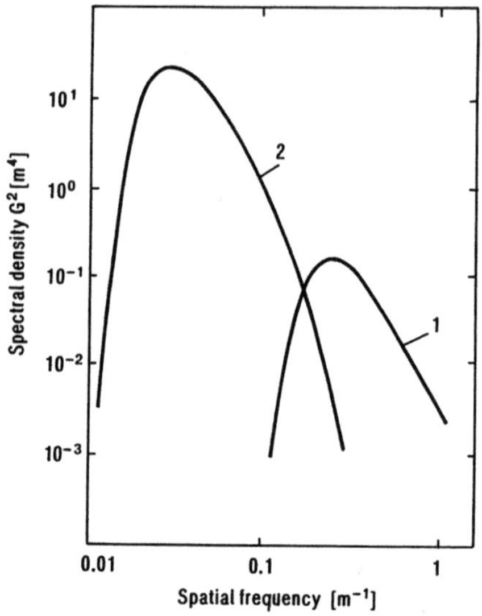


**Fig. 9.7.** Spectral density of fully developed windy surface elevations at $v = 5$ m/s (curve 1) and swells at $h_{sw} = 0.5$ m and $T_{sw} = 8$ s (curve 2)

object plane generated, respectively, by the real unit power source and fictitious source located at the receiving point, and having the same aperture and directional diagram as the receiver.

It is significant that, as in viewing in the stochastic medium (Sect. 8.5), the functions $E_s(\vec{r}, \vec{r}_1)$ and $E_{rec}(\vec{r}, \vec{r}_1)$ are random, and depend on the rough sea surface realization.

The power input averaged over the surface realization ensemble is, naturally, most interesting. By expressing $P_{VS}(\vec{r}_1')$ in terms of the PSF $a(\vec{r}, \vec{r}_1)$ for the averaged system PSF $\langle a(\vec{r} - \vec{r}_1)\rangle$ and averaged signal power, we obtain

$$\langle P_{VS}(\vec{r}_1')\rangle = P_0 \int\int_{-\infty}^{\infty} A(\vec{r})\langle a(\vec{r} - \vec{r}_1)\rangle \, d\vec{r} \, .$$

The Fouier spectrum $\langle a(\vec{v})\rangle$, which corresponds to the averaged PSF $\langle a(\vec{r})\rangle$, is the averaged OTF of the vision system.

In the general case, the expression for $\langle a(\vec{v})\rangle$ is rather complicated [9.37, 38]. We shall confine our consideration to the simple, but important case of the sea surface illuminated by the sun. In this instance, the averaged OTF is [9.37]

$$\langle a(\vec{v})\rangle = \frac{\Sigma_{rec}\Omega_{rec}t_{sur}}{\pi m^2} \int\int_{-\infty}^{\infty} \varphi_{rec}(\vec{v}, \vec{v}H + \vec{v}' z_0/m) S(\vec{v}') \times \theta\left[(\vec{v} - \vec{v}')z_0 \frac{m-1}{m}, \vec{v}' z_0 \frac{m-1}{m}, \vec{r}_d\right] \times e^{-i(\vec{v} - \vec{v}')\cdot(\vec{r}_d + \Delta\vec{r})} \, d\vec{r}_d \, d\vec{v}' \, . \tag{9.4.7}$$

Here, $t_{sur}$ is the sea surface transmission coefficient, $z_0$ the object depth, $H$ the receiver altitude, $\varphi_{rec}(\vec{v}, \vec{p})$ the spatial-angular Fourier spectrum of the receiver aperture and directional diagram, and $\theta(\cdot)$ the characteristic two-point function for sea surface slopes.

As follows from (9.4.7), in the general case, the surface impact on the transferred image pattern is nonlocal, and may be expressed in terms of the correlation characteristics of the rough sea surface. For example, (9.4.7) includes the characteristic two-point function of slopes. This may cause the so-called backscattering enhancement effect [9.36], as radiation passes twice (there and back) through the correlated areas of the rough surface. Recently, it has been shown that, under some perfect conditions, this effect is rather essential and should be taken into account, for examples, in problems of remote sensing [9.43].

In the case of a sufficiently large angular difference $\Delta\vec{n} = \vec{n}_s - \vec{n}_{rec}$, or signals received from depths which exceed the depth of maximum light focusing due to the rough surface, correlation effects are insignificant, and the backscattering enhancement effect can be neglected. In this case, the expressions for the vision system OTF $\langle a(\vec{v})\rangle$ and MTF $\langle T(\vec{v})\rangle = \langle a(\vec{v})\rangle / \langle a(\vec{v} = 0)\rangle$ essentially reduce. At $|\Delta\vec{r}| \gg r_c$, where $r_c$ is the correlation radius of the surface slopes, the functions $E_s(\vec{r}, \vec{r}_1)$ and $E_{rec}(\vec{r}, \vec{r}_1)$ are statistically independent, and the system OTF in viewing through the rough sea surface may be represented as a product of the transfer functions of the optical system, water layer and rough surface. In particular, assuming the function $\theta[(\vec{v} - \vec{v}')z_0(m-1)/m, v'z_0(m-1)/m, \vec{r}_d]$ in (9.4.7) to be independent of $\vec{r}_d$, the MTF expression becomes [9.37].

$$T(\vec{v}) = \frac{\varphi_{rec}(\vec{v}, \vec{v}(H + z_0/m))S(\vec{v})\theta(0, \vec{v}z_0(m-1)/m)}{\varphi_{rec}(0,0)S(0)}. \tag{9.4.8}$$

Here, the function $\theta(0, \vec{v}z_0(m-1)/m)$ describes the contrast degeneration due to rough sea. It has been studied in [9.35] and assumes the form

$$\theta\left(0, \vec{v}z_0\frac{m-1}{m}\right) = \exp\left[-0.5v^2\sigma_\theta^2 z_0^2\left(\frac{m-1}{m}\right)^2\right] \tag{9.4.9}$$

for the case of normal distribution of rough surface slopes having the variance $\sigma_\theta^2$.

At $|\Delta\vec{r}| \lesssim r_c$, the picture is much more intricate [9.38]. The transfer coefficient $\langle a(\vec{v} = 0)\rangle$ is a nonmonotonic function of $\Delta\vec{r}$. At $|\Delta\vec{r}| \ll r_c$, the quantity $\langle a(\vec{v} = 0)\rangle$ is twice as large as at $|\Delta\vec{r}| \gg r_c$ (the backscattering enhancement effect). The statistical correlation of the functions $E_s$ and $E_{rec}$ results in a complex system OTF. In this case, the contrast may be inverted at some spatial frequencies, and other interesting and important effects may also occur [9.37, 38]. A comprehensive OTF analysis requires reliable data on the correlation functions of slopes.

In viewing through a rough sea surface, shot noises and system transport noises may be encountered, as well as image fluctuations induced by the rough surface. They may be essentially correlated for different picture elements. Then, the SNR may be calculated by the formula given in Sect. 5.1.5. The variance of these noises depends significantly on the space and time averaging. Pioneering studies of this factor as applied to self-luminous object image transfer have been made in [9.41], as well as in [9.42] for observation in sunlight. It has been shown, for example, that the SNR of the image of an underwater object may increase with viewer altitude because of better averaging of the fluctuations caused by the rough sea surface. However, the study of these aspects and the problem of image quality in viewing through a rough surface is far from complete.

## 9.5 The Range of Optical Pulsed Location in Sea Water

In Sect. 5.5, we compared location and television target detection systems in order to show that the range of location in the known direction to the object (echo-ranging) may be taken as the upper limit in the detection range for TV systems. This case is encountered, for example, in sea-bed location. In the following analysis, we shall discuss this situation in more detail, and evaluate the limiting location depths.

The signal-to-noise ratio which characterizes the possibility of object detection is described by (5.5.1). The energy $\Delta W$ of the valid signal (Fig. 5.8), and the energy $W_b$ of the background component for the time $\Delta t_{el}$ specified by the receiver time resolution, depend on the optical distance $\tau$ to the object. Therefore, basing ourselves on (5.5.1) and using the condition $\delta = \delta^*$, where $\delta^*$ is the threshold value of the signal-to-noise ratio (Sect. 5.1), we could find the limiting location range $\tau^*$. However, we shall proceed in another manner.

We write (5.5.1) as

$$\delta = \frac{1}{\sqrt{1/\delta_{sh}^2 + 1/\delta_{tr}^2}}, \tag{9.5.1}$$

where

$$\delta_{sh} = \frac{\Delta W}{W_b}\sqrt{CW_b}, \tag{9.5.2}$$

and

$$\delta_{tr} = \frac{\Delta W}{W_b\sqrt{\alpha_2}} \tag{9.5.3}$$

are, respectively, the signal-to-noise ratio with only shot noises ($\alpha_2 CW_b \ll 1$), or with only system transport noises ($\alpha_2 CW_b \gg 1$). The relations (9.5.1–3) provide an instructive physical analysis. The values of $\tau_{sh}^*$ and $\tau_{tr}^*$ calculated using (9.5.2, 3) from the conditions $\delta_{sh} = \delta^*$ and $\delta_{tr} = \delta^*$ specify the limiting location ranges in the presence of one of the noises. Naturally, the real location range is determined by the minimum of these values.

In studying the visibility ranges in Sect. 9.3.2, the quasi-single scattering approximation was used (Table 9.4) to estimate the energy $W_b$ of the background component, which is a sum of the background noise $W_{\mathrm{BSN}}$ and external noise $W_{\mathrm{ext}}$. In the case of location at very large distances, the asymptotic formulas at $t \to \infty$ given in Sect. 4.8.3, rather than the quasi-single scattering approximation, should be used for the BSN. The quantity $\Delta W$ may be calculated in the small-angle diffusion approximation (Sect. 4.5). It is very important to choose an optimum receiver time resolution $\Delta t_{\mathrm{el}}$, and initial pulse width $\Delta t_{\mathrm{s}}$. Clearly, the smaller $\Delta t_{\mathrm{s}}$, the greater the SNR value. It makes no sense, however, to choose a very small time $\Delta t_{\mathrm{s}}$ since, in the scattering medium the pulse is spread by $\Delta t_{\mathrm{m}} \approx \sqrt{V_t}$ (Sect. 4.8.5), where $V_t$ is the variance of the $\delta$-pulse time spread. Therefore, the condition

$$\Delta t_{\mathrm{s}} \leqslant \Delta t_{\mathrm{m}} \approx \Delta t_{\mathrm{el}} \tag{9.5.4}$$

is optimum for the greatest signal-to-noise ratio to be obtained, and will be regarded as valid. In Table 9.5, the values of $\varepsilon v \Delta t_{\mathrm{m}}$, where $v$ is the light velocity in the medium, are given as SADA calculated by the formulas of Sect. 4.8.5 [9.44] to locate an object of radius $R = 0.15$ m and $R \to \infty$ at $\Omega_{\mathrm{s}} = \Omega_{\mathrm{rec}} = 6.10^{-5}$ sr.

First, we consider the location of the sea-bed, or some other large object whose size is greater than the diameter of the light spot on it. The analysis indicates that, for the realistic source energy, the location range is such that $\Delta W \gg W_{\mathrm{BSN}}$ may always be assumed. In this case, where no external noise exists, the location range $\tau^*_{\mathrm{sh}}$ may be roughly approximated by the relation

$$\tau^*_{\mathrm{sh}} = \frac{1}{\gamma} \ln(0.05\varepsilon\sqrt{AM}/\delta^*)\,, \tag{9.5.5}$$

where $\gamma = 1 - \Lambda + \sqrt{\Lambda(1-\Lambda)\beta_2/2}$ is the asymptotic attenuation coefficient, $A$ the object albedo,

$$M = W_0 S_\lambda \Sigma_{\mathrm{rec}} \Omega_{\mathrm{rec}}/e\,, \tag{9.5.6}$$

$W_0$ the source energy, $S_\lambda$ the spectral sensitivity of the receiver, and $e$ the electron charge.

**Table 9.5.** Values of $\varepsilon v \Delta t_m$.

| | | $R = 0.15$ m | | | | $R \to \infty$ | | | |
|---|---|---|---|---|---|---|---|---|---|
| | | $\tau$ | | | | $\tau$ | | | |
| $\Lambda$ | $\beta_2$ | 5 | 10 | 20 | 40 | 5 | 10 | 20 | 40 |
| 0.6 | 0.03 | 0.004 | 0.025 | 0.31 | 0.87 | 0.13 | 0.45 | 1.3 | 2.6 |
| | 0.1 | 0.007 | 0.044 | 0.75 | 2.0 | 0.39 | 1.2 | 2.4 | 3.7 |
| 0.7 | 0.03 | 0.006 | 0.043 | 0.33 | 1.0 | 0.15 | 0.54 | 1.6 | 3.3 |
| | 0.1 | 0.010 | 0.078 | 1.10 | 2.5 | 0.46 | 1.4 | 3.1 | 4.8 |
| 0.8 | 0.03 | 0.008 | 0.071 | 0.36 | 1.2 | 0.17 | 0.63 | 2.0 | 4.5 |
| | 0.1 | 0.014 | 0.14 | 1.10 | 3.2 | 0.54 | 1.8 | 4.1 | 6.6 |

In the presence of external background of radiance $I_{ext}$, when $W_{ext} \gg \Delta W$, an approximate $\tau^*_{sh}$ dependence on the optical medium characteristics and the location system parameters is

$$\tau^*_{sh} \approx \frac{1}{2\gamma} \ln \frac{10^{-3} \varepsilon^{5/2} A \sqrt{W_0 M v}}{\delta^* \sqrt{I_{ext}}} . \tag{9.5.7}$$

We now consider the quantity $\tau^*_{tr}$. Taking into account that $\sqrt{\alpha_2} = k^*/\delta^*$ (5.1.8), where $k^*$ is the threshold contrast of the system, and assuming $\delta_{tr} = \delta^*$ at $\tau = \tau^*_{tr}$, from (9.5.3), we have

$$\frac{\Delta W}{W_b} \approx k^* \ll 1 . \tag{9.5.8}$$

When there is no external noise, where $W_b = W_{BSN}$, the range $\tau^*_{tr}$ is large and, taking into account (9.5.8), may approximately be calculated by

$$\tau^*_{tr} = \frac{1}{2\sqrt{\Lambda(1-\Lambda)\beta_2/2}} \ln \frac{230(1-\Lambda)^{3/2} A}{k^* \Lambda \beta_2} . \tag{9.5.9}$$

In the case where there is external noise

$$\tau^*_{tr} \approx \frac{1}{2\gamma} \ln(1.2 \times 10^{-3} \varepsilon^3 A W_0 v / k^* I_{ext}) . \tag{9.5.10}$$

In Table 9.6, the values of $\tau^*$ are given as calculated by (9.5.5, 7, 9, 10) for the following medium and system characteristics: $\varepsilon = 0.15\ \mathrm{m}^{-1}$, $A = 0.2$, $k^* = 0.03$, $\delta^* = 3$, $M = 1.7 \times 10^{12}\ \mathrm{m}^2$. At $W_{ext} \neq 0$, the radiance of the external noise on the photodetector is set to be $I_{ext} = 10^{-3}\ \mathrm{Wm^{-2}sr^{-1}}$, which is in approximate conformity with the background lighting on a cloudy day, if the light filter with the transmission band $\Delta\lambda = 1$ nm is used. The value of $M$ adopted in calculations may be realized, for example, at $W_0 = 1$ J, $S_\lambda = 6.5 \times 10^{-2}\ \mathrm{AW}^{-1}$, $\Sigma_{rec} = 0.14\ \mathrm{m}^2$, $\Omega_{rec} = 6.10^{-5}$ sr.

As seen from Table 9.6, if $W_{ext} = 0$, then $\tau^*_{sh} \ll \tau^*_{tr}$, i.e. the real location range is limited to shot noises and $\tau^* = \tau^*_{sh}$. In the case of external noises, $\tau^*_{sh}$ essentially

**Table 9.6.** Limiting optical depths in laser see-bed location.

| | $\Lambda$ | | | | | |
|---|---|---|---|---|---|---|
| | 0.6 | | 0.7 | | 0.8 | |
| $\tau^*$ | $\beta_2 = 0.03$ | $\beta_2 = 0.1$ | $\beta_2 = 0.03$ | $\beta_2 = 0.10$ | $\beta_2 = 0.03$ | $\beta_2 = 0.1$ |
| $\tau^*_{sh}(W_{ext} = 0)$ | 17 | 15 | 21 | 19 | 31 | 26 |
| $\tau^*_{tr}(W_{ext} = 0)$ | 83 | 40 | 85 | 40 | 88 | 41 |
| $\tau^*_{sh}(W_{ext} \neq 0)$ | 15 | 18 | 19 | 17 | 26 | 23 |
| $\tau^*_{tr}(W_{ext} \neq 0)$ | 17 | 16 | 22 | 20 | 31 | 28 |

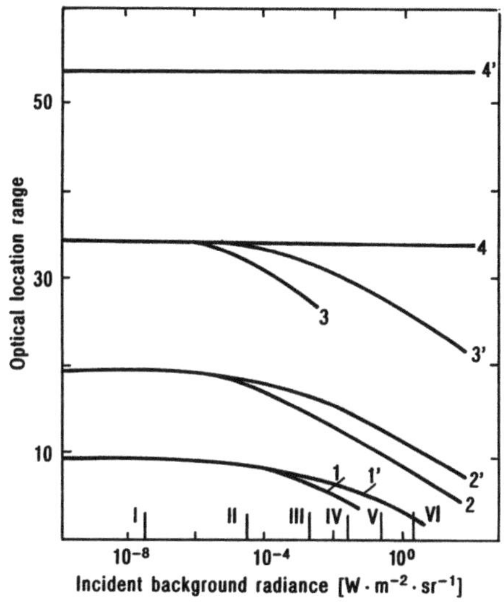


**Fig. 9.8.** $\tau^*$ versus $I_{ext}$ at $\Lambda = 0.7$, $\beta_2 = 0.06$, $\varepsilon = 0.15\ m^{-1}$ and $W_0 = 10^{-4}$ J (curves 1, 1′), 1 J (curves 2, 2′), $10^6$ J (curves 3, 3′), $\infty$ (curves 4, 4′); $k^* = 0.3$ (curves 1–4), 0.03 (curves 1′–4′). Marks on the abcissa correspond to I, clear night sky ($\Delta\lambda = 100$ nm); II, full moon ($\Delta\lambda = 100$ nm); III, dense clouds in daytime ($\Delta\lambda = 1$ nm); IV, sun in zenith ($\Delta\lambda = 1$ nm); V, sun in zenith ($\Delta\lambda = 10$ nm); VI, sun in zenith ($\Delta\lambda = 100$ nm)

decreases to become $\tau^*_{sh} \lesssim \tau^*_{tr}$ for the given medium and system parameters. This means that, the parameter $M$ being somewhat increased due to growing $W_0$, $S_\lambda$, $\Sigma_{rec}$, etc., the transport system noises may prevail, and the location range will be limited by the threshold contrast $k^*$.

The dependence of $\tau^*$ on the medium parameters $\Lambda$ and $\beta_2$ is illustrated by the data in Table 9.6. The effect of the extinction coefficient, object albedo $A$ and location system parameters ($W_0, S_\lambda$, etc.) is easily analysed using the relations (9.5.5, 7, 9, 10). As seen, the $\tau^*$ dependence on these parameters is logarithmic, i.e., their wide variation changes the limiting optical depth of location slightly. A similar property was mentioned in Sect. 9.3.3 concerning the limiting range of vision (9.3.9).

The quantity $\tau^*$ is dependent on the proportion between the pulse energy $W_0$ and external noise energy $W_{ext}$ (Fig. 9.8). The results have been derived for $\Lambda = 0.7$, $\beta_2 = 0.06$, $\varepsilon = 0.15\ m^{-1}$ and the system parameters mentioned above. The quantity $\Delta t_{el} = \Delta t_m$ ranged from 10 ns to 140 ns, depending on the location depth. Roman figures show the level of the external noise $I_{ext}$ for different conditions of viewing and different $\Delta\lambda$, as well as the spectral photodetector filter half-width. The location range $\tau^*$ is seen to increase with the pulse energy $W_0$. At low external background level, $\tau^*$ is independent of $I_{ext}$ and $k^*$ since shot noises prevail here. As $I_{ext}$ increases, the location range decreases and starts to depend on the threshold contrast $k^*$, i.e. on the level of system transport noises.

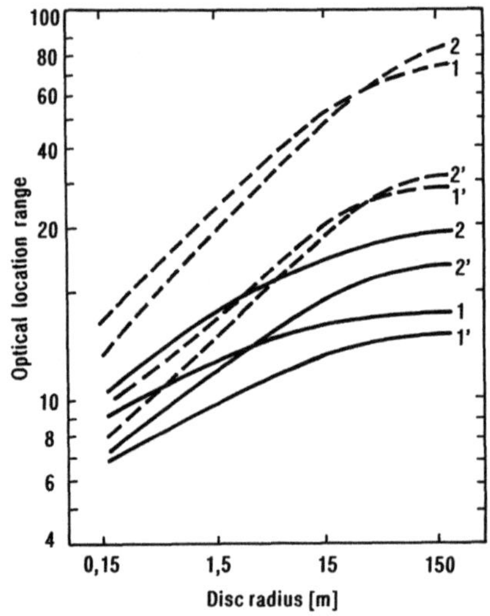


**Fig. 9.9.** Optical location range $\tau^*$ at $\varepsilon = 0.15\ m^{-1}$ for a finite object. $A = 0.2$; $\Lambda = 0.6$ (curves 1, 1′), 0.8 (curves 2, 2′); $\beta_2 = 0.03$ (curves 1, 2), 0.1 (curves 1′, 2′); $M = 3.5 \cdot 10^2 m^2$ (*solid lines*) and $M \to \infty$ (*dashed lines*)

The quantity $k^*$, naturally, governs the limiting range $\tau^*$ at $W_0 \to \infty$. For example, if $W_0 \to \infty$, then $\tau^* \to 35$ at $k^* = 0.3$ and $\tau^* \to 53$ at $k^* = 0.03$.

Figure 9.9 illustrates the range of location of a finite object. The range $\tau^*$ in sea water as a function of the round screen radius at $M = 3.5 \times 10^{12}\ m^2$ with predominant shot noises is depicted by the solid lines, and that at $M \to \infty$, when system transport noises prevail, by the dashed lines. The data were derived from [9.45], and taking into account the relations given in Sect. 5.5. As seen, the location range increases with object size. The $\tau^*$ gradient depends on the proportion between the size of the object and the light spot on it generated by multiple scattering in the medium.

In conclusion, we shall compare the limiting depths of object detection in the sea by viewing, and by optical pulsed location.

In viewing, the range of visibility of Sekky's 30 cm dia disc is (9.2.7)

$$z_d = \frac{1}{\varepsilon + \alpha} \ln \frac{A - r}{rk^*}, \tag{9.5.11}$$

where $r$ is the sea albedo, and $\alpha$ the vertical extinction coefficient for irradiance. The range of visibility of an infinite object, for example of the sea-bed, is

$$z_\infty = \frac{1}{2\alpha} \ln \frac{A - r}{rk^*}. \tag{9.5.12}$$

By way of example, we consider viewing in extremely pure sea water. Let $\varepsilon = 0.07\ m^{-1}$, $\kappa = 0.014\ m^{-1}$, $\beta_2 = 0.03$. In this case, $\alpha \approx 0.017$. Setting $k^* = 0.03$, at

$A = 1; 0.2; 0.05$ equation (9.5.12) gives the range of sea-bed visibility as 205 m, 154 m and 91 m, respectively. For pulsed location ($W_0 = 1$ J, $M = 3.5 \times 10^{12}$ m$^2$), the sea-bed location range is 300–400 m. According to (9.5.11), at $A = 1$, Sekky's disk can be viewed in sea water up to a depth of 80 m. The pulsed locator will detect it at a depth of up to 160 m. Thus, the range of optical pulsed location in sea water in the known direction to the object (laser echo-ranging) may achieve magnitudes 2–3 times greater than the range of visual detection.

# 10 Image Quality Problems in Photographic Layers and Luminescent Screens

Let us consider two more specific light-scattering materials, such as a photographic layer and a luminescent screen. These objects are used for image recording and are components of many imaging systems. They are notable for their multistage imaging. This aspect, as well as the other main characteristics which determine image quality in photographic layers and luminescent screens, is elaborated in monographs [10.1, 2], the data of which are basic to this chapter. Another distinctive feature of photographic layers and luminescent screens is the fact that they allow close packing of the scatterers. However, this point is not covered in the present book. The details of light scattering in close-packed particles are given in [10.3].

## 10.1 Optical Parameters of a Photographic Layer

### 10.1.1 Undeveloped Layer

In the simplest case, a photographic layer is a plane-parallel layer of a gelatin matrix with implanted light-sensitive microcrystals (MC) of halogen silver. The light-sensitive layer in photographic paper is 6 to 8 μm thick, while it is 5 to 30 μm in photographic films. In a number of situations, much greater thicknesses are required to record the tracks of elementary particles and holograms. The MC sizes range from $5 \times 10^{-3}$ to 5 μm, depending on the designated purpose of the material. Nuclear emulsion is characterized by a volumetric concentration $C_V$ of about 0.5; $C_V$ for ordinary emulsion is from 0.1 to 0.2 [10.4].

The efficiency factors of extinction $Q_{\text{ext}}$, scattering $Q_{\text{s}}$ and absorption $Q_{\text{a}}$ of individual MCs, single scattering albedo as well as the forward/back ratio $F/(1-F)$ and the assymetry parameter $g$ at $\lambda = 436$ nm and complex refractive index $m = 1.54 - \text{i} \times 10^{-3}$ for different MC sizes $2a$ and the parameter $\rho = 2\pi a/\lambda$ computed by the Mie formulas are listed in Table 10.1. The conversion to the elementary volume parameters is achieved using

$$\varepsilon = \frac{3}{4}\frac{C_V Q_{\text{ext}}}{a}, \qquad \sigma = \frac{3}{4}\frac{C_V Q_{\text{s}}}{a}, \qquad \kappa = \frac{3}{4}\frac{C_V Q_{\text{a}}}{a}. \tag{10.1.1}$$

The single scattering albedo and scattering indicatrix for individual particles and the elementary volume coincide.

**Table 10.1.** Scattering characteristics of spherical halogen silver MCs at $\lambda = 0.436$ μm.

| $2a$ [μm] | $\rho$ | $Q_{ext}$ | $Q_s$ | $Q_a$ | $\Lambda$ | $g$ | $\frac{F}{1-F}$ |
|---|---|---|---|---|---|---|---|
| 0.002 | 0.022 | 4.3 − 5[a] | 6.5 − 8 | 4.3 − 5 | 1.50 − 3 | 0 | 1 |
| 0.006 | 0.067 | 1.4 − 4 | 5.2 − 6 | 1.3 − 4 | 3.88 − 2 | 0 | |
| 0.010 | 0.11 | 2.6 − 4 | 4.0 − 5 | 2.2 − 4 | 0.157 | 2.9 − 3 | 1 |
| 0.014 | 0.16 | 4.6 − 4 | 1.6 − 4 | 3.1 − 4 | 0.337 | 4.9 − 3 | 1 |
| 0.02 | 0.22 | 1.1 − 3 | 6.5 − 4 | 4.4 − 4 | 0.594 | 0.01 | 1 |
| 0.06 | 0.67 | 5.5 − 2 | 5.3 − 2 | 1.6 − 3 | 0.971 | 0.09 | 1.3 |
| 0.10 | 1.1 | 0.365 | 0.361 | 0.004 | 0.990 | 0.26 | 2.2 |
| 0.14 | 1.6 | 1.003 | 0.997 | 0.006 | 0.994 | 0.61 | 12 |
| 0.2 | 2.2 | 2.361 | 2.351 | 0.010 | 0.996 | 0.62 | 10 |
| 0.4 | 4.4 | 4.021 | 3.998 | 0.023 | 0.994 | 0.71 | 9.7 |
| 0.6 | 6.7 | 1.930 | 1.893 | 0.037 | 0.981 | 0.48 | 4.5 |
| 1.0 | 11 | 2.608 | 2.547 | 0.061 | 0.977 | 0.72 | 7.1 |
| 1.4 | 15 | 2.681 | 2.572 | 0.110 | 0.959 | 0.74 | 11.5 |
| 1.8 | 20 | 2.157 | 2.064 | 0.092 | 0.957 | 0.76 | 14 |

[a] It should be $4.3 \cdot 10^{-5}$

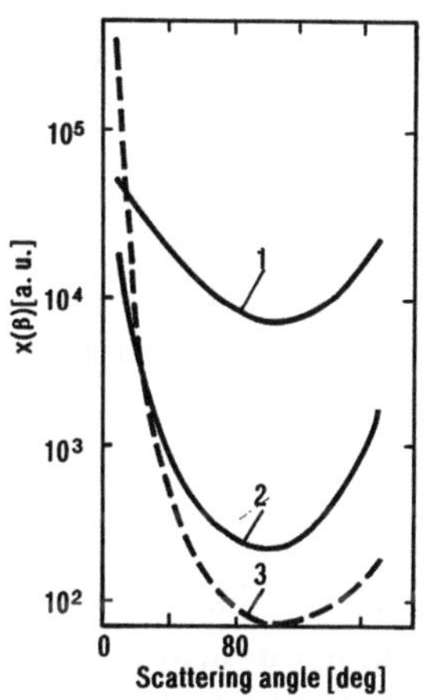


**Fig. 10.1.** Indicatrices of undeveloped photoemulsions (curves 1, 2) and luminophor ZnS:Ag particles (curve 3) at $a = 0.15$ μm (curve 1); 0.475 (curve 2); 0.25–1.5 (curve 3)

As seen from the table, for small $\rho$, $Q_{ext} \approx Q_a$, whereas at $\rho \lesssim 0.3$ the forward and backscattering is, in fact, the same. As $\rho$ goes up, the single scattering albedo increases, and from $\rho > 0.6$, it exceeds 0.95. At the same time, the parameters $F/(1-F)$ and $g$, which tend to grow, oscillate (the case is different for polydispersed emulsions). The quantity $Q_{ext}$ does not achieve its asymptotic limit of 2 until $\rho = 20$.

Figure 10.1 shows experimental $x(\beta)$ values for different mean microcrystal sizes $2a$ [10.5]. The spectral dependence of the optical layer density $\tau_0 = \varepsilon z_0$ for

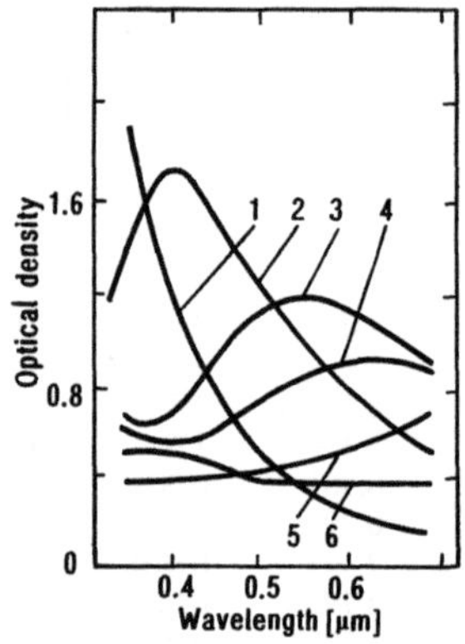


**Fig. 10.2.** Spectral variation of the optical density of emulsion layers with cubic MCs of length 0.22 μm (curve 1); 0.32 (curve 2); 0.47 (curve 3); 0.55 (curve 4); 0.67 (curve 5); 0.83 (curve 6)

different $a$ is illustrated in Fig. 10.2 [10.6]. As $a$ goes up, the maximum in $\varepsilon$ moves to long wavelengths. At $2a \gtrsim 0.83$ μm, light extinction is, in fact, no longer selective. Such behavior is typical of layers of any MC shape. Nowadays, flat (plates) and core-shell MCs are widely used to obtain a fast and high-resolution photographic layer [10.7, 8].

When the photographic material is sensitized, absorption is generally increased several fold, while scattering is almost unchanged. More detailed information on the optical properties of photographic materials can be found in [10.1]. The methods of measuring the optical properties of undeveloped layers are discussed in [10.1, 9].

### 10.1.2 Exposed Developed Layer

In contrast to an undeveloped layer, the optical properties of an exposed photographic layer essentially change upon chemical and photographic treatment. The developed fixed photographic layer contains silver particles distributed in gelatin. Their ability to absorb or scatter the light is different from that of initial MCs. The developed picture forming particles are referred to as the grains. The grain concentration in a photographic layer depends on the exposure and development conditions (it generally decreases with depth). The geometric thickness of such a layer is less, or equal to that of an undeveloped layer.

The shape, structure and size of the undeveloped silver grains depend on the shape, size and structure of light-sensitive MCs, matrix properties, exposure time and the incident irradiance of the layer, the developer, development conditions and time.

Table 10.2 and Figs. 10.3, 4 show the optical characteristics of silver grains obtained in [10.1]. As follows from the table, the scattering indicatrix is more extended upon chemical than physical development. Figures 10.3, 4 show the

**Table 10.2.** Scattering indicatrix parameters for silver grains in gelatin at $\lambda = 0.5$ μm.

| Development | Parameter | 2a [μm] | | | | | | | |
|---|---|---|---|---|---|---|---|---|---|
| | | 0.1 | 0.14 | 0.2 | 0.4 | 0.8 | 1.2 | 2 | 3 |
| Chemical | $\frac{F}{1-F}$ | 1.6 | 2.7 | 8.3 | 22.5 | 30 | 32 | 36 | 25 |
| | $g$ | 0.16 | 0.34 | 0.59 | 0.82 | 0.88 | 0.9 | 0.91 | 0.88 |
| Physical, after fixing | $\frac{F}{1-F}$ | 1 | 1.7 | 2.9 | 3.5 | 4.5 | 4 | 4.1 | 2.7 |
| | $g$ | 0 | 0.2 | 0.36 | 0.5 | 0.58 | 0.56 | 0.58 | 0.47 |

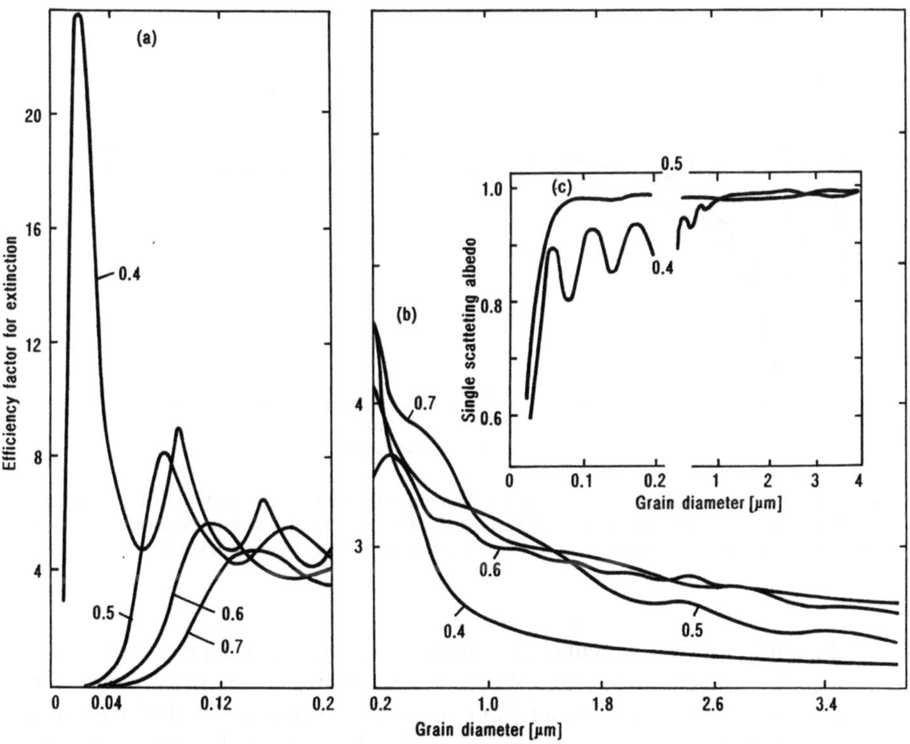


**Fig. 10.3.** Photographic layer $Q_{ext}$ (a, b) and $\Lambda$ (c) as functions of the silver grain diameter under physical development after fixing. The numbers at the curves are $\lambda$ in μm

changes in the extinction efficiency factor and the single-scattering albedo in the photomaterial upon physical and chemical development. Here, the $\varepsilon$ vs $a$ curve is seen to be smoother for chemical development. As $a$ increases, the quantity $\Lambda$ rapidly tends to unity in the case of physical development, and slowly approaches the values from 0.58 to 0.63 in chemical treatment.

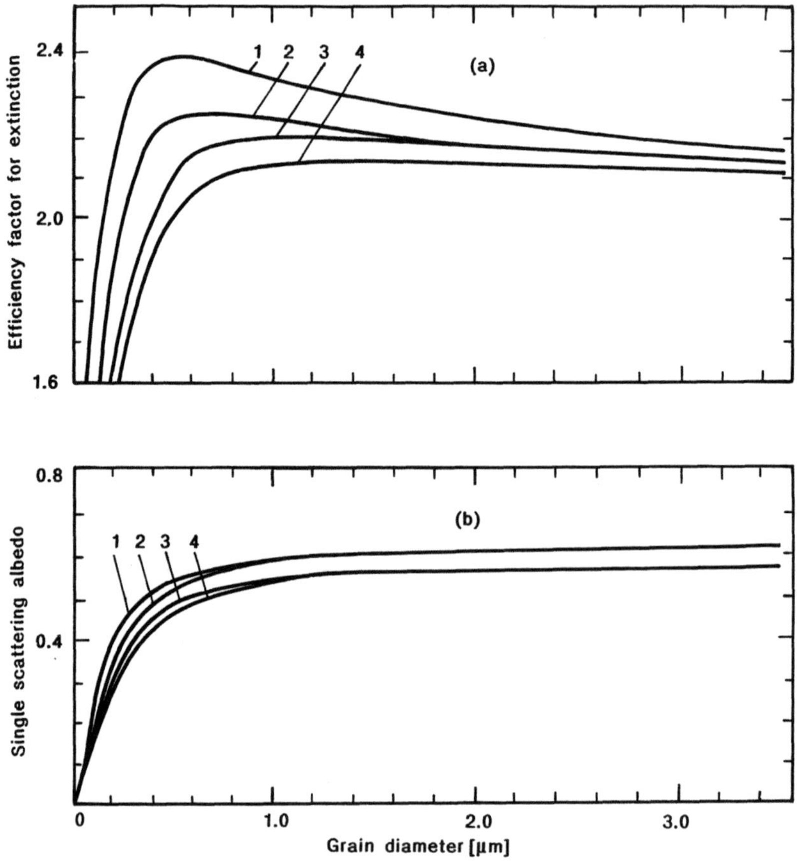


**Fig. 10.4.** $Q_{ext}$ (a) and $\Lambda$ (b) versus silver grain diameter under chemical development within 5 min for different ratios of the initial crystal volume $V_{cr}$ to the grain volume after development $V_{gr}$: $V_{gr}/V_{cr} = 1$ (curves 1, 2); 1.48 (curves 3, 4); $\lambda = 0.5$ μm (curves 2, 4); 0.6 μm (curves 1, 3)

Changing from the optical characteristics of grains to the elementary volume parameters requires a knowledge of the concentration of grains in the elementary volume which depends, first and foremost, on the exposure at the layer point which is of interest. As the irradiance generally decreases with increasing depth, the concentration of developed particles also decreases. We may show (see, for example, [10.1]) that the extinction coefficient at a depth $z$ in the layer is

$$\varepsilon = \int_0^\infty C_{ext}(a)\Phi(a)\,da$$

$$= N \int_0^\infty C_{ext}(a) f(a)\,da \int_0^\infty \beta(a,\eta)\left[1 - e^{-N_p} \sum_{i=0}^{\eta-1} \frac{(N_p)^i}{i!}\right] d\eta\,. \tag{10.1.2}$$

Here, $\Phi(a)$ is the grain size distribution function; $\eta$ denotes the MC quantum sensitivity, i.e., the minimum number of absorbed quanta responsible for MC development; $f(a)$ and $\beta(a, \eta)$ are the normalized MC size and sensitivity distribution functions; $N_p = Q_a^{un}\Sigma^{un}Et_e/hf$ is the mean number of MC absorbed photons at a depth $z$ ($E$ the irradiance, $t_e$ the exposure time, $Q_a^{un}$ the efficiency factor of absorption of an undeveloped particle, $\Sigma^{un}$ its cross-sectional area, $h$ the Planck constant and $f$ the irradiating light frequency); $C_{ext} = Q_{ext}\Sigma$ defines the extinction cross-section of a grain ($\Sigma$ the grain cross-sectional area); $N$ is the maximum possible number of grains per unit volume.

For monodispersed emulsion with $\eta$ constant, the expression (10.1.2) assumes the form

$$\varepsilon = C_{ext}N\left[1 - e^{-N_p}\sum_{i=0}^{\eta-1}\frac{(N_p)^i}{i!}\right]. \tag{10.1.3}$$

As $\lg N_p$ increases, the quantity $\varepsilon$ first increases slightly, then grows sharply to slowly approach the limit. In scientific photography, the function $\varepsilon(\lg N_p)$ is referred to as the characteristic curve of an elementary volume. The effect of various factors on the characteristic curve is analysed elsewhere [10.1, 10–12].

Naturally, the $\sigma$ and $\kappa$ of the developed layer are also defined by (10.1.3), but have $C_s$ or $C_a$ rather than $C_{ext}$. The differences in the light scattering and absorption characteristics of developed and undeveloped particles are used in liquid fixation-free sensitometry to estimate the development adjacency effect due to light scattering by the MC and grain mixture [10.13].

## 10.2 Modulation Transfer Function of Photographic Materials

In contrast to the majority of image transfer systems, which are linear, imaging in photographic materials is generally nonlinear. One can distinguish three stages, such as exposure, development and observation. The first stage is linear scattering and absorption of the incident (superimposed) radiation on the photomaterial surface, which causes image spread deep in the light-sensitive layer. At the second nonlinear stage, metal silver is produced. Both the developed and undeveloped materials scatter and absorb radiation. This third stage, as well as the first, seems to be linear.

### 10.2.1 Optical and Photographic Modulation Transfer Functions

The modulation transfer function (MTF) is the most complete characteristic of the quality of photographic material, as well as of other media. However, when the nonlinearity of image formation and the multiplicity of the processing stages is considered, its significance is somewhat different. The MTF is expressed in terms of the effective exposure (or irradiance) introduced into scientific photography by *Friezer* [10.4].

Such an approach assumes that at any point $\vec{r}$ in the photomaterial plane, the exposure $H_{\text{eff}}(\vec{r}) = E_{\text{eff}}(\vec{r})t_e$ may be found, where $E_{\text{eff}}$ is the irradiance and $t_e$ the exposure time, which provides the optical density $\mathscr{D}$ equal to the optical density $\mathscr{D}(\vec{r})$ at the point $\vec{r}$ in the case of uniform illumination of a large area. Here, as adopted in scientific photography, the optical density $\mathscr{D} = \log[1/t(\tau_0)]$, where $t(\tau_0)$ is the transmission coefficient of a layer [10.1]. The exposure $H_{\text{eff}}(\vec{r})$ and, hence, the irradiance $E_{\text{eff}}(\vec{r}) = H_{\text{eff}}(\vec{r})/t_e$, at which the condition $\mathscr{D} = \mathscr{D}(\vec{r})$ is realized, are termed effective.

The photographic modulation transfer function of a photographic layer is defined as the ratio of the effective irradiance (exposure) to the superimposed irradiance modulation as a function of the spatial frequency of the superimposed irradiance

$$T_p(\nu) = \frac{M_{\text{eff}}(\nu)}{M_{\text{su}}(\nu)} = \frac{E_{\text{eff}}^{\max}(\nu) - E_{\text{eff}}^{\min}(\nu)}{E_{\text{eff}}^{\max}(\nu) + E_{\text{eff}}^{\min}(\nu)} \Bigg/ \frac{E_{\text{su}}^{\max}(\nu) - E_{\text{su}}^{\min}(\nu)}{E_{\text{su}}^{\max}(\nu) + E_{\text{su}}^{\min}(\nu)}, \tag{10.2.1}$$

where $M_{\text{eff}}(\nu)$ and $M_{\text{su}}(\nu)$ are the effective and superimposed irradiance modulations; $E_{\text{eff}}^{\max}(\nu)$ and $E_{\text{eff}}^{\min}(\nu)$, $E_{\text{su}}^{\max}(\nu)$ and $E_{\text{su}}^{\min}(\nu)$ are the maximum and minimum effective and superimposed exposures, respectively.

The function $T_p(\nu)$ is related to the blackening quantity. For the optical densities which conform to the linear section on the characteristic curve (CC), we have [10.4]

$$T_p(\nu) = \frac{10^{\Delta\mathscr{D}(\nu)/\gamma_{\text{CC}}} - 1}{10^{\Delta\mathscr{D}(\nu)/\gamma_{\text{CC}}} + 1} \frac{1}{M_{\text{su}}(\nu)} = \frac{10^{\Delta \lg H_{\text{eff}}(\nu)} - 1}{10^{\Delta \lg H_{\text{eff}}(\nu)} + 1} \frac{1}{M_{\text{su}}(\nu)}, \tag{10.2.2}$$

where $\Delta\mathscr{D} = \mathscr{D}_{\max} - \mathscr{D}_{\min}$, $\mathscr{D}_{\max}$ and $\mathscr{D}_{\min}$ stand for the maximum and minimum optical densities in the harmonic superimposed exposure distribution, and the gamma $\gamma_{\text{CC}}$ is the maximum CC gradient.

At small $\Delta \lg H_{\text{eff}}$, when $M_{\text{su}} \lesssim 0.4/\gamma_{\text{CC}}$, (10.2.2) reduces to

$$T_p(\nu) = \frac{\Delta\mathscr{D}(\nu)}{0.86\gamma_{\text{CC}} M_{\text{su}}(\nu)}. \tag{10.2.3}$$

Expression (10.2.3) is approximately valid not only for the linear gamma characterized CC fragment, but also for the other fragments. In this case, instead of $\gamma_{\text{CC}}$, formula (10.2.3) must include the real CC gradient $g_{\text{CC}}$ [10.4], so that

$$T_p(\nu) = \frac{\Delta\mathscr{D}(\nu)}{0.86 g_{\text{CC}} M_{\text{su}}(\nu)}. \tag{10.2.4}$$

The relations (10.2.2–4) connect the photographic modulation transfer function and the optical blackening density. However, $T_p(\nu)$ depends not only on the photographic layer properties, but also on the light scattering in it.

Consider materials whose parameters $\varepsilon$, $\Lambda$, $x(\beta)$ do not change with the undeveloped layer depth. In this case, following [10.14]

$$T_p(\nu) = \frac{\int_0^{z_0} \varepsilon_f(z)\,dz \int_{-\infty}^{\infty} S(x,z)\cos(\nu x)\,dx}{\int_0^{z_0} \varepsilon_f(z)\,dz \int_{-\infty}^{\infty} S(x,z)\,dx}. \tag{10.2.5}$$

Here, $\varepsilon_f(z)$ is the exposure derivative of the extinction coefficient of the developed layer at the point $H = \langle H \rangle t(z)$, where $\langle H \rangle$ is the surface average of the superimposed exposure, and $t(z)$ the ratio of scalar irradiances at a depth $z$ and above the layer surface; $S(x, z)$ is the line spread function at depth $z$. In contrast to previous chapters, the function $S(x, z)$ defines the distribution of the scalar irradiance (2.1.4), rather than the irradiance (2.1.8).

The mean value theorem allows (10.2.5) to be written as

$$T_p(\nu) = \frac{\varepsilon_f^u}{\varepsilon_f^l} T_0(\nu) . \tag{10.2.6}$$

In (10.2.6), $\varepsilon_f^u$ and $\varepsilon_f^l$ are the effective mean $\varepsilon_f(z)$ in the upper and lower integrals in (10.2.5),

$$T_0(\nu) = \frac{\int_{-\infty}^{\infty} \langle S(x) \rangle \cos(\nu x)\, dx}{\int_{-\infty}^{\infty} \langle S(x) \rangle\, dx} , \tag{10.2.7}$$

where

$$\langle S(x) \rangle = \frac{1}{z_0} \int_0^{z_0} S(x, z)\, dz .$$

The characteristic $\langle S(x) \rangle$, which is the layer thickness average of the spread function, will be termed the integral line spread function, while $T_0(\nu)$ is the optical MTF of a photographic layer. It depends only on the light scattering and absorption in the undeveloped layer.

The coefficient $\varepsilon_f^u/\varepsilon_f^l$ in (10.2.6) is a function of $\nu$. However, for weakly absorbing layers where $\varepsilon_f(z)$ changes slowly with depth, this dependence is weak. At $\varepsilon_f(z) = \text{const}$, we have

$$T_p(\nu) \equiv T_0(\nu) , \tag{10.2.8}$$

i.e., the MTF in terms of the thickness average of the scalar and effective irradiances coincides.

Such a situation holds, for instance, in a uniformly developed layer exposed to electrons, where the linear range of $\varepsilon$ vs $H$ is wide [10.4]. In the case of light exposure, (10.2.8) may be satisfied in uniform development (i.e., such that all MCs, irrespective of their distance from the layer surface, are subject to equal development conditions) and a slight change occurs in $t(z)$ with depth, since the linear range of $\varepsilon$ vs $H$ is small. We return to (10.2.5) and express $\varepsilon_f(z)$ in terms of the elementary CC gradient $g_{CC}(z) = \partial\varepsilon(z)/\partial \lg H$. Now,

$$T_p(\nu) = \frac{\int_0^{z_0} g_{CC}(z) T(\nu, z)\, dz}{\int_0^{z_0} g_{CC}(z)\, dz} . \tag{10.2.9}$$

Using the mean value theorem, we arrive at

$$T_p(\nu) = \frac{g_{CC}^u}{g_{CC}^l} \langle T(\nu) \rangle . \tag{10.2.10}$$

Here, $g^{u}_{CC}(z)$ and $g^{l}_{CC}(z)$ are the mean values of $g_{CC}(z)$ in the upper and lower integrals of (10.2.9), whereas

$$\langle T(v) \rangle = \frac{1}{z_0} \int_0^{z_0} T(v, z)\, dz \qquad (10.2.11)$$

is the layer thickness-averaged MTF.

For equal CCs of all elementary layers and a sufficiently large transmission coefficient of the undeveloped layer, $T_p(v) \equiv \langle T(v) \rangle$. As seen from (10.2.6, 10), the expressions for $T_0(v)$ and $T(v)$ coincide at $t(z)$ independent of $z$.

It is worth considering the relationship between $T_p(v)$ and $T_0(v)$ [or $\langle T(v) \rangle$] in the case of nonuniform development of the exposed layer, i.e., when the developer ability to change exposed MCs into grains reduces with depth. In such a situation, blackening appears only at the layer surface, where the initial radiation scattering is not large. The MCs deep in the layer do not contribute to blackening, and, therefore, the densitometric $T_p(v)$ turns out to be high enough. Meanwhile, in calculating $T_0(v)$ and $\langle T(v) \rangle$, light scattering over the entire emulsion thickness is taken into account, thus resulting in $T_0(v)$ being much less than the appropriate $T_p(v)$ at any $v$.

We have so far considered the MTF for a homogeneous layer. The data on the MTF for multilayer photographic materials are cited, for instance, in [10.15].

### 10.2.2 Empirical and Approximate MTF Formulas

Many attempts have been made to find analytical expressions for the MTF. For example, the following empirical formulas are cited in [10.4, 16–18]:

$$T_p(v) = [1 + (a_1 v)^2]^{-1/2}\,, \qquad T_p(v) = e^{-(a_2 v)^2}\,,$$

where $a_1, a_2$ are the empirical constants. The integral line spread function is most often described as a sum of two exponents [10.4]. In this case, the MTF model is

$$T_p(v) = \frac{\rho_1}{1 + (k_1 v)^2} + \frac{1 - \rho_1}{1 + (k_2 v)^2}\,. \qquad (10.2.12)$$

Here, the parameters $k_1$ and $k_2$ stand for spreading due to low- and high-order scattering; $\rho_1$ is the fraction of low-order scattering.

In strongly scattering layers, at $\rho_1 = 0$,

$$T_p(v) = [1 + (k_2 v)^2]^{-1}\,. \qquad (10.2.12')$$

In weakly scattering layers, when $k_1 = 0$,

$$T_p(v) = \rho_1 + (1 - \rho_1)[1 + (k_2 v)^2]^{-1}\,. \qquad (10.2.12'')$$

For optically thick strongly scattering layers, [10.19] offers the following formula to relate the MTF to the optical layer constants

$$T_0(v) = \frac{1 - e^{-\tau_0} + T(v)}{1 - e^{-\tau_0} + T(0)}\,, \qquad (10.2.13)$$

where

$$T(\nu) = \frac{\Lambda}{4\pi m^2}\left\{\frac{e^{-m\tau_0}(1 + e^{-2m(\tau_0+d_2)})(1 + e^{-2md_2})}{1 - e^{-2m\tau_e}}\left[\eta_1 e^{-2md_2}(e^{-(1+m)\tau_e} - e^{-2m\tau_e})\right.\right.$$
$$+ \eta_2(1 - e^{-(1+m)\tau_e}) + \eta_1(1 - e^{-m\tau_e}) + \eta_2(1 - e^{-m\tau_e})$$
$$\left.\left. - \frac{6m^2}{1 - m^2}(1 - e^{-\tau_0})\right]\right\},$$

$$\eta_1 = \frac{3}{1-m} - x_1\,, \qquad \eta_2 = \frac{3}{1+m} - x_1\,,$$

$$m = \sqrt{\omega^2 + \gamma^2}\,, \qquad \gamma = \sqrt{(1-\Lambda)(3 - x_1)}\,.$$

Here, $x_1$ is the first coefficient in the Legendre polynomial expansion of the indicatrix $\omega = \nu/\varepsilon$, $\tau_e = \tau_0 + d_1 + d_2$, $d_i = [2/\Lambda(3 - x_1)](1 + A_i)/(1 - A_i)$, and $A_i$ is the surface albedo. The subscript $i = 1, 2$ stands for the upper and lower surfaces, respectively.

Expression (10.2.13) was derived in the diffusion approximation (Sects. 4.3 and 6.1.3). Its accuracy grows with both the optical thickness of the layer, and with decreasing indicatrix extension and absorption.

### 10.2.3 MTF Dependence on the Optical and Emulsion Parameters of Photographic Materials

Let us consider the MTF variation with the optical constants of the photographic material (optical thickness, single scattering albedo and scattering indicatrix), as well as emulsion and technological factors (mean size and polydispersion of MCs, amount of halogen silver per unit surface). The dependence of the MTF on these parameters is shown in Fig. 10.5. These data are obtained by measurements taken using the optical model of photographic layers [10.20, 21], and plotted against the dimensionless frequency $\omega = \nu/\varepsilon$. The increase in the optical layer thickness $\tau_0 = \varepsilon z_0$ (Fig. 10.5a) results in a decrease in the MTF value for all spatial frequencies. At first glance, this seems to imply that the contrast transfer deteriorates with increasing $z_0$. In fact, in passing to dimensional spatial frequencies the situation might occur where an optically thick layer, owing to the higher value of $\varepsilon$, transfers lower frequencies more effectively than an optically thin layer. This effect is discussed in more detail in Sect. 6.1.4.

The increasing contribution of true absorption to light extinction, i.e., $\Lambda$ decrease (for example, with radiation wavelength or dye injection) improves the transfer of all spatial frequencies (Fig. 10.5b). This is due to the fact that the total amount of scattered light is reduced, becoming more and more concentrated about the beam axis as $\Lambda$ decreases.

Figure 10.5c illustrates the MTF variation for different scattering indicatrices. In Fig. 10.5d, the MTF of the photographic material (layer of emulsion taken from the CN-2 film and diluted in water) is plotted for different optical depths $\tau$ from the surface. As seen from this figure, with growing $\tau$, the $T_0(\omega, \tau)$ transforma-

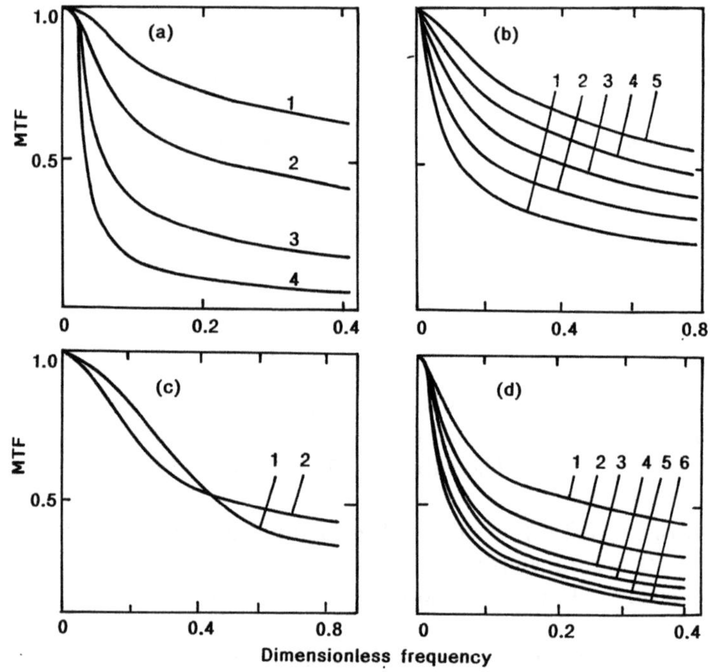


**Fig. 10.5.** Modulation transfer function (MTF) of model layers (a–c) and of emulsion film CN-2 (d). In (a) $\Lambda \approx 1$, $g = 0.87$, $\tau_0 = 1.8$ (curve 1); 3.2 (curve 2); 6.4 (curve 3); 13 (curve 4). In (b) $\tau_0 = 9.2$; $g = 0.87$; $\Lambda \approx 1$ (curve 1); 0.95 (curve 2); 0.9 (curve 3); 0.85 (curve 4); 0.77 (curve 5). In (c) $\tau_0 = 13$, $\Lambda \approx 1$, $g = 0.34$ (curve 1); 0.87 (curve 2). In (d) $\tau_0 = 13$, $\tau = 1.6$ (curve 1); 3.2 (curve 2); 4.8 (curve 3); 6.4 (curve 4); 7.4 (curve 5); 13 (curve 6)

tion becomes slower. This is because the light scattering intensity at large depth is small and affects only slightly the MTF of a $\tau$ and, hence, $\tau_0$ thick layer.

The results presented provide information on MTF transformation and its sensitivity to changes in the optical parameters which are, in turn, determined by the emulsion factors. In producing photographic materials, such characteristics are, generally, applied as layer thickness, MC density and mean size, as well as the MC size distribution variation coefficient. The increase in the geometric thickness $z_0$ and MC density which results in the growth of $\varepsilon$, increases the optical thickness as $\tau_0 = \varepsilon z_0$. The $\tau_0$ influence on the MTF was discussed above. The effect of the MC size on the MTF can be seen from

$$T_p(\nu) = \left[1 + \frac{0.96}{4\pi^2}\frac{a^2\nu^2}{C_V}\right]^{-1} . \tag{10.2.14}$$

This follows from (10.2.12′) and holds for the layers of monodispersed emulsions at $2a$ of 0.2 to 0.9 μm in the range of $\nu \leqslant 100\ \mathrm{mm}^{-1}$ in the case of optimum coating for the given MC size, when $C_V \gtrsim 0.07$ [10.22].

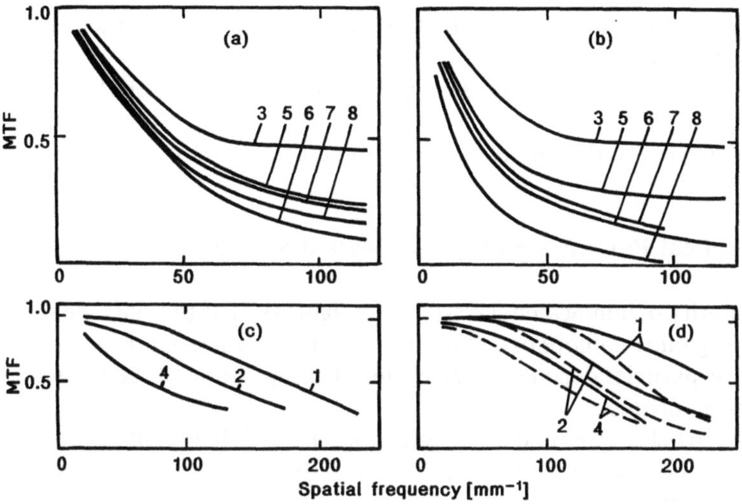


**Fig. 10.6.** MTFs of commercial films. In (a) Photo-32 (MC mean area $\langle \Sigma \rangle = 0.5\ \mu m^2$); (b, c) Photo-250 ($\langle \Sigma \rangle = 1.5\ \mu m^2$); (d) UV-4 (*solid lines*, $\langle \Sigma \rangle = 0.07\ \mu m^2$) and UVSh (*dashed lines*); $\lambda = 0.313\ \mu m$ (curve 1); 0.365 (curve 2); 0.390 (curve 3); 0.405 (curve 4); 0.46 (curve 5); 0.55 (curve 6); 0.6 (curve 7); 0.665 (curve 8)

The effect of optical and emulsion parameters on the MTF is discussed in [10.23–26]. The MTFs of a great number of commercial photographic films are presented in monograph [10.27]. We shall consider one aspect and analyse the MTF as a function of the exposure radiation wavelength for some commercial photographic materials [10.28, 29].

In Fig. 10.6 [10.28, 29], the MTFs of several fine- and coarse-grain films are plotted. In the former case, the MTF is a nonmonotonic function of $\lambda$. Such behavior is attributable to scattering selectivity. In the material in question, a light extinction maximum occurs at $\lambda = 0.52$ to $0.53\ \mu m$ [10.28], and the MTF values at all spatial frequencies are less than in other spectra. At small wavelengths, the layer MTF improves as the halogen silver absorption increases sharply. For large wavelengths, where the self-absorption of halogen silver and sensitizers is low, the radiation extinction is mainly due to scattering, and the MTF is again degraded. In coarse-grain films, light scattering is no longer selective since the single scattering albedo decreases with decreasing wavelength. Therefore, for all spatial frequencies, the MTF value is seen to grow monotonically with the decreasing wavelength. A similar dependence is encountered for films designed for ultraviolet radiation recording (Fig. 10.6d). It is worth emphasizing the availability of a low-frequency region with $T_p(\nu) \approx 1$. In the ultraviolet, the single-scattering albedo is low, and the effect of multiple-scattered light is insignificant. The image is distorted here due to low-order scattering, the spread function being narrow. The reduction in the contrast transfer at high spatial frequencies is probably caused by the angular divergence of the experimental exposure radiation.

Information on image transfer in the photographic materials is detailed in [10.1, 27, 30, 31].

We now proceed to an analysis of luminescent screens.

## 10.3 Optical Parameters of a Luminescent Screen

The majority of luminescent layers scatter light because they include particles of solid inorganic luminophors, phosphor crystals. The size, shape and structure of phosphor particles depend on the chemical composition of the matter, and the production process.

The particles deposited on a screen may be sufficiently uniform and round-shaped in one case, have smaller particles on their surfaces in another or, finally, be large aggregates.

To obtain images of different colours, which depend on the exciting electron energy, barrier (lamellar) luminophors with multilayer particles have been employed in recent years. Depending on the screen purpose, the mean size of the particles (aggregates) may range from one to tens of micrometers.

A luminescent layer includes, besides phosphor crystal particles, a small number of different substances for production purposes. In a finished layer, they may cover the luminophor particles or some of the faceplate surface, form clots and fill all or some of the space between the luminophor particles.

Information on the molecular mass, density and refractive index of some luminophor arrays and binders is cited in [10.32].

The luminescent layer thickness is closely related to the problem of maximum luminous efficiency and depends on the type of exciting radiation: the stronger its absorption, the thinner the layer. In cathode-ray tubes (CRT) the electron penetration amounts to some micrometers, the layer thickness lying between 3 and 15 μm. The penetration of radiaton through the photo- and $\gamma$-rays excited screens is much larger, and the layer thickness may reach hundreds of micrometers.

The luminescent layer thickness is generally characterized by the specific load (weight) $m$, or the luminophor mass per unit layer surface, rather than by its geometric size $z_0$. Depending on the physico-chemical luminophor properties and the layer production, the volumetric concentration of particles (packing) $C_V$ ranges from 0.3 to 0.5.

Before discussing scattering and absorption characteristics, we must consider the details of exciting radiation penetration through the layer, since these are mainly responsible for the elementary volume parameters. When luminescence is excited by photons or $\gamma$-quanta, the radiation penetrates at depths far in excess of the phosphor crystal grain size. Consequently, a large number of particles turn out to be uniformly exposed to light. Therefore, we can speak of the elementary volume parameters in a general sense, and relate the volume and particle characteristics. In this case, the energy of the photon interaction with the medium will not change, whereas their total number decreases due to partial absorption.

If the luminescence is excited by $\alpha$- or $\beta$-particles, the latter, in interacting with the medium atoms, penetrate almost micrometer distances which are comparable to the phosphor crystal sizes, or are sometimes even less. The penetration range is a function of the particle energy and the material properties. At sufficiently high initial energy of a particle, the probability of its being absorbed on interacting with an atom is low. The interaction generally involves, only partial energy transfer to an atom, and changes the particle direction. Therefore, when the particles travel through the material, their total number does not change, in contrast to their velocity distribution, their mean velocity decreasing.

In this case, the scattering and absorption characteristics of the exciting radiation depend only on the properties of the medium atoms. The radiation in the screen surface layer is reduced. Therefore, it is important to estimate the depth $R_s$ of particle penetration through the screen. In the case of electron excitation, it is the depth at which their mean energy decreases to as little as 50 eV.

The empirical formula for $R_s$ is suggested in [10.33]

$$R_s = \frac{0.1 W_0^{1.5}}{\rho}, \tag{10.3.1}$$

where $R_s$ is given in μm, $\rho$ (material density) is in g/cm$^3$ and $W_0$ (initial energy of electrons), in keV. The formula can be applied for $W_0$ between 0.5 and 300 keV.

If the luminescent screen is bombarded by $\alpha$-particles, they penetrate, on account of greater mass, at much smaller distances than electrons of the same energy. The electron scattering indicatrix is given by [10.34]

$$\frac{\sigma(\beta)}{\sigma(0)} = \left[1 + 0.0096 \frac{W \sin^2(\beta/2)}{Z^{2/3}}\right]^{-1}. \tag{10.3.2}$$

Here, the number of electrons at any scattering angle is normalized to their number at $\beta = 0$; $Z$ is the effective atomic number equal to the arithmetic mean of the ordinal numbers of all chemical elements forming a material molecule. As follows from (10.2.6), at large electron energies, the scattering indicatrix is essentially extended in the forward direction. In penetrating through the medium, electrons lose their velocity and energy $W$, and their indicatrix gradually transforms to become spherical. The lower the atomic number, the more anisotropic the indicatrix.

The optical parameters of an elementary volume in the case of light excitation can be found in [10.2]. Figure 10.1 illustrates the scattering indicatrix of the particles of the widespread luminophor ZnS · Ag at $\lambda = 546$ nm. It is worth recalling the indicatrix of luminescence, which originates in the phosphor crystal particles which absorb exciting radiation. The luminescence radiance emitted by a small volume is equal in all directions (the Lommel law).

**Table 10.3.** $\sigma_m$ (numerator) and $\kappa_m$ (denominator) coefficients for some luminophors [10.2].

| Luminophor | λ [nm] 250 | 300 | 350 | 400 | 500 | 750 |
|---|---|---|---|---|---|---|
| $Al_2O_3$:Ce | 19/6.5 | 21.1/7.83 | 24.3/5.15 | 30.1/1.26 | 32.8/0.04 | 41.5/1.05 |
| ZnS·CdS:Cu | 1.4/1.2 | 1.0/2.08 | 1.12/1.6 | 1.64/0.08 | 1.8/0.08 | 1.64/0.08 |
| ZnS:Ag | 14.2/1.0 | 3.67/11.6 | 9.47/5.96 | 14.6/1.19 | 15.4/0.74 | 16.4/0.63 |
| $YVO_4$:Eu | 3.2/1.28 | 2.48/2.68 | 3.16/0.4 | 3.04/0.2 | 2.84/0.16 | 2.4/0.16 |
| ZnS·CdS:Ag | 0.52/1.4 | 0.24/1.88 | 0.28/2.56 | 0.84/0.4 | 1.0/0.16 | 1.24/0.16 |
| ZnS:Ag | 1.92/1.0 | 1.16/2.76 | 1.92/2.0 | 2.44/0.32 | 2.52/0.16 | 2.48/0.16 |
| $Y_2O_3$:Eu | 3.4/0.32 | 3.4/0.24 | 3.48/0.16 | 3.44/0.16 | 3.24/0.16 | 3.48/0.16 |

In practice, weighed coefficients are convenient. They are related to the elementary volume parameters as

$$\sigma_m = \sigma z_0/m , \qquad \kappa_m = \kappa z_0/m , \qquad \varepsilon_m = \varepsilon z_0/m , \tag{10.3.3}$$

where $z_0$ is the geometric thickness of the layer, and $m$ its specific weighed amount. The quantities $\sigma_m$ and $\kappa_m$ are usually measured in $cm^2/mg$. Table 10.3 shows the spectral dependence of the absorption and scattering weighed coefficients of the luminophors. All the luminophors possess low absorption in the visible spectrum. The essential growth of $\kappa_m$ in the ultraviolet is attributable to the absorption band of luminescence-exciting radiation. In the visible spectrum, the $\sigma_m$ of all luminophors changes slightly with the wavelength, or remains constant. For the ultraviolet, all the luminophors have minimum $\sigma_m$. This is explained by the presence of the absorption band.

The spectral dependences of $\sigma_m$ and $\kappa_m$ for a contrast luminophor at different fractions $\Delta$ of the contrast matter in the total weight of powder are plotted in Fig. 10.7. The contrast luminophor consisted of the luminophor "white mixture" [1 mass fraction of the luminophor (ZnS:Ag) and 1.2 mass fraction of the luminophor (ZnS·CdS:Ag)], with the addition of the same luminophor treated by $Bi_2S_3$ contrast material. As seen from Fig. 10.7, the treatment of the luminescent screen results in an increase in its absorption in the visible spectrum. The dependence of $\kappa_m$ on $\lambda$ is most pronounced in the ultraviolet. As $\Delta$ increases, $\kappa_m$ reduces, i.e., the increase in the contrast material content reduces its photo-

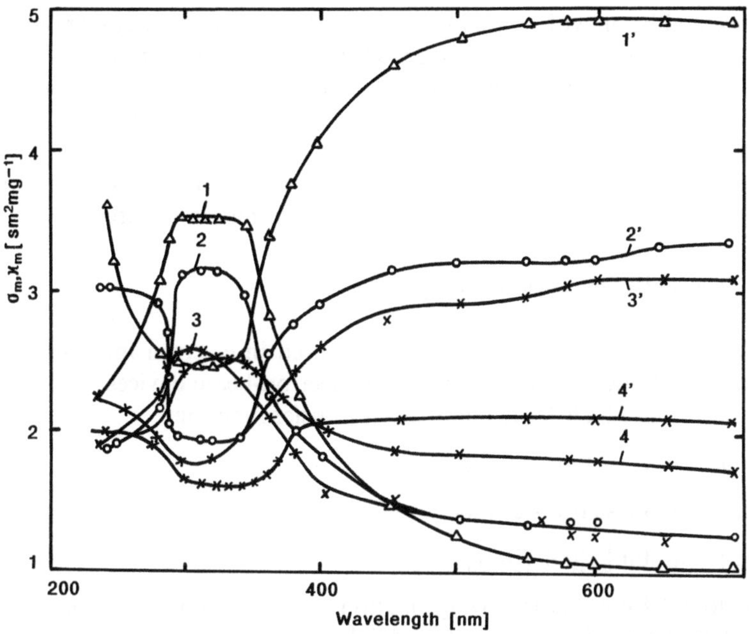


**Fig. 10.7.** Spectral variation of optical characteristics of contrast luminophors at $\Delta = 0$ (curves 1, 1'); 0.2 (curves 2, 2'); 0.4 (curves 3, 3'); 1 (curves 4, 4'). Curves 1'–4' show $\sigma_m$, curves 1–4, $\kappa_m$. From [10.35]

excitation ability. Scattering in the luminescent layer diminishes throughout the spectrum as the material fraction in it increases.

Although $\kappa_m(\lambda)$ and $\sigma_m(\lambda)$ are distinctly spectral for all the luminophors, the sum $\kappa_m(\lambda) + \sigma_m(\lambda) = \varepsilon_m(\lambda)$ is almost independent of $\lambda$. This is because of the predominance of large particles in the screens, for which the scattering theory provides spectrally neutral $\varepsilon_m$ (Sect. 2.2).

## 10.4 Modulation Transfer Function of Luminescent Screens

In luminescent screens, imaging undergoes two stages. A certain spatial structure is first projected onto a screen due to exciting radiation. This structure is partially spreaded thanks to the exciting radiation scattering. The light induced in the exposed areas is also scattered, thus smearing still further the visualized image pattern. On the luminescent screen, the image can be observed both from the side of exciting radiation (in reflected light) and from the opposite side (in transmitted light). The exciting radiation, which carries the object pattern, is incident normal to the screen as a parallel beam. The observation runs at a certain angle. In this connection, as mentioned in Chap. 5, a somewhat specific definition of the modulation transfer function can appropriately be used. Let a monodirectional excited beam with spatial sinusoidal (at spatial frequency $\omega$) power distribution be incident normal to a screen. Then, the MTF $T_I(\omega, \vec{n})$ is the ratio of the contrast of spatially distributed luminescent screen radiance in the $\vec{n}$-direction to the initial contrast of the excited radiation power distribution. Here, $\vec{n}$ is considered to be the direction of observation.

As in the general case, the MTF calculation is rather intricate, and will be analysed theoretically only for some particular luminescent screens.

### 10.4.1 Nonscattering Luminescent Screens

Nonscattering screens (glasses) are widely used in electron microscopy, high-resolution cathode-ray tubes, and special electron-beam devices.

As shown in [10.36], the modulation transfer functions

$$T_I(\omega, \vec{n}) = \sqrt{T_c^2(\omega) + T_s^2(\omega)} \tag{10.4.1}$$

as well as the phase transfer functions

$$\varphi(\omega) = \arctan[T_s(\omega)/T_c(\omega)] \tag{10.4.2}$$

can be found from the following relations. When observing from the side of incident exciting radiance (in reflected light),

$$\begin{aligned} T_{c,\mathrm{ref}}(\omega) &= \frac{e^{\mathscr{D}_1}\mathscr{D}_1^2 \cos C + C\mathscr{D}_1 e^{\mathscr{D}_1} \sin C - \mathscr{D}_1^2}{(\mathscr{D}_1^2 + C)(e^{\mathscr{D}_1} - 1)}, \\ T_{s,\mathrm{ref}}(\omega) &= \frac{-e^{\mathscr{D}_1}\mathscr{D}_1^2 \sin C + C\mathscr{D}_1 e^{\mathscr{D}_1} \cos C - C\mathscr{D}_1}{(\mathscr{D}_1^2 + C)(e^{\mathscr{D}_1} - 1)}, \end{aligned} \tag{10.4.3}$$

and when observing from the opposite side (in transmitted light),

$$\begin{aligned} T_c(\omega) &= \frac{e^{\mathscr{D}}\mathscr{D}^2 - \mathscr{D}^2 \cos C + C\mathscr{D} \sin C}{(\mathscr{D}^2 + C^2)(e^{\mathscr{D}} - 1)}, \\ T_s(\omega) &= \frac{-C\mathscr{D}e^{\mathscr{D}} + \mathscr{D}^2 \sin C + C\mathscr{D} \cos C}{(\mathscr{D}^2 + C^2)(e^{\mathscr{D}} - 1)}, \end{aligned} \tag{10.4.4}$$

where $\mathscr{D} = \tau_0/\cos\theta - \tau_{\mathrm{ex},0}$, $\mathscr{D}_1 = -\tau_0/\cos\theta - \tau_{\mathrm{ex},0}$, $C = \omega\tau_0 \operatorname{tg}\theta$ ($\theta$ being the observation angle), $\tau_0 = \kappa z_0$, $\tau_{\mathrm{ex},0} = \kappa_{\mathrm{ex}} z_0$, $\omega = \nu/\kappa$. Here and below, the subscript "ex" on the optical parameters applies to exciting light; the absence of a subscript implies luminescence parameters.

Let us consider the modulation transfer functions calculated using (10.4.4). Figure 10.8 shows the strong MTF dependence on the observation angle. Thus, at $\theta = 0$, the contrast is unity at any spatial frequencies. Observation at some angle $\theta$ results in a lower contrast and spatial frequency dependent phase shift of the picture observed. The shift due to the asymmetric form of the spread function may also be observed at $\theta = 0$ if the screen is excited at some angle to the normal. The contrast is degraded with increasing $\theta$, i.e., when observing the

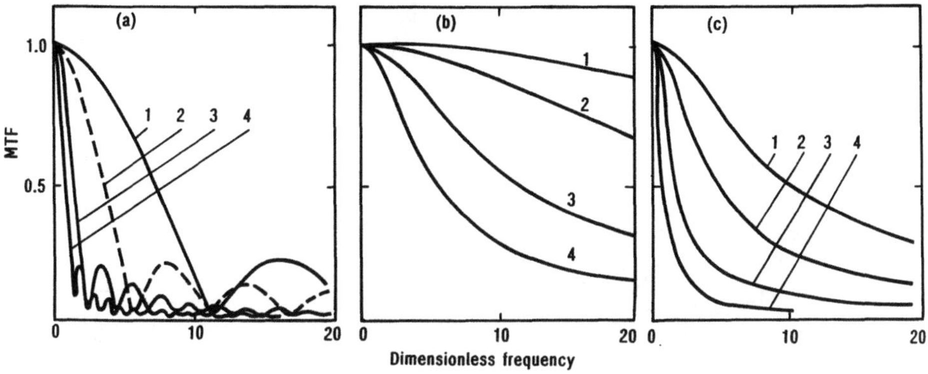


**Fig. 10.8.** Radiance MTF of a nonscattering luminescent screen at $\theta = 5°$ (curve 1), $10°$ (curve 2), $25°$ (curve 3), $45°$ (curve 4). (a) $\tau_{ex,0} = 1$, $\tau_0 = 1$. (b) $\tau_{ex,0} = 20$, $\tau_0 = 1$. (c) $\tau_{ex,0} = 20$, $\tau_0 = 5$

side view of the picture. This is connected with the spread widening. The nonmonotonic behavior of the MTF curves for small optical thicknesses $\tau_{ex,0}$ and $\tau_0$ (Fig. 10.8a) seems to be of interest. The pulsation frequency is seen to increase with the observation angle owing to the fact that the specific form of the spread function terminates abruptly as zero at $x = 0$ and $x = z_0 \operatorname{tg} \theta$. This jump is specifically large at small $\tau_{ex,0}$ and $\tau_0$.

As the optical thickness grows, the geometric thickness being constant, the luminescent area reduces. It is subtended to the screen side on which the exciting radiation is incident. This results in a "peaked" spread function, and higher image quality.

The image phase shift occurs at small optical frequencies and increases to $\pi/2$ as they grow. It changes rapidly at large observation angles and small optical thicknesses $\tau_{ex,0}$ and $\tau_0$. Like the MTF, the PTF at small $\tau_{ex,0}$ and $\tau_0$ is a nonmonotonic functon of the spatial frequency.

The above distinctions are similar in observation from the side of incident exciting radiation. However, the parameters being equal, the MTF for reflection is always better than that for transmission.

Given the radiance MTF $T_I(\omega, \theta)$ of the screen, the ordinary MTF $T(\omega)$ (of radiant emittance) can be found from the obvious relationship $T(\omega) = \int_{2\pi} T_I(\omega, \theta) \cos\theta \, d\Omega$. The spread function of the radiant emittance of the screen can be found in the same way. It is evident that the spread function so obtained is symmetric about the point $x = 0$, whereas the PTF vanishes. In this case, the transmission bandwidth of the spatial frequencies is essentially reduced.

### 10.4.2 Screens Weakly Absorbing Exciting Radiation

As the energy of the luminescence exciting particles grows, the absorbing power of all the materials generally decreases. Thus, slow electrons are absorbed by

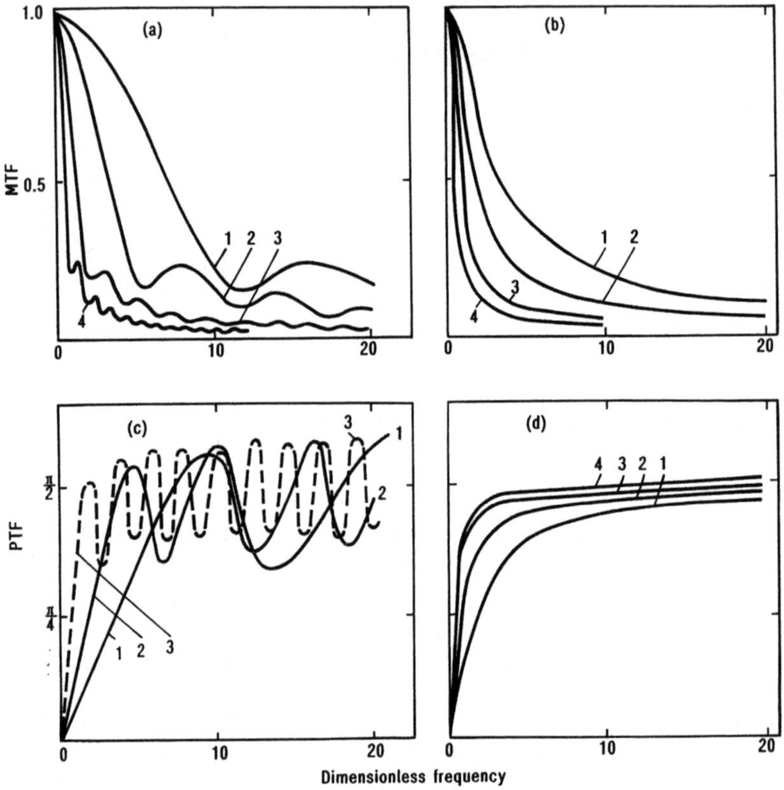


**Fig. 10.9.** Radiance MTF (a, b) and PTF (c, d) of a nonscattering luminescent screen weakly absorbing exciting radiation at $\theta = 5°$ (curve 1), $10°$ (curve 2), $25°$ (curve 3), $45°$ (curve 4). In (a, c) $\tau_0 = 1$; in (b, d) $\tau_0 = 5$

layers tenths of a micrometer thick, whereas they penetrate 2 μm deep at an accelerating voltage of about 20 kV. High-power X-rays and $\gamma$-quanta can travel large distances in the material without being attenuated. Therefore, the study of imaging in a luminescent medium weakly absorbing exciting radiation is important for an analysis of the operation of radiographic and X-ray screens.

Consider first the simplest case of a nonscattering screen. Here, $\tau_{ex,0} = 0$, and in the formulas for $T_c(\omega)$, $T_s(\omega)$, $T_{c,ref}(\omega)$, $T_{s,ref}(\omega)$ in the previous section, $\mathscr{D} = \tau_0/\cos\theta$, $\mathscr{D}_1 = -\mathscr{D}$. The results of MTF and PTF calculations are plotted in Fig. 10.9.

The radiant emittance of such a screen, provided it is sufficiently transparent for luminescence, can be analysed in terms of the approximate MTF expression [10.37]

$$T(v) = \frac{1}{vz_0}(1 - e^{-vz_0}) . \tag{10.4.5}$$

As follows from (10.4.5), the MTF changes slightly at large $v$. Therefore, the screen resolution determined by the contrast threshold will be high. The MTF depends on the product of $v$ and $z_0$, i.e., on the angular frequency $vz_0$. Hence, as the screen thickness is increased several times, the image emittance modulation will be the same for the spatial frequencies that are similarly decreased.

For the case of a light scattering screen, the MTF may be found in [10.38]. For the optical transfer function $S(\omega)$ ($T(\omega) = S(\omega)/S\ (\omega = 0)$), we have

$$S(\omega) = \frac{3 - x_1}{2\pi m^2}\left\{\frac{\sinh[m(\tau_0 + d_1)]}{\sinh[m(\tau_0 + d_1 + d_2)]}[\cosh[m(\tau_0 + d_2)] - \cosh(md_2)] + 1 - \cosh(m\tau_0)\right\},$$

$$m = \sqrt{\gamma^2 + \omega^2} \tag{10.4.6}$$

for a transmitting screen, and

$$S_{\text{ref}}(\omega) = \frac{(3 - x_1)\sinh(md_1)}{2\pi m^2 \sinh[m(\tau_0 + d_1 + d_2)]}[\cosh[m(\tau_0 + d_2)] - \cosh(md_2)] \tag{10.4.7}$$

for a reflecting screen.

An analysis of the relations in the present section shows that the MTF of scattering and nonscattering weakly absorbing screens for excitng radiation are similar. At the same time, a number of specific features can be revealed in the image transfer with a scattering screen. Essential changes in the optical thickness $\tau_0$ only change the MTF form slightly. It is especially distinct with appreciable luminescent light absorption, when the scattering of luminophor particles is low. The image quality characteristics are mainly determined by the light spread in the surface screen sections close to the observer.

### 10.4.3 Screens Strongly Absorbing Exciting Radiation

First, we consider a screen which does not scatter luminescent light. The image observed is concentrated on an extremely thin surface layer. The radiance spread function is essentially narrow and the MTF is, in fact, of unity for all spatial frequencies. This is true both of radiant emittance and radiance when observing in reflected light.

The radiant emittance distribution over the lower layer boundary (i.e., the spread function for a transmitting screen has some width), and the MTFs depend on the screen parameters. As shown in [10.2], in this case

$$T(\omega) = K_0(\tau_0\sqrt{1 + \omega^2})/K_0(\tau_0), \tag{10.4.8}$$

where $K_0(\tau_0)$ is the McDonald function. Curves calculated using (10.4.8) are plotted in Fig. 10.10. The image quality is seen to degenerate considerably with increasing optical thickness of the screen.

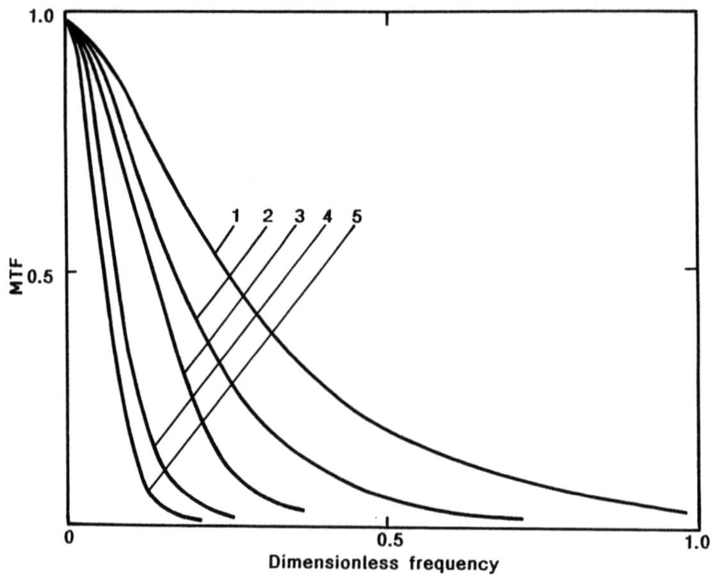


**Fig. 10.10.** Radiant emittance MTF of a nonscattering screen strongly absorbing exciting radiation at $\tau_0 = 0.5$ (curve 1), 1 (curve 2), 2 (curve 3), 5 (curve 4), 10 (curve 5)

Let us compare the transfer characteristics of luminescent screens with weak and strong absorption of exciting radiation. An analysis of (10.4.5, 8) suggests that in the former case, the MTF decrease is essential at low frequencies and insignificant at high frequencies with growing spatial frequency. The spread function has long tails and sharp peaks. In the latter case, as $\omega$ increases, the MTF changes slightly at very small frequencies, its decrease being rapid at large spatial frequencies.

For a light scattering screen, the results of [10.38] can be used to find the OTF for the radiation density

$$S(\omega) = \frac{3 - x_1}{2\pi m}\left[\frac{\sinh[m(\tau_0 + d_1)]\sinh[m(\tau_0 + d_2)]}{\sinh[m(\tau_0 + d_1 + d_2)]} - \sinh(m\tau_0)\right], \quad (10.4.9)$$

for a transmitting screen and

$$S_{\text{ref}}(\omega) = \frac{(3 - x_1)\sinh(md_1)\sinh[m(\tau_0 + d_2)]}{2\pi m \sinh[m(\tau_0 + d_1 + d_2)]}, \quad (10.4.10)$$

if the observation is from the side of exciting radiation incidence. Here, the same notation is used as in (10.4.6, 7).

These expressions demonstrate that the MTF is of the same form as for a nonscattering screen. In reflected light, the image quality is much better than in transmitted light.

### 10.4.4 Modulation Transfer Function of Cathode-Ray Screens

The previous section provided information on the specific features of imaging on a cathode-ray screen when the electron penetration is small as compared to the screen thickness. Let us now consider the MTF, taking into account the electron penetration depth.

As shown in [10.39], the optical transfer function in transmitted light is

$$S(\omega) = \frac{1}{2\pi}\left[\Gamma(\omega, d_2)\frac{\sinh[m(\tau_0 + d_1)]}{\sinh[m(\tau_0 + d_1 + d_2)]} - \Gamma(\omega, d_2 = 0)\right], \qquad (10.4.11)$$

whereas for observing in reflected light,

$$S_{\mathrm{ref}}(\omega) = \frac{1}{2\pi}\Gamma(\omega, d_2)\frac{\sinh(md_1)}{\sinh[m(\tau_0 + d_1 + d_2)]}. \qquad (10.4.12)$$

Here,

$$\Gamma(\omega, d_2) = \frac{\sinh[m(\tau_0 + d_2)]}{m\varepsilon R_s} - \int_0^1 (1-u)^{1/n}\cosh\left[m\varepsilon R_s\left(\frac{\tau_0 + d_2}{\varepsilon R_s} - u\right)\right]$$
$$\times \exp\left[-\frac{\sigma_{\mathrm{ex}}\beta_{2,\mathrm{ex}}}{4}\omega^2\varepsilon^2 R_s^3\frac{u^3}{3}\right]du, \qquad (10.4.13)$$

where $u = z/R_s$, and $n \approx 2$ is the empirical coefficient. In (10.4.13), the parameters $R_s$, $n$, $\sigma_{\mathrm{ex}}$ and $\beta_{2,\mathrm{ex}}$ apply to electrons, while the remainder are optical characteristics of the luminescent light scattering screen. In addition, $R_s \leqslant z_0$ is assumed.

The above equations provide a numerical analysis of the MTFs of cathode-ray screens. For transmitting screens, the MTF reduces with increasing thickness, while no changes are, in fact, observed in the case of reflection. The explanation is that the luminescent light scattering smears the image when it is observed in transmitted light, but almost does not distort it, when observed in light escaping from the excitation side of the screen. The reflecting screen resolution is 3–4 times as large as that of transmitting screens.

The MTF and resolution are weak functions of the accelerating voltage. For electron energies from 5 to 40 keV, the changes are less than 15%. This points to the predominant contribution of luminescent light scattering, rather than of electron diffusion, to image degradation.

### 10.4.5 Influence of Technological Screen Parameters on MTF

The general dependence of image characteristics on optical screen parameters mentioned above can be applied to real examples. A knowledge of the way in which the particular physical and technological parameters influence the optical thickness and single scattering albedo provides a guide to the MTF and resolution variation which depends on one parameter or another.

Some fragmentary data are available on the dependence of imaging characteristics on screen parameters. A great deal of such information is held in laboratories which are developing the production of optimum screens. Moreover, the data frequently reported describe the properties of an imaging device as a whole, a screen being only one of its numerous elements. This obscures the role played by the screen itself.

The optical thickness of real luminescent screens is usually characterized by a specific weighed amount of luminophor used. Numerous experimental data [10.40–47] show that the MTF and resolving power $R$ of the screen improve with a decrease in the specific weighed amount. It seems that $R$ might increase to become ultimately large with decreasing thickness. However, the specific features of thin screen production impose constraints on the specific weighed amount and, hence, to $R$. In addition, a sufficiently high absolute radiant emittance of the screen is required in practice. Therefore, some optimum specific weighed amount must exist.

The greatest number of experimental data apply to the effect of the type of luminophor used [10.36, 41, 42, 45, 47–50]. Their analysis suggests that, other conditions being equal, the influence of the original luminophor granularity is predominant. When equal amounts of screen coating material are used, fine-grained luminophors possess greater scattering, and the MTF of fine-grained screens is lower than in the case of coarse-grained luminophors. At the same time, a small number of fine particle layers can decrease the layer thickness and, thus, improve the MTF. The number of layers being equal, the image quality on fine-grained screens is higher. The dependence is inverse for equal geometric thickness.

The choice and improvement of screen production provides essential MTF enhancement. Thus, an increase in the binder concentration of more than 0.3% degrades the MTF [10.43]. This is due to the formation of a coarser luminophor coating.

Technological admixtures which change the optical properties of luminophor particles contribute much to MTF improvement [10.51–53]. A small amount of absorbing pigment may essentially improve imaging characteristics, with an insignificant reduction in the screen radiant emittance.

Information on the MTF of luminescent screens is detailed in [10.2].

It should be noted that in photographic and luminescent layers, particles are close-packed. This results in specific light scattering which is described in [10.3].

# List of Symbols and Abbreviations

| Symbol | Units | Meaning and first occurrence |
|---|---|---|
| $A$ | | Surface albedo (Sect. 3.4) |
| $A(\vec{r})$ | | Reflection coefficient (albedo) distribution over surface (Sect. 5.3.1) |
| $A_{ob}$ | | Object albedo (Sect. 5.4.2) |
| $A_b$ | | Background albedo (Sect. 5.4.2) |
| $a$ | m | Particle radius (Sect. 2.2) |
| $a(\vec{r})$ | $m^{-2}$ | Point spread function (PSF) of system (Sect. 5.2.1) |
| $a(\vec{v})$ | | Optical transfer function (OTF) of system (Sect. 5.2.1) |
| $B(\vec{R}, \vec{n})$ | $Wm^{-3}sr^{-1}$ | Emission coefficient (Sect. 2.3) |
| $B'(\vec{\tau}, \vec{n})$ | $Wm^{-2}sr^{-1}$ | Source function (Sect. 2.4) |
| $C_a$ | $m^2$ | Absorption cross-section (Sect. 2.2) |
| $C_{ext}$ | $m^2$ | Extinction cross-section (Sect. 2.2) |
| $C_s$ | $m^2$ | Scattering cross-section (Sect. 2.2) |
| $C_V$ | | Volume concentration of scatterers (Sect. 8.1.2) |
| $d$ | | Extrapolated length in Milne problem (Sect. 4.3.1) |
| $E$ | $Wm^{-2}$ | Irradiance (Sect. 2.1) |
| $F$ | | Portion of light scattered into forward hemisphere by a volume element or individual particle (Sect. 2.2) |
| $F_{\beta_0}$ | | Fraction of light scattered over angle range $0 \leqslant \beta \leqslant \beta_0$ (Sect. 4.4.2) |
| $\vec{F}(\vec{R})$ | $Wm^{-2}$ | Flux density vector (Sect. 2.1) |
| $f$ | $s^{-1}$ | Spectral frequency (Sect. 5.6.3) |
| $f(a)$ | $m^{-1}$ | Particle size distribution (Sect. 2.2) |
| $g$ | | Asymmetry parameter of scattering indicatrix (Sect. 2.2) |
| $H(\vec{r})$ | $Jm^{-2}$ | Exposure (Sect. 7.2.2) |
| $I(\vec{R}, \vec{n})$ | $Wm^{-2}sr^{-1}$ | Radiance (Sect. 2.1) |
| $I_{em}(\vec{n})$ | $Wsr^{-1}$ | Radiant intensity |
| $I_{ext}$ | $Wm^{-2}sr^{-1}$ | External background radiance (Sect. 9.5) |
| $I_n$ | $m^{-2}sr^{-1}$ | Normalized radiance (localized source) (Sect. 2.4) |
| $I_n$ | $sr^{-1}$ | Normalized radiance (infinitely wide source) (Sect. 3.1) |
| $I^s$ | $Wm^{-2}sr^{-1}$ | Scattered light radiance (Sect. 4.5.7) |
| $J_0, J_1$ | | Bessel functions, zero and first order (Sect. 4.1) |
| $j(\vec{R}, \vec{n})$ | $Wm^{-2}sr^{-1}$ | Angular spectrum of coherence function (Sect. 7.2.1) |
| $k$ | | Contrast |
| $k_0$ | | True (initial) object contrast (Sect. 5.2) |
| $k^*$ | | Threshold contrast (Sect. 5.1.1) |
| $k_{st}$ | | Observed contrast of Johnson stripped test object image (Sect. 5.4.3) |
| $M_{r,2n}$ | | Moments of radial irradiance distribution (Sect. 4.1) |
| $M_{\theta,2n}$ | | Moments of angular radiance distribution (Sect. 4.4.4) |
| $M_{u,k}$ | | Moments of time radiance (or irradiance) distribution (Sect. 4.8.1) |

| | | |
|---|---|---|
| $m$ | | Complex refractive index (Sect. 2.2) |
| $N$ | $m^{-3}$ | Number of particles per unit volume (Sect. 2.1) |
| $N$ | | Number of photons (Sect. 5.1) |
| $n$ | | Real part of complex refractive index |
| $n'$ | | Imaginary part of complex refractive index |
| $\vec{n}$ | | Solid angle vector (Sect. 2.1) |
| $\vec{n}_{\perp}$ | | $\vec{n}$ projection onto plane normal to beam axis (Sect. 4.4.1) |
| $\vec{n}_0$ | | Unit vector of incident beam direction (Sect. 3.1.1) |
| $\vec{n}_s$ | | Unit vector of source axis direction (Sect. 4.8.3) |
| $\vec{n}_{rec}$ | | Unit vector of receiver axis direction (Sect. 4.8.3) |
| $P$ | W | Radiant flux (Sect. 2.1) |
| $P_a$ | W | Absorbed flux (Sect. 2.2) |
| $P_s$ | W | Scattered flux (Sect. 2.2) |
| $P_0$ | W | Radiation flux from source (Sect. 4.1) |
| $P_{VS}$ | W | Valid signal power (Sect. 5.3.1) |
| $P_b$ | W | Background radiation power (Sect. 5.3.3) |
| $P_i(\cos\beta)$ | | Legendre polynomials (Sect. 2.2) |
| $P_{det}$ | | Detection probability (Sect. 5.1.4) |
| $P_f$ | | False alarm probability (Sect. 5.1.4) |
| $\vec{p}$ | $rad^{-1}$ | Angular frequency (Sect. 4.4.3) |
| $Q_a$ | | Absorption efficiency (Sect. 10.1) |
| $Q_{ext}$ | | Extinction efficiency (Sect. 2.2) |
| $Q_s$ | | Scattering efficiency (Sect. 8.1.2) |
| $q$ | | Scattering indicatrix parameter (Sect. 3.2) |
| $q_{eq}$ | | Scattering indicatrix parameter for similar media (Sect. 3.6.2) |
| $q$ | $gm^{-3}$ | Liquid water content (Sect. 8.1.2) |
| $\vec{R}$ | m | Point radius-vector (Sect. 2.1) |
| $R_c$ | m | Correlation radius (Sect. 4.4.7) |
| $\vec{r}$ | m | $\vec{R}$ projection onto $z = \text{const}$ plane (Sect. 4.1) |
| $r(\mu, \mu_0, \psi)$ | | Reflection function for semi-infinite medium (Sect. 3.3) |
| $r(\mu, \mu_0)$ | | Azimuth-averaged reflection function (Sect. 3.3.1) |
| $r(\mu_0)$ | | Reflection coefficient for directional incident beam (layer albedo) (Sect. 3.3.2) |
| $r$ | | Reflection coefficient for diffuse illumination (spherical albedo) (Sect. 3.3.2) |
| $r_0(\mu, \mu_0, \psi)$ | | $r(\mu, \mu_0, \psi)$ at $\Lambda = 1$ (Sect. 3.3.1) |
| $r_0^{(1)}(\mu, \mu_0, \psi)$ | | $r_0(\mu, \mu_0, \psi)$ for single scattering (Sect. 3.3.1) |
| $r(\tau_0, \mu, \mu_0, \psi)$ | | Reflection function for layer of optical thickness $\tau_0$ (Sect. 3.6.2) |
| $S_1, S_2$ | | Downward and upward irradiance coefficient (infinitely wide source) (Sect. 3.7.1) |
| $S(\vec{r}, z)$ | $m^{-2}$ | Irradiance coefficient (narrow beam source), medium point spread function (PSF) (Sect. 4.1) |
| $S(z) \equiv S_1$ | | Irradiance coefficient due to downward flux (infinitely wide source) (Sect. 4.1) |
| $S(\vec{v})$ | | Medium optical transfer function (Sect. 4.1) |
| $S_m$ | | Meteorological visibility range (MVR) (Sect. 8.1.2) |
| $S_{\lambda}$ | $AW^{-1}$ | Spectral response of photocathode (Sect. 5.1.1) |
| $s$ | | Scattering media parameter equal to $\sqrt{\Lambda(1-\Lambda)\beta_2/2}$ or $\sqrt{\Lambda(1-\Lambda)(1-g)}$ (Sect. 3.2) |
| $T(\vec{v})$ | | Modulation transfer function (MTF) (Sect. 5.2.1) |
| $T_p(\vec{v})$ | | Photographic MTF (Sect. 10.2.1) |
| $t$ | s | Time (Sect. 4.8.1) |
| $t_e$ | s | Exposure time (Sect. 7.1.1) |
| $t_{el}$ | s | Recording time for image cell (Sect. 5.1.1) |
| $t(\tau_0, \mu, \mu_0, \psi)$ | | Transmission function of thick layer (Sect. 3.4) |
| $t(\tau_0, \mu_0)$ | | Transmission coefficient for directional beam (Sect. 3.4) |
| $t(\tau_0)$ | | Transmission coefficient for diffuse illumination (Sect. 3.4) |
| $U(\tau)$ | | Heaviside unit function (Sect. 4.3.3) |

| | | |
|---|---|---|
| $U(\vec{r}, \Sigma)$ | | Unit function over area (Sect. 5.1.5) |
| $u$ | | Dimensionless time (Sect. 4.8.1) |
| $\langle u \rangle$ | | Mean photon delay time (Sect. 4.8.1) |
| $u(\mu)$ | | Escape function in Milne problem (Sect. 3.2) |
| $u_0(\mu)$ | | $u(\mu)$ at $\Lambda = 1$ (Sect. 3.2) |
| $V$ | $m^3$ | Volume (Sect. 2.2) |
| $V_N$ | | Variance of photon number (Sect. 5.1.1) |
| $V_r$ | $m^2$ | Variance of PSF, mean square radial photon deviation from beam axis in $z$ plane (Sect. 4.1) |
| $V_\rho$ | | Variance of PSF in nondimensional coordinates equal to $\varepsilon^2 V_r$ (Sect. 4.2) |
| $V_\theta$ | $rad^2$ | Variance of angular radiance distribution, mean square angular photon deviation from beam axis (Sect. 4.4.4) |
| $V_\theta^{as}$ | $rad^2$ | $V_\theta$ for asymptotic regime (Sect. 4.5.2) |
| $V_{\rho\theta}$ | $rad^2$ | Variance of angular spatial radiance distribution (narrow beam source (Sect. 4.5.3) |
| $V_u$ | | Variance of photon paths (Sect. 4.8.1) |
| $V_v$ | | Parameter of OPF bandwidth equal to $V_r^{-1}$ (Sect. 6.1.3) |
| $v$ | $ms^{-1}$ | Light velocity in medium (Sect. 2.1) |
| $W_0$ | J | Pulse energy (Sect. 4.8.3) |
| $\Delta W$ | J | Valid signal energy (Sect. 5.4.2) |
| $W_b$ | J | Background energy (Sect. 5.4.2) |
| $W_{ext}$ | J | Energy from external sources (Sect. 9.3.2) |
| $w$ | $Wm^{-2}$ | Scalar irradiance (Sect. 2.1) |
| $w^*$ | $Jm^{-3}$ | Radiant density (Sect. 2.1) |
| $x$ | m | Cartesian coordinate (Sect. 4.1) |
| $x(\beta)$ | | Scattering indicatrix (Sect. 2.2) |
| $x_{eq}(\beta)$ | | Scattering indicatrix of similar media (Sect. 3.6.1) |
| $x_i$ | | Coefficient in Legendre polynomial expansion (Sect. 2.2) |
| $y$ | m | Cartesian coordinate (Sect. 4.1) |
| $z$ | m | Cartesian coordinate, depth (Sect. 3.1) |
| $z_d$ | m | Sekky's disk visibility depth (Sect. 9.2.3) |
| $\alpha$ | | Generalized parameter of approximate transfer theory (Sect. 3.2) |
| $\alpha_2$ | | Video system parameter (Sect 5.1.2) |
| $\beta$ | rad | Scattering angle (Sect. 2.2) |
| $\beta_2$ | $rad^2$ | Second angular moment of scattering indicatrix (Sect. 4.4.4) |
| $\Gamma_{\varepsilon\varepsilon}(\vec{r}', z)$ | $m^{-2}$ | Correlation function (Sect. 4.4.7) |
| $\Gamma(\vec{R}_1, \vec{R}_2, t_d)$ | | Mutual coherence function (Sect. 7.2.1) |
| $\gamma$ | | Asymptotic attenuation coefficient (Sect. 3.2) |
| $\gamma_{CC}$ | | Maximum gradient of characteristic curve (CC) (Sect. 10.2.1) |
| $\delta$ | | Signal-to-noise ratio (SNR) (Sect. 5.1.1) |
| $\delta^*$ | | Threshold signal-to-noise ratio (Sect. 5.1.2) |
| $\delta(\cdot)$ | | Dirac function (Sect. 2.2) |
| $\varepsilon$ | $m^{-1}$ | Extinction coefficient (Sect. 2.2) |
| $\varepsilon_{eq}$ | $m^{-1}$ | Extinction coefficient for similar media (Sect. 3.6.1) |
| $\xi$ | | SADA parameter (Sect. 4.5.2) |
| $\eta$ | | Portion of light scattered at small angles (Sect. 3.6.1) |
| $\eta_{VS}$ | | Valid signal power transfer coefficient (Sect. 5.3.1) |
| $\eta_{BSN}$ | | BSN power transfer coefficient (Sect. 5.3.3) |
| $\theta$ | rad | Polar angle (Sect. 3.1) |
| $\kappa$ | $m^{-1}$ | Absorption coefficient (Sect. 2.2) |
| $\Lambda$ | | Single scattering albedo (Sect. 2.2) |
| $\Lambda_{eq}$ | | Single scattering albedo for similar media (Sect. 3.6.2) |
| $\lambda$ | m | Radiation wavelength (Sect. 2.2) |
| $\mu$ | | Cosine of polar angle (Sect. 3.1) |
| $\mu_0$ | | Cosine of polar angle of incident beam (Sect. 3.1) |
| $\vec{v}$ | $m^{-1}$ | Spatial frequency (Sect. 4.1) |
| $v_x, v_y$ | $m^{-1}$ | Vector $\vec{v}$ components (Sect. 4.1) |

| | | |
|---|---|---|
| $\nu^*$ | | Dimensionless angular frequency equal to $\nu z$ (Sect. 6.1.1) |
| $\zeta$ | | Generalized parameter of approximate transfer theory (Sect. 4.7) |
| $\vec{\rho}$ | | $\vec{\tau}$-projection onto the $z = \text{const}$ plane (Sect. 4.1) |
| $\Sigma$ | $m^2$ | Area (Sect. 2.1) |
| $\Sigma_{el}$ | $m^2$ | Image cell area (Sect. 5.4.2) |
| $\Sigma_{ob}$ | $m^2$ | Object area (Sect. 5.4.2) |
| $\Sigma_{rec}$ | $m^2$ | Receiver area (Sect. 4.6) |
| $\sigma$ | $m^{-1}$ | Scattering coefficient (Sect. 2.2) |
| $\sigma_{eq}$ | $m^{-1}$ | Scattering coefficient for similar media (Sect. 3.6.1) |
| $\sigma(\beta)$ | $m^{-1}sr^{-1}$ | Volume scattering function (Sect. 2.2) |
| $\vec{\tau}$ | | Optical radius-vector (Sect. 2.4) |
| $\tau$ | | Optical depth (Sect. 3.1) |
| $\tau_z, \tau_y$ | | Radius-vector $\vec{\tau}$ components (Sect. 4.1) |
| $\tau_0$ | | Optical thickness of $z_0$ thick layer (Sect. 3.1) |
| $\tau_{eq}$ | | Optical depth of similar media (Sect. 3.6.2) |
| $\tau_\kappa$ | | Absorbing optical thickness (Sect. 4.4.4) |
| $\tau_{ex,0}$ | | Optical thickness for exciting light (Sect. 10.4.1) |
| $\varphi$ | rad | Azimuth of $\vec{R}$ point in cylindrical coordinate system with $z$ axis (Sect. 4.1) |
| $\varphi(\mu)$ | | Angular radiance distribution for asymptotic regime (Sect. 3.2) |
| $\varphi(\vec{\nu})$ | | Phase transfer function (PTF) (Sect. 5.2.1) |
| $\varphi_{rec}(\vec{n})$ | | Angular directional diagram of receiver sensitivity (Sect. 4.8.3) |
| $\varphi_s(\vec{n})$ | | Angular directional diagram of source emittance (Sect. 4.8.3) |
| $\Psi(\Sigma_{ob}, \Sigma_{el})$ | | $\Psi$-function of optimum detection theory (Sect. 5.4.2) |
| $\psi$ | rad | Azimuthal angle of $\vec{n}$ direction in spherical coordinate system (Sect. 3.1) |
| $\Omega$ | sr | Solid angle (Sect. 2.1) |
| $\Omega_{rec}$ | sr | Solid angle of receiver (Sect. 4.6) |
| $\Omega_s$ | sr | Solid angle of source (Sect. 4.5.1) |
| $\vec{\omega}$ | | Dimensionless spatial frequency (Sect. 4.1) |
| $\omega^*$ | | Dimensionless angular frequency equal to $\omega\tau$ (Sect. 6.1.1) |
| BSN | | Back-scattering noise, 115 |
| CC | | Characteristic curve, 312 |
| CRT | | Cathode ray tubes, 318 |
| DA | | Diffusion approximation, 65 |
| H-G | | Henyey-Greenstein, 12 |
| LSF | | Line spread function, 164 |
| LTC | | Limited time coherence, 219 |
| MC | | Microcrystals, 306 |
| MTF | | Modulation transfer function, 155 |
| MVR | | Meteorological visibility range, 242 |
| OTF | | Optical transfer function, 69 |
| PSF | | Point spread function, 69 |
| PTF | | Phase transfer function, 156 |
| RFIPH | | Reference-free image plane holography, 219 |
| SAA | | Small-angle approximation, 65 |
| SADA | | Small-angle diffusion approximation, 65 |
| SNR | | Signal-to-noise ratio, 143 |
| TV | | Television, 142 |
| VS | | Valid signal, 167 |

# References


1.1. S.Q. Duntley: J. Opt. Soc. Amer. **53**, 214–233 (1963)
1.2. O.A. Sokolov: Deep-Sea Res. **10**, 70–72 (1963)
1.3. V.V. Sharonov: *Measurement and Calculation of Remote Subject Visibility* (Gostekhizdat, Moscow, Leningrad 1947)
1.4. H. Koshmider: Beit. Phys. Atmosph. **12**, 171–187 (1925)
1.5. I.M. Levin: Izv. AN SSSR, Fiz. Atm. Okeana, **5**, 62–76 (1969)
1.6. D.M. Bravo-Zhivotovsky, L.S. Dolin, A.G. Luchinin, V.A. Saveliev: Izv. AN SSSR, Fiz. Atm. Okeana, **5**, 672–684 (1969)
1.7. D.M. Bravo-Zhivotovsky, L.S. Dolin, I.M. Levin, A.G. Luchinin, V.A. Saveliev: Vopr. Radioelektron., Ser. Tekh. Televid., Vyp. **3**, 35–46 (1972)
1.8. D.M. Bravo-Zhivotovsky, L.S. Dolin, I.M. Levin, A.G. Luchinin, V.A. Saveliev: Vopr. Radioelektron., Ser. Tekh. Televid., Vyp. **3**, 38–49 (1973)
1.9. D.M. Bravo-Zhivotovsky, L.S. Dolin, I.M. Levin, A.G. Luchinin, V.A. Saveliev: Izv. AN SSSR, Fiz. Atm. Okeana, **7**, 1143–1152 (1971)
1.10. H. Frieser: *Photographic Information Recording* (Focal, London 1975)
1.11. A.P. Ivanov: *Physical Principles of Hydrooptics* (Nauka i Tekhnika, Minsk 1975) pp. 349–405
1.12. A.P. Ivanov, V.A. Loyko: *Optics of a Photographic Layer* (Nauka i Tekhnika, Minsk 1983)
1.13. A.P. Ivanov, K.G. Predko: *Optics of a Luminescent Screen* (Nauka i Tekhnika, Minsk 1984)
1.14. V.E. Zuev, M.B. Kabanov: *Transfer of Optical Signals in Earth's Atmosphere* (*with Noise*) (Sov. Radio, Moscow 1977)
1.15. O.A. Sokolov: *Underwater Vision* (Gidrometeoizdat, Leningrad 1974)
1.16. N.G. Jerlov: *Marine Optics* (Elsevier, Amsterdam 1976) Chap. 12
1.17. S. Chandrasekhar: *Radiative Transfer* (Dover, New York 1960)
1.18. B. Davison: *Neutron Transport Theory* (Clarendon, Oxford 1958)
1.19. V.V. Sobolev: *A Treatise of Radiative Transfer* (D. van Nostrand, New York 1963)
1.20. V.V. Sobolev: *Light Scattering in Planetary Atmospheres* (Pergamon, Oxford 1975)
1.21. H.C. van de Hulst: *Multiple Light Scattering* (*Tables, Formulae and Applications*), Vol. 1, 2 (Academic, New York 1980)
1.22. A. Ishimaru: *Wave Propagation and Scattering in Random Media*, Vol. 1 (Academic, New York 1978)
1.23. A.S. Monin (ed.): *Optics of the Ocean*, Vol. 1, 2 (Nauka, Moscow 1983)
1.24. *International Dictionary of Physics and Electronics* (D. van Nostrand, New York 1956)
1.25. *International Lighting Vocabulary*, 1(1) (Comm. Int. Eclairage Publ. 1957)
1.26. *American Institute of Physics Handbook* (McGraw-Hill, New York 1963) Sect. 6
1.27. D.A. Roberts: "Radiometry/Photometry: Concepts and Terms", in *Encyclopedia—The Optical Industry and Systems Directory* (1973–1974) pp. H77–H81
1.28. E. Ingeistam: Optik **B18**, 657–658 (1961)
2.1. S. Chandrasekhar: *Radiative Transfer* (Dover, New York 1960)
2.2. H.C. van de Hulst: *Multiple Light Scattering* (*Tables, Formulae and Applications*), Vol. 1, 2 (Academic, New York 1980)
2.3. D.A. Roberts: "Radiometry/Photometry: Concepts and Terms", in *Encyclopedia—The Optical Industry and Systems Directory* (1973–1974) pp. H77–H81
2.4. V.V. Sobolev: *A Treatise of Radiative Transfer* (D. van Nostrand, New York 1963)
2.5. V.V. Sobolev: *Light Scattering in Planetary Atmospheres* (Pergamon, Oxford 1975)
2.6. A. Ishimaru: *Wave Propagation and Scattering in Random Media*, Vol. 1 (Academic, New York 1978)

2.7. C.F. Bohren, D.R. Hufman: *Absorption and Scattering of Light by Small Particles* (John Wiley and Sons, New York 1983)
2.8. K.S. Shifrin: *Scattering of Light in a Turbid Medium* NASA Report No. TTF-447 (Washington, D.C., 1951)
2.9. H.C. van de Hulst: *Light Scattering by Small Particles* (Wiley, New York 1957)
2.10. M. Kerker: *The Scattering of Light and Other Electromagnetic Radiation* (Academic, New York 1969)
2.11. R.G. Newton: *Scattering Theory of Waves and Particles* (McGraw-Hill, New York 1966)
2.12. G. Mie: Ann. Phys. **25**, 377–445 (1908)
2.13. K.S. Shifrin, I.L. Zelmanovich: *Tables on Light Scattering, Vol. 2. Tables of Scattering Matrices and Scattering Field Components* (Gidrometeoizdat, Leningrad 1968)
2.14. I.L. Zelmanovich, K.S. Shifrin: *Tables on Light Scattering, Vol. 4. Light Scattering by Polydisperse Systems* (Gidrometeoizdat, Leningrad 1971)
2.15. D. Deirmendjian: *Electromagnetic Scattering on Spherical Polydispersion* (Elsevier, New York 1969)
2.16. W.M. Irwine: Communication from the Observatory at Leiden B.A.N. **17**, N 3, 176–184 (1963)
2.17. E.P. Zege, A.A. Kohanovsky: Izv. AN SSSR, Fiz. Atm. Okeana, **24**, 695–701 (1988)
2.18. D. Schnerman (ed.): *Light Scattering by Irregularly Shaped Particles* (Plenum, New York 1980)
2.19. A.P. Prishivalko, V.A. Babenko, V.N. Kuzmin: *Light Scattering and Absorption by Inhomogeneous and Anisotropic Spherical Particles* (Nauka i Tekhnika, Minsk 1984)
2.20. A.R. Holt: Radio Science **17**, 929–945 (1982)
2.21. O.N. Volkovitsky, L.N. Pavlova, A.G. Petrushin: *Optical Properties of Crystal Clouds* (Gidrometeoizdat, Leningrad 1984)
2.22. L.L. Foldy: Phys. Rev. **67**, 107–119 (1945)
2.23. M. Lax: Phys. Rev. **85**, 621–629 (1962)
2.24. A.P. Ivanov, V.A. Loyko, V.P. Dik: *Light Propagation in Media with Dense Scatterers* (Nauka i Tekhnika, Minsk 1988)
2.25. M.M. Gurevich: Trudy GOI **6**, 1–18 (1931)
2.26. V.P. Rvachev: *Methods of Scattering Media Optics in Physics and Biology* (Byelorus. Univ., Minsk 1978)
2.27. F. Ya. Sidko, I.A. Terskov, N.S. Eroshin, I.A. Nemchenko, V.A. Zakharova: "Light Scattering Media Spectroscopy as Applied to Biophysical Studies of Unicellular Organisms and Their Suspensions", in *Theoretical and Applies Light Scattering Problems* (Nauka i Teknika, Minsk 1971) pp. 361–375
2.28. A.Ya. Khairullina: "Study of Biological Cells by Light Scattering Methods", in *Light Propagation in a Dispersed Medium* (Nauka i Tekhnika, Minsk 1982) pp. 275–292
2.29. A.G. Bezrukova, I.K. Vladimirskaya: Tsitologiya **24**, 508–521 (1982)
2.30. B. Davison: *Neutron Transport Theory* (Clarendon, Oxford 1958)
2.31. A.P. Ivanov: *Optics of Scattering Media* (Nauka i Tekhnika, Minsk 1969)
2.32. A.P. Ivanov, V.A. Loyko: *Optics of a Photographic Layer* (Nauka i Tekhnika, Minsk 1983)
2.33. A.P. Ivanov: *Physical Principles of Hydrooptics* (Nauka i Tekhnika, Minsk 1975) pp. 349–405
2.34. K.M. Case, P.F. Zweifel: *Linear Transport Theory* (Addison-Wesley, Mass. 1967)
2.35. I.M. Levin: Izv. AN SSSR, Fiz. Atm. Okeana, **5**, 62–76 (1969)
3.1. S. Chandrasekhar: *Radiative Transfer* (Dover, New York 1960)
3.2. B. Davison: *Neutron Transport Theory* (Clarendon, Oxford 1958)
3.3. V.V. Sobolev: *A Treatise of Radiative Transfer* (D. van Nostrand, New York 1963)
3.4. V.V. Sobolev: *Light Scattering in Planetary Atmospheres* (Pergamon, Oxford 1975)
3.5. H.C. van de Hulst: *Multiple Light Scattering* (*Tables, Formulae and Applications*), Vol. 1, 2 (Academic, New York 1980)
3.6. K.M. Case, P.F. Zweifel: *Linear Transport Theory* (Addison-Wesley, Mass. 1967)
3.7. T.A. Germogenova: "Development of Numerical Methods of Solution of the Radiative Transfer Equation", in *Light Propagation in a Dispersed Medium* (Nauka i Tekhnika, Minsk 1982) pp. 105–118
3.8. J. Lenoble (ed.): Standard Procedures to Compute Atmospheric Radiative Transfer in a Scattering Atmosphere, Vol. 1 (National Center for Atmospheric Research, Boulder, Colorado 1977)

3.9. Y. Fouquart, W.M. Irwine, J. Lenoble: *Standard Procedures to Compute Atmospheric Radiative Transfer in a Scattering Atmosphere, Vol.* 2 (National Center for Atmospheric Research, Boulder, Colorado 1980)
3.10. G.I. Marchuk, G.A. Mikhailov, M.A. Nazaraliev, R.A. Darbinyan, B.A. Kargin, B.S. Elepov: *Monte-Carlo Methods in Atmospheric Optics* (Springer, Berlin–Heidelberg 1980)
3.11. G.I. Marchuk, G.A. Mikhailov: "Monte-Carlo Solution of Radiative Transfer Theory Problems", in *Theoretical and Applied Light Scattering Problems* (Nauka i Tekhnika, Minsk 1971) pp. 43–58
3.12. A. Ishimaru: *Wave Propagation and Scattering in Random Media*, Vol. 1 (Academic, New York 1978)
3.13. M.V. Maslennikov: Trudy Inst. Mat. AN SSSR, Vyp. 97, 3–131 (1968)
3.14. V.A. Ambartsumyan: Izv. AN SSSR, Ser. Geograf. Geofiz. **3**, 97–104 (1942)
3.15. L.M. Romanova: Izv. AN SSSR, Ser. Geofiz., N 9, 1294–1296 (1962)
3.16. E.P. Zege: Izv. AN SSSR, Fiz. Atm. Okeana **7**, 121–132 (1971)
3.17. V.N. Loskutov: Vestn. Leningr. Univ., N 13, 143–149 (1969)
3.18. N.V. Konovalov: "Asymptotic Characteristics of Monochromatic Radiation Fields in the Problems on an Inhomogeneous Layer of Large Optical Sizes. 2. Calculation of Basic Parameters and Functions"; Preprint, Inst. Appl. Math. Acad. Sci. Moscow USSR, No. 14 (1974)
3.19. V.G. Schellenberg: Pure and Appl. Geophys. **58**, N 3 (1964)
3.20. E.P. Zege, L.I. Chaikovskaya: Izv. AN BSSR, Ser. Fiz. Mat. Nauk, N 2, 68–75 (1982)
3.21. E.P. Zege: "Some Results of the Asymptotic Radiative Transfer Theory for Absorbing Anisotropically Scattering Media as Applied to Ocean and Cloud Optics"; Preprint, Inst. Phys. Acad. Sci. BSSR, N 274 (Minsk 1982)
3.22. E.P. Zege: "Design Calculations of Light Fields at Multiple Scattering", in *Light Propagation in a Dispersed Medium* (Nauka i Tekhnika, Minsk 1982) pp. 84–105
3.23. O.V. Bushmakova, E.P. Zege, I.L. Katsev: Dokl. AN BSSR **15**, 309–311 (1971)
3.24. E.P. Zege, O.V. Bushmakova, I.L. Katsev, N.V. Konovalov: Zh. Prikl. Spektrosk. **30**, 900–907 (1979)
3.25. V.A. Ambartsumyan: Dokl. AN SSSR **38**, 229–232 (1943)
3.26. E.G. Yanovitsky: "Approximate Formulas for the Intensity of Diffusely Reflected Atmospheric Radiation", in *Astrometry and Astrophysics* (Naukova Dumka, Kiev 1972) pp. 63–75
3.27. E.P. Zege: Izv. AN SSSR, Fiz. Atm. Okeana **19**, 927–936 (1983)
3.28. B.M. Golubitsky, I.M. Levin, M.V. Tantashev: Izv. AN SSSR, Fiz. Atm. Okeana **10**, 1235–1238 (1974)
3.29. L.S. Anikanov, S.Yu. Ermolayev: Vestn. Leningr. Univ., N 7 Vyp. 2, 132–137 (1977)
3.30. G.V. Rozenberg: "Optical Properties of Thick Inhomogeneous Scattering Layers", in *Spectroscopy of Scattering Media* (AN BSSR, Minsk 1963) pp. 5–36
3.31. T.A. Germogenova: "Some Formulas to Solve the Transfer Equation in the Plane Layer Problem", in *Spectroscopy of Scattering Media* (AN BSSR, Minsk 1963) pp. 36–41
3.32. T.A. Germogenova: Zh. Vych. Mat. Mat. Fiz., No. 6, 1001–1019 (1961)
3.33. H.C. van de Hulst: Bull. Astron. Inst. Nether. **20**, 77–86 (1968)
3.34. V.V. Ivanov: Astron. Zh. **53**, 589–595 (1976)
3.35. S.D. Gutshabash: Izv. AN SSSR, Fiz. Atm. Okeana **14**, 912–919 (1978)
3.36. Kh. Domke, V.V. Ivanov: Astron. Zh. **52**, 1034–1036 (1976)
3.37. E.G. Yanovitsky: Astron. Zh. **48**, 323 (1971)
3.38. E.G. Yanovitsky: "Light Scattering in an Anisotropic Atmosphere", in *Light Propagation in a Dispersed Medium* (Nauka i Tekhnika, Minsk 1982) pp. 36–54
3.39. T.A. Germogenova: Zh. Vych. Mat. Mat. Fiz. **14**, 1528–1535 (1974)
3.40. N.V. Konovalov: "On the Accuracy of the Transfer Equation Solution for a Plane Layer of a Large Optical Thickness with Strongly Anisotropic Scattering"; Preprint, Inst. Appl. Math. Acad. Sci. USSR, N 94 (Moscow, 1975)
3.41. E.P. Zege, I.L. Katsev: "Temporal Asymptotic Solutions of the Radiative Transfer Equation and Their Application"; Preprint, Inst. Phys. Acad. Sci. BSSR (Minsk, 1973)
3.42. E.P. Zege, I.L. Katsev: Zh. Prikl. Spektrosk. **31**, 327–332 (1979)
3.43. T.A. Sushkevish: "Allowance for Strong Scattering Anisotropy in the Problems on Radiation Propagation with a Monodirectional Source"; Preprint, Inst. Appl. Math. Acad. Sci. USSR, N 132 (Moscow 1979)
3.44. J.E. Hansen: J. Atmosph. Sci. **26**, 478–487 (1969)

3.45. W.J. Wiscombe: J. Atmosph. Sci. **34**, 1408–1422 (1977)
3.46. V.V. Sobolev: Astron. Zh. **20**, 14–22 (1943)
3.47. J.H. Joseph, W.J. Wiscombe: J. Atmosph. Sci. **33**, 2452–2459 (1976)
3.48. O. Aspelund: "A New Method of Boltzmann Equation Solution in a Transport Approximation", in *Physics of Nuclear Reactors* (Atomizdat, Moscow 1959) pp. 433–442
3.49. I.L. Katsev: Izv. AN SSSR, Fiz. Atm. Okeana **10**, 425–430 (1974)
3.50. A.M. Samson: Inzh. Fiz. Zh. **1**, 65–73 (1958)
3.51. O.V. Bushmakova, L.P. Bass, E.P. Zege: Izv. AN SSSR, Fiz. Atm. Okeana **8**, 1114–1119 (1972)
3.52. H.R. Gordon: Appl. Opt. **12**, 2803–2804 (1973)
3.53. B.M. Golubitsky, I.M. Levin: Izv. AN SSSR, Fiz. Atm. Okeana **16**, 1051–1058 (1980)
3.54. L.S. Dolin, V.A. Saveliev: Izv. AN SSSR, Fiz. Atm. Okeana **7**, 505–510 (1971)
3.55. B.V. Ermakov, Yu. A. Iliinsky: Izv. Vuz., Ser. Radiofiz. **12**, 694–701 (1969)
3.56. O.A. Avaste, V.S. Atroshenko: Izv. AN SSSR, Ser. Geofiz. N 3, 507–510 (1960)
3.57. I.W. Basbridge, S.E. Orchard: Astrophys. J. **149**, 655 (1967)
3.58. V.V. Sobolev: *Courses in Theoretical Astrophysics* (Nauka, Moscow 1967)
3.59. E.P. Zege: "Two-Stream Approximation in Radiative Transfer Theory"; Preprint, Inst. Phys. Acad. Sci. BSSR (Minsk 1971)
3.60. E.S. Kuznetsov: Dokl. AN SSSR **37**, 237–244 (1942)
3.61. V.I. Khalturin: Izv. AN SSSR, Fiz. Atm. Okeana (Deposited in VINITI N 2956–82, 1982)
3.62. A.A. Gershun: "Light Propagation through a Plane Light Scattering Layer", in *Selected Papers on Photometry and Lighting Engineering* (Fizmatgiz, Moscow 1958) pp. 68–85
3.63. A.P. Ivanov: *Physical Principles of Hydrooptics* (Nauka i Tekhnika, Minsk 1975) pp. 349–405
3.64. D. Deirmendjian: *Electromagnetic Scattering on Spherical Polydispersion* (Elsevier, New York 1969)
4.1. G.I. Marchuk, G.A. Mikhailov, M.A. Nazaraliev, R.A. Darbinyan, B.A. Kargin, B.S. Elepov: *Monte-Carlo Methods in Atmospheric Optics* (Springer, Berlin–Heidelberg 1980)
4.2. G.I. Marchuk, G.A. Mikhailov: "Monte-Carlo Solution of Radiative Transfer Theory Problems", in *Theoretical and Applied Light Scattering Problems* (Nauka i Tekhnika, Minsk 1971) pp. 43–58
4.3. A.P. Ivanov, I.D. Sherbaf: Opt. Spektrosk. **18**, 698–703 (1965)
4.4. A.P. Ivanov, S.A. Makarevich, I.D. Sherbaf: Izv. AN SSSR, Fiz. Atm. Okeana **4**, 879–887 (1968)
4.5. A.P. Ivanov, I.D. Sherbaf: "Searchlight Beam Scattering in a Turbid Medium", in *Actinometry and Optics of the* Atmosphere (Nauka, Moscow 1964) pp. 267–275
4.6. A.P. Ivanov: *Physical Principles of Hydrooptics* (Nauka i Tekhnika, Minsk 1975) pp. 349–405
4.7. A.P. Ivanov: *Optics of Scattering Media* (Nauka i Tekhnika, Minsk 1969)
4.8. G.V. Rozenberg (ed.): *A Searchlight Beam in Atmosphere* (AN SSSR, Moscow 1960)
4.9. A.I. Gribanov: *Methods of Calculating Visibility at Directional Illuminance* (Gosenergoizdat, Moscow 1955)
4.10. I.L. Katsev: Izv. AN SSSR, Fiz. Atm. Okeana **10**, 425–430 (1974)
4.11. V.M. Kochetkov: Izv. AN SSSR, Fiz. Atm. Okeana **7**, 1163–1171 (1971)
4.12. G.A. Korn, T.M. Korn: *Mathematical Handbook for Scientists and Engineers* (McGraw-Hill, New York 1961)
4.13. I.L. Katsev: Izv. AN SSSR, Fiz. Atm. Okeana **17**, 478–486 (1981)
4.14. L.M. Romanova: Izv. AN SSSR, Fiz. Atm. Okeana **4**, 311–319 (1968)
4.15. L.M. Romanova: Izv. AN SSSR, Fiz. Atm. Okeana **4**, 1185–1195 (1968)
4.16. L.M. Romanova: Izv. AN SSSR, Fiz. Atm. Okeana **7**, 1153–1164 (1971)
4.17. E.B. Breshenkova, V.V. Orlov: Atom. Energ. **10**, 175–177 (1961)
4.18. Yu.N. Drozhzhinov, O.P. Moskalev: Dokl. AN SSSR **193**, 1271–1273 (1970)
4.19. T.A. Germogenova: "Problems with Localized Sources in the Radiative Transfer Theory"; Preprint, Inst. Appl. Math. Acad. Sci. USSR, N 23 (Moscow 1971)
4.20. T.A. Germogenova: *Local Properties of Transfer Equation Solution* (Nauka, Moscow 1986) pp. 204–209
4.21. B. Davison: *Neutron Transport Theory* (Clarendon, Oxford 1958)
4.22. A.M. Weinberg, E.P. Wigner: *The Physical Theory of Neutron Chain Reactors* (University of Chicago Press 1959)
4.23. K.M. Case, P.F. Zweifel: *Linear Transport Theory* (Addison-Wesley, Mass. 1967)

4.24. S. Glasstone, M.C. Edlund: *The Elements of Nuclear Reactor Theory* (London 1952)
4.25. M.V. Maslennikov: Trudy Inst. Mat. AN SSSR, Vyp. 97, 3–131 (1968)
4.26. E.P. Zege, I.L. Katsev: "Temporal Asymptotic Solutions of the Radiative Transfer Equation and Their Application"; Preprint, Inst. Phys. Acad. Sci. BSSR (Minsk, 1973)
4.27. L.S. Dolin: Dokl. AN SSSR **260**, 1344–1347 (1981)
4.28. L.S. Dolin: Izv. AN SSSR, Fiz. Atm. Okeana **18**, 840–849 (1982)
4.29. O.V. Bushmakova, E.P. Zege, I.L. Katsev: Izv. AN SSSR, Fiz. Atm. Okeana **8**, 711–719 (1972)
4.30. E.P. Zege, I.L. Katsev, A.M. Lazaruk: Opt. Spektrosk. **37**, 1126–1133 (1974)
4.31. L.M. Romanova: Izv. AN SSSR, Fiz. Atm. Okeana **7**, 410–421 (1971)
4.32. E.P. Zege: "Design Calculations of Light Fields at Multiple Scattering", in *Light Propagation in a Dispersed Medium* (Nauka i Tekhnika, Minsk 1982) pp. 84–105
4.33. E.P. Zege, I.L. Katsev: "Radiation Scattering in Plane Layers with Optical Thickness-Dependent Characteristics", in *Proc. All-Union Conf. Opt. Rad. Propag. Dispersed Medium* (Gidrometeoizdat, Moscow 1978) pp. 67–70
4.34. H. Frieser: *Photographic Information Recording* (Focal, London 1975)
4.35. G. Haaze, H. Muller: Optic B **17**, 1–25 (1960)
4.36. A.K. Kolesov: Vestn. Leningr. Univ., Ser. Mat., Mekh., Astron., Vyp. 1, 103–110 (1959)
4.37. Yu. A. Tsirlin, A.E. Pargamanik, A.R. Daich: Opt. Spektrosk. **12**, 304–310 (1962)
4.38. P.M. Morse, H. Feshback: *Methods of Theoretical Physics*, Part II (McGraw-Hill, New York 1953)
4.39. M.C. Wang, E. Guth: Phys. Rev. **84**, 1092–1111 (1951)
4.40. H.S. Snyder, W.T. Scott: Phys. Rev. **76**, 220–225 (1949)
4.41. W. Both: Zh. Phys. **54**, 167–178 (1929)
4.42. B. Rossi, K. Greisen: Rev. Mod. Phys. **13**, 240–309 (1941)
4.43. L.M. Romanova: Izv. AN SSSR, Ser. Geofiz. **8**, 1108–1112 (1962)
4.44. L.M. Romanova: Opt. Spektrosk. **13**, 429–435 (1962)
4.45. L.S. Dolin: Izv. Vuz., Ser. Radiofiz. **7**, 380–383 (1964)
4.46. H. Bremmer: Radio Sci. **68** D., 967–981 (1964)
4.47. L.S. Dolin, V.A. Saveliev: Izv. Vuz., Ser. Radiofiz. **22**, 1310–1317 (1979)
4.48. L.S. Dolin: Izv. AN SSSR, Fiz. Atm. Okeana **16**, 55–64 (1980)
4.49. D.M. Bravo-Zhivotovsky, L.S. Dolin, A.G. Luchinin, V.A. Saveliev: Izv. AN SSSR, Fiz. Atm. Okeana **5**, 160–167 (1969)
4.50. B.M. Golubitsky, I.M. Levin, M.V. Tantashev: Izv. AN SSSR, Fiz. Atm. Okeana **10**, 798–801 (1974)
4.51. H.R. Gordon, W.R. McLuney: Appl. Opt. **14**, 413–416 (1975)
4.52. I.L. Katsev: Izv. AN SSSR, Fiz. Atm. Okeana **18**, 884–885 (1982)
4.53. I.L. Katsev: Izv. AN BSSR, Ser. Fiz.-Mat. Nauk, N 1, 66–70 (1980)
4.54. I.L. Katsev: Izv. AN SSSR, Ser. Fiz. Atm. Okeana **19**, 1172–1179 (1983)
4.55. L.S. Dolin: Dokl. AN SSSR **277**, 77–80 (1984)
4.56. J. Kuza, A. Ishimaru: Appl. Opt. **25**, 4328–4385 (1986)
4.57. W.G. Tam, A. Zargecki: Appl. Opt. **19**, 2822–2827 (1980)
4.58. A. Zargecki, A.W. Gerste-Siegfild: Appl. Opt. **26**, 3000–3004 (1987)
4.59. J.W. McLean, D.R. Grawford, C.L. Hindman: Appl. Opt. **26**, 2053–2054 (1987)
4.60. A.N. Valentyuk: Issled. Zemli iz Kosmosa, No. 3, 91–99 (1987)
4.61. H.C. van de Hulst: *Multiple Light Scattering* (*Tables, Formulae and Applications*), Vol. 1, 2 (Academic, New York 1980)
4.62. I.L. Katsev: Opt. Spektrosk. **60**, 1245–1250 (1986)
4.63. I.L. Katsev: Opt. Spektrosk. **64**, 599–603 (1988)
4.64. L.S. Dolin: Izv. Vuz., Ser. Radiofiz. **7**, 471–478 (1964)
4.65. L.S. Dolin: Izv. Vuz., Ser. Radiofiz. **23**, 332–338 (1980)
4.66. L.S. Dolin: Izv. Vuz., Ser. Radiofiz. **11**, 840–849 (1968)
4.67. V.V. Sobolev: *Light Scattering in Planetary Atmospheres* (Pergamon, Oxford 1975)
4.68. I.Ya. Pomeranchuk: *Collection of Scientific Works*, Vol. 2 (Nauka, Moscow 1972)
4.69. L.S. Dolin: Izv. Vuz., Ser. Radiofiz. **9**, 67–71 (1966)
4.70. A.N. Valentyuk, K.G. Predko: Izv. AN SSSR, Fiz. Atm. Okeana **14**, 226–229 (1978)
4.71. A.G. Luchinin: Izv. Vuz., Ser. Radiofiz. **14**, 1925–1927 (1971)
4.72. A.G. Luchinin, V.A. Saveliev: Izv. Vuz., Ser. Radiofiz. **12**, 256–264 (1969)
4.73. A.G. Luchinin: Izv. AN SSSR, Fiz. Atm. Okeana **10**, 1312–1317 (1974)

4.74. V.S. Remizovich, D.B. Rogozkin, M.I. Ryazanov: "Narrow Beam Light Propagation in Turbid Atmosphere with Regard for Photon Fluctuations at Multiple Scattering", in *Collected Abstracts, 6th All-Union Symp. Laser Rad. Propag. Atm.*, Vol. 1 (Tomsk 1981) pp. 155–158
4.75. V.S. Remizovich, D.B. Rogozkin M.I. Ryazanov: Izv. Vuz., Ser. Radiofiz. **25**, 891–898 (1982)
4.76. L.S. Dolin: Izv. AN SSSR, Fiz. Atm. Okeana **19**, 400–404 (1983)
4.77. V.S. Remizovich, D.B. Rogozkin, M.I. Ryazanov: Izv. AN SSSR, Fiz. Atm. Okeana **19**, 1053–1061 (1983)
4.78. E.P. Zege, I.N. Polonsky, L.I. Chaikovskaya: Izv. AN SSSR, Fiz. Atm. Okeana **23**, 486–492 (1987)
4.79. A.S. Monin (ed.): *Optics of the Ocean*, Vol. 1, 2 (Nauka, Moscow 1983)
4.80. V.S. Remizovich, Sh.A. Shekhmametiev: "Light Propagation in the Scattering Medium with Large Inhomogeneities at Oblique Incidence"; Preprint, Moscow Engn. Physics Inst., 047–88 (Moscow 1988)
4.81. V.S. Remizovich, D.B. Rogozkin. M.I. Ryazanov: Izv. AN SSSR, Fiz. Atm. Okeana **22**, 736–742 (1986)
4.82. D. Deirmendjian: *Electromagnetic Scattering on Spherical Polydispersion* (Elsevier, New York 1969)
4.83. L.S. Dolin: Izv. Vuz., Ser. Radiofiz. **206**, 300–309 (1983)
4.84. E.P. Zege, I.L. Katsev, I.N. Polonsky: Opt. Atm. **1**, N 11, 19–27 (1988)
4.85. E.P. Zege, I.L. Katsev, I.N. Polonsky: "One Approach to the Solution of the Transfer Equation with an Arbitrary Scattering Indicatrix", in *Collected Abstracts, 6th All-Union Conf. Laser Rad. Propag. Dispersed Medium*, Vol. 1 (Obninsk-Barnaul 1988) pp. 172–174
4.86. B.V. Ermakov, Yu. A. Iliinsky: Izv. Vuz., Ser. Radiofiz. **12**, 694–701 (1969)
4.87. L.S. Dolin, V.A. Saveliev: Izv. AN SSSR, Fiz. Atm. Okeana **7**, 505–510 (1971)
4.88. D.B. Rogozkin: Izv. AN SSSR, Fiz. Atm. Okeana **23**, 366–375 (1987)
4.89. L.S. Dolin, O.V. Kopelevich, I.M. Levin, V.I. Feigels: Izv. AN SSSR, Fiz. Atm. Okeana **24**, 1218–1222 (1988)
4.90. O.V. Kopelevich: "Factors Determining Optical Properties of Sea Water", in *Optics of Ocean, Vol. 1. Physical Optics of Ocean* (Nauka, Moscow 1983) pp. 166–208
4.91. H.R. Gordon: Appl. Opt. **12**, 2803–2804 (1973)
4.92. B.M. Golubitsky, I.M. Levin: Izv. AN SSSR, Fiz. Atm. Okeana **16**, 1051–1058 (1980)
4.93. V.V. Sobolev: *A Treatise of Radiative Transfer* (D. van Nostrand, New York 1963)
4.94. L.M. Romanova: "Nonstationary Light Field in Turbid Media", in *Theoretical and Applied Light Scattering Problems* (Nauka i Tekhnika, Minsk 1971) pp. 74–92
4.95. I.N. Minin: "Theory of Nonstationary Radiation", in *Theoretical and Applied Light Scattering Problems* (Nauka i Tekhnika, Minsk 1971) pp. 59–73
4.96. D.I. Nagirner: Astrofiz. **10**, 445–469 (1974)
4.97. I.L. Katsev, E.P. Zege: Astrofiz. **10**, 219–225 (1974)
4.98. I.N. Minin: Vestn. Leningr. Univ., Vyp.3, 131–141 (1959)
4.99. I.L. Katsev: Dokl. AN BSSR **13**, 118–121 (1969)
4.100. I.L. Katsev: Zh. Prikl. Spektrosk. **11**, 85–91 (1969)
4.101. E.P. Zege, I.L. Katsev, I.D. Sherbaf: Izv. AN SSSR, Fiz. Atm. Okeana **9**, 937–946 (1973)
4.102. O.V. Bushmakova, E.P. Zege, I.L. Katsev: Izv. AN SSSR, Fiz. Atm. Okeana **10**, 250–257 (1974)
4.103. L.M. Romanova: Izv. AN SSSR, Fiz. Atm. Okeana **5**, 463–472 (1969)
4.104. E.P. Zege, I.L. Katsev: Dokl. AN BSSR **13**, 687–689 (1969)
4.105. E.P. Zege, I.L. Katsev: Izv. AN SSSR, Fiz. Atm. Okeana **8**, 945–953 (1972)
4.106. E.P. Zege, I.L. Katsev: "Light Field in a Scattering Medium due to a Pulsed Laser Source", in *Collected Abstracts, 2nd All-Union Symp. Laser Rad. Propag. Atm.* (Tomsk 1973) pp. 49–51
4.107. E.P. Zege, A.P. Ivanov, I.L. Katsev, B.A. Kargin, G.A. Mikhailov: Izv. AN SSSR, Fiz. Atm. Okeana **9**, 1054–1062 (1973)
4.108. E.P. Zege, A.P. Ivanov, I.L. Katsev, B.A. Kargin, S.V. Kuznetsov, G.A. Mikhailov: "Some Problems of Optical Pulse Location in Natural Scattering Media", in *Proc. 10th All-Union Conf. Radiowave Propagation* (Nauka Moscow 1972) pp. 337–341
4.109. E.P. Zege, A.P. Ivanov, I.L. Katsev: Izv. AN SSSR, Fiz. Atm. Okeana **7**, 750–757 (1971)
4.110. A. Ishimaru: J. Opt. Soc. Amer. **68**, 1045–1048 (1978)
4.111. K. Furutsu: J. Opt. Soc. Amer. **70**, 360–365 (1980)

4.112. S. Ito, K. Furutsu: J. Opt. Soc. Amer. **70**, 356–374 (1980)
4.113. E.A. Bucher, R.M. Lerner: Appl. Opt. **12**, 2401–2414 (1973)
4.114. I.L. Katsev: "Nonstationary Scattering of Localized Pulsed Sources and Supershort Pulses", in *Light Propagation in a Dispersed Medium* (Nauka i Tekhnika, Minsk 1982) pp. 67–83
4.115. I.L. Katsev, A.P. Ivanov: Izv. AN BSSR, Ser. Fiz.-Tekh., N 4, 102–108 (1968)
4.116. A.M. Belyantsev, L.S. Dolin, V.A. Saveliev: Izv. Vuz., Ser. Radiofiz. **10**, 489–497 (1967)
4.117. L.M. Romanova: Izv. AN SSSR, Ser. Fiz. Atm. Okeana **6**, 489–498 (1970)
4.118. E.A. Bucher: Appl. Opt. **12**, 2391–2400 (1973)
4.119. R.A. Elliot: Appl. Opt. **22**, 2670–2681 (1983)
4.120. Yu.A. Goldin, V.A. Gashko, T.T. Karlson et al.: "Experimental Studies of Nonstationary Light Fields in Ocean", in *Hydrophysical and Hydrooptical Researches in the Atlantic and Pacific Oceans* (Nauka, Moscow 1974) pp. 228–241
4.121. V.K. Gavrikov, V.G. Korenev: Izv. AN SSSR, Fiz. Atm. Okeana **17**, 763–765 (1981)
4.122. V.G. Korenev: "On Temporal Dispersion of an Optical Pulse in a Limited Volume of Artificial Fog", in *Collected Abstracts, 6th All-Union Symp. Laser Rad. Propag. Atm.* (Tomsk 1981), pp. 162–164
4.123. E.P. Zege, I.L. Katsev, A.I. Kolesnik: Izv. AN SSSR, Fiz. Atm. Okeana **24**, 1163–1169 (1988)
4.124. E.V. Genin, G.P. Kokhanenko, V.A. Krutikov, D.S. Mezhevoy: "Pulse Depolarization and Formation of Two-Scale Time Radiation Structure", in *Collected Abstracts, 6th All-Union Conf. Laser Rad. Propag. Dispersed Medium,* Vol. 1 (Obninsk-Barnaul 1988) pp. 83–84
4.125. E.P. Zege: "Some Results of the Asymptotic Radiative Transfer Theory for Absorbing Anisotropically Scattering Media as Applied to Ocean and Cloud Optics"; Preprint, Inst. Phys. Acad. Sci. BSSR, N 274 (Minsk 1982)
5.1. A. Marechal, M. Francon: *Diffraction Structure des Images* (Editions de la Revue d'Optique Theorique et Instrumentale, Paris 1960)
5.2. E.L. O'Neill: *Introduction to Statistical Optics* (Addison-Wesley, London 1963)
5.3. J.W. Goodmen: *Introduction to Fourier Optics* (McGraw-Hill, New York 1968)
5.4. A. Papoulis: *Systems and Transforms with Applications in Optics* (McGraw-Hill, New York 1968)
5.5. I.M. Levin: Izv. AN SSSR, Fiz. Atm. Okeana, **5**, 62–76 (1969)
5.6. D.M. Bravo-Zhivotovsky, L.S. Dolin, A.G. Luchinin, V.A. Saveliev: Izv. AN SSSR, Fiz. Atm. Okeana, **5**, 672–684 (1969)
5.7. D.M. Bravo-Zhivotovsky, L.S. Dolin, I.M. Levin, A.G. Luchinin, V.A. Saveliev: Vopr. Radioelektron., Ser. Tekh. Televid., Vyp. **3**, 35–46 (1972)
5.8. D.M. Bravo-Zhivotovsky, L.S. Dolin, I.M. Levin, A.G. Luchinin, V.A. Saveliev: Vopr. Radioelektron., Ser. Tekh. Televid., Vyp. **3**, 38–49 (1973)
5.9. D.M. Bravo-Zhivotovsky, L.S. Dolin, I.M. Levin, A.G. Luchinin, V.A. Saveliev: Izv. AN SSSR, Fiz. Atm. Okeana, **7**, 1143–1152 (1971)
5.10. A. Rose: *Vision Human and Electronic* (Plenum, New York 1973)
5.11. J.M. Lloyd: *Thermal Imaging Systems* (Plenum, New York 1975)
5.12. J.A. Hall: "Signal and Noise in Image Transfer", in *Solid State Imaging*, ed. by P.G. Jespers, F. van de Wiele, M.H. White (Noordhoff, Leyden 1976)
5.13. Yu.P. Leonov: *Theory of Statistical Solutions and Psychophysics* (Nauka, Moscow 1977)
5.14. I.L. Katsev: Izv. AN SSSR, Fiz. Atm. Okeana **17**, 725–731 (1981)
5.15. M.M. Miroshnikov: *Theoretical Principles of Optico-Electronic Devices* (Mashinostroyeniye, Leningrad 1983)
5.16. F.A. Rosell, R.H. Willson: "Recent Psychophysical Experiments and the Display Signal-to-Noise Ratio Concept", Chap. 5, in *Perception of Displayed Information*, ed. by L.M. Biberman (Plenum, New York 1973)
5.17. F. Perren: Usp. Fiz. Nauk **28**, 307–344 (1962)
5.18. G.A. Korn, T.M. Korn: *Mathematical Handbook for Scientists and Engineers* (McGraw-Hill, New York 1961)
5.19. A.S. Drofa, I.L. Katsev: Meteorology and Hydrology, N 1, 101–109 (1981)
5.20. A.L. Usachev: Izv. AN SSSR, Fiz. Atm. Okeana **17**, 1107–1109 (1981)
5.21. L.S. Dolin, V.A. Saveliev: Izv. AN SSSR, Fiz. Atm. Okeana **15**, 717–723 (1979)
5.22. J.A. Hall: "Application of Solid State Image Receivers", in *Solid State Imaging*, ed. by P.G. Jespers, F. van de Wiele, M.H. White (Noordhoff, Leyden 1976)
5.23. J. Kuza, A. Ishimaru: Appl. Opt. **25**, 4328–4385 (1986)

5.24. A.L. Usachev: Izv. AN SSSR, Fiz. Atm. Okeana **17**, 1107–1109 (1981)
5.25. D.M. Bravo-Zhivotovsky, L.S. Dolin, A.G. Luchinin, V.A. Saveliev: "Modulation Transfer Functions of Sea Water", in *Hydrophysical and Hydrooptical Researches in the Atlantic and Pacific Oceans* (Nauka, Moscow 1974) pp. 213–217
5.26. L.P. Volnistova, A.S. Drofa: Izv. AN SSSR, Fiz. Atm. Okeana **21**, 50–57 (1985)
5.27. A.P. Ivanov, V.A. Loyko: *Optics of a Photographic Layer* (Nauka i Tekhnika, Minsk 1983)
5.28. K.G. Predko, M.P. Znachenok: Zh. Prikl. Spektrosk. **10**, 1019–1021 (1969)
5.29. G.G. Gribakin, G.A. Istomin: "Light Conditions in a Photographic Layer and Image Formation", in *Theoretical and Applied Light Scattering Problems*, ed. by B.I. Stepanov (Nauka i Teknika, Minsk 1971) pp. 432–453
5.30. G.A. Istomin: J. Signal Amer. **B1**, 7–21 (1973)
5.31. Z.L. Petrushkina, E.G. Ivakina: Zh. Nauch. Prikl. Fotograf. Kinematograf. **16**, 375–377 (1971)
5.32. A.P. Ivanov, K.G. Predko: *Optics of a Luminescent Screen* (Nauka i Tekhnika, Minsk 1984)
5.33. D.M. Bravo-Zhivotovsky, L.S. Dolin, A.G. Luchinin, I.M. Levin: "Theory of Optical Image Transfer in Water", in *Optics of Ocean, Vol.* 1. *Physical Optics of Ocean* (Nauka, Moscow 1983) pp. 96–113
5.34. I.M. Levin: "Underwater Television", in *Optics of Ocean, Vol.* 2. *Applied Optics of Ocean* (Nauka, Moscow 1983) pp. 187–199
5.35. I.F. Elistratov, I.M. Levin, T.N. Lomonosova: "Polarization of Backscattered Radiation as Applied to the Problem on Vision in Turbid Media", in *Light Scattering in Earth's Atmosphere* (Nauka, Alma-Ata 1972) pp. 161–164
5.36. M.P. Vanyukov, E.V. Nilov, A.A. Chertkov: Opt.-Mekh. Prom., No. 6, 50–55 (1970)
5.37. V.M. Arpishkin, V.P. Vanyukov, P.P. Muratov, E.V. Nilov, G.G. Yarovaya: Opt.-Mekh. Prom., N 9, 1–3 (1967)
5.38. I.I. Kalinin, O.A. Yutskevich: "Receiver Sensitivity Modulation Suppression of Backscattering Noise in Pulsed Location", in *Collected Abstracts, 3rd All-Union Symp. Laser Sound. Atm.* (Tomsk 1974) pp. 275–279
5.39. E.V. Ivakin, A.I. Kitsak: Opt. Spektrosk. **52**, 99–102 (1982)
5.40. K.A. Stetson: J. Opt. Soc. Amer. **57**, 1060–1062 (1967)
5.41. J. Johnson: "Analysis of Image Forming Systems", in *Proc. Image Intesif. Symp.* (U.S. Army Engng. Res. Develop. Lab., Fort Belvoir, Virginia, October 1958)
5.42. A.A. Vaskovsky, S.S. Romanov: Opt.-Mekh. Prom., N 9, 13–16 (1987)
5.43. T.H. Couttenay, J.F. Boulter, H. Henshall: Infrared Phys. **16**, 95 (1976)
5.44. I.N. Matveev, V.V. Protopopov, I.N. Troitsky, N.D. Ustinov: *Laser Detection* (Mashinostroyeniye, Moscow 1984)
5.45. S.I. Vavilov: *Eye and Sun* (Nauka, Moscow 1981)
5.46. S.V. Kravkov: The *Eye and its Operation* (AN SSSR, Moscow 1950)
5.47. *Physiology of Sensor Systems, Pt.* 1. *Sight Physiology* (Nauka, Leningrad 1971)
5.48. M.A. Ostrovskaya: Opt.-Mekh. Prom **36**, 45–54 (1969)
5.49. D.A. Roberts: "Radiometry/Photometry: Concepts and Terms", in *Encyclopedia—The Optical Industry and Systems Directory* (1973–1974) pp. H77–H81
5.50. E. Ingeistam: Optik **B18**, 657–658 (1961)
5.51. *Physical Encyclopedia*, Vol. 2 (Sovetskaya Entsyklopediya, Moscow 1962) pp. 94–97
5.52. V.E. Demidov: *How Do We See What We See* (Znaniye, Moscow 1979)
5.53. H.H. Ory: Statistical Detection Theory of Threshold Visual Performance", in *Rand Corp. Memor. RM-5992-PR* (Santa Monica 1969)
5.54. F.M. Bagrash, L.G. Kerr, J.P. Thomas: Vision Research **11**, 625–634 (1971)
5.55. H. Frieser: *Photographic Information Recording* (Focal, London 1975)
5.56. P.G. Jespers, F. Van de Wiele, M.H. White (eds.): *Solid State Imaging* (Noordhoff, Leyden (1976)
5.57. J.A. Hall: "Comparison of Solid State and Electrovacuum Image Receivers", in *Solid State Imaging*, ed. by P.G. Jespers, F. van de Wiele, M.H. White (Noordhoff, Leyden 1976)
5.58. A. Rose: J. Opt. Soc. Amer. **38(2)**, 196–208 (1958)
5.59. R.D. Hudson, Jr.: *Infrared System Engineering* (Wiley-Interscience, New York 1969)
5.60. D.F. Barbe: "Solid State Infrared Image Receivers", in *Solid State Imaging*, ed. by P.G. Jespers, F. van de Wiele, M.H. White (Noordhoff, Leyden 1976)

6.1. G.I. Marchuk, G.A. Mikhailov, M.A. Nazaraliev, R.A. Darbinyan, B.A. Kargin, B.S. Elepov: *Monte-Carlo Methods in Atmospheric Optics* (Springer, Berlin–Heidelberg 1980)
6.2. B.A. Kargin, S.V. Kuznetsov, G.A. Mikhailov: Izv. AN SSSR, Fiz. Atm. Okeana **15**, 1027–1035 (1979)
6.3. B.A. Kargin: "Statistical Modelling Algorithm for Calculating the Modulation Transfer Function of Atmosphere", in *Remote Sensing of Environment in Siberia and Far East* (Nauka, Novosibirsk 1982) pp. 169–174
6.4. A.S. Drofa, A.L. Usachev: Izv. Vuz. SSSR, Ser. Radiofiz. **26**, 408–414 (1983)
6.5. D.M. Bravo-Zhivotovsky, L.S. Dolin, A.G. Luchinin, V.A. Saveliev: Izv. AN SSSR, Fiz. Atm. Okeana **5**, 160–167 (1969)
6.6. I.L. Katsev: Izv. AN SSSR, Fiz. Atm. Okeana **17**, 478–486 (1981)
6.7. T.A. Sushkevich, V.F. Khokhlov: "Approximation of Functions by an Exponential Sum"; Preprint, Inst. Appl. Math. Acad. Sci. USSR, N 80 (Moscow 1975)
6.8. E.P. Zege: "Design Calculations of Light Fields at Multiple Scattering", in *Light Propagation in a Dispersed Medium* (Nauka i Tekhnika, Minsk 1982) pp. 84–105
6.9. O.V. Bushmakova, E.P. Zege, I.L. Katsev: Izv. AN SSSR, Fiz. Atm. Okeana **8**, 711–719 (1972)
6.10. E.P. Zege, I.L. Katsev: Zh. Prikl. Spektrosk. **44**, 854–860 (1986)
6.11. H. Frieser: *Photographic Information Recording* (Focal, London 1975)
6.12. A.P. Ivanov, V.A. Loyko: *Optics of a Photographic Layer* (Nauka i Tekhnika, Minsk 1983)
6.13. L.S. Dolin, V.A. Saveliev: Izv. Vuz., Ser. Radiofiz. **22**, 1310–1317 (1979)
6.14. E.P. Zege, I.L. Katsev: "Radiation Scattering in Plane Layers with Optical Thickness-Dependent Characteristics", in *Proc. All-Union Conf. Opt. Rad. Propag. Dispersed Medium* (Gidrometeoizdat, Moscow 1978) pp. 67–70
6.15. O.V. Bushmakova, E.P. Zege, I.L. Katsev: "Optical Transfer Function of an Inhomogeneous Layer", in *Collected Abstracts, 3rd All-Union Symp. Laser Sound. Atm.* (Tomsk 1974) pp. 189–194
6.16. O.V. Bushmakova, E.P. Zege, I.L. Katsev: "Object Vision Through a Cloud", in *Collected Abstracts, 3rd All-Union Symp. Laser Rad. Propag. Atm.* (Tomsk 1975) pp. 30–32
6.17. L.P. Volnistova, M.I. Vorobiev, A.S. Drofa: Izv. AN SSSR, Fiz. Atm. Okeana **17**, 1031–1038 (1981)
6.18. V.V. Belov, V.E. Zuev, G.M. Krekov: Izv. AN SSSR, Fiz. Atm. Okeana **18**, 963–969 (1982)
6.19. V.V. Belov, B.D. Borisov, V.N. Genin, M.V. Kabanov, G.M. Krekov: Izv. AN SSSR, Fiz. Atm. Okeana **18**, 1303–1311 (1982)
6.20. V.E. Zuev, V.V. Belov, B.D. Borisov, V.N. Genin, M.V. Kabanov, G.M. Krekov: Dokl. AN SSSR **268**, 321–324 (1983)
6.21. V.E. Zuev, V.V. Belov, G.M. Krekov: Dokl. AN SSSR **269**, 1353–1356 (1983)
6.22. L.P. Volnistova, A.S. Drofa: Izv. AN SSSR, Fiz. Atm. Okeana **21**, 50–57 (1985)
6.23. I.L. Katsev: Izv. AN SSSR, Ser. Fiz. Atm. Okeana **19**, 1172–1179 (1983)
6.24. A.B. Gavrilovich, P.Ya. Ganich, A.P. Ivanov: Dokl. AN BSSR **22**, 220–222 (1978)
6.25. A.L. Usachev: Izv. AN SSSR, Fiz. Atm. Okeana **17**, 1107–1109 (1981)
6.26. T.A. Sushkevich, I.V. Mishin: Dokl. AN SSSR **269**, 60–64 (1982)
6.27. T.A. Sushkevich: "Numerical Methods of Solving the Transfer Equation with a Complex Function"; Preprint, Inst. Appl. Math. Acad. Sci. USSR, No. 136, Moscow (1980)
6.28. E.P. Zege, I.L. Katsev, S.A. Makarevich: Izv. AN SSSR, Ser. Fiz.-Mat. Nauk, No. 3, 61–66 (1986)
6.29. A.S. Drofa, A.L. Usachev: Izv. AN SSSR, Fiz. Atm. Okeana **21**, 603–608 (1985)
6.30. I.V. Mishin, T.A. Sushkevich: "A Mathematical Model of the Optical Transfer Function of the Underlying Surface-Atmosphere System"; Preprint, Inst. Appl. Math. Acad. Sci. Moscow USSR, No. 76 (1979)
6.31. A.A. Ioltukhovsky, I.V. Mishin, T.A. Sushkevich: "Allowance for Nonlinear Distortions of the Radiance Field in Three-Dimensionally Inhomogeneous Atmosphere Bounded by the Underlying Surface with a Homogeneous Albedo"; Preprint, Inst. Appl. Math. Acad. Sci. Moscow, USSR, No. 76 (1981)
6.32. A.A. Ioltukhovsky, I.V. Mishin, T.A. Sushkevich: "Allowance for Nonlinear Distortions of the Radiance Field in Three-Dimensionally Inhomogeneous Atmosphere Bounded by the Underlying Surface with an Inhomogeneous Albedo"; Preprint, Inst. Appl. Math. Acad. Sci. Moscow, USSR, No. 63 (1982)

6.33. G.A. Korn, T.M. Korn: *Mathematical Handbook for Scientists and Engineers* (McGraw-Hill, New York 1961)
6.34. L.S. Dolin, V.A. Saveliev: Izv. AN SSSR, Fiz. Atm. Okeana **15**, 717–723 (1979)
6.35. A.S. Drofa: Izv. AN SSSR, Fiz. Atm. Okeana **20**, 939–946 (1984)
6.36. D.M. Bravo-Zhivotovsky, L.S. Dolin, I.M. Levin, A.G. Luchinin, V.A. Saveliev: Izv. AN SSSR, Fiz. Atm. Okeana, **7**, 1143–1152 (1971)
7.1. E.N. Leith, J. Upatnieks: J. Opt. Soc. Amer. **52**, 1123–1130 (1962)
7.2. E.N. Leith, J. Upatnieks: J. Opt. Soc. Amer. **54**, 1295–1301 (1964)
7.3. E.N. Leith, J. Upatnieks: J. Opt. Soc. Amer. **53**, 1377–1381 (1963)
7.4. Yu.N. Denisyuk: Dokl. AN SSSR **144**, 1275–1278 (1962)
7.5. K.A. Stetson: J. Opt. Soc. Amer. **57**, 1060–1062 (1967)
7.6. E. Spitz: Comp. Rend. **264 B**, 1449–1452 (1967)
7.7. A.W. Lohman, C.A. Shuman: Opt. Commun. **7**, 93–97 (1973)
7.8. A.D. Galpern: Opt. Spektrosk. **35**, 142–147 (1973)
7.9. A.D. Galpern: Opt. Spektrosk. **37**, 965–972 (1974)
7.10. A.D. Galpern, A.S. Sudarushkin: Opt. Spektrosk. **40**, 357–363 (1976)
7.11. V.K. Polyansky, L.V. Kovalsky, O.V. Angelsky: Ukr. Fiz. Zh. **25**, 1679–1682 (1980)
7.12. H.S. Caufield: J. Opt. Soc. Amer. **58**, 276–277 (1968)
7.13. A.P. Ivanov, A.P. Chaikovsky, A.A. Kumeisha, V.N. Shcherbakov: Zh. Prikl. Spektrosk. **28**, 518–525 (1978)
7.14. A.P. Ivanov, A.A. Kumeisha: Dokl. AN BSSR **25**, 108–111 (1981)
7.15. E.V. Ivakin, A.I. Kitsak: Opt. Spektrosk. **52**, 99–102 (1982)
7.16. I.S. Klimenko, E.G. Matinyan, G.V. Skrotsky: Dokl. AN SSSR **211**, 571–573 (1973)
7.17. I.S. Klimenko, N.I. Shushlebina, E.G. Shikhalev: Opt. Spektrosk. **54**, 324–327 (1983)
7.18. I.S. Klimenko, T.G. Matinyan: Opt. Spektrosk. **31**, 776–779 (1971)
7.19. J.W. Goodman, W.H. Huntly, D.W. Jackson, M. Lehmann: Appl. Phys. Lett. **8**, 311–313 (1966)
7.20. A.P. Ivanov, A.P. Chaikovsky, V.N. Shcherbakov: Dokl. AN BSSR **23**, 979–982 (1979)
7.21. A.P. Ivanov: "Methods of Laser Sounding", in *Quantum Electronics and Laser Spectroscopy* (Nauka i Tekhnika, Minsk 1974) pp. 381–405
7.22. V.E. Zuev: *Laser-Meteorologist* (Gidrometeoizdat, Leningrad 1974)
7.23. J.W. Goodman: Proc. IEEE **53**, 1688–1700 (1965)
7.24. P.A. Bakut, N.D. Ustinov, I.N. Troitsky, K.N. Sviridov: Zarubezh. Radioelektron., No. 9, 3–30 (1976)
7.25. I.S. Klimenko: *Focused Image Holography* (Nauka, Moscow 1985)
7.26. E.V. Ivakin, A.I. Kitsak: Zh. Prikl. Spektrosk. **44**, 43–47 (1986)
7.27. V.G. Sinchenko: Opt. Spektrosk. **59**, 391–396 (1985)
7.28. M. Born, E. Wolf: *Principles of Optics* (Pergamon, Oxford 1968)
7.29. L.S. Dolin: Izv. Vuz., Ser. Radiofiz. **7**, 471–478 (1964)
7.30. L.S. Dolin: Izv. Vuz., Ser. Radiofiz. **23**, 332–338 (1980)
7.31. Yu.N. Barabanenkov: Zh. Eksp. Teor. Fiz. **56**, 1262–1272 (1969)
7.32. L.S. Dolin: Izv. Vuz., Ser. Radiofiz. **11**, 840–849 (1968)
7.33. Yu.N. Barabanenkov, Yu.A. Kravtsov, S.M. Rytov, V.I. Tatarsky: Usp. Fiz. Nauk **102**, 3–42 (1970)
7.34. Yu.N. Barabanenkov: Usp. Fiz. Nauk **117**, 49–78 (1975)
7.35. G.V. Rozenberg: Usp. Fiz. Nauk **121**, 97–138 (1977)
7.36. G.I. Ovchinnikov: Radiotekh. Radioelektron. **18**, 2044–2050 (1973)
7.37. A. Ishimaru: *Wave Propagation and Scattering in Random Media*, Vol. 2 (Academic, New York 1978)
7.38. A.P. Ivanov, A.Ya. Khairullina, A.P. Chaikovsky: "Statistical Optics Methods as Applied to Study Scattering Media Parameters"; Preprint, Inst. Phys. Acad. Sci. Minsk, BSSR (1980)
7.39. J.W. Goodman: J. Opt. Soc. Amer. **57**, 493–502 (1967)
7.40. J.W. Goodman, R.B. Miles, R.B. Kimball: J. Opt. Soc. Amer. **58**, 609–613 (1968)
7.41. I.L. Katsev: "Image Quality in Holographic Systems of Observation Through a Scattering Medium"; Preprint, Inst. Phys. Acad. Sci. Minsk, BSSR (1985)
7.42. I.N. Troitsky, N.D. Ustinov: *Statistical Theory of Holography* (Radio i Svyaz, Moscow 1981)
8.1. L. Elterman: "UV, Visible and IR Attenuation to 50 km"; Report AFCRL-68-0153 (Air Force Cambridge Research Laboratories Bedford, Mass. 1968)

8.2. R.A. McClatche, R.W. Fenn, J.E.A. Selby, F.E. Volz, G.S. Garing: "Optical Properties of the Atmosphere (Revised)"; Report AFCRL-72-0497, (Air Force Cambridge Research Laboratories, Bedford, Mass. 1972)
8.3. G. Hänel, R. Bullrich: Beitr. Phys. Atm. **B51**, 129–138 (1978)
8.4. E.P. Shettle, R.W. Fenn: "Models for the Aerosols of the Lower Atmosphere and the Effects of Humidity Variations on Their Optical Properties"; Report AFCRL-79-0214, (Air Force Cambridge Research Laboratories, Bedford, Mass. 1979)
8.5. K.Ya. Kondratiev, D.V. Pozdnyakov: *Aerosol Models of the Atmosphere*" (Nauka, Moscow 1981)
8.6. K.Ya. Kondratiev, N.I. Moskalenko, V.F. Terzi: Dokl. AN SSSR **253**, 1354–1356 (1980)
8.7. V.E. Zuev, G.M. Krekov: *Optical Models of the Atmosphere* (Gidrometeoizdat, Leningrad 1986)
8.8. G.M. Krekov, R.F. Rakhimov: *Optical Models of Atmospheric Aerosol* (Tomsk 1986)
8.9. Report of the World Meteorological Association, Scientific Advisory Council/Radiation Commission of the Intl. Assn. of Meteorology and Atmospheric Physics. (IAMAP) "Meeting of Experts on Aerosols and Their Climatic Effects (Williamsburg, Virginia, USA, 28–30 March 1983)
8.10. R. Renndorf: J. Opt. Soc. Amer. **47**, 176–182 (1957)
8.11. V.E. Zuev, M.V. Kabanov: *Optics of Atmospheric Aerosol* (Gidrometeoizdat, Leningrad 1987)
8.12. K.Ya. Kondratiev, M.A. Prokofiev: Izv. AN SSSR, Fiz. Atm. Okeana **20**, 339–348 (1984)
8.13. K.S. Shiftin, V.M. Volgin, B.N. Volkov, O.A. Ershov, A.B. Smirnov: Issl. Zeml. Kosm., No. 4, 21–30 (1985)
8.14. V.G. Tsverava: Meteorolog. Gidrolog., No. 5, 34–39 (1974)
8.15. A.B. Gavrilovich, V.P. Busygin, I.A. Slabkaya: Izv. AN BSSR, Ser. Fiz.-Mat. Nauk, No. 3, 97–103 (1983)
8.16. E.L. Aleksandrov, K.B. Yudin: Meteorolog. Gidrolog., N 12, 57–61 (1979)
8.17. V.E. Zuev: *Propagation of Visible and IR Waves in the Atmosphere* (Sov. Radio, Moscow 1970)
8.18. A.P. Ivanov, A.Ya. Khairullina: Izv. AN SSSR, Fiz. Atm. Okeana **3**, 537–543 (1967)
8.19. D. Deiŕmendjian: *Electromagnetic Scattering on Spherical Polydispersion* (Elsevier, New York 1969)
8.20. E.P. Zege, A.A. Kohanovsky: Izv. AN SSSR, Fiz. Atm. Okeana, **24**, 695–701 (1988)
8.21. A.Ya. Khairullina, A.P. Ivanov: Izv. AN SSSR, Fiz. Atm. Okeana **13**, 503–505 (1968)
8.22. P.Ya. Ganich, A.P. Ivanov: Zh. Prikl. Spektrosk. **12**, 280–285 (1970)
8.23. V.E. Zuev, M.V. Kabanov, B.A. Saveliev: Dokl. AN SSSR **175**, 327–330 (1967)
8.24. A.P. Ivanov: *Physical Principles of Hydrooptics* (Nauka i Tekhnika, Minsk 1975) pp. 349–405
8.25. V.V. Sharonov: *Measurement and Calculation of Remote Subject Visibility* (Gostekhizdat, Moscow, Leningrad 1947)
8.26. O.A. Sokolov: *Underwater Vision* (Gidrometeoizdat, Leningrad 1974)
8.27. N.G. Jerlov: *Marine Optics* (Elsevier, Amsterdam 1976) Chap. 12
8.28. A.A. Ivanoff: *Introduction á L'océanographie* (Librairie Vuibert, Paris 1975)
8.29. V.A. Gavrilov: *Visibility in the Atmosphere* (Gidrometeoizdat, Leningrad 1966)
8.30. I.L. Katsev: Izv. AN BSSR, Ser. Fiz.-Mat. Nauk, N 2, 93–98 (1984)
8.31. J. Kuza, A. Ishimaru: Appl. Opt. **25**, 4328–4385 (1986)
8.32. A.S. Drofa, A.L. Usachev: Izv. AN SSSR, Fiz. Atm. Okeana **16**, 933–938 (1980)
8.33. L.P. Volnistova, M.I. Vorobiev, A.S. Drofa: Izv. AN SSSR, Fiz. Atm. Okeana **17**, 1031–1038 (1981)
8.34. A.S. Drofa, I.L. Katsev: Meteorolog. Gidrolog., N 11, 101–109 (1981)
8.35. I.L. Katsev: Izv. AN SSSR, Ser. Fiz. Atm. Okeana **19**, 1172–1179 (1983)
8.36. L.S. Dolin: "Spatial Fluctuations of the Image Radiance in Viewing Through the Medium with Optical Characteristics", in *Optics of the Ocean and the Atmosphere* (Elm, Baku 1983) pp. 153–158
8.37. I.L. Katsev: Izv. AN SSSR, Fiz. Atm. Okeana **23**, 84–90 (1987)
8.38. I.L. Katsev: Izv. AN SSSR, Fiz. Atm. Okeana **17**, 725–731 (1981)
9.1. A.P. Ivanov: *Physical Principles of Hydrooptics* (Nauka i Tekhnika, Minsk 1975) pp. 349–405
9.2. N.G. Jerlov: *Marine Optics* (Elsevier, Amsterdam 1976) Chap. 12
9.3. O.V. Kopelevich: "Factors Determining Optical Properties of Sea Water", in *Optics of Ocean, Vol. 1. Physical Optics of Ocean* (Nauka, Moscow 1983) pp. 166–208

9.4. A.A. Ivanoff: *Introduction á L'océanographie* (Librairie Vuibert, Paris 1975)
9.5. K.S. Shifrin: *Introduction to Optics of the Ocean* (Gidrometeoizdat, Leningrad 1983)
9.6. O.V. Kopelevich: "A Simple Model for Optical Sea Water Properties", in *Optics of the Ocean, Vol. 1. Physical Optics of the Ocean* (Nauka, Moscow 1983) pp. 208–236
9.7. A.N. Dorogin, O.V. Kopelevich, I.M. Levin, V.I. Feigels: "Correlations Between Some Hydrooptical Characteristics", in *Collected Abstracts, 10th Plenum on Optics of the Ocean* (S.I. Vavilov State Optical Inst., Leningrad 1988) pp. 136–137
9.8. V.I. Mankovsky: Okeanolog. **17**, 750–753 (1978)
9.9. I.M. Levin: Izv. AN SSSR, Fiz. Atm. Okeana **16**, 926–932 (1980)
9.10. E. Kornstein, H. Wetzstein: Electron. **41**, 21, 140–150 (1968)
9.11. N.V. Vershinsky: *Underwater Television* (Gosenergoizdat, Moscow 1960)
9.12. R.F. Zhukov, A.A. Kondatovich, S.D. Mogilny, B.I. Shchipko: *Underwater Searching Systems, Instruments and Devices* (Voenizdat, Moscow 1972)
9.13. M. Kawasaki: Technocrat **8**, 25–34 (1975)
9.14. I.M. Levin: "Underwater Television", in *Optics of Ocean, Vol. 2. Applied Optics of Ocean* (Nauka, Moscow 1983) pp. 187–199
9.15. I.F. Elistratov, I.M. Levin, T.N. Lomonosova: Izv. AN SSSR, Fiz. Atm. Okeana **6**, 736–739 (1970)
9.16. C.D. Gilbert, L.S. Pernicka: Appl. Opt. **6**, 741–746 (1967)
9.17. I.M. Levin: Tekh. Sredstv Svyaz., Ser. Tekh. Televid., Vyp.4, 97–106 (1981)
9.18. P.Ya. Ganich, A.P. Ivanov: Zh. Prikl. Spektrosk. **12**, 653–656 (1970)
9.19. A.P. Ivanov, P.B. Boiko: Zh. Prikl. Spektrosk. **10**, 846–851 (1969)
9.20. N.E. Johnson: Res. Develop. **19**, 28–31 (1968)
9.21. "Television Systems for Underwater Explorations", in *Radioelectronics* 1974: *Review of Foreign Publications Vol.* 3 (Moscow 1975) pp. 63–67
9.22. D.M. Bravo-Zhivotovsky, L.S. Dolin, I.M. Levin, A.G. Luchinin, V.A. Saveliev: Izv. AN SSSR, Fiz. Atm. Okeana, **7**, 1143–1152 (1971)
9.23. D.M. Bravo-Zhivotovsky, L.S. Dolin, A.G. Luchinin, V.A. Saveliev: Izv. AN SSSR, Fiz. Atm. Okeana **5**, 160–167 (1969)
9.24. D.M. Bravo-Zhivotovsky, L.S. Dolin, A.G. Luchinin, V.A. Saveliev: "Modulation Transfer Functions of Sea Water", in *Hydrophysical and Hydrooptical Researches in the Atlantic and Pacific Oceans* (Nauka, Moscow 1974) pp. 213–217
9.25. N.T. Jura: Appl. Opt. **10**, 114–118 (1971)
9.26. N.T. Jura: Appl. Opt. **12**, 1061–1066 (1973)
9.27. N.T. Jura: Appl. Opt. **12**, 108–115 (1973)
9.28. S.A. Danley: "Underwater Visibility and Photography", in *Optical Aspects of Oceanography* (Academic, London 1974) pp. 135–149
9.29. D.M. Bravo-Zhivotovsky, L.S. Dolin, A.G. Luchinin, I.M. Levin: "Theory of Optical Image Transfer in Water", in *Optics of Ocean, Vol. 1. Physical Optics of Ocean* (Nauka, Moscow 1983) pp. 96–113
9.30. I.M. Levin: "Limiting Visibility Range of Self-Luminous Underwater Objects", in *Optical Methods of Ocean and Internal Water Basin Exploration* (AN Est. SSR, Tallin 1980) pp. 152–155
9.31. W.I. Pierson, L. Moskowitz: J. Geophys. Res. **69**, 5181–5190 (1964)
9.32. A.G. Luchinin, I.A. Sergievskaya: Izv. AN SSSR, Fiz. Atm. Okeana **18**, 850–858 (1982)
9.33. A.G. Luchinin, I.A. Sergievskaya: Izv. AN SSSR, Fiz. Atm. Okeana **22**, 773–776 (1986)
9.34. C. Cox, W. Munk: J. Mar. Res. **13**, 198–227 (1954)
9.35. Yu.-A.R. Mullamaa: Izv. AN SSSR, Fiz. Atm. Okeana **11**, 199–205 (1975)
9.36. A.G. Luchinin: Izv. AN SSSR, Fiz. Atm. Okeana **15**, 770–775 (1979)
9.37. A.G. Luchinin: Izv. AN SSSR, Fiz. Atm. Okeana **17**, 732–736 (1981)
9.38. A.G. Luchinin: Izv. AN SSSR, Fiz. Atm. Okeana **18**, 330–333 (1982)
9.39. S.V. Dotsenko: Izv. AN SSSR, Fiz. Atm. Okeana **18**, 408–415 (1982)
9.40. V.L. Veber: Izv. Vuz., Ser. Radiofiz. **22**, 989–997 (1979)
9.41. V.L. Veber, L.S. Dolin: Izv. AN SSSR, Fiz. Atm. Okeana **17**, 1168–1177 (1981)
9.42. V.L. Veber, A.G. Luchinin: Izv. AN SSSR, Fiz. Atm. Okeana **19**, 631–638 (1983)
9.43. D.V. Bunkin, D.V. Vlasov, D.M. Mirkamilov: *Physical Principles of Earth Surface Laser Aerosounding* (Fan, Tashkent 1987)

9.44. E.P. Zege, I.L. Katsev, A.I. Kolesnik: Izv. AN SSSR, Fiz. Atm. Okeana **24**, 1163–1169 (1988)
9.45. A.I. Kolesnik, A.P. Ivanov: Dokl. AN BSSR **30**, 601–604 (1986)
10.1. A.P. Ivanov, V.A. Loyko: *Optics of a Photographic Layer* (Nauka i Tekhnika, Minsk 1983)
10.2. A.P. Ivanov, K.G. Predko: *Optics of a Luminescent Screen* (Nauka i Tekhnika, Minsk 1984)
10.3. A.P. Ivanov, V.A. Loyko, V.P. Dik: *Light Propagation in Media with Dense Scatterers* (Nauka i Tekhnika, Minsk 1988)
10.4. H. Frieser: *Photographic Information Recording* (Focal, London 1975)
10.5. O.N. Kozakov, Yu.S. Andreev: Zh. Nauch. Prikl. Fotograf. Kinematograf. **22**, 124–126 (1977)
10.6. C.R. Berry: J. Opt. Soc. Amer. **52**, 888–895 (1962)
10.7. Yu.A. Breslav, V.V. Peisakhov: Zh. Nauch. Prikl. Fotogr. Kinematogr. **30**, 150–157; 228–236 (1985)
10.8. D.G. Nizhner, V.M. Belous, K.V. Chibisov: Zh. Nauch. Prikl. Fotogr. Kinematogr. **32**, 391–392 (1987)
10.9. V.V. Berdnik, A.P. Ivanov, V.A. Loyko: Dokl. AN BSSR **32**, 219–222 (1988)
10.10. A.I. Veitsman, K.V. Vendrovsky: Zh. Nauch. Prikl. Fotogr. Kinematogr. **29**, 143–159 (1984)
10.11. V.V. Andreyanov, Yu.G. Oransky: Tekh. Kino Televid., N 7, 22–29 (1983)
10.12. G.A. Desyatnik, A.P. Ivanov, V.A. Loyko, V.M. Shvarts: Zn. Nauch. Prikl. Fotogr. Kinematogr. **31**, 428–433 (1986)
10.13. V.V. Berdnik, A.P. Ivanov, V.A. Loyko: Zh. Nauch. Prikl. Fotogr. Kinematogr. **32**, 210–212 (1987)
10.14. A.N. Valentyuk, K.G. Predko: Dokl. AN BSSR **26**, 801–804 (1982)
10.15. O.N. Kozakov, Yu.S. Andreev: "Modulation Transfer Function of Multilayer Photographic Material", in *Light Sensitive Materials and Their Treatment* (Gosniikhimfotoproyekt, Moscow 1980) pp. 126–129
10.16. K. Sayanagi: J. Opt. Soc. Amer. **47**, 566–567 (1957)
10.17. L.O. Hendeberg: Ark. Fys. **16**, 417–456 (1960)
10.18. H.F. Gilmore: J. Opt. Soc. Amer. **57**, 75–80 (1967)
10.19. V.A. Loyko: J. Signal Amer. **8**, 299–304 (1980)
10.20. A.P. Ivanov, V.A. Loyko: "Optical Image Recording with Photographic Layers"; Preprint, Inst. Phys. Acad. Sci. BSSR, N 105 (Minsk 1976)
10.21. A.P. Ivanov, V.A. Loyko: J. Signal Amer. **3**, 23–33 (1975)
10.22. V.D. Katsman, K.V. Vendrovsky: Zh. Nauch. Prikl. Fotogr. Kinematogr. **27**, 377–378 (1982)
10.23. J.J. de Palme, J. Casper: Photogr. Sci. Eng. **16**, 181–191 (1972)
10.24. M. Belder, J. Kerf, S.J. Jesper, R. Verbrugghe: J. Opt. Soc. Amer. **55**, 1261–1268 (1965)
10.25. A.L. Kuznetsova, Yu.A. Anokhin, T.P. Kozban: Zh. Nauch. Prikl. Fotogr. Kinematogr. **24**, 290–292 (1979)
10.26. V.M. Nefedchenkov, L.M. Rutskaya, A.I. Veitsman, Z.Ya. Shevchenko: Zn. Nauch. Prikl. Fotogr. Kinematogr. **21**, 33–37 (1976)
10.27. Yu.N. Gorokhovsky, V.P. Baranova: *Properties of Black and White Photographic Films* (Nauka, Moscow 1970)
10.28. Yu.N. Gorokhovsky, I.M. Davydkin, T.A. Lomachenkova: Zh. Nauch. Prikl. Fotogr. Kinematogr. **22**, 420–425 (1977)
10.29. I.M. Davydkin, T.A. Lomachenkova: Zh. Nauch. Prikl. Fotogr. Kinematogr. **26**, 58–60 (1981)
10.30. A.I. Veitsman, K.V. Vendrovsky: *Photographic Structuremetry* (Iskusstvo, Moscow 1982)
10.31. Yu.A. Anokhin, A.Yu. Potekhin: Trudy GOI **64**, 49–55 (1987)
10.32. *Handbook for a Chemist, Vol. 2. Main Properties of Inorganic and Organic Compounds* (Khimiya, Leningrad 1964)
10.33. K.F. Kovalenko: Elektron. Tekh., Ser. 1, N 1, 3–11 (1972)
10.34. D. Bohm: *Quantum Theory* (Prentice Hall, New York 1952)
10.35. M.P. Znachenok, N.P. Kupreichik, K.G. Predko: Zh. Prikl. Spektrosk. **18**, 648–652 (1973)
10.36. K.G. Predko: "Optical Image Reproduction with Scattering Luminescent Layers"; Preprint, Inst. Phys. Acad. Sci. Minsk, BSSR (1969)
10.37. A.P. Ivanov, K.G. Predko: Elektron. Tekh., Ser. 4, No. 9, 66–71 (1973)
10.38. O.V. Bushmakova, E.P. Zege, I.L. Katsev: Izv. AN SSSR, Fiz. Atm. Okeana **8**, 711–719 (1972)
10.39. A.N. Valentyuk, K.G. Predko: "Relationship Between the Characteristics of Photo- and Cathode-Luminescent Images on Crystal Phosphorescent Screens", in *Collected Abstracts of the 25th Conf. Luminescence (Crystal Phosphores)*, (Lvov State Univ., Lvov 1978) p. 236

10.40. I.A. Dibrov, S.A. Selitskaya, T.P. Zavodova: Elektron. Tekh., Ser. 4, No. 4, 227–235 (1968)
10.41. V.N. Vishnevsky, R.G. Gnyp, M.S. Pidziraylo, N.A. Stefansky: Vestn. Lvov Univ., Ser. Fiz., No. 15, 83–88 (1980)
10.42. V.N. Vishnevsky, R.G. Gnyp, N.S. Pidziraylo, N.A. Stefansky: Elektron. Tekh., Ser. 4, No. 3, 261–264 (1968)
10.43. G.Ya. Kossov, S.A. Selitskaya: Elektron. Tekh., Ser. 4, No. 1, 44–49 (1969)
10.44. G.E. Giakoumakis, C.D. Nomicos, P.C. Buthymiou: Canad. J. Phys. **57**, 2190–2193 (1979)
10.45. K.G. Predko: Elektron. Tekh., Ser. 4, No. 1, 50–60 (1969)
10.46. A.P. Ivanov, K.G. Predko: Izv. AN BSSR, Ser. Fiz.-Mat. Nauk, No. 3, 74–78 (1972)
10.47. M.M. Butslov, O.N. Kashina, A.F. Klenov: "Electrophoretic Luminescant Screens for Image Converters", in *Phys. Electron.* (Nauka, Moscow 1976) pp. 39–43
10.48. K.G. Predko: "Characteristics of Optical Image Recording on Electron-Sensitive Layers", in *Light Propagation in a Dispersed Medium* (Nauka i Tekhnika, Minsk 1982) pp. 239–258
10.49. L.V. Aleksandrov, I.A. Dibrov, A.M. Tutantseva, A.G. Yakovlev: Elektron. Tekh., Ser. 4, No. 6, 39–45 (1973)
10.50. M.M. Butslov, B.M. Stepanov, S.D. Fanchenko: *Image Converters in Scientific Research* (Nauka, Moscow 1978)
10.51. N.P. Soshin: Izv. AN SSSR, Ser. Fiz. **40**, 2364–2371 (1976)
10.52. N.P. Soshin: Izv. AN SSSR, Ser. Fiz. **43**, 1212–1217 (1979)
10.53. I.E. Ponomarenko, N.N. Filimonov: Elektron. Tekh., Ser. 4, No. 4, 23–30 (1981)

# Subject Index

**O. Sild, K. Haller,** Tartu (Eds.)

# *Zero-Phonon Lines*

## *and Spectral Hole Burning in Spectroscopy and Photochemistry*

1988. IX, 183 pp. 64 figs. Hardcover DM 72,– ISBN 3-540-19214-X

**Contents:** Zero-Phonon Lines in the Spectroscopy and Photochemistry of Impurity-Doped Solid Matter. – Homogeneous Broadening of Zero-Phonon Lines in the Impurity Spectra of Crystals and Glasses. – On the Theory of Stationary and Time-Dependent Zero-Phonon Lines. – Zero-Phonon Lines in the Spectra of Small Impurity Molecules as an Indicator of Electron-Phonon Interactions in Crystals. – Zero-Phonon Lines in the Spectra of Polyatomic, Including Biogenetic, Molecules. – Spectral Hole Burning (SHB): Scientific and Practical Applications. – Broadening of Zero-Phonon Lines of Impurities in Glasses. – Zero-Phonon Lines and Time-and-Space-Domain Holography of Ultrafast Events. – Zero-Phonon Lines and Anomalous Photoelectric Properties of Ruby. – Zero-Phonon Lines in Superradiance. – Transient Phenomena in Mössbauer Gamma Radiation. – Index of Contributors.

**A. Hasegawa,** AT&T Bell Laboratories, Murray Hill, NJ

# *Optical Solitons in Fibers*

2nd enl. ed. 1990. XII, 79 pp. 25 figs. Softcover DM 48,– ISBN 3-540-51747-2

This first edition was published as hardcover edition of the Series “Springer Tracts in Modern Physics” Vol. 116)

Already after six months high demand made a new edition of this textbook necessary. The most recent developments associated with two topical and very important theoretical and practical subjects are combined: **Solitons** as analytical solutions of nonlinear partial differential equations and as lossless signals in dielectric **fibers**. The practical implications point towards technological advances allowing for an economic and undistorted propagation of signals revolutionizing telecommunications. Starting from an elementary level readily accessible to undergraduates, this pioneer in the field provides a clear and up-to-date exposition of the prominent aspects of the theoretical background and most recent experimental results in this new and rapidly evolving branch of science.

Springer-Verlag
Berlin
Heidelberg
New York
London
Paris
Tokyo
Hong Kong
Barcelona

MIX
Papier aus verantwortungsvollen Quellen
Paper from responsible sources
FSC® C105338

If you have any concerns about our products,
you can contact us on
**ProductSafety@springernature.com**

In case Publisher is established outside the EU,
the EU authorized representative is:
**Springer Nature Customer Service Center GmbH**
**Europaplatz 3, 69115 Heidelberg, Germany**

Printed by Libri Plureos GmbH
in Hamburg, Germany